IEE History of Technology Series 7

Series Editor: Brian Bowers

BRITISH TELEVISION
The formative years

Other volumes in this series

Volume 1	Measuring instruments—Tools of knowledge P. H. Sydenham
Volume 2	Early radio wave detectors V. J. Phillips
Volume 3	A history of electric light and power B. Bowers
Volume 4	The history of electric wires and cables R. M. Black
Volume 5	An early history of electricity supply J. D. Poulter
Volume 6	Technical history of the beginnings of radar S. S. Swords

BRITISH TELEVISION
The formative years

R.W. BURNS

Peter Peregrinus Ltd in association with the Science Museum, London

Published by: Peter Peregrinus Ltd., London, United Kingdom

British Library Cataloguing in Publication Data

Burns, R.W.
British television, the formative years.
—(IEE history of technology series; 7)
1. Television—Great Britain—History
I. Title II. Series
621.388'00941 TK6638.G7

ISBN 0-86341-079-0

Printed in England by Short Run Press Ltd., Exeter

Contents

Acknowledgments

Many persons and organisations have aided me in my researches on the history of television and in the preparation of this book. I am most grateful indeed for all the contributions and help which I have received.

In the first place I wish to thank most sincerely both Professor J O'Neill, director of Trent Polytechnic, and Councillor F Riddell, chairman of the Polytechnic Council, Trent Polytechnic, for their efforts which led to me enjoying a period of study leave to pursue my writings and for making available to me library, photocopying and study facilities.

Some of my research work was carried out in the USA following the award of a travelling scholarship generously provided by Radio Rentals Ltd. Mr S L McCrearie, service director of this company, has for many years taken an active interest in promoting Baird's image and I wish to express my gratitude to him for his kindnesses and cooperation.

My chief debt is to the British Broadcasting Corporation and the Post Office. Both of these bodies have quite splendid archival collections, at the BBC's Written Archives Centre, Caversham and at the Post Office Archives, London respectively. The staff at these centres have always given me every assistance and it has been a joy to research in conditions where knowledge, experience and expertise are so freely available. Particular mention must be made of Miss M S Hodgson, formerly the BBC's Written Archives Officer and Mrs S C McNamara, formerly of the Post Office's Record Office for their kindly tolerance to my many requests. Their respective successors, Mrs J Kavanagh and Mrs J Farrugia, have also provided excellent services which I gratefully acknowledge. Apart from these I am agreatly indebted to the BBC and the Post Office for allowing me to quote extensively from their primary written source material.

The late Mrs B Hance, formerly company historian, and Mr R Rodwell, company historian of Marconi Historical Archives, and Mr C J Somers, formerly archivist of Ferranti Ltd, have also given me valuable assistance. So too have the librarians of Trent Polytechnic, who, over a period of many years, have processed my numerous requests for information.

I should like to express my appreciation of the courtesy and helpfulness of the staffs of the libraries and archives mentioned above and also of the staffs of the libraries of the Institution of Electrical Engineers, the Science Museum, the Patent Office, the Nottinghamshire County Council, the University of Nottingham, the University of Strathclyde and of the Mitchell Library, Glasgow. The staff of the Public Record Office, Kew have also been most helpful.

For reasons given in the preface my book contains many quotations from written material. I am especially grateful to the following persons for giving me permission to use extracts from their organisations publications and/or private papers: the editors of The Daily Telegraph, The Daily Mail, The Birmingham Post and Mail, the Folkstone Herald, the Evening Telegraph and Post, the Hastings and St Leonards Observer, The Sunday Express, The Times, The Guardian, the Glasgow Herald, The New York Times, the Daily Express, The London Standard, Electronics and Wireless World, Nature, the Journals of the Royal Television Society, The Engineer, the Journals of the Institute of Physics, and the Institute of Electrical and Electronics Engineers; Sir Cifford Cornford, chairman of A C Cossor Ltd.; Mr B C Owers, secretary of The Rank Organisation; Mr R Price, publisher, Odhams Leisure Group Ltd.; Mr O P Sutton, director of the British Radio and Electronic Equipment Manufacturer's Association; and Mr G C Arnold, chairman of P W Publishing Ltd.

Of the many persons, friends and colleagues in the UK and abroad who have assisted me with advice, information and/or documents I should like to mention: Mr R M Herbert, Mr W C Fox, Dr P Waddell, Mr E Peach, Mr T N Rimington, Mrs D Douglas, Dr A G Keller, Mr W P Lucas, Dr R F Thumwood, Mr P V Reveley, Mr K Geddes, Mr R V Mills, Mr P A Bennett, Mrs B Ryan, Mrs E H Day, Mr T Ivall, Mrs P Holt, Mr A Pinsky and Mr G Schindler. Professor M Baird very kindly let me use some extracts from his father's unpublished autobiographical notes. To all these individuals I wish to express my gratitude and appreciation.

Especially warm thanks are due to those persons who gave me – in all cases most willingly – permission on behalf of their organisation to use photographs and drawings. To the following I am deeply indebted: Mr S L McCrearie, service director of Radio Rentals Ltd.; Ms R Pearce of Granard Communications; Mr G D Speake, deputy director of research, GEC Research (Marconi Research Centre); Mr W P Lucas, formerly administration manager, Thorn EMI Central Research Laboratories; Mr D P Leggatt, chief engineer, external relations, BBC; the controller of Her Majesty's Stationary Office; Mr B C Owers, secretary, The Rank Organisation; Mr A Kemp, editor of the Glasgow Herald; Mr V Yates, company archivist and historian, Selfridges; Mrs E Romano of the Photo Centre, AT and T Bell Laboratories; Mr A Pinsky, of the information department, RCA; and, Mr R M Herbert and Professor D Shoenberg.

Very considerable efforts were made to locate the copyright holders of all the photographs and written material used but in some instances because of either the deaths of the original copyright holders, and the unknown consequential transfers of the copyrights, or the unknown whereabouts of these persons, success was not achieved. To these copyright holders I offer my apologies for the non-inclusion of their names.

Preface

On the 29th May 1934 the first meeting was held of the Television Committee under the chairmanship of Lord Selsdon, a former Postmaster General.

The terms of reference of the committee were:

> To consider the development of television and to advise the BBC on the relative merits of the several systems and on the conditions – technical, financial and general – under which any public service of television should be provided.

Subsequently the Television Committee submitted its report to the Postmaster General, Sir H. Kingsley Wood, on 14th January 1935. The report included a number of recommendations, the most important of which was:

> High definition television has reached such a standard of development as to justify the first steps being taken towards the early establishment of a public television service of this type.

This service was initially based on the utilisation of equipment provided by the Marconi–EMI Television Co. Ltd. and by Baird Television Ltd. and commenced on 2nd November 1936 at Alexandra Palace, London. The London Station was the world's first, public, regular, high definition television station and operated on an alternate basis with the television systems of the above mentioned companies from one site. After a trial period the Marconi–EMI Television Co. Ltd. system was adopted and the last transmission by the Baird Television Ltd. system took place on 30th January 1937.

The establishment of the Television Committee followed a period of concern by J. L. Baird and the directors of Baird Television Ltd. about the endeavours, in the field of television, of EMI Ltd. These endeavours effectively started in 1931 when Electric and Musical Industries was formed by the amalgamation of the Gramophone Company and the Columbia Graphophone Company. Because the latter company was associated with the Radio Corporation of America the new company of EMI Ltd. had a link with the American company. This link upset Baird and his directors who wished to establish in the United Kingdom a purely British system of television based on the pioneer work which had been undertaken by Baird from around 1923.

Baird was the first person in the world to demonstrate a rudimentary form of television; this he did on 25th October 1925. He worked tirelessly to advance his basic ideas on television but was considerably frustrated in his efforts. Much lobbying had to be embarked upon by Baird and his supporters in an effort to achieve their objective. Unfortunately for them the one organisation which could contribute to their aim – the British Broadcasting Corporation – was not impressed by low definition television.

With the emergence of a competitor – EMI Ltd. – Baird Television Ltd. made determined efforts to match their rival's advances. However, the 240-line television system of Baird Television Ltd. was generally inferior to the 405-line television system of Marconi–EMI Television Co. Ltd. and the latter system was adopted for British television.

The development of the 405-line television system in a period of less than four years was quite remarkable bearing in mind that electronics generally, as the subject is known today, was then (1931–1935) in its infancy. Notwithstanding the achievement, which placed the United Kingdom in the forefront of television advancement anywhere in the world, the general public, for various reasons, did not immediately embrace the new media form. When the London Station closed down for the war period on 2nd September 1939 the total number of television receiving sets sold was less than 20 000.

This book is concerned with the history to 1939 of British television for home reception. It is based predominantly on written primary source material, that is on private letters, memoranda, committee minutes and reports and on published articles, papers, editorials, reports and letters in the technical and non-technical press.

Great care has been taken to ensure that an unbiased, accurate history has been written. Fahie, who wrote definitive histories on electric telegraphy to the year 1837 and on wireless telegraphy from 1839 to 1899, took considerable pains when writing not to include 'the fictions with which many historians embellish their facts'. He accomplished his purpose by quoting extensively from primary sources. The present author has followed Fahie's practice: numerous references are given at the end of each chapter and the total number of references quoted is approximately 900.

This approach has been followed because controversy, largely based on ignorance, still exists about the contributions to the development of television of Baird and of EMI Ltd. Histories based on the spoken word should always be treated with some circumspection. People's memories are not always infallible; sometimes a person's recollection is coloured by what he/she has been told or has read and it is possible that an oral account may be unintentionally distorted because a person stresses the importance of his/her contribution – real or imagined – in the saga of events.

An endeavour has been made to write a balanced history rather than a purely technical history. Thus this book considers the factors – technical, financial and general – which led to the establishment of the world's first, all-electronic, public, regular, high definition television broadcasting service.

R. W. Burns

Personalities associated with the development of pre-war television

The Baird Companies

J. L. Baird	Pioneer of television, born 1888, died 1946
W. L. Day	Co-founder of Television Ltd.
O. G. Hutchinson	Baird's business partner
S. A. Moseley	Author, journalist and publicist: Baird's close friend
Sir Edward Manville	Chairman of Baird Television Development Company Ltd.
Lord Ampthill	Chairman of Baird International Television
T. W. Bartlett	Secretary of Baird Television Development Company
W. W. Jacomb	Chief Engineer, 1928–
A. G. D. West	Chief Engineer of Baird Television Ltd., 1933–
Ostrer Brothers (Isidore, Mark)	Chairman, and Managing Director of Gaumont-British
Sir Harry Greer	Chairman of Baird Television Ltd.
B. Clapp T. H. Bridgewater J. Denton D. R. Campbell J. D. Percy A. F. Birch J. C. Wilson H. Barton-Chapple	Technical assistants
W. C. Fox	Press Officer
H. Bradley	Television Programme Producer
Dora Jackson	Baird's personal secretary

The General Post Office (at 1931)

Sir Evelyn P. Murray	Secretary
F. W. Phillips	Assistant Secretary (later, 1935, Director of Communications)
L. Simon	Director of Telegraphs and Telephones
J. W. Wissenden	Senior Staff Officer
H. Napier	Private secretary to the Postmaster General
C. O. L. Leigh-Clare	Private secretary to the Assistant Postmaster General
Col. A. S. Angwin	Superintending Engineer (later Assistant Engineer-in-Chief, and then Deputy Engineer-in-Chief)
Col. Sir T. F. Purves	Engineer-in-Chief
E. H. Shaughnessy	Assistant Engineer-in-Chief
Lt. Col. A. G. Lee	Assistant Engineer-in-Chief (later Engineer-in-Chief)
A. J. Gill	Assistant Staff Engineer (later Staff Engineer)
J. V. Roberts	Staff Officer
(Sir Stephen Tallents	Public Relations Officer, 1933–1938)
Sir William Mitchell-Thomson (later Lord Selsdon)	Postmaster General, 1924–1929
H. B. Lees-Smith	Postmaster General, 1929–1931
Sir Kingsley Wood	Postmaster General, 1931–1935
Major G. C. Tryon	Postmaster General, 1935–1940

The British Broadcasting Corporation

Mr (later Sir John) C. W. Reith	Managing Director BBCo., Director General, 1927–1938
Capt. P. P. Eckersley	Chief Engineer
W. Gladstone Murray	Assistant Controller Information
Admiral Sir Charles Carpendale	Controller
N. Ashbridge	Assistant Chief Engineer, (later Chief Engineer)
H. Bishop	Assistant Chief Engineer
D. C. Birkinshaw	Superintendent Engineer, television service
E. Robb	Producer, 30-line television, 1932–1935
L. Sieveking	Co-producer of the first television play
G. Goldsmith	Director of Business Relations
G. Cock	Director of Television, 1936–
C. Madden	Programme Producer

Electric and Musical Industries

I. Shoenberg (later Sir Isaac)	Director of Research Laboratories and of the Patent Department
G. E. Condliffe	Manager of the Research Laboratories
Dr L. F. Broadway	Leader of the group which developed cathode-ray tubes for receivers
Dr J. D. McGee	Leader of the group which developed the emitron camera, and other camera tubes
A. D Blumlein	Leader of the group which developed the circuits for the 405-line system
C. O. Browne	Worked on 150-line television system; in charge of the design of cameras and studio equipment
C. S. Agate	Designed commercial receivers
Dr L. Klatzow	Worked on photosensitivity
Dr H. G. Lubszynski	Worked on super-emitron camera tube
Dr E. L. C. White	With Blumlein and Hardwick, designed the circuits associated with the 405-line system
N. E. Davis (Marconi)	Designed EMI and Alexandra Palace transmitters
A. Clark	Chairman
L. Sterling	Managing Director

Chapter 1

The birth of television, 1923–1926

The early, post 1914–18, history of television in the United Kingdom is essentially the history of John Logie Baird's work in this field, for until 1931 Baird had no effective competition from any individual or industrial research group in this country.

In 1923, the Scottish inventor, who probably did more to advance the early progress of television than any other individual anywhere, embarked on his future life's work at the age of 35 years in quite an unexpected way. Baird's quest to find a solution to the television problem was not the outcome of a long standing interest or hobby – unlike the work of Marconi in the field of radio communication, which represented a continuation of his youthful electrical experiments[1] – but rather it came about while Baird was convalescing in Hastings.

John Logie Baird was born in 1888 – 14 years after the birth of Marconi – in the town of Helensburgh,[2] a small seaside resort situated approximately 22 miles north-west of Glasgow.

Like the famous Italian inventor, Baird was brought up in a comfortable, middle class, professional household. His father was a minister of the local church and an intellectual of some merit. He had entered Glasgow University in 1860, and, by dint of a great deal of hard work and self-help (he supplemented his small financial resources by engaging in tutoring) he obtained an MA with honours in classics and then a BD.[3] In 1868 he was chosen to open a new church in Helensburgh. One of the subsequent members of the congregation was a Jessie Inglis, who came from a wealthy family of shipbuilders, and in 1878 the Reverend John Baird and she were married. Her dowry enabled them to buy The Lodge,[4] a square stone house as the church had no manse, and it was there that John Logie Baird spent the first 26 years or so of his life. (Fig. 1.1).

Towards the end of his life Baird drafted some autobiographical notes which were found after his death by his close friend, S.A. Moseley. In these, the television pioneer lucidly and graphically described his early upbringing and schooling; first in 'an extraordinary Dickensian menage, kept by a Mr Porteous and his wife' and then at a Miss Johnston's preparatory school, where a most formidable middle-aged spinster ruled the pupils by the same method of inconsiderate and indiscriminate corporal punishment as Mr Porteous. For Baird this was a most miserable time: 'I was terrorised and the years spent at that school were among the most unhappy of my life' he wrote.[3]

But if his early school days were unpleasant, John Logie, as with Marconi, received comfort, love and understanding from his mother. 'Of my mother, I find it difficult to write', he observed. 'She was the only experience I have had of pure, unselfish devotion. Her whole life was taken up in looking after others, particularly after myself . . .'[3]

Fig. 1.1 *The birthplace of J. L. Baird. The manse in Helensburgh*

His father, on the other hand, seemed a ponderous, forbidding figure, with a large beard through which he boomed at his children in the manner of an oracle. He was 47 years older than his youngest son, John Logie, in an era when any man over 50 had become an elderly, pompous figure. As a consequence, no doubt, the young Baird found his father's efforts to joke with him frightening and embarrassing, although the minister was famed for his humour throughout the west of Scotland.[3] Here again a similarity exists between the two inventors, for Gugielmo's father was 48 years old when the radio pioneer was born.

Baird's days at Miss Johnstone's school came to an end when a new school, 'Larchfield', was opened in Helensburgh. The Scot described this as a bad school run by three public schoolmen fresh from Oxford and Cambridge. 'They made it an imitation of their public schools,' he wrote, 'and a very poor imitation it was, with all the worst features and none of the best'.[3] No form of science was taught at Larchfield,[4] but nevertheless Baird acquired an interest in practical science which led him to engage in projects and experiments with telephones, cameras, gliders and electricity in his spare time.

This interest in science seems to have greatly influenced Baird's post-Larchfield education, for he rejected his father's request to enter the ministery and enrolled at the Royal Technical College, Glasgow, in 1906[4] to follow a course in electrical engineering. Eight years later Baird was awarded an Associateship of the College.

Baird does not seem to have found Professor Angus Maclean's course particularly enlightening; indeed his autobiographical notes show that he found much of the material he studied quite boring.

> On leaving Larchfield I went to the Royal Technical College in Glascow, filled with zeal and enthusiasm and feeling quite sure I would distinguish myself. I found it not so easy. There were plenty of other youths there filled with zeal and determination. How these youths worked. They were, for the most part, working men, bright lads out to make a career for themselves. They were not the intellectual cream (those won scholarships and went to the University): nevertheless they were doughty competitors. Nothing could approach the frenzied concentration with which they absorbed learning. There was no pretence at social life; there was no time for it.
>
> The first year I was there I learned a good deal that was very useful and interesting; the remaining years were, I think, almost entirely a waste of time. I learned, with great pains and boredom, masses of formulae and tedious dates, of which much was never used and soon forgotten. But what I learned in the first year remained with me all my life and has been of great value.[5]

The first year course was common to all engineering programmes of study and aimed to provide a sound basis in elementary science for the studies of the succeeding years. An examination of the course curriculum shows that the subject timetables for the second and third years of the mechanical engineering and electrical engineering programmes were almost identical 'as modern industrial conditions required that those who had adopted one branch should have a knowledge of the other'.

Curriculum for Electrical Engineering (E) and Mechanical Engineering (M) students

First Year	Second Year	Third Year
Natural Philosophy	Mathematics	Mechanics
Mathematics	Motive Power Engineering	Electrical Engineering
Inorganic Chemistry	Electrical Engineering	Motive Power Engineering
Drawing	Mechanics	Natural Philosophy

First Year	Second Year	Third Year
Physical Laboratory	Engineering Metallurgy	Electrical Engineering Laboratory (E only)
Chemical Laboratory	Drawing	Graphics
Surveying	Electrical Engineering Laboratory	Fuels
Additional Physical or Chemical Laboratory	Motive Power Engineering Laboratory	Motive Power Engineering Laboratory
	Mechanics Laboratory	Mechanics Laboratory (M only)
		Drawing

This fact has an important bearing on Baird's work on television, for he had a penchant for designing and inventing devices which had a mechanical basis rather than an electrical foundation. Baird displayed considerable ingenuity and innovativeness in the fields of optics and mechanics and produced many patents on aperture disc, lens disc and mirror drum scanning mechanisms, but only a few on electronic components or systems. Electronics was not Baird's forte. *Prima facie*, it would seem that neither electric telegraphy nor wireless telegraphy formed part of the diploma programme.

Attendance on the first, second and third years of the courses in the engineering disciplines qualified a student for the award of a diploma in engineering science, while attendance on an appropriate fourth year course enabled a student to become an Associate of the College. However, candidates for this award had to have completed at least two years practical works experience in the related industry before entering the fourth or Associateship year of their course.[6]

Baird spent three periods in industry. First at Halley's Industrial Motors Ltd., Yoker[7] (from May 1910 to February 1911); then with Argylls Ltd. of Alexandria, Dumbartonshire (from March 1911 to September 1911), where he was an Improver in their car drawing office working on engine and chassis details, and finally at Brash and Russell (Electrical and General Engineers) of Glasgow[8] (from October 1911 to May 1913) where he was employed in the design and layout of high and low tension switchboards and switchgear. Baird's testimonials refer to his industry: he is a 'very industrious, sober and efficient workman' (Halley's); 'he is persevering and industrious and anxious to get on' (Argylls); he is 'obliging and attentive to his work' (Brash and Russell).

These periods gave Baird an insight into factory working conditions, although the soul destroying, monotonous drudgery of the work he engaged in at Halley's Industrial Motors was hardly pertinent to the studies of a student of electrical engineering. During one of these spells in industry, at Argylls, the young Baird met a person who was later to play an important role in advancing the Baird system of television, namely Oliver George Hutchinson.

Baird was 26 when he left the Royal Technical College. He tried to enlist in August 1914 and when he was declared unfit for service entered Glasgow

University as a BSc degree student.[3] Possession of the associateship award of the RTC entitled a holder to take the appropriate final year degree examination of the university after a period of six months attendance. Baird spent an enjoyable session at the university but did not sit the examinations.

Subsequently he obtained work as an assistant mains engineer, at 30 shillings per week, with the Clyde Valley Electrical Power Company. This job entailed the supervision of the repair of any electrical failure in the Rutherglen area of Glasgow, whatever the weather, day or night, and as a consequence Baird had to live in lodgings in the area.[3] For a while he stayed at 17 Blairbeth Road, and later lodged with a family called Sommerville at 2 Millar Terrace.[3]

Throughout his life Baird suffered from chills, colds and infleunza which necessitated lengthy periods of convalescence. His studies at the Royal Technical College 'were continually interrupted by long illnesses' and his numerous absences from his employment as an assisant mains engineer because of illness militated against any promotion in the company. Because of this he disliked the job and eventually resigned. It was a wise decision. Baird's notes tell how he came to the turning point in his life.

> If I had remained travelling along the straight road of an engineering career, I should either be dead by now or a hopeless, broken-spirited object. To break my career seemed to those about me the act of an irresponsible madman, the throwing away of all my expensive training. If the choice was between slavery and madness, I preferred madness — there seemed no middle course. 'Are there no ways but these alone, madman or slave must man be one?' It seemed so in my case. If I remained an engineer I saw nothing before me but a vista of grey days, of unrelieved drudgery. Coughing and shivering through the winters, what hope to force my way through the mob of lusty competitors?[3]

Essentially Baird was a man of ideas, of imagination: as a boy he had installed electric lighting in The Lodge at a time when such an event could make news in the local press: he had constructed a small telephone system so that he could easily contact his friends: he had tried to make a flying machine and had fabricated selenium cells, and, as an assistant mains engineer, he had used the energy resources of the Clyde Valley Electrical Power Company in an attempt to make diamonds. Later events were to show that he possessed an inherent genius not only for adapting and developing the inventions and innovations of others but for originating and pursuing what, *prima facie*, must have seemed unlikely business enterprises. His subsequent life showed that he possessed an indomitable spirit for overcoming difficulties with a tenaciousness and relentlessness which could not be satisfied by a humdrum, boring and intellectually deflating existence as an assistant mains engineer.

Thus the break with routine came in 1919. Now he was free to make his fortune as he dictated.

Actually, Baird's departure from the CVEP Company was hastened by his entreprenurial exploits during the period 1917–19. In 1917 boot polish was difficult to obtain. Baird siezed the opportunity to enliven his existence by registering a company at 196 St Vincent Street[8] and employing girls to fill cardboard boxes with his own boot polish. (There is an unsupported story that his locker at the office where he worked was filled with tins of polish bought in Clydebank and Yoker in an attempt to corner the market for his own product.) This venture seems to have escaped the notice of his employers, but the next did not.

Baird had always suffered, and always did suffer, from cold feet, and, on the principle of capitalising on one's deficiencies, he devised an understock – consisting of an ordinary sock sprinkled with borax – and arranged its commercial exploitation with such a degree of business acumen and skill that, when he sold the enterprise 12 months later, he had made roughly £1600; a sum of money which would have taken him 12 years to earn as an engineer with the Clyde Valley Electrical Power Company.[3] He employed the first woman bearers of sandwich boards seen in Glasgow and also constructed a large model of a tank – plastered with posters about the efficacy of the Baird Understock in providing comfort for soldiers' feet – which was trundled about the streets of the city.[8]

Solid scent was added as a sideline to Baird's main business. However, all of this was too much for the managers of the CVEP Company, and their erring engineer was given an ultimatum; either he had to give up his business interests or he had to leave the company. Baird chose the latter alternative.[4]

By 1919 the future television pioneer appeared to be on the threshold of a lucrative commercial life. Unfortunately, continuous good health was not a blessing which had been bestowed on Baird, and, during the winter of 1920–21, he suffered a cold which entailed an absence of six weeks away from his venture. He decided to sell out and try his luck in the Caribbean.

Godfrey Harris, a school boy friend (with whom Baird had engaged in attempts to fly), had gone out to Trinidad and had sent Baird glowing accounts of the possibilities there. 'I was full of 'optimism,' wrote Baird, 'and set out blithely for the West Indies, taking a cheap passage in a cargo boat so as to keep as much as possible of my capital intact. I had three trunks filled with samples of cotton and other goods to sell to the natives. However, I began to feel a little doubtful of the prospects during the voyage.'[3]

Shortly after his arrival at Port of Spain, his doubts were confirmed and Baird conceivd the idea of establishing a jam factory – the island teamed with citrus fruits, guavas seemed plentiful, and sugar was produced in great quantity. His attempts to make and sell the jam were unsuccessful; there did not seem to be any adequate markets for his products in Port of Spain and Baird decided to return to London and endeavour to establish a market there. 'I bought a large cask and a number of kerosene tins and packed them with mango chutney, guava jelly, marmalade and tamarind syrup.'

On returning to London in September 1920, Baird opened a small shop at 166 Lupus Street[3] and set about trying to find a sales outlet for his wares, but his produce was sub-standard, and, in in desperation, he sold his total stock to a sausage maker, to mix with other material which went into his sausages, for £15.

By now Baird's capital had dwindled to approximately £200, in spite of extreme economy, and his financial position was causing him some concern. He tried answering advertisements in *The Times* under 'Business Opportunities' but without any beneficial effect.

Then he bought two tons of Australian honey at a give-away price and did a brisk business from Lupus Street selling it in 28-pound tins. Baird added the sale of fertilisers and coir-fibre dust to his interests and his enterprise prospered. An illness, unfortunately, caused him to remain in bed for several weeks, 'the business meanwhile going to bits'[3] and when his cold did not improve he went to Buxton to convalesce and sold his undertaking to a friend of Harold Pound for £100 and £200 in shares in an oil company. (Baird met Pound in Port of Spain: he subsequently accompanied Baird back to England and played a part in several of the Scot's commercial exploits.)

Baird was ill for nearly six months but finally recovered sufficiently to return to London and settle in lodgings in Pembroke Crescent. He had little more than £100 in his pocket and no source of income, he urgently needed money and commenced searching through the papers for a likely opening.

An advertisement in *The Grocer* offering two tons of resin soap at an amazingly low price caught Baird's eye and so once more another business venture was initiated. The year was 1922 and Baird's Speedy Cleaner was soon being sold to hotels, boarding-houses, ship's chandlers and street barrow boys. The business flourished so much that Baird later imported large quantities of soap from France and Belgium and formed a limited liability company with two associates of Harold Pound.[4]

Soon a competitor appeared. Baird was selling his soap at 18 shillings per hundredweight, but his rival, O.G. Hutchinson (whom he had met at the Argyle Motor Works), was undercutting Baird's price by two shillings per hundredweight. Baird's Speedy Cleaner and Hutchinson's Rapid Washer both catered for the same market and Baird felt that if he could not beat Hutchinson's price the best course of action was to consider a merger of the two businesses. He therefore invited his rival, a jovial young Irishman, to meet him to discuss a suitable business arrangement.

> We sat long into the night, drinking old brandy and settling the last details of our merger. I felt ill when he saw me off at Leicester Square Tube station. Next morning I had a high temperature and a terrific cold. I had left my old lodgings and was staying in a cheap residential hotel (bed and breakfast, full board Sunday, thirty shillings). My bedroom was a converted conservatory and bitterly cold. I got rapidly worse. Hutchinson appeared with a bottle of eau-de-cologne and was thoroughly alarmed at my state. The doctor was called in and I got steadily worse. He, too, became concerned and told me that I must get out of London at once, or he would not answer for my recovery. 'Mephy' (a friend from childhood)

> was in Hastings and next day I packed my bag and set off to join him. The business had prospered and I had fulfilled a long-held ambition by forming it into a limited company with £2000 authorised capital. My co-directors, two young business men to whom Pound had introduced me, bought out my shares, leaving me with a sum of roughly £200.[3]

Television development was not initially in Baird's mind when he settled in Hastings, for in his notes Baird related how, when his health improved, he tried to invent a pair of boots with pneumatic soles so that people could walk with the same advantage that a car gains from its pneumatic tyres.

> I bought a pair of very large boots, put inside them two partially inflated balloons, very carefully inserted my feet, laced up the boots, and set off on a short trial run. I walked a hundred yards in a sucession of drunken and uncontrollable lurches, followed by a few delighted urchins. Then the demonstration was brought to an end by one of the balloons bursting. More thought was needed.[3]

Also while staying in Hastings, Baird had tried to invent a glass safety razor, which would not rust or tarnish, but without success. His financial state was giving him some concern, and, as his health precluded the possibility of him establishing a business or engaging in salaried work, he reasoned that he would have to invent something which would give him a source of income.

Baird's biographers – his wife, Margaret, and staunch friend, S.A. Moseley – state that Baird thought out a complete system of television while walking over the cliffs to Fairlight Glen. Unfortunately, neither author mentions explicitly why Baird should have considered this topic. Margaret Baird has, however, written that, during her future husband's period of convalescence, some time was spent by him browsing in the public library. As Baird was trained as an electrical engineer, it seems highly likely that he would have taken at least a cursory look at the popular and semipopular technical magazines available at the library – particularly as sound broadcasing had not long been in operation in England. If this is so, then the possibility exists that he read an article on 'A development in the problem of television' by Langer in the *Wireless World and Radio Review* issue of 11th November 1922.[9] (The month Baird moved to Hastings is not known, but he was there in the winter of 1922–23.)

In his article Langer described the rudimentary principles of television and proposed a solution based on the use of oscillating mirror scanners, a selenium cell, a 'light-valve' of the string galvanometer type and a suitable amplifier. The most important aspect of Langer's paper concerned the results which he had obtained from some experiments 'to ascertain the limit of speed of change which a selenium cell was capable of recording'. These results[9] showed that selenium cells were capable of responding to variations in light at a frequency of 10 000 Hz; that with an increase of speed the sensitivity of the cell was greatly reduced; and that by the use of valve amplifiers it was possible to compensate largely for the loss in sensitivity at high speeds.

From his observations Langer considered that the selenium cell was applicable to experiments 'for enabling objects at a distance to be seen by the eye by means of electrical circuits or by wireless, in much the same manner as the ear is enabled by means of these methods to hear sounds at a distance'.

Langer concluded his article by saying that he had endeavoured to indicate the lines along which a solution to the problem could be found. 'I feel that it may be of interest to other experiments', he wrote, 'to have these suggestions put forward. The solution of the problem from a wireless point of view must be looked for as a logical outcome of television by line wires.'

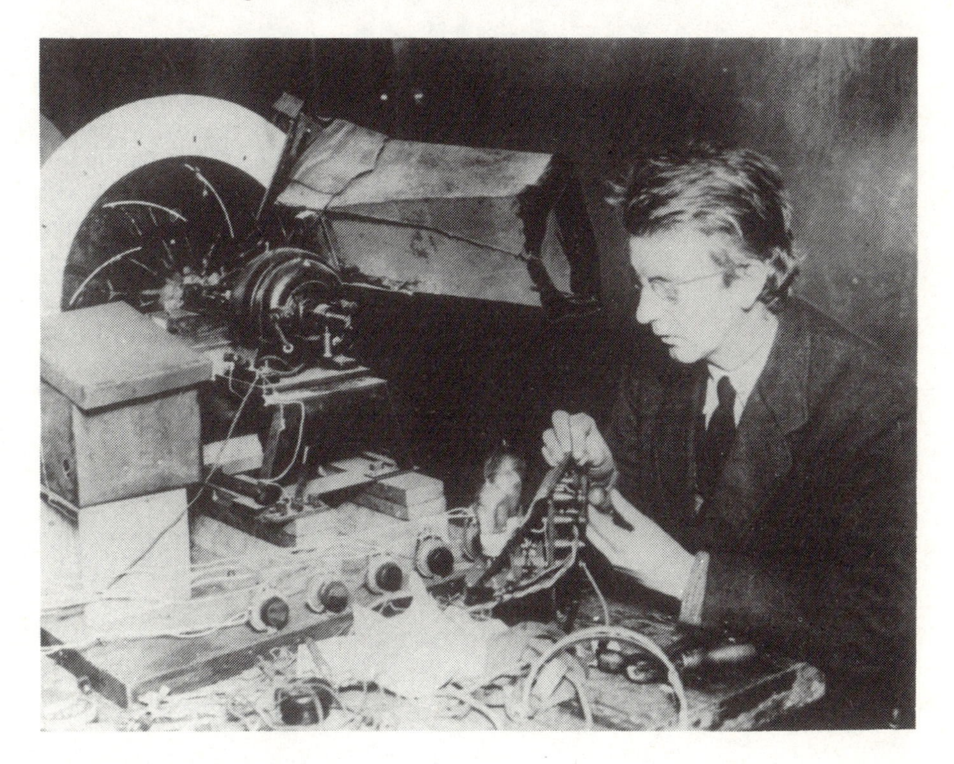

Fig. 1.2 *Baird at work on his early television apparatus*

Now Baird, in his youth, had tried to make selenium cells while living with his father and mother in their manse in Helensburgh. However, when one day some hot selenium scarred one of his hands he abandoned his project and diverted his energies elsewhere.[10] It has been reported that Baird carried out some investigations, in 1913 or 1914, on television while staying with relatives at 17 Coldingham Avenue, Yoker, and succeeded in transmitting shadowy images from one room to another.[8] These episodes show that Baird had an early interest in light cells; an interest which could well have been rekindled by Langer's paper.

A further point of circumstantial evidence which suggests that Baird read the *Wireless World* articles comes from his views on the applications of television. The *Hastings and St Leonards Observer* reported in January 1924:

> A Scotsman has come south, in fact he has come to Hastings, and this particular Scotsman is now engaged (Fig. 1.2) upon perfecting an invention which at some not very distant date may enable people to sit in a cinema and see on the screen the finish of the Derby at the same moment as the horses are passing the post, or may be the Carpentier-Dempsey fight . . .

This prospect for the future was similar to Langer's, for he wrote:

> Personally I look forward with confidence to the time when we shall not only speak with, but also see those with whom we carry on telephone or wireless telephone conversation, and the distribution of cinematograph films will be superseded by the direct transmission from a central cinema.[9]

It is interesting to note that Baird's first patent, 222,604 dated 25th June 1923 (see Fig. 1.3), is concerned with a system of reproducing television images which the inventor used to show large screen television at the London Coliseum on 28th July 1930.

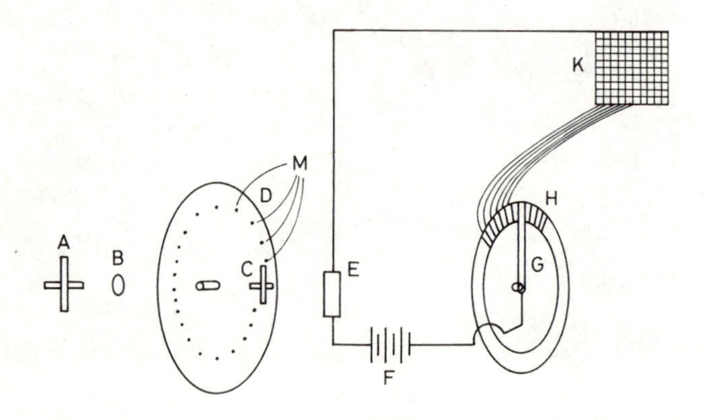

Fig. 1.3 *Baird's first patent 222,604, was taken out on 26th July 1923. It describes a method using a screen of small lamps to reproduce a televised scene: the concept formed the basis for Baird's large screen display, of images of the Derby, at the Coliseum on 28th July 1930*

A further significant point which lends support to the previously mentioned premise concerns a statement made by Langar in the first paragraph of his paper: 'The problem of television has been already partly solved by the methods adopted by Professor Korn.' If the premise is true, then Baird presumably would have wanted to refer to Korn's work. This he did, for Margaret Baird has written[4] that Baird found a musty and torn copy of a book in German called 'Handbuch der Phototelegraphie' (1911), by A. Korn and B. Glatzel in the public library. Korn and Glatzel's work was the only pre-1922 publication which gave a clear description, with diagrams, of some of the early suggestions and schemes for distant vision. (A published lecture which Baird gave in January 1927, indicates that he was familiar with the inventions of the television pioneers.)[11]

Thus it seems reasonable to suppose that Baird's renewed interest in 'seeing by electricity' was triggered off by an article in a semi-popular technical magazine and that his subsequent choice of the Nipkow disc scanner for his work resulted from a perusal of the appropriate chapter of Korn and Glatzel's book. Certainly Langer's paper was optimistic in tone and many 1922–23 readers with only a slight knowledge of the advances which had taken place in radio communications and valve circuits would probably have formed the opinion that the accomplishment of television was near at hand.

Baird, however, realised the difficult nature of the task: 'The only ominous cloud on the horizon,' he wrote, 'was that, in spite of the apparent simplicity of the task, no one had produced television.'[11]

It is interesting to recall that Marconi was led to pursue his life's work when he read an obituary describing Hertz's experiments, while holidaying at Biellese in the Italian Alps. He subsequently said that, as a result of the article, the idea of wireless telegraphy using Hertzian waves suddenly came to him. In later life he observed:

> The idea obsessed me more and more and, in these mountains of Biellese, I worked it out in imagination. I did not attempt any experiments until we returned to the Villa Grifone in the autumn, but then two large rooms at the top of the house were set aside for me by my mother. And there I began experiments in earnest.[1]

Like Baird, the young Italian inventor considered the solution to his posed problem to be essentially simple and seems to have been surprised to learn that it had not been solved by others. 'My chief trouble,' he noted, 'was that the idea was so elementary, so simple in logic, that it seemed difficult for me to believe that no one else had thought of putting it into practice. Surely, I argued, there must be much more mature scientists than myself who had followed the same line of thought and arrived at an almost similar conclusion.'[1]

A further point of similarity concerns the approach which the two inventors adopted to solve their problems. Neither Baird nor Marconi had any highly original suggestions to put forward at the outset of their investigations and both experimenters modelled their schemes on the ideas of others. Marconi's earliest transmitter was still essentially a coil and spark-gap as used by Hertz (although the design of the spark-gap had been slightly changed to incorporate an improvement due to Righi), and a curved metal reflector which was situated behind it to direct the radiation to the receiver. Baird's earliest distant vision apparatuses were based on proposals which had been advanced by Nipkow and others in the 19th and early 20th century.

P. P. Eckersley, a former Chief Engineer of the BBC and an opponent of Baird's low definition system, wrote in 1960:

> Baird is to be honoured . . . among those who see past immediate technical difficulties to an eventual achievement; Marconi did much the same with radio. Neither Baird nor Marconi were pre-eminently inventors or physicists; they had, however, that flair for picking about on the scrapheap of unrelated discoveries and assembling the bits and pieces to make something work and so revealing possibilities if not finality.[12]

Neverthless, notwithstanding their lack of profound intellect, their inability to emulate Clerk Maxwell or Lord Kelvin in original thought, Baird and Marconi suceedeed where others had failed because they possessed qualities of patience, of concentration and of an overwhelming desire to succeed which enabled them to pursue their objectives with an indefatigable resolve.

Both Baird and Marconi commenced their investigations at opportune times, for, in addition to the ideas which had been put forward by others, the technology existed for narrowband television broadcasting in the one case and narrowband wireless communication in the other.

Also Marconi and Baird had the luck, or the intuition, to select the right notions to try out. Of course, in some instances, particularly when Baird commenced his experiments in 1923, the choice of components was also decided by factors of cost, ease and simplicity of construction and of availability.

Marconi and Baird began their experiments in private residences. Marconi had two large rooms at the top of the Villa Grifone set aside for him by his mother: Baird made use of various rooms which he rented when staying in Hastings and elsewhere.

Baird hoped to realise a quest which had engaged the attention of scientists, engineers and inventors from 1873. For, in that year, Willoughby Smith communicated a discovery[13] to the Society of Telegraph Engineers which appeared to provide, a few years later, a means for 'seeing by electricity'. Smith's disclosure concerned the photoconductive property of selenium, a characteristic of the element which was utilised in many of the early schemes for television until the development of suitable amplifiers and photoemissive cells made selenium cells obsolete in the 1920–1930 decade.

Prior to 1873, several systems had been advanced for transmitting images of line drawings and printed sheets, commencing with the inventions of Alexander Bain and Frederick Bakewell in 1843[14] and 1848[15], respectively, and culminating in the operational picture telegraph schemes of Caselli (1862), Meyer (1869), and d'Arlincourt (1872).[16]. However, with all these systems, the lack of a suitable photoelectric transducer had limited their application to the reproduction, at a distance, of images of pictures and the like which could be drawn in varnish on a conducting sheet or which could be suitably prepared so that a scanning stylus and associated apparatus could discriminate between the picture and background.

During the period 1843–1872, the only known effect which related changes of light intensity to changes in an electric current was that discovered by Becquerel in 1839 in his investigations on the electrochemical effects of light. Becquerel's observations demanded the use of a highly sensitive galvanometer and consequently the effect was not suitable for incorporation into a distant-vision scheme.

The photoconductive property of selenium was easily demonstrated – 'its sensibility to light is extraordinary, that of a mere lucifer match being sufficient to effect its conductive powers' – and in the decade following Willoughby Smith's communication there was an expectation among scientists and others that 'seeing by electricity' would soon be a reality. This expectation was based not only on the results that had been achieved in the field of picture telegraphy, together with Smith's disclosure, but also on the invention of the telephone, by Alexander Graham Bell in 1876, which enabled 'hearing by electricity' to be readily implemented. The simplicity of Bell's device and the lack of effort involved in its development possibly stimulated inventors to attempt the transmission of moving images by electrical means, for numerous suggestions for 'telectroscopes' were put forward in the 10-year period after Smith's announcement (see Table 1.1).

However, the problem of 'seeing by electricity' was of an altogether different order of complexity compared with the problem of 'hearing by electricity', and success eluded the above workers. An early indication that selenium was not an ideal photoconductive material was given by Sale in a communication to the Royal Society in 1873.[17] Sale's experiments indicated that instantaneous changes of light intensity on a selenium bar did not cause instantaneous changes of resistance in the material. This property of selenium was to limit television development for many years, and even when selenium cells were employed by Korn in low-speed picture telegraphy systems, in the first decade of the 20th century, a special circuit had to be devised by him to overcome partially this effect.

Nevertheless, the work of the 19th century television pioneers was not wholly unproductive, and, by the end of the century, some of the basic system components needed to implement a television scheme had been proposed. The basic principles of scanning in particular were well understood and the scanners of Nipkow (1884), Weiller (1889) and Brillouin (1891) were later successfully utilised by scientists and engineers in the period 1925–1936.

Also, the development effort which had enabled practical picture telegraph systems to be demonstrated and introduced into the public service (albeit for short periods) had given inventors an understanding of the principles and difficulties of synchronisation. This understanding was applied to the problem of television; indeed, the use of line synchronising pulses in modern television can easily be traced back to the work of Bain and Bakewell in 1843 and 1848, respectively.

Table 1.1 *Dates (and names of inventors) of some 'distant-vision' proposals for the period 1875–1925*

1878 de Paiva	1904 von Jaworśky and Frankenstein	1923 von Mihaly
1879 Perosino	1906 Lux	1923 Jenkins
1879 Senlecq	1906 Rignoux	1923 Westinghouse
1880 Carey	1906 Dieckmann and Glage	1923 Dauvillier
1880 Ayrton and Perry	1907 Rosing	1923 Zworykin
1880 Le Blanc	1908 Campbell Swinton	1923 Western Electric
1880 Sawyer	1910 Ekstrom	1923 Stephenson and Walton
1880 Middleton	1910 Schwierer	1923 Baird
1881 Bidwell	1911 Rosing	1923 Nisco
1882 Senlecq	1911 Campbell Swinton	1923 Robb and Martin
1882 Lucas	1914 Lavington-Hart	1924 Seguin and Seguin
1884 Nipkow	1915 Voulgre	1924 Blake and Spooner
1889 Weiller	1917 Nicolson	1924 Belin
1891 Brillouin	1920 Baden-Powell	1924 Baird
1891 Sutton	1921 Whiston	1924 d'Albe
1897 Szczepanik	1921 Schoultz	1925 Whitten
1902 Coblyn	1922 Belin	1925 Baird
1904 Belin and Belin	1922 Valensi	

After the turn of the century the work and notions of Rosing (1907) and Campbell Swinton (1911), Fig. 1.4, on the utilisation of cathode-ray tubes in distant-vision schemes – which subsequently led to the experiments of Zworykin (1923), Fig. 1.5, and the production of the iconoscope and emitron by the RCA and EMI, respectively, several years later – provided the new ideas which were needed to achieve an all-electronic solution to the television problem.

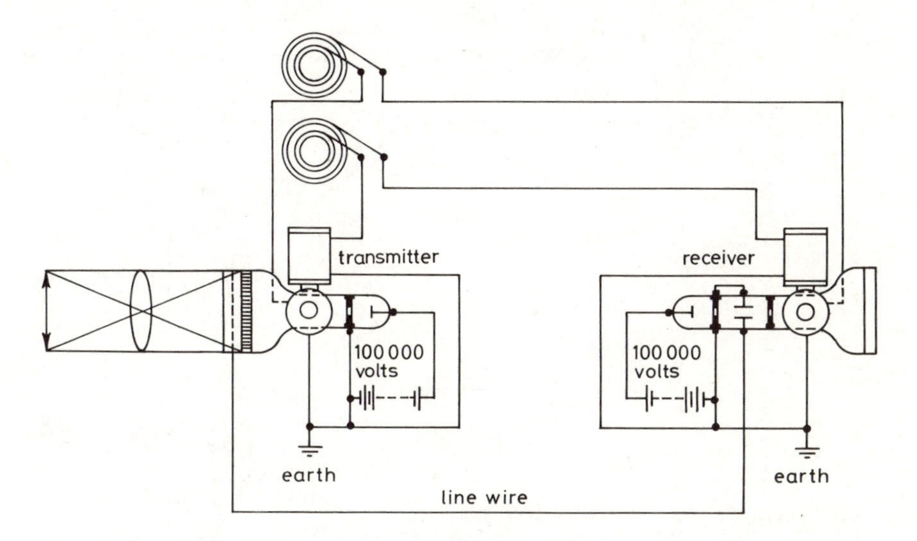

Fig. 1.4 *In his Presidential address to the Röntgen Society in 1911, Campbell Swinton described his scheme for television using non-mechanical scanning means at both the transmitter and receiver. His ideas influenced McGee and Tedham, of Electric and Musical Industries, in the early 1930s*

In addition, de Forest's invention of the audion (1907), Hallwach's demonstration of the photoelectric effect (1888) and the detailed investigations of Elster and Geitel (1889–1913) on photoelectricity were important contributions that were to play a vital part in the progress of television.

Consequently, when Baird, in 1923, decided to apply his inventive abilities to the problem of a practical television scheme, the problem seemed to him to be comparatively simple. Two optical exploring devices rotating in synchronism, a light-sensitive cell and a controlled varying light source capable of rapid variations in light flux were all that were required, and these appeared to be already, to use a Patent Office term, known to the art.

Baird's principal contemporaries in this challenge were C. F. Jenkins of the USA and D. von Mihaly of Hungary. Other inventors were patenting their ideas on television at this time (1923), but only Jenkins, Mihaly, Baird and a few others were pursuing a practical study of the problem based on the utilisation of mechanical scanners.

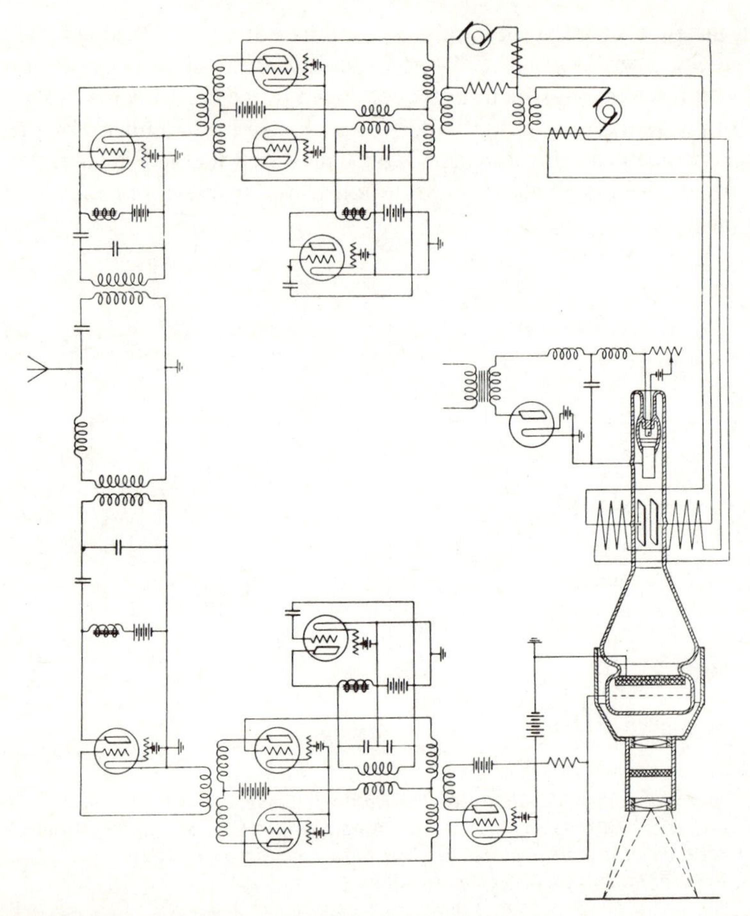

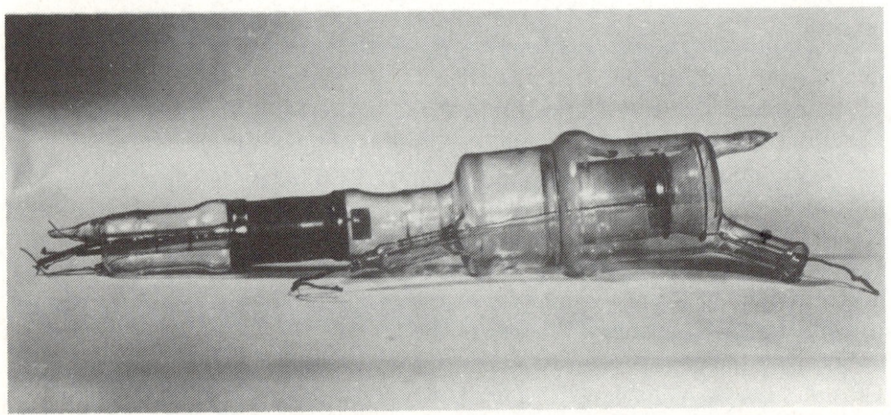

Fig. 1.5 *V. K. Zworykin, of the American Westinghouse Company, patented his television scheme in 1923. The circuit and electronic pick-up tube are shown, together with a photograph of the first tube made by Zworykin in 1923*

The approaches of the three inventors to their tasks were individualistic. Jenkins was a well known inventor and a person of considerable means. He had produced important inventions in the field of cinematography and was able to design and manufacture apparatus of some complexity. His early rotary scanners consisted either of specially ground prismatic discs or costly lensed discs, see Figs. 1.6 and 1.7. Mihaly, an experienced patent expert and engineer, used an oscillating mirror scanner, together with tuning forks and phonic motors for synchronising purposes, (Fig. 1.8). Baird's aproach necessarily had to be entirely different to those of his contemporaries. He had little money, no laboratory facilities for the construction and repair of equipment, no access to specialist expertise, and no experience of research and development work in electrical engineering. He had to carry out his experiments in the unsuitable conditions of private lodgings. Still, undaunted by the formidable difficulties which faced him, he commenced his experiments in 1923 by collecting bits and pieces of scrap materiual and assembling them into a system. The constraints imposed by his financial state severely limited the type of investigations he could carry out, but nevertheless he pursued his objective with dogged determination, ingenuity and resourcefulness. A Nipkow disc could be made from a cardboard hat box, the apertures could be formed using a knitting needle or a pair of scissors, electric motors could be bought cheaply and bull's-eye lenses could be obtained at low cost from a cycle shop.

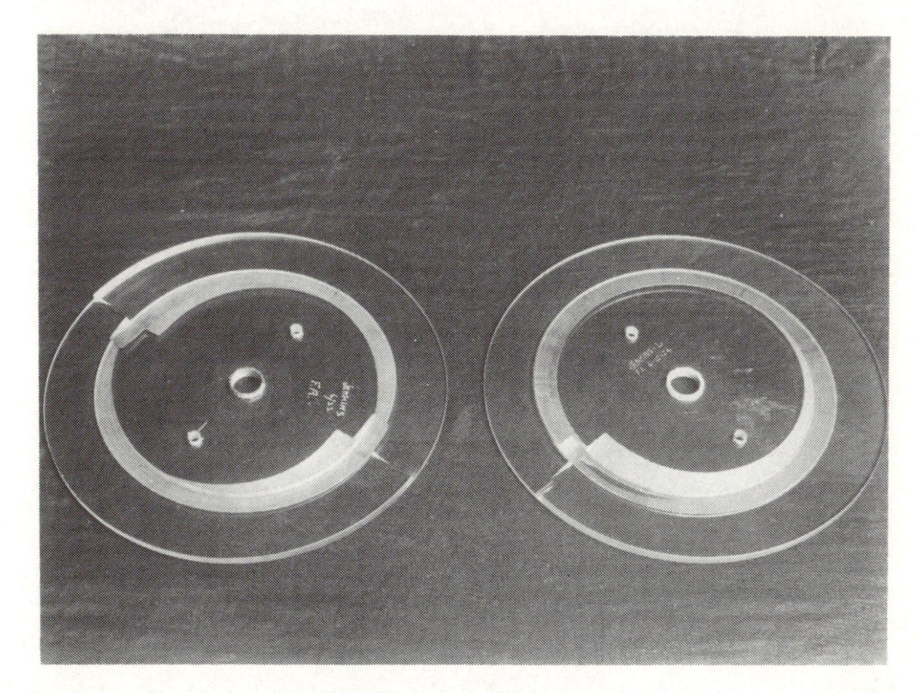

Fig. 1.6 *C. F. Jenkins, of the USA, employed specially ground prismatic discs as his scanning elements*

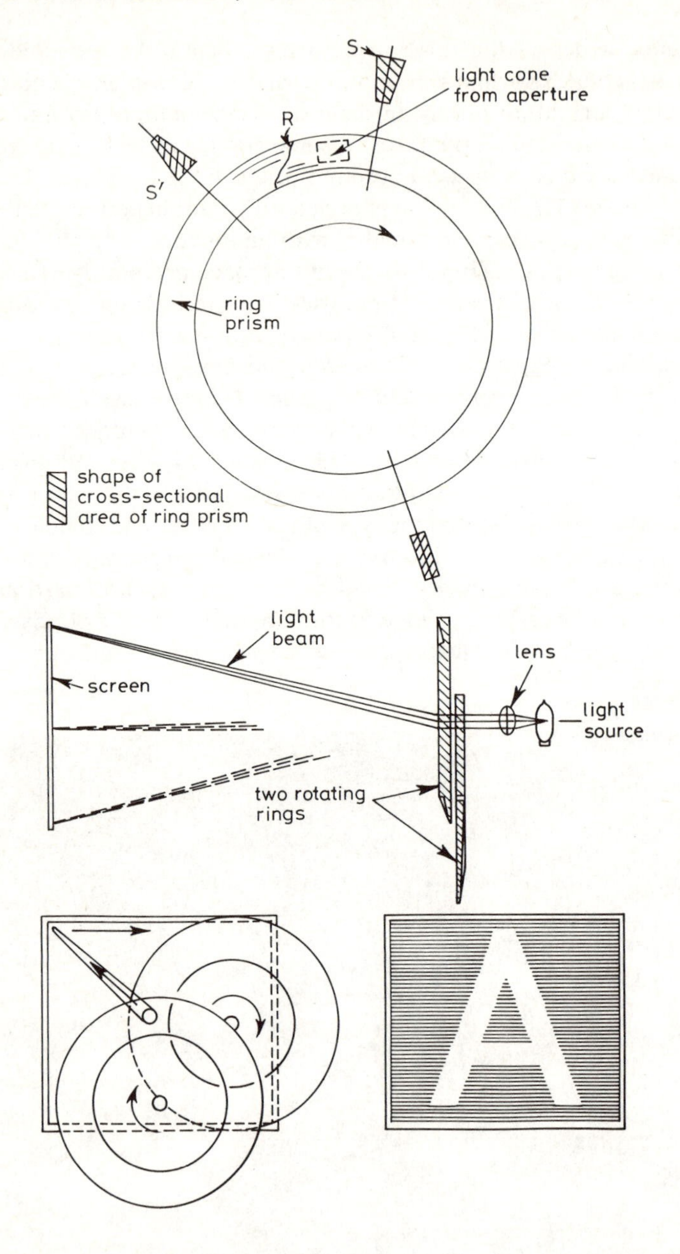

Fig. 1.7 *Diagrams showing (a) the variation of the cross-sectional shape of the ring prism, and (b) the use of two discs to scan a raster*

According to Moseley (Baird's staunchest supporter from 1928), the first televisor Baird devised 'had the ingenuity of Heath Robinson and a touch of Robinson Crusoe'.[2] Baird described it as having the saving grace of simplicity.

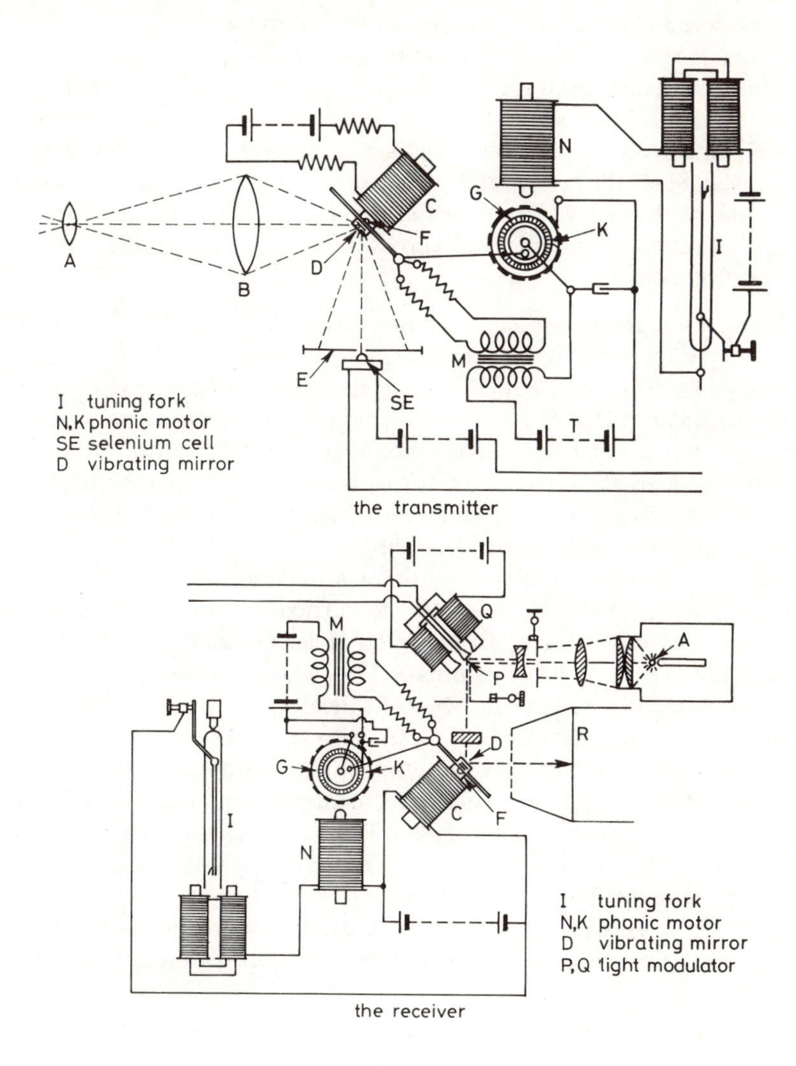

Fig. 1.8 *D. von Mihaly's television transmitter and receiver utilised mirrors for scanning and phonic motors and tuning forks for synchronisation*

A circular cardboard disc cut from the hat box and pieced with two spirals of small holes formed the Nipkow disc, a darning needle served as a spindle, and bobbins supplied the means of revolving the disc with the use of an electric motor.

> On one side a powerful electric lamp shone through the bull's-eye lens on to a little cardboard cross, whose shadow was cast onto the disc. To one side of the cardboard disc was another of tin. This had a large number of little serrations around its edge, was mounted on the spindle of a small but high speed electric fan motor, and stood in the path of light.[2]
>
> On the other side of the cardboard disc was a selenium cell. The interrupted light falling on to this cell generated a current which went to a neon lamp fixed behind the same disc but at the opposite edge from the cell, which was connected to the lamp through an amplifier. The lamp glowed when the cell was illuminated and went out when it was in the shade, so that when the apparatus revolved it was possible to see in one half of the disc the shadow of a cross (used as an object) on the other half two feet away.

Of the original machine nothing now remains. Baird's recollection was that it was sold for £2 in order to pay the rent.

Moseley's description of Baird's first' televisor cannot be taken as a definitive account: he stated that Baird used a neon lamp in his televisor, but a May 1924 description by Baird of a system corresponding to the transmitting arrangement mentioned by Moseley (and shown to the press at the beginning of the year) referred to a receiving disc having lamps arranged in the same staggered formation as the holes in the transmitter disc.

Baird's movements in 1923 are not known in detail. Apart from staying in Hastings there is some evidence that he carried out experimental work in Folkestone, Tunbridge Wells, London and Helensburgh. His Folkestone association only came to light in February 1961 when a reference to Baird by a columnist reminded a resident, a Mr P. A. Bennett of 21 Shorncliffe Crescent, that the inventor had boarded in West Terrace, Folkestone.[18] It seems that Baird's landlandy objected to his cluttering up the bedroom with apparatus and he was asked to leave. Following a recommendation, Baird next acquired accmmodation with T. C. Gilbert & Co. Ltd., at 26 Guildhall Street (Bennett was then a director of the company). The company's business as electrical contractors brought it into the wireless experimenting circle and Baird was welcomed. He was provided with bench space in the workshop and given certain facilities, including the use of some Western Electric audiofrequency amplifiers.[19]

Bennett has stated:

> Baird stayed here perhaps two or three months, and during that period was in touch with someone in the Tunbridge Wells area, who was experimenting with a light cell . . . When he left Folkestone I feel certain he went into that area, and in retrospect it seems to me that he went to Hastings some time after that.

Baird was certainly back in Hastings towards the end of 1923. He had for some time, according to the *Hastings and St Leonards Observer* (19th January,1924), been working in his own laboratory near Hastings station, but when the premises were required for use as a shop he transferred his intricate machine to Linton Crescent (number 21), where Mephy, his friend, lived.

Mr V. R. Mills of 35 Ghyllside Drive, Hastings, gave Baird some useful assistance while he was in Hastings. Mills has written:[21]

> I cannot put an exact date to Baird's first visit to my people's house which was no. 122 Hughenden Road . . . My mother answered the door and then told me that there was a very strange man to see me. On meeting me Baird said that he had heard about my wireless experiments. (There had been something in the paper about my gear, station G5QM.) He told me that he was working on television and I do remember asking him what television was. He than told me about the signal which he was getting from his selenium cell. I then saw his apparatus in Queen's Avenue, (no. 8), and helped him just for the fun of it until we managed to get a picture. At this time I had quite a good amplifier made by Stirling Telephones Ltd. (cost about £30). This used bright power triodes, I think LS1 and LS2s. I lent him all sorts of things free of cost. Mother did not charge him anything and when he left she just said 'God will repay'.

Baird worked at 8 Queens Avenue for a period until an explosion induced the landlord to seek another tenant.[22] From Hastings the inventor next moved to Frith Street in Soho, London.[23]

During the early part of 1923 Baird had no source of income, although there was a constant drain on his small capital reserves for the essentials of life and the needs of his ideas. By June 1923 his financial state was becoming precarious and he inserted an advertisement in the 'Personal' column of *The Times*, (27th June 1923):

> Seeing by Wireless – Inventor of apparatus wishes to hear from someone who will assist (not financially) in making working models. Write Box S686 The times, EC4[24]

Baird thought it best not to ask directly for money and afterwards maintained that this was the wiser course.

The advertisement was answered by three men. 'One hadn't any money. One, a relative of Sir John Reith, sent an expert (Captain West) to investigate. The third man, Will Day, did not see my apparatus but he said he would take a sporting chance . . .', wrote Baird.

Day, a successful person in the wireless and cinema businesses, bought a one-third interest in Baird's invention for £200[25] and as a consequene Day's name was included on the first four patents produced by Baird. Later Day acquired a further share of one-sixth from Baird. He had to bear all the costs of patenting and developing the inventions, but, as he noted in some autobiographical notes: 'I would have signed away my immortal soul for two hundred pounds and I was not going to quibble over the terms of a legal document.'

The exact sequence of experiments and investigations which Baird carried out in 1923 is not known, but Mills has written that during his association with Baird he always used a Nipkow disc scanner in one of its many different forms, either at the transmitting end or at the receiving end of the television link. Baird utilised disc scanners for many years, and even when the London Station was established in 1936 one of the essential items of equipment provided by Baird Television Ltd. was a spotlight scanner of the Nipkow disc type.

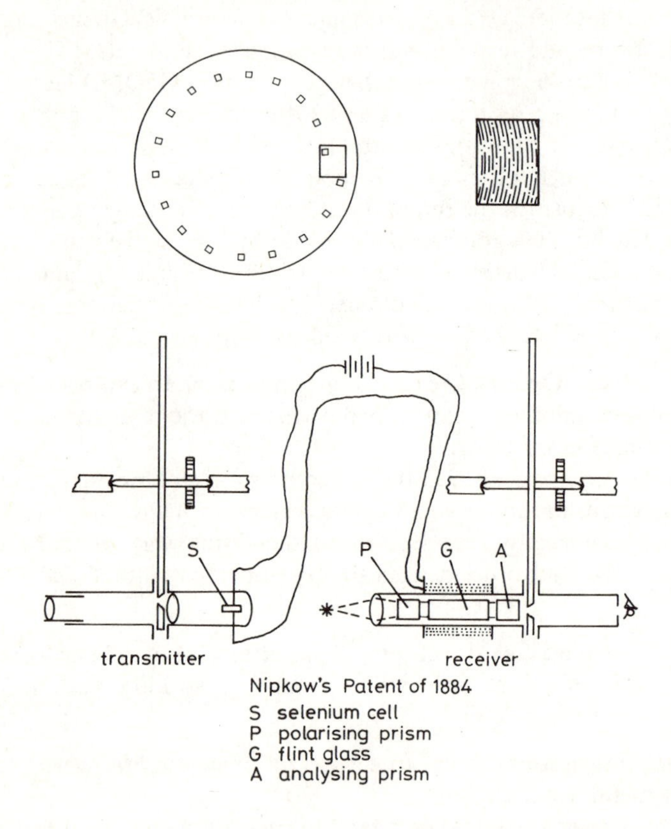

Fig. 1.9 *P. Nipkow was the first person to propose the use of apertured discs for scanning in a 'seeing by electricity' scheme. He patented his method in 1884*

Nipkow's invention[26] had an inherent limitation which precluded its use for a purpose which Baird held to be of some importance, namely, cinema television. Fig. 1.9 illustrates the principle of operation of a Nipkow disc scanner. In its simplest form the Nipkow disc consists of a thin, flat, circular piece of metal, or other suitable material, pierced by a number of small apertures, equal in number to the number of scanning lines required, and arranged to lie along a single turn spiral. Given a disc of diameter D and a number n of scanning lines, the circumferential distance between any two apertures is $\pi D/n$ and the radial distance separating the first and nth holes is $\pi D/nk$ where k is the aspect ratio (assuming horizontal line scanning). Thus for a disc 30 inches in diameter the size of the aperture scanned is only approximately 3 inches by 2.4 inches, for an aspect ratio of 5:4 and a sequential scan of 30 lines. Hence while such a disc could be employed to analyse an image of the object or scene to be televised it was not possible to contemplate its used as an image synthesiser in a large screen system, bearing in mind the controllable light sources then available.

Barid's solution, which he put forward in his first patient[27] (222, 604), was based on an idea advanced by Ayrton and Perry, by Redmond, by Middleton, and by Carey in 1880 and by Senlecq in 1881 – the idea of a mosaic of reproducing elements. The patent, titled 'A system of transmitting views, portraits and scenes by telegraphy or wireless telegraphy' describes the use of an analysing Nipkow disc and a receiver mosaic of incandescent lamps. Baird envisaged the disc to be provided with a series of 18 small holes, each 1/8 inch in diameter, and each circumferentially separated from the next by 1 inch. At the receiver a rotating brush commutator was to be employed to switch the received signal to the appropriate lamps of the mosaic, whence the varying brightness of the lamps would reproduce the image, and persistence of vision would cause the whole reproduced image to appear simultaneously on the screen of lamps, see Fig. 1.3.

Baird actually tried to implement this scheme in 1923, for the first known published report of his work, which was given in *Chambers Journal* (November 1923), describes some of the characteristics of the aparatus.[28] The aperture disc scanner at the transmitter was 20 inch in diameter and the image frame measured 2 inch square. A rotation speed of 20 revolutions per second was adopted for the frame scanning rate and at the receiver the signals were taken to the fulcrum of an arm with a copper brush at the end, which rotated around a ring of tiny contacts. These contacts were connected in sequence to a number of lamps, of only 1/8th inch in diameter mounted in a picture frame. No indication was given in the report of the number of lamps, or holes employed: nor was any reference made to the performance of the system. The reporter of the above mentioned article seemed to have an optimist's outlook for he/she wrote: '. . . . we may shortly be able to sit at home in comfort and watch a thrilling run at an international football match, or the finish of the Derby'.

It is interesting to note that the American Telephone and Telegraph Company employed a similar system to provide the visual equivalent of a public address system at a demonstration given in April 1927. The essential difference between the two equipments lay in the design of the receiver light source: AT & T utilised a special type of multi-electrode glow discharge lamp, whereas Baird, with no laboratory or manufacturing facilities at his disposal, was obliged to consider the use of easily obtainable electric lamps.

By utilising a screen of lamps, the individual lamps may have a considerable light decay time constant, whereas with a single source of illumination the source must vary instantaneoulsy with changes of applied voltage. However, the above scheme would have required *n* times *n* lamps for a square picture scanned by *n* lines, but by mounting the lamps spirally on a rotating disc, Baird was able to reduce the number of lamps needed for domestic viewing to *n*. Each lamp was positioned on the receiver disc in the same place as the corresponding aperture in the transmitter disc and the lamps connected to a commutator at the centre of the disc.

This arrangement was described by Baird in a paper published in May 1924,[29] but as with most of Baird's early writings there was an obvious absence of practical details in the description. Nevertheless, the apparatus, or a similar version to it, was seen by a resident of Hastings who described the crude images it produced. (Fig. 1.10)

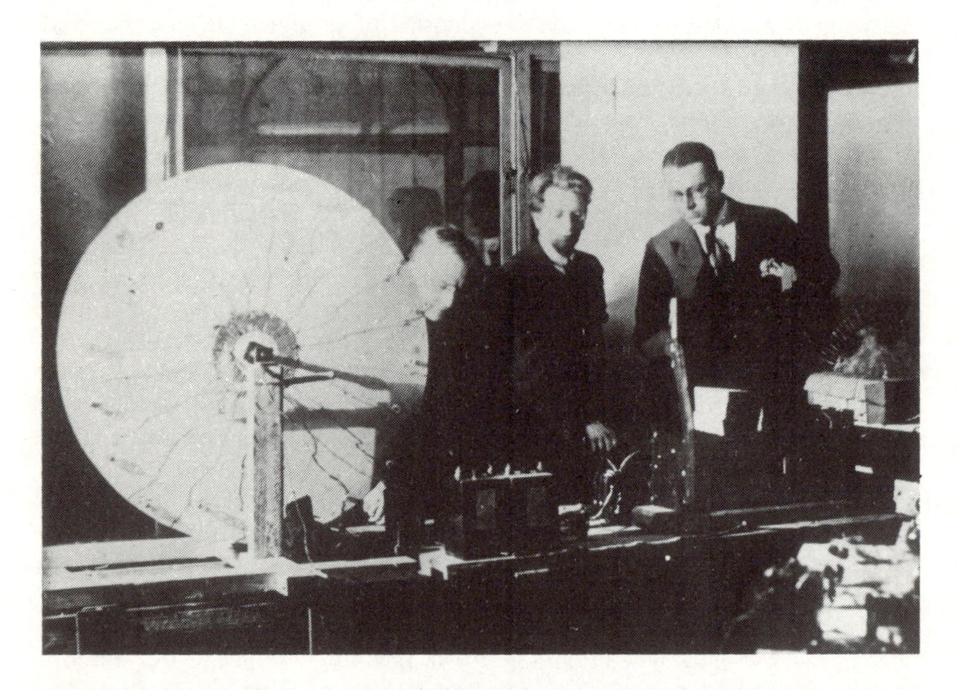

Fig. 1.10 *J. L. Baird demonstrating his apparatus to W. Le Queux (left) and C. Frowd (right) in his laboratory in Queens Arcade, Hastings (early 1924)*

Baird was at this time in urgent need of money. He therefore gave a demonstration of his apparatus to the press and managed to get a mention in the *Daily News* (15th January 1924). A friend of Baird's father saw the *Daily News* paragraph, and mentioned it to him with the consequence that Baird received a much required present of £50.[2]

Another person who probably saw the few lines in the *Daily News* was Mr

Odhams of Odhams Press. He has recalled how he met Baird but seemed to be under the impression that he wrote to Baird following *The Times* advertisement. The dates given in Odhams's account do not agree with the data of this advertisement but do accord with the *Daily News* report.

Odhams wrote to Moseley about his negotiations with Baird many years later, as follows:[2]

> I have pleasure in recalling the story of my pleasant relations with Mr Baird. There appeared in *The Times* an advertisement headed 'Seeing by Wireless' inviting co-operation in the then unknown science of television. The world had just been startled by the practical uses of broadcast sound: What if here was the counterpart – broadcast seeing? I replied to the advertisement the same day and on the 14th was rung up from Hastings by Mr Baird who wanted to call and see me at an early date. An appointment was made for the 16th, and as the unexpected result of a long talk Mr Baird wrote me on the following day:
>
> Re: 'Television Invention'
>
> Dear Mr. Odham,
>
> I have been considering our conversation of yesterday, and will be very pleased to let you have a 20% interest in my Television invention, and its developments, in consideration of your paying the cost of experimental apparatus – not exceeding £100 (one hundred pounds) – and giving introductions likely to prove advantageous.
>
> You will understand that I am anxious to have a decision as early as possible – but I can leave the option open for seven days.
>
> I enclose another cutting giving some further information (the technical details are not accurate).
>
> Trusting the above will be satisfactory.
>
> Yours faithfully,
> JOHN L. BAIRD
>
> W. Odham, Esq.,
> Long Acre
> LONDON
>
> Another appointment was made for the 18th and I was able to have with me a Mr West, the Assistant Chief Engineer of the British Broadcasting Company, then a comparatively young concern. The interview must have taken at least an hour, I being the silent listener to much technical talk. The only thing I can remember understanding was the enormous difference in the waves of light and sound constituted the difficulty of television as compared with broadcasting.
>
> I had another visit from Mr Baird a few days later, and he told me that

he was hung up for various apparatus. These, by the courtesy of Mr. West, I was able to obtain, viz: three DER valves, three LS5 power valves, two RI inter-valve transformers, one three valve note amplifier.

'I hope', wrote Mr West in sending them, 'this will be sufficient for Mr Baird's requirements, but if he requires any more I hope you will ring me and I might be able to send round this afternoon further apparatus.'

Thus equipped, Mr Baird, who had been delayed by a bout of influenza, advised me later that the apparatus was working properly, and a few days after Mr West went down to Hastings to see it. As far as I can recollect, he reported to me that undoubtedly the claims made for television were scientifically sound but that it would be a considerable time before they would be likely to become commercially available. The same view was taken by Mr F. H. Robinson, the Editor of the *Broadcaster*, some months later after a visit to Hastings at my suggestion.

He said his claims were soundly substantiated by the experiments which I saw . . . Mr Baird's invention is, of course, extremely wonderful, but at present I cannot see anything spectacular in it.

With this expert evidence before me, I had to consider the desirability of taking or refusing the handsome offer made to me by Mr Baird in his original letter, and I decided not to avail myself of it because, having already put in over fifty years of strenuous work, I felt it would be unwise to embark on what would evidently be a long-drawn out period of further anxious toil.

Baird later wrote:

Mr Odhams was very charming. he gave me his time and entertained me with a respect and consideration which were as balm to the soul of a struggling inventor instead of being regarded as a dangerous crank. 'Well now, Mr Odhams', I said, 'what kind of demonstration would convince you?' He said: 'If you could put a machine next door, seat someone in front of it, and then on the screen in this room show his face – not a shadow but a face – then I am certain you would get all the money you would want. I am anxious to help and I have discussed this with West and Robinson, but we can see no future for a device which only sends out shadows.[3]

Notwithstanding the lack of financial support from Odhams, Baird did receive help in another way. Both W. Surrey Dane, subsequently a joint manager at Odhams, and John Dunker, editorial chief, were interested in Baird and his work and gave him much needed publicity and encouragement. A few months later the editor of the *Wireless World and Radio Review* was able to mention: 'A good deal of popular interest has been aroused by the experiments in television recently conducted by Mr Baird.'[2]

The concept of television had a popular appeal and in February 1924 the *Radio Times* carried an article[30] headed: 'Seeing the world from an armchair.

When television is an accomplished fact'. After describing certain aspects of wireless broadcasting, such as international broadcasting, the transmission of wireless waves, fading and so forth, the writer posed the question 'What will be the next stage?' He went on 'The answer seems to be television. We have encircled the earth with our music and speech, will the next year enable us to see around the earth with our eyes? Eminent scientists have progressed far along the road at the end of which will be discovered the secret of television, or simply, seeing by wireless.'

The writer, whose name is not known, described Fournier D'Albe's vision of the future:

> It is highly probable, he, (d'Albe), is reported to have said, that we shall be able to sit in, say, the Albert Hall and actually watch the Derby or the 'Varsity Boat Race, or a Naval Review, or a prize fight in America, or, for that matter, a battle. I mean, watch a moving picture of any of these things on a screen, at the moment they are happening . . . As we know now that wireless waves can be relayed almost indefinitely, I see no reason why in ten years time we should not be able to see what is happening on the other side of the globe. It is only a matter of effort in research, and if the public interest is there the effort will be there.

Baird's future colleague Moseley and his business partner Hutchinson were to excite this interest in large measure. The writer ended his article by stating that J. L. Baird had succeeded in transmitting the outlines of objects and that C. F. Jenkins had reproduced his moving hand on the screen. 'These experiments indicate the miraculous linking up of the whole earth by wireless in the not distant future.'

One well-known person who lived in Hastings at this time was the novelist William Le Queux. He was very interested in radio communications and had carried out some radio experiments in Switzerland in 1924 with Dr Petit Pierre and Mr Max Amstutz.[31] Le Queux's fame and interests resulted in him being elected first President of the Hastings Radio Society in 1924.[32] The inaugural lecture to the newly formed society was given by Baird on 28th April 1924 when he talked on 'Television'. The report[33] of the lecture in the local press indicated that Baird was still using selenium cells.

Le Queux attended various demonstrations and he was willing and eager to help Baird but unfortunately all his money was tied up in investments in Switzerland. He did, however, write, an article[34] for the *Radio Times*, in April 1924, with the title 'Television – a fact'. The article is important as it gives some indication of the progress which Baird was making at this time. After mentioning the successful transmission of outline images by Jenkins and Baird, Le Queux wrote:

> In both cases, however, the receiving and transmitting machines were mechanically coupled. Mr Baird has now succeeded in overcoming the great synchronising difficulty and has successfully transmitted images

> between two totally disconnected machines, synchronism being accomplished with perfect accuracy by comparatively simple and inexpensive apparatus.

Baird was still experimenting with small lamps in his receiver but there is no doubt that these could produce crude images.

> My fingers, moved up and down in front of the transmitting lens were clearly seen moving up and down on the receiving disc, and so forth. It remains now to transmit detailed images and a machine to do this has already been designed. A public demonstration will probably be given shortly and then those who listen to broadcasting will be amazed at being able to actually see by wireless. Soon we shall be able to hear and see a thousand miles away.

Le Queux seemed to be using his novelist's imagination to whet public interest.

The machine alluded to in this article was also hinted at by Baird in his first technical article[29] published in May, 1924: 'I am now engaged on aparatus capable of giving a certain amount of detail, considerable modifications have been made, the large revolving disc and lamps described below have been dispensed with.' The details of this new machine were not disclosed until he wrote a further paper, in January 1925, for the same periodical, as he wished to confine himself in the above article to a description of his earlier experiments.

In these experiments, Baird, like most other experimenters, used a selenium cell. As noted previously the time lag of this type of cell had proved a source of constant concern to distant-vision workers. Various attempts had been made to compensate the lag, and Baird, in his early experiments, gave the subject much thought. His patent 235, 619[35] of 12th March 1924 describes a method which he was to employ for about a year in order to overcome the difficulty.

Baird outlined the application of this invention in his first paper as follows:

> In the transmitter an image of the object to be transmitted is focused on a disc rotating at a speed of approximately 200 r.p.m.. The disc is perforated by a series of holes staggered around the circumference. In the experimental apparatus described four sets of five holes were thus arranged: in proximity to this disc revolved a serrated disc at some 2000 r.p.m., and on the other side of this and in line with the focused image of the object to be transmitted there was a single selenium cell connected to a valve amplifier.

Baird's rotating serrated disc was a light chopper, a device which is still used in certain applications. It enables the picture information to be sent by means of a modulation of a carrier signal and its utilisation eased the problem of low frequency amplification.

Baird wrote about his method as follows:

> Selenium is instantaneous in its response to light, that is to say, the instant light falls on it, it begins to change its resistance. Time, however, is

> required for this effect to reach its maximum, and this property of selenium known as its 'chemical inertia', or 'time lag', is fatal to any system which depends simply upon passing the image over a selenium cell, as in passing along a light strip of the image, the resistance is very much lower at the end of the strip than at the beginning, and again in passing along a dark strip the resistance increases as the strip is traversed.
>
> The use of the rapidly revolving serrated disc overcomes this, as the actual resistance of the cell at any instant is not of consequence, it is the pulsations which are transmitted. To make the matter clearer it might be said that light was turned into sound. Loud for the high-lights, low for darker areas and complete silence for darkness.

Although Baird did not patent this important idea until 12th March 1924, a photograph published in the 19th January 1924 issue of the *Hastings and St Leonards Observer* clearly shows the serrated disc incorporated in his apparatus.

Baird gave much thought in 1924 and 1925 to the problem of improving the unsatisfactory performance of selenium cells. He was at this time faced with a real dilemma. The photoconductive selenium cells which he was utilising were moderately sensitive, easy to manufacture but suffered from the serious drawback mentioned above and thereby gave rise to blurred television images. On the other hand, photoemissive cells had a much superior transient response but their sensitivity was extremely low, and most probably below the capacity of Baird's earliest amplifiers. Baird chose to persevere with the use of selenium cells and noticed that when such a cell was exposed to a change of light flux the current would change rapidly at first and then more slowly. By employing a rate of change of current signal in addition to the cell's current he tried to minimise the effect of the lag.

This solution to the problem of reducing the inertia effect of selenium cells was put forward as a claim in Baird's patent 270,222 of 21st October 1925.[36]

Prior to this date many investigations had been undertaken by research workers on the dependence of the inertia effect of such cells on the duration and strength of the illumination, on the heat treatment of the selenium, on the thickness of the photoconductive layer, on the previous illumination, on the colour of the incident light, on the purity of the selenium and the nature of the electrodes, on the voltage applied to the cell, and on the cell's temperature.

Of particular interest to television experimenters was the behaviour of a selenium cell exposed to intermittent illumination. This point had been studied by Nisco, Glatzel, Romanese, Bellati, and Majoranda.[37]

How much of this work was known to Baird in 1923–1924 is a matter for conjecture, but as he was living in the non-university town of Hastings at that time it seems unlikely that he had easy access to the learned journals in which the findings of the above workers were published.

Various attempts had been made by experimenters to minimise the undesirable property of the selenium cell during the development of picture telegraph systems and of these those of Szczepanik (1895), Korn (1906), Zavada (1911), and Cox (1921) were of some importance.

Baird's solution to the problem had a characteristic simplicity and consisted in adding to the output current of the selenium cell a current proportional to the first derivative of the output current. Fig. 1.11 taken from the patent shows the result of adding these two currents together. In this patent the inventor described several circuit arrangements, for accomplishing the desired effect, using passive circuit elements. It appears strange that the method was not advanced and utilised by the early developers of picture telegraphy and this fact further illustrates Baird's ability to propose a simple, realistic solution to a problem which had for many years been considered difficult, if not intractable.

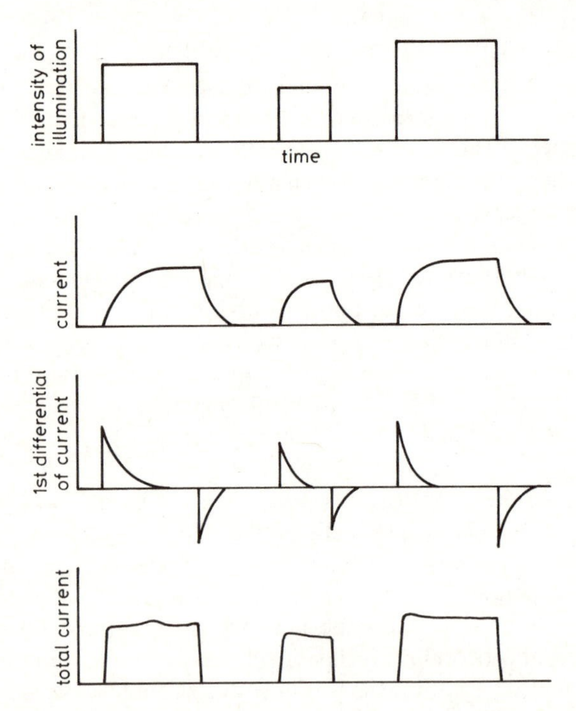

Fig. 1.11 *Baird's patent 270,222 describes a method for ameliorating the effect produced by the utilisation of selenium cells in television system. The combination of the first derivative of a cell's current and the current itself leads to an improved response*

A similar solution was employed by Dr F. Gray, of the American Telephone and Telegraph Company, and was described by him in a memorandum,[38] dated 27th February 1926, on 'Correcting for the time lag in photo-sensitive cells. Gray wrote:

> We may consider the light intensity as the integral of a series of sine and cosine components. For good transmission, the cell must transform these

> components into alternating electric current components of the same relative amplitude and phase relation.
>
> Cells that exhibit time lag do not meet this requirement. For certain cells, however, the error in phase relation of the transformed components is negligible over the frequency region concerned in picture transmission. Such cells behave as if they transformed low frequency components into current much more effectively than they transform the high frequency components.
>
> It is therefore possible to approximately correct for the time lag of such cells by introducing in the circuit an element that will equalise the components in the proper manner. This can be done either by an element that will attenuate low frequencies more than high frequencies, or by an element that will amplify high frequencies more than low frequencies.

Gray used a simple *CR* circuit in his amplifiers to enhance the gain at the high frequencies.

In his early articles and patents Baird did not give any details of the amplifiers which he employed. Bridgewater, who at one time worked for Baird before transferring to the BBC (he later became Chief Engineer, BBC Television), has written:[7]

> It has been said of him: 'An amplifier was just a necessary and rather unimportant box to him, and one amplifier was very like another. 'Undoubtedly an oversimplification, but electronics was not Baird's strong point. Baird was principally interested in optics and mechanics and it is significant that he sought, initially, for a solution to the difficulty of using selenium cells by recourse to a mechanical rather than an electrical device.
>
> However he fully realised the importance of amplifiers and many years later referred to Sir Ambrose Fleming's work on valves in the following way: 'His invention of the thermionic valve revolutionised wireless communications and is in my opinion by far the most valuable invention of the century.'

One of the major problems which had to be solved in any system of television by wireless using Nipkow discs, Weiller drums or other mechanical scanners at the transmitting and receiving stations was the need to maintain the two scanners in synchronism. Baird worked on this problem in 1923–24 and on the 17th March 1924 applied for a patent[39] with the title 'A system of transmitting views portraits and scenes by telegraphy or wireless telegraphy'. Baird devised a suitable scheme in a characteristic way; his solution was the simplest, most robust and probably the cheapest which could be used – but it worked.

He wrote:

> To obtain isochronism an alternating current generator is coupled to the shaft of the transmitter and current from it controls the speed of a synchronous motor driving the receiving machine. To obtain synchronism

the driving mechanism is rotated about the spindle of the receiver until the image comes correctly into view.

While this method enabled the transmitting and receiving discs to be mechanically decoupled, the solution, of necessity, involved the transmission of low frequency alternating currents, in addition to the signal currents which were of a higher frequency. Baird was not at this time working on the problem of transmitting both the signal and synchronising currents as a modulation on a carrier wave. This had to wait several years for a solution. There is no doubt that Baird's method worked, as Le Queux was able to confirm in his *Radio Times* article.

The above patent further illustrates Baird's gift for reducing a problem to its simplest statement and then finding an equally simple solution. Not for him were the complicated, fragile and expensive mirror scanning systems of Szczepanik and Mihaly or the more robust but still costly mirror drums of Weiller. What could be simpler, stronger or cheaper than a Nipkow disc cut from an old hat box and pierced with a darning needle? Baird was short of finance and had to find answers to his questions using almost literally the inventors' beloved sealing wax, string and glue (Fig. 1.12). He could not have engaged in research work on cathode-ray oscilloscope systems but had to make do with the meagre and crude facilities at his disposal. It is to his everlasting credit that he achieved any results at all when faced with such daunting difficulties.

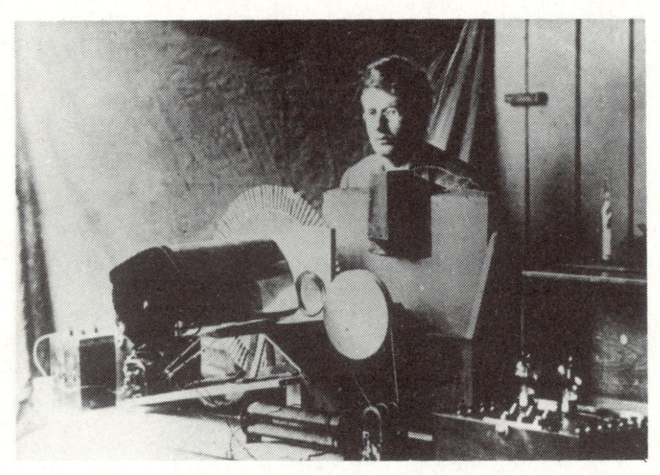

Fig. 1.12 *J. L. Baird with his apparatus. Note the serrated disc which was employed as a light chopper (late 1923)*

However, old electric motors and generators could be obtained from a scrap metal merchant for a few shillings, and so Baird, probably using his experience as an assistant mains engineer with the Clyde Valley Electricity Company during the war years, tackled the problem of synchronisation and solved it, together with many others, so that by early May 1924 *Amateur Wireless* was incautious enough to state 'so much progress has been made with television that we may regard it already as an accomplished fact . . . the main problem has been solved . . .'.[40]

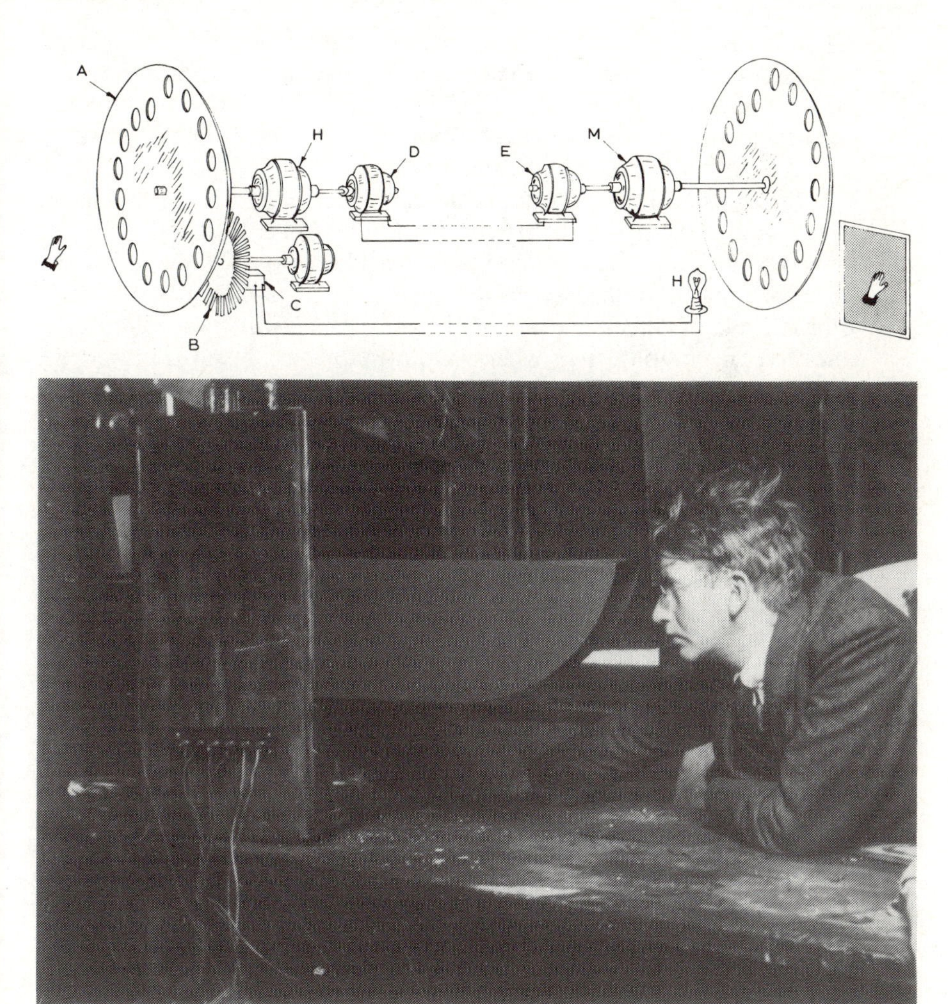

Fig. 1.13 *(a) Baird's television system, as described by him in his paper published in the Wireless World* and *Radio Review* (January 1925). *(b) Baird viewing an image produced by one of his early televisors. (c. 1925)*

From Hastings Baird moved to a small attic at 22 Frith Street, London, found for him by his associate Will Day. Here he continued his researchers which he described in a paper published in a January issue of the *Wireless World and Radio Review* for 1925,[41] see Figs. 1.13.

Baird was at this time attempting to transmit an image of an actual object, not by having a source of light behind it, but by reflecting light from it. This was a problem of quite a different order of difficulty to the transmission of shadowgraphs, as Baird explained.

> The distinction between the transmission of shadowgraphs and that of actual objects is much greater than is at first apparent. With shadowgraphs the light-sensitive cell is only called upon to distinguish between total darkness on the one hand and the full power of the source of light, possibly several thousand candle power, on the other.
>
> In the transmission, however, of actual objects even where only black and white are concerned, the cell has to distinguish between darkness and the very small light, usually indeed only a small fraction of a candle power, reflective from the white part of the object. The apparatus has therefore to be capable of detecting changes of light, probably at least a thousand times less in intensity than when shadowgraphs are being transmitted. There are other optical problems which render the transmission of the actual light much more difficult of the two, but the above is sufficient indication of the difficulty of the step.
>
> The apparatus now used consists of a revolving disc containing a single spiral of lenses, and behind this, a serrated shutter revolving at high speed. The object to be viewed is placed in front of the disc which revolves at about 500 revolutions per minute. The arrangement will be better understood by referring to the accompanying drawing [Fig. 1.13b]. The hand is the object being transmitted. As the disc revolves, light from every part of the hand falls in succession on to the light-sensitive cell 'C' after having been interrupted at very high frequency by the serrated disc 'B'. This interupted light, as it falls upon the cell, gives rise to a pulsating current in its circuit. The pulsating current after amplification is transmitted to the receiving machine, which consists of a disc containing a spiral of lenses exactly similar to the disc at the transmitting end, and revolving in synchronism with it. Behind this disc, in a position corresponding to that of the light-sensitive cell at the transmitting station, is a lamp which is fed by the received current. The varying point from this lamp traverses a ground glass screen and reproduces there the hand which is being transmitted.

Unfortunately Baird gave no clue as to the nature of this lamp or of the light-sensitive cell except to mention that it was neither a photoemissive nor a selenium cell, but 'a colloidal (fluid) cell of my own invention . . .'. He did mention the method of synchronisation – the same as quoted in his patent of 17th March 1924 (236, 478).

Baird was very modest in reporting the success he had achieved:

> The letter 'H' for example, can be clearly transmitted, but the hand, moved in front of the transmitter, is reproduced only as a blurred outline. A face is exceptionally difficult to send with the experimental apparatus, but, with

> careful focussing, a white oval, with dark patches for the eyes and mouth, appears at the receiving end and the mouth can be clearly seen opening and closing.

The sequence and detail of the continuous experiments that Baird pursued relentlessly during the 1923–1926 period are not on record, for Baird kept no regular notes.[42] He did not keep a laboratory notebook and the very few articles which he published were usually of an elementary character. J. D. Percy, who later worked with Baird, has said:

> . . . it is true, I think, to say, that his mind worked too quickly for the satisfactory recording of his myriads of ideas. Explanation to those of us who worked with him was amplified only by rough sketches on the backs of old envelopes, by scrawling diagrams on walls (the Long Acre Laboratories were literally covered in these) or even occasionally on the tablecloths of restaurants and inns.

However, there are the occasional press notices, the memories of a few of his colleagues and some general recollections by Baird himself written nearly 20 years later. His patents, of which he took out a considerable number from mid-1923 onwards, also record some of his thoughts.

Apart from the occasional notices in the newspapers the general public first became aware of television in April 1925. Selfridge's had been on the look out for an attraction for their Birthday Week and George Selfridge had visited Baird. The consequence was that Baird was offered twenty pounds a week to give three demonstrations of television a day to the public in Selfridge's store (Fig. 1.14). Baird, of course, was not so concerned with the publicity value of this enterprise: rather it was a case that he could not refuse the weekly cheque without which it would have been difficult for him to have carried on.[25]

The explanation for the rather surprising meeting between the large store owner and the struggling inventor is that it came about because Baird's neighbour in Hastings was a person called Bosdari who was a friend of Selfridge. Bosdari consequently knew of Selfreidge's desire to have a special feature in his Oxford Street shop and told him of Baird's work.[10]

> It was in Frith Street, in September 1925, that there was a knock at my door one day and there stood Bosdari and George Selfridge. I showed them in. Mr Selfridge asked me if I would demonstrate my machine in his shop during the birthday celebrations for £20 a week. [Baird's account is slightly inaccurate: the month was April, not September.]

Baird spent three weeks demonstrating to long queues of spectators, most of them shoppers, but with a few scientists who had come especially to see the show. They were handed leaflets by shop assistants which read:

> Television is to light what telephoning is to sound. It means the instantaneous transmission of a picture so that the observer at the receiving end

can see, to all intents and purposes, what is a cinematographic view of what is happening at the 'sending' end . . .

The apparatus here demonstrated is, of course, absolutely 'in the rough' – the question of finance is always an important one for the inventor. But it does, undoubtedly transmit an instantaneous picture. The picture is flickering and defective, and at present only simple pictures can be sent successfully; but Edison's first phonograph rendered that 'Mary had a little lamb' in a way that only hearers who were 'in the secret' could

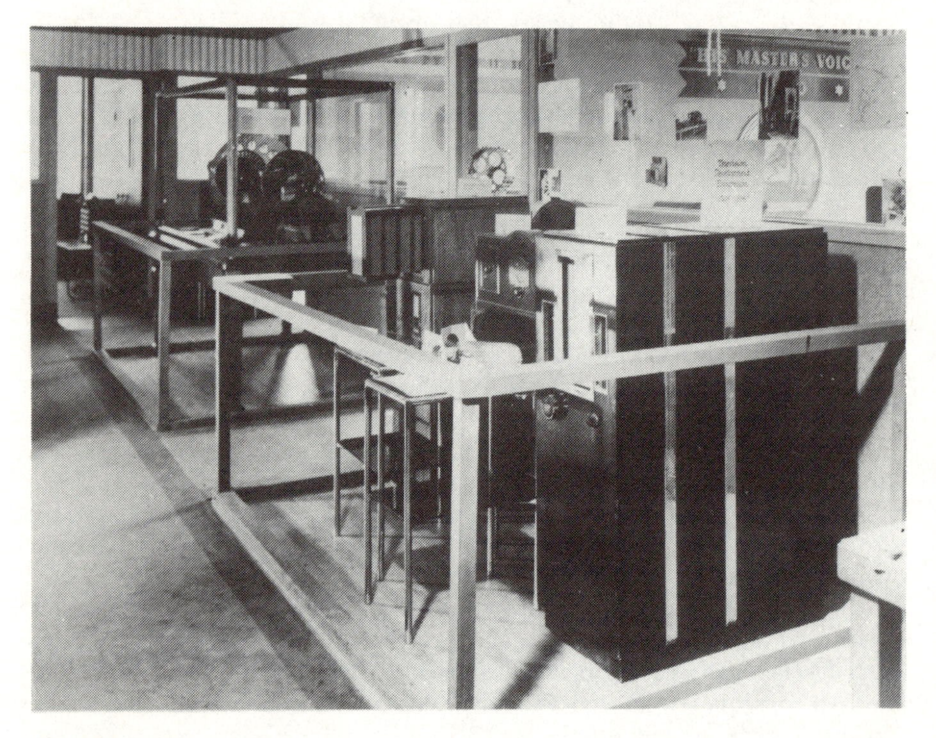

Fig. 1.14 *Selfridge's store in Oxford Street, London, featured aspects of Baird's work in 1925, 1928, 1932 and 1939. The Television Exhibition which opened in February 1939 included the apparatus used by Baird in 1925 (shown on the left of the photograph)*

understand and yet, from that first result has developed the gramophone of today . . .

We should perhaps explain that we are in no way financially interested in this remarkable invention, the demonstrations are taking place here only because we know that our friends will be interested in some thing that should rank with the greatest inventions of the century.

Selfridge and Company Ltd.

A description of the apparatus which Baird used for the demonstration is given in *Nature* (4th April 1925).[43] Unfortunately the writer seemed uncertain about some aspects of the equipment. He referred to 'a selenium *or other* photo-electric cell' and 'a disc with lenses *or* holes' in his description. It would seem unlikely that Baird experimented with each of these possibilities during the hectic three weeks he spent at the store. Moseley has described how the strain of giving three shows a day with the rickety apparatus was too much for Baird and he was ill for several weeks afterwards.[2]

The article in *Nature* does, however, include an important new fact, namely, that Baird was now using a neon lamp in his receiver rather than the arrangement described in his 1924 article. Otherwise Baird's television apparatus was very similar to his earlier arrangement which he mentioned in the *Wireless World and Radio Review*.

The crudeness of the images hinted at in Selfridge's handout was confirmed in the report published in *Nature*: 'Mr Baird has overcome many practical difficulties, but we are afraid that there are many more to be surmounted before ideal television is accomplished.'

Baird had few competitors in 1924–25 in this country, and consequently, whenever television was mentioned in the press, Baird's name tended to be coupled with the report. He was the only experimenter who was able to give demonstrations of his equipment and to show the transmission of crude outlines. However, during the month that Baird gave his Selfridge shows, Dr Fournier d'Albe, who had made a name for himself by his invention of the optophone – an instrument which allowed blind persons to read – was giving a private demonstration of his television apparatus at his laboratory at Kingston upon Thames.[44] He patented his ideas in January 1924 in a patent[45] titled: 'Telegraphic transmission of pictures and images' and a few months later (October), adapted the equipment to record and reproduce sounds.[46] D'Albe's method was doomed to failure. Whereas Baird had adopted, along with most other investigators, a sequential scanning arrangement, d'Albe experimented with a system for transmitting information about the luminosity of the elements of a picture simultaneously. His apparatus, therefore, belongs to the same group as those of Carey and Lux.

The general arrangement of d'Albe's original method is shown in Fig. 1.15. An essential feature was the inclusion of a 'means for breaking up the light'. This consisted of seven transparent cylinders each divided into seven sections which were split up by opaque parts as indicated by black lines. Each section had twice as many sections as the preceeding section and the cylinders rotated at different speeds. The effect of this was to produce alternate 'brightenings and darkenings in the light transmitted across it'. The light was then collected by a lens and projected onto a selenium cell. 'The medley of electrical impulses produced in the selenium' was subsequently 'converted into sound by any of the methods usual for this purpose' and transmitted to the receiver. Here they excited a loudspeaker and 'the medley of sounds' was analysed by a set of resonators equal in number to the number of patches in the original picture or image, each resonator responding to only one pitch. The resonators were so constructed that they produced a luminous patch on a screen when a note of their own pitch was contained in the acoustic output of the loudspeaker. This was achieved by having reeds, to which were attached small mirrors, fixed to the resonators so that when the resonators sounded the reeds vibrated in sympathy with them. The light reflected from the reeds was then directed on to a screen in such a way as to correspond to the original patches of the picture transmitted.

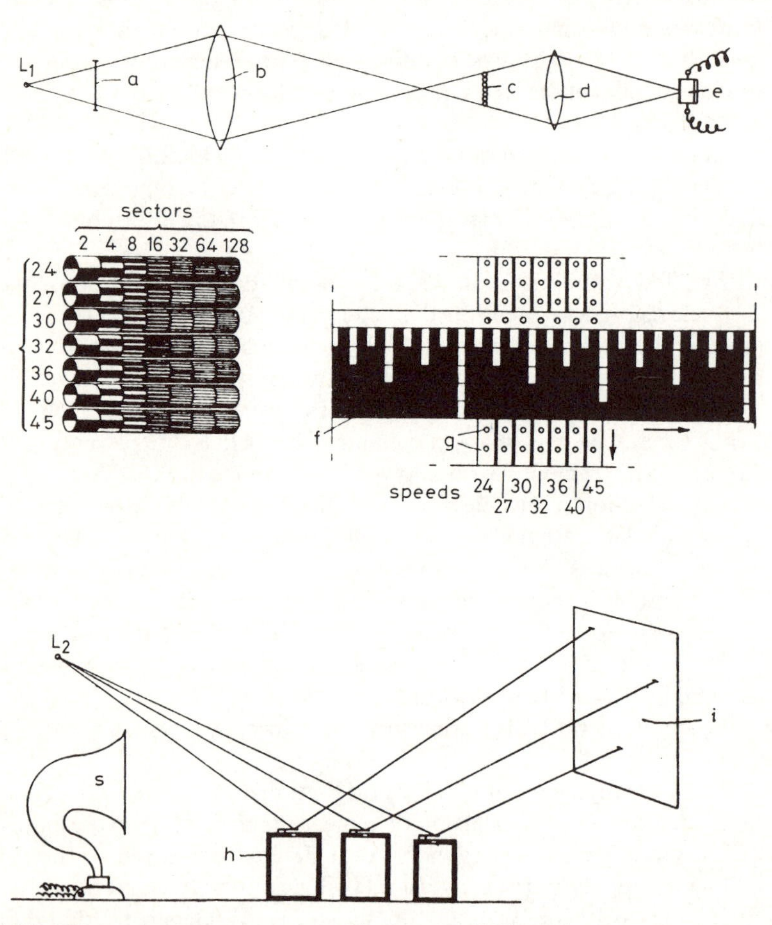

Fig. 1.15 *Diagrammatic representation of Dr Fournier d'Albe's television scheme, patented in January 1924*

The apparatus which d'Albe demonstrated on 18th April 1925 represented a further stage in his development of a system which would allow the elements of a picture to be simultaneously transmitted. The account given in *Nature* described the system as follows:

> An image of the object to be transmitted was projected on a revolving siren disc provided with thirty concentric circles of holes. An image of the disc was in turn, projected on a transmitting screen studded with thirty small selenium tablets, arranged so that each tablet was exposed to a different audio-frequency of intermittent light produced by the disc. The selenium tablets were connected in parallel with a two-valve amplifier, and the sound produced in a loud speaker at the receiving station was allowed to act upon thirty compound resonators, each of which responded to its own note when it occurred in the medley of sound transmitted. The response manifested itself by the appearance of a luminous patch on a ground glass screen reproducing an element of the original object both as to position and intensity. As the response occurs within a twentieth of a second, it is claimed that the apparatus transmits some six hundred signals per second.

Unfortunately the report did not indicate whether the image of the object or transparency was successfully transmitted. The writer, however, did mention that 'the complete transmission of an object such as a changing face requires at least ten thousand signals per second' and that therefore 'there is still a considerable gap to be filled'.

D'Albe hoped to do this by increasing the number of resonators and their selectivity, or, in the last resort, by transmitting over more than one wire or on more than one radio wavelength.

It seems surprising that a person of d'Albe's ability should have pursued the above method. Either he was not aware of, for example, Shelford Bidwell's calculations[47] on the number of elements necessary to achieve good television or he chose to ignore them. D'Albe's system was severely restricted, by the response of the loudspeaker used, to audiofrequencies, whereas Baird's use of the Nipkow disc did allow, by an increase in the number of holes used and an increase in the rotational speed of the disc, an extension to high definition television. More than ten years after Baird's initial use of the disc it was still being employed in the world's first, public, regular high definition television system as a film scanner. This example shows, again, Baird's considerable insight into the difficulties associated with seeing by electricity. The solutions which he adopted in the 1923–26 period were probably the only ones which could lead to a practical demonstration of television with the technology known at that time.

Baird's demonstrations at Selfridge's produced a number of unexpected gifts for him. Shortly after his show he was visited by a representative from Hart Accumulators and a few days later received £200 worth of batteries. 'It was a bright spot in the darkness of anxious days', wrote Baird.[3]

GEC too made him a present of £200 worth of valves soon after this. The gifts, of course, were vitally welcome and together with £500 put up by a cousin named Inglis enabled Baird to continue his researches. Baird was most grateful for the GEC donation and later said '. . . for hard headed business men to give £200 of goods to a dilapidated and penniless crank in a garrett is a phenomena worth recording'.

Shortly after the demonstrations at Selfreidge's store the first of the companies associated with the Scottish inventor was formed. This was called Television Ltd., and was registered on 11th June 1925.

The objective[48] of the company was 'to purchase or otherwise acquire from any person or persons lawfully entitled to dispose thereof the whole or any part of the right, title and interest in, or appertaining to the inventions relating to 'A system of transmitting views, portraits and scenes by telegraphy or wireless telegraphy' comprised in and covered by the registered patents no. 222,604 and no. 230,576; also the provisional specifications nos. 4,800, 6,363 and 6,774 of 1924; and nos. 48 and 911 of 1925 . . .'.

Baird ('electrical engineer') and Day ('merchant') were the subscribers of 20 founders shares and the nominal capital of the company was £3000. This was made up of 2900 ordinary shares of £1 each and 2000 founders shares of 1 shilling each. The registered office of Television Ltd. was initially at 22 Frith Street, W1, and the first directors were Baird and Day. Later the company moved to Motorgraph House, Upper St Martins Lane, WC (15th February 1926), and then to 133 Long Acre, WC (9th February 1928).

Following the establishment of Television Ltd., Baird and Day entered into an agreement with the firm (on 12th June 1925), the pertinent points of which were:

> 1 The Vendors shall sell and the Company shall purchase the whole of the Vendors right, title and interest in or benefit of the Patents and Inventions . . .
>
> 2 The consideration for the said sale shall be the sum of £100 which shall be paid and satisfied by the allottment to the Vendors or their respective nominees of 2000 Founders shares of 1s. each in the capital of the company – (1000 shares to J. L. Baird and 1000 shares to W. E. L. Day) – and £500 to be paid by the company after £1500 in shares had been sold and paid for at which time the directors, J. L. Baird and W. E. L. Day may draw upon the company for all or any part of this sum (£500) as they mutually agree.

The precise sequence of experiments which Baird carried out during the months following the Selfridge demonstrations is not known, but on 2nd October 1925 he noticed that, when he viewed the head of the dummy which he used as an object, in his receiver, there was light and shade: he had achieved crude television by reflected light with tone graduation.

Later, in 1931, while giving a broadcast from New York, Baird described this important occasion:

> In 1925 television was still regarded as something of a myth. No true television had ever been shown – only crude shadows. At that time I was

working very intensively in a small attic laboratory in the Soho district of London. Things were very black; my cash resources were almost exhausted and as, day by day, success seemed as far away as ever I began to wonder if general opinion was not after all correct, and that television was in truth a myth. But one day – it was, in fact, the first Friday in October – I experienced the one great thrill which research work has brought me.

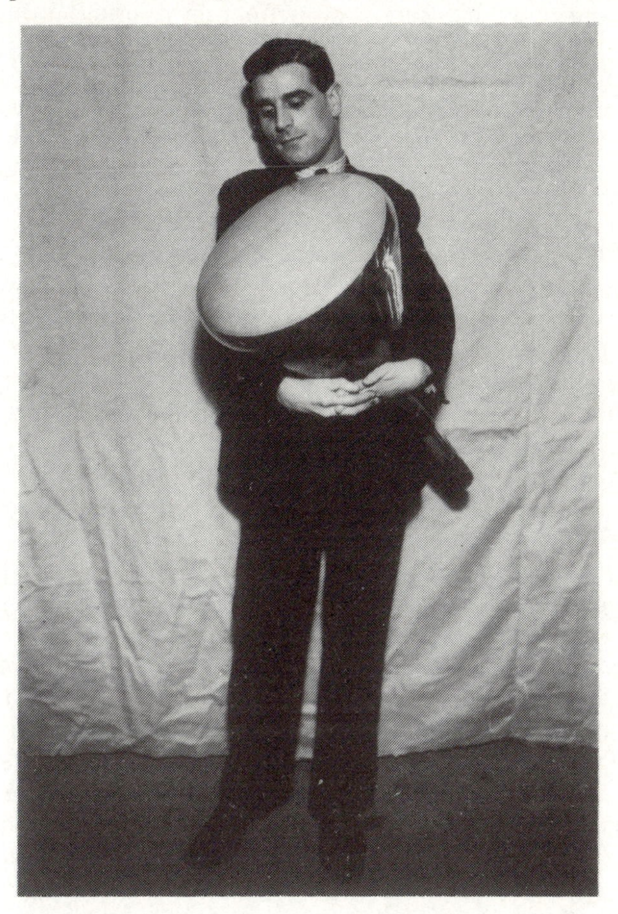

Fig. 1.16 *'I was vastly excited and ran downstairs to obtain a living object. The first person to appear was the office boy, a youth named William Taynton, and he rather reluctantly consented to subject himself to the experiment.' Taynton was televised in October 1925*

The dummy's head . . . suddenly showed up on the screen not as a mere smudge of black and white, but as a real image with details and gradations of light and shade.

I was vastly excited and ran downstairs to obtain a living object (Fig. 1.16). The first person to appear was the office boy from the floor below, a youth named William Taynton, and he, rather reluctantly, consented to subject himself to the experiment. I placed him before the transmitter and went into the next room to see what the screen would show. The screen was entirely blank, and no effort of tuning would produce any result. Puzzled, and very disappointed, I went back to the transmitter and there the cause of the failure became at once evident. The boy, scared by the intense white light, had backed away from the transmitter. In the excitement of the moment I gave him half-a-crown, and this time he kept his head in the right position. Going again into the next room I saw his head on the screen quite clearly. It is curious to consider that the first person in the world to be seen by television should have required a bribe to accept the distinction.

The success which Baird accomplished on the 2nd October 1925 (see Figs. 1.17) posed a dilemma for him. He urgently needed publicity and funds, but feared that there was a danger of television being exploited and developed by powerful interests. Three months were to elapse before Baird had the courage to demonstrate his results. As he himself said '. . . I was extremely nervous in case while I waited someone else achieved television – terrified that someone would copy my work, and particularly frightened that big wireless concerns would take up television research and use my work as a guide . . .'

Fortunately, during this period of indecision Baird met,[2] by chance, Captain O. G. Hutchinson, with whom he had had talks in 1922 about a merger of their soap interests.

Hutchinson became Baird's business partner towards the end of 1925 (the date is not known precisely) and together they embarked on a programme of expansion which was to cause Baird and his work to become alienated from certain sections of the scientific community – principally the British Broadcasting Corporation – because of the publicity methods adopted by Hutchinson.

The appointment of Hutchinson enabled Baird to concentrate on laboratory work. As a consequence, he was able to devote all his energy and inventive skills to the furtherance of television without being encumbered by the need to attend to business matters: he was able to demonstrate a number of applications of his basic scheme before any other person or industrial organisation succeeded in doing so.

Hutchinson's first task was to obtain some much needed extra financial assistance for Baird's efforts. This he did by reorganising the nominal capital of Television Ltd., and persuading various persons and bodies to take up shares in the company. Day's interest in Television Ltd. was purchased by Hutchinson, and Day resigned his directorship on 16th December 1925.[48]

The special resolutions which were needed to accomplish the reshaping of the company were passed on 15th February 1926 and confirmed on 4th March 1926. They provided that:

1 the 2900 Ordinary Shares in the company shall in future be called and known as First Participating Cummulative Preference Shares;

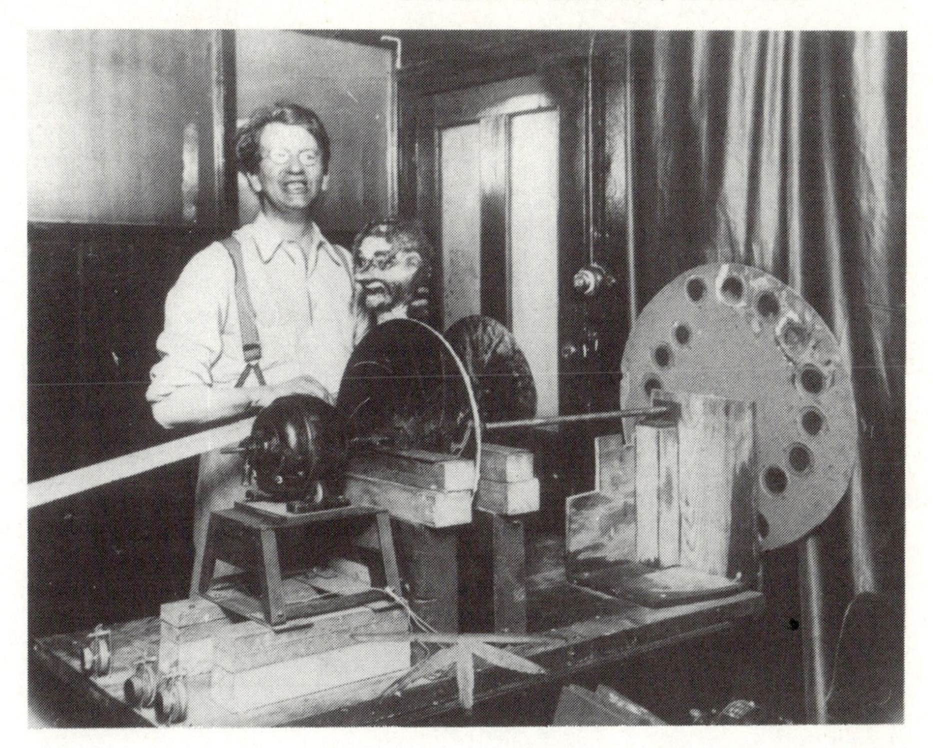

Fig. 1.17 *(a) Baird with the television apparatus which he presented to the Science Museum.*
(b) Baird television apparatus of 1925, now in the Science Museum, London

2 the 2000 Founders Shares in the capital of the company shall in future be called and known as Ordinary Shares;
3 the capital of the company be increased to £9050 by the creation of 6000 new First Participating Cummulative Preference Shares of £1 each and 1000 new Ordinary Shares of 1s each;
4 the rights specified in the articles of association of the company be attached to the said First Participating cummulative Preference Shares and to the Ordinary Shares respectively.

Thus the nominal capital of Television Ltd. changed as follows:

Pre-4th March 1926	Post-4th March 1926
£3000	£9050
2900 ordinary shares of £1 each	8900 FPCP shares of £1 each
2000 founder shares of 1 shilling each	3000 ordinary shares of 1 shilling each

By 7th September 1926, 7320 FPCP shares and the 3000 ordinary shares had been taken up. Some of the holdings of these shares by certain individuals are given below:

	Ordinary shares	FPCP shares
Baird	1125	
Hutchinson	625	
Broderip	625	10
Baird Hutchinson Broderip (jointly)	625	
Bishopgate Nominees Ltd.	2000	
G. A. Inglis		500
J. D. Inglis		750

Altogether 43 persons or organisations had an interest in Television Ltd. less than one year after Hutchinson became Baird's business partner. The Inglis' referred to in the above list were wealthy shipbuilders and were cousins of Baird — his mother's maiden name was Inglis. Broderip was an associate of Hutchinson and was a colliery proprietor.

Changes took place in the directorate of Television Ltd. following Day's departure. The directors on 3rd July 1928 were Baird, Hutchinson and Broderip: two years later, on 12th March 1930, two other names were added, namely J. D. Inglis and H. J. Whitcomb.

Larger premises were acquired and in February 1926 Baird moved his apparatus from the small room he rented at 22 Frith Street to Motorgraph House in Upper St Martins Lane.

Prior to this transfer Baird and his friends had arrived at a solution to the dilemma which faced them after the successful crude televising of William Taynton. A compromise was proposed — only selected individuals would be given a demonstration of television. Invitations were sent out to members of the Royal Institution, and to *The Times* to represent the press, '. . . . these would give dignity and importance to the occasion'.[2]

References

1 JOLLY, W. P.: 'Marconi' (Constable, London, 1922)
2 MOSELEY, S. A.: 'John Baird' (Odhams, London, 1952)
3 BAIRD, J. L.: autobiographical notes quoted in Moseley, *op. cit.*
4 BAIRD, M.: 'Television Baird' (HAUM, South Africa, 1974)
5 BAIRD, J. L.: autobiographical notes quoted in M. Baird, *op. cit.*
6 Calender for the year 1914–15 of the Royal Technical College, Glasgow
7 BRIDGEWATER, T. H.: 'Baird and television', *Journal of the Royal Television Society*, **9**, No. 2, pp. 60–68
8 KEMPSELL, A.: 'The man with many dreams', *Helensburgh Times*, 3rd September 1975, p. 15
9 LANGER, N.: 'A development in the problem of television', *The Wireless World and Radio Review*, 11th November 1922, pp. 197–210
10 WALLER, C.: 'This new wonder television', *Sunday Express*, 26th February 1939
11 BAIRD, J. L.: 'Television', *Journal of Scientific Instruments*, 1927, **4**, pp. 138–143
12 ECKERSLEY, P. P.: quoted in MOSELEY, S. A., *op. cit.*, pp. 250–251
13 SMITH, W.: letter to Latimer Clark, *Journal of the Society of Telegraph Engineers*, 1873, **2,** p. 31
14 BAIN, A.: 'Certain improvements in producing and regulating currents and improvements in electric time pieces and in electric printing and signal telegraphs', British patent no. 9745, 27th November 1843
15 BAKEWELL, F. C.: 'Electric telegraphs', British patent no. 12352, 2nd June 1849
16 BOYER, J. M. J.: 'La transmission telegraphique des images et des photographique', Paris, 1914
17 SALE, Lt.: 'The action of light on the electrical resistance of selenium', *Proc. Roy. Soc.*, 1873, **21**, pp. 283–285
18 Anon.: 'When John Baird experimented in Folkeston', *Folkeston and Hythe Herald*, 25th February 1961
19 BENNETT, P. A.: private communication to R. W. Burns
20 Anon.: 'Pioneer of television research', *Folkeston and Hythe Herald*, 3rd June 1961
21 MILLS, V. R.: private communication to R. W. Burns
22 Anon.: 'Inventor injured', *Hastings and St Leonards Observer*, 26th July 1924
23 WATKINS, H.: 'The first public demonstration', *London Home Magazine*, May 1962, p. xii
24 Personal Column, *The Times*, 27th June 1923
25 TILTMAN, R. F.: 'Baird of television' (Seeley service, London, 1933)
26 NIPKOW, P.: 'Elektrisches Teleskop', German patent no. 30105, 6th January 1884
27 BAIRD, J. L., and DAY, W. E. L.: 'A system of transmitting views, portraits and scenes by telegraphy or wireless telegraphy', British patent no. 222 604, 26th July 1923
28 Anon.: 'Seeing by wireless', *Chambers Journal*, November 1923, pp. 766–767
29 BAIRD, J. L.: 'An account of some experiments in television', *The Wireless World and Radio Review*, 7th May 1924, pp. 153–155

30 Anon.: 'Seeing the world from an armchair. When television is an accomplished fact', *Radio Times*, 15th February 1924, p. 301
31 Anon.: report in the *Hastings and St Leonards Observer*, 26th April 1924
32 Anon.: report in the *Hastings and St Leonards Observer*, 12th April 1924
33 Anon.: 'Television', *Hastings and St Leonards Observer*, 3rd May 1924
34 LE QUEUX, W.: 'Television – a fact', *Radio Times*, 25th April 1924, p. 194
35 BAIRD, J. L., and DAY, W. E. L.: 'A system of overcoming the time lag in a selenium or other light sensitive cell used in a television or like system', British patent no. 235 619. 12th March 1924
36 BAIRD, J. L.: 'Improvements in or relating to television systems and apparatus', British patent no. 270 222, 21st October 1925
37 BARNARD, G. P.: 'The selenium cell: its properties and applications' (Constable, London, 1930)
38 GRAY, F.: internal memorandum, 27th February 1926, A T and T Co., Warren Record Centre, USA
39 BAIRD, J. L., and DAY, W. E. L.: 'A system of transmitting views, portraits and scenes by telegraphy or wireless telegraphy', British patent no. 236 978, 17th March 1924
40 Editorial, *Amateur Wireless*, May 1924
41 BAIRD, J. L.: 'Television. A description of the Baird system by its inventor', *Wireless World and Radio Review*, 21st January 1925, pp. 533–535
42 PERCY, J. D.: 'The founding of British television', *Journal of the Television Society*, 1950, pp. 3–16
43 Anon.: a report, *Nature*, 4th April 1925, pp. 505–506
44 Anon.: a report, *Nature*, 25th April 1925, p. 613
45 D'ALBE, E. E. F.: 'Telegraphic transmission of pictures and images', British patent no. 233 746, 15th January 1924
46 D'ALBE, E. E. F.: 'Improvements in apparatus for recording and reproducing sound,' British patent no. 247 629, 23rd October 1924
47 BIDWELL, S.: 'Telegraphic photography and electric vision', *Nature*, 4th June 1908, pp. 105–106
48 Company file, Television Ltd., Public Record Office

Chapter 2

The first demonstration, 1926

The chosen date for the first public demonstration of television was 26th January 1926. *The Times* report for 28th January was the only press statement obtained first hand of this historic event.

> Members of the Royal Institution and other visitiors to a laboratory in an upper room in Frith Street, Soho, on Tuesday saw a demonstration of apparaus invented by Mr J. L. Baird . . .
>
> For the purpose of the demonstration the head of a ventriloquists' doll was manipulated as the image to be transmitted, though the human face was also reproduced. First on a receiver in the same room as the transmitter, and then on a portable receiver in another room, the visitors were shown recognisable reception of the movements of the dummy head and of a person speaking. The image as transmitted was faint and blurred, but substantiated a claim that through the 'Televisor', as Mr Baird has named his apparatus, it is possible to transmit and reproduce instantly the details of movement, and such things as the play of expression on the face.[1]

So many members of the Royal Institution accepted Baird's invitation that when they foregathered in Frith Street they had to climb the stairs and enter the attic 'laboratory' in batches of six at a time. While *The Times* reporter was the only national newspaper journalist present on the above occasion, another demonstration had been given a few days earlier, on 7th January 1926, to a reporter of the *Daily Express*.[2] This consisted of the transmission by wireless of a moving picture of Captain O. G. Hutchinson. The *Daily Express*, and *Evening Standard*, dutifully observed that this transmission took place not many yards away from the Soho house where in 1889 another British pioneer, William Friese-Green, showed the world's first moving pictures to a few friends.

Looking back at Baird's achievement it may seem that his choice of transmitting and receiving Nipkow disc scanners, neon lamp and electrically compensated selenium cell and his method of obtaining synchronism between two scanners based on the use of a synchronous motor and a.c. generator was obvious and that it really was surprising that no one had experimented with this particular combination of system components before 1923. The possibilities for component selection and invention, however, were very great.

Almost every conceivable type of scanner and display device were suggested in the period 1877–1936:[3] there were vibrating mirrors, rocking mirrors, rotating mirrors, mirror polyhedra, mirror drums, mirror screws, and mirror discs; there were lens discs, lens drums, circles of lenses, lenticular slices, reciprocating lenses, lens cascades, and eccentrically rotating lenses; there were rocking prisms, sliding prisms, reciprocating prisms, prism discs, prism rings, electric prisms, lens prisms, and rotating prism pairs; there were apertured discs, apertured bands, apertured drums, vibrating apertures, intersecting slots, multispiral apertures, and ancillary slotted discs; there were cell banks, lamp banks, rotary cell discs, neon discs, corona discs, convolute neon tubes, tubes with bubbles in them; there were cathode-ray tubes, Lenard tubes, X-ray tubes, tubes with fluorescent screens, gas screens, photoelectric matrices, secondary emitting surfaces, electroscope screens and Schlieren screens.

However, of all the many different mechanical scanners listed here the Nipkow disc was the most robust, the most versatile in operation, the simplest and the cheapest. It could be made from cardboard or metal, in a workshop or home, by the skilled or the unskilled, for a few pence or many pounds.

Nipkow's invention was greatly developed after 1926: the scanner was used in 30, 50, 60, 90, 120, 180, and 240-line television systems. It was run in air and in vacuum; it was constructed with single spirals and spiral segments; it was provided with round holes, square holes, diamond shaped holes and holes of various other shapes; it was designed to produce orthodox scanning, discontiguous scanning, graduated scanning, overlapping scanning, and when pairs of discs were employed, interlaced scanning, and bilateral scanning; it was produced in apertured hole and lensed form; it was used by amateurs and professionals alike. It was employed for several years in telecine equipment and enabled films to be shown with a degree of clarity and brilliance which surpassed that which could be shown by apparatus which used the emitron camera.

The Nipkow disc was utilised by the American Telephone and Telegraph Company for their ambitious and well-engineered field tests in 1927, and during the succeeding years, the Nipkow disc formed the essential part of the early television schemes of Mihaly, Karolous, Fernseh A. G. and Telefunken in Germany, of Sanabria, Jenkins, GEC and others in the USA, of Baird, Marconi and HMV in Great Britain, of Faeber in Czechoslovakia and of several others.

Baird's achievement may be weighed by considering the work of the Admiralty Research Laboratory (ARL) on television. The objective of the Admiralty in conducting experiments in this field was 'for spotting at sea with the use of aeroplanes'. A university-trained research scientist and others commenced the investigation in 1923. In January 1925, Dr C. V. Drysdale, the superintendent of the laboratories, described the problem as difficult but felt that it could be solved with 'money and staff'. Approximately 17 months later he had modified his opinion and referred to the extreme difficulty of finding a practical solution. An inspection of ARL's television equipment, which included a photoelectric cell made by the National Physical Laboratory (NPL), was undertaken by two representatives of the Air Ministry on 27th May 1926, and during this an image of an object consisting of a grid of three bars of cardboard, each about 0.25 inch (6.35 mm) in width and 0.25 inch (6.35 mm) apart, was transmitted. Although transmitted light was employed the object 'could just be recognised at the receiving end, but the reproduction was very crude'.

News of Baird's success travelled fast, for two days after the demonstration to members of the Royal Institution, the Press Photonachtrichten Dienst of Berlin wrote to the General Post Office requesting details of the Baird system.[4] (Later Baird, Hutchinson and Moseley were to make many visits to Germany to advance their companies. They were always treated well and with respect and much to Baird's and Moseley's disgust they found negotiations with the German broadcasting authorities easier and more immediately helpful than with either the BBC or GPO).

Captain Hutchinson, the dashing, romantic, impetuous Irishman,[5] as Moseley called him, did not waste any time in taking advantage of Baird's achievement. 'So far are the experiments completed', he told the *Evening Standard* reporter,[6] 'that we are now having made 500 television receiving machines, which will not cost more than £30. Each of these, if connected to an ordinary wireless set, will enable the purchaser to 'look-in', just as now he can 'listen-in'.

Unfortunately, Hutchinson's publicity remarks were to cause difficulties for the Baird companies before the end of the decade. His methods in both business and public relations were such as to stir Campbell Swinton to write privately to Eckersley in 1928 and make scathing and defamatory remarks about Baird and Hutchinson which had they been published might have been held to be libellous.[7]

However, Baird was still very short of money and no doubt Hutchinson thought that he could fan the flames of public interest by making sweeping statements to a gullible public.

The prospect which faced Baird in 1926 was much more daunting than that which faced Marconi in 1895–96. Marconi had the advantage that his invention had an immediate application in military and naval operations, and, when his demonstrations before Service officers proved successful, his future seemed assured. Additionally, Marconi obtained valuable support for a short but critical period from W. Preece, the Engineer-in-Chief of the General Post Office, who had himself been experimenting on signalling through space without wires.

On the other hand, Baird's invention had no immediate application to warfare or safety, and he received no patronage from the one body which could assist him, the BBC. There is no readily available evidence to show that Baird approached the Department of Scientific and Industrial Research for a development grant: possibly he wished to remain independent of this organisation, and so, in 1926, the only source of money for his purpose seemed to be the general public. However, the general public had to be given an indication that investment in the new form of broadcasting was worthwhile.

One of the constant arguments used for the development of television in

Britain was the appeal to national pride. Moseley, who did not meet Baird until 1928,[8] used this point on many occasions in his letters to the BBC, the Prime Minister, the Postmaster General[9] and others. Hutchinson seems to have given the lead, for in his interview with the *Evening Standard* reporter he mentioned:

> Four nations have been competing in the race for television, America, France, Germany and England. In America they have merely succeeded in transmitting shadowgraphs. Germany and France have only talked a lot and prophesied. Mr Baird is the first, and only, inventor to demonstrate instantaneous vision between two entirely separate machines with no wire connecting them.[6]

Hutchinson was certainly keen to start broadcasting television, notwithstanding the faint and blurred image,[1] and consequently applied on 4th January 1926 to the Postmaster General (PMG) on behalf of Television Ltd. for a licence to transmit from London, Glasgow, Manchester and Belfast.[10] Baird told a *Daily Telegraph* reporter[11] that as the law stood at that time there was no need for a licence to broadcast movement, but to have matters in order they had applied to the PMG.

This was a new problem for the Post Office and so the Secretary sought the advice of its solicitor, Sir Raymond Woods, on whether the company was bound under the existing wireless acts to obtain a licence from the PMG.[12] The legal aspects of television transmission were to exercise the attention of Sir Raymond Woods for the whole period from 1926 until 1940 in various ways. For the immediate problem Woods advised that in his opinion the transmission of vision by itself could not be regarded as the transmission of a message or a communication but it would appear to be quite impractical for the company to transmit vision by itself.

> They would I should imagine be compelled to send some kind of signal to the person receiving the vision indicating either some step in the process or the time when the process was about to commence or the nature or kind of vision which was being sent, for example, the name of the person whose photograph was transmitted.[13]

Thus the position was that a licence would be required.

While this point was being clarified by the Post Office, Hutchinson again wrote to the Secretary (on 11th January 1926) saying:

> The Television Company was hoping to hear from you stating that there was no objection as we have several important demonstrations to give. Also we are having made 500 receiving sets and until we know officially that there is no objection we cannot offer these to our clients.[14]

The receiving sets seem to have been an inspiration of Hutchinson's imagination as none was available for sale until 1930.[15] No doubt his impetuosity was endeavouring to speed the bureaucratic machine.

The Secretary (GPO), replied to Hutchinson's letters of the 4th and 11th January and asked for further information regarding the proposed operating procedure, the points between which it was hoped to work the apparatus, the power and wavelength to be employed, the method of securing synchronisation between the sending and receiving apparatus and the means of communicating the subject matter of a 'vision' to a receiving station.[16] This letter was followed by a request from the Secretary to the Engineer-in-Chief (GPO), suggesting that 'it would be better for a representative of the GPO to visit Baird in order to clear up one or two points.[17]

Following the demonstration at Frith Street to members of the Royal Institution, arrangements were made for a private demonstration to be given to a Mr E. G. Stewart.[18] Stewart's very interesting report, written in April 1926, only came to light in 1948 when L. Hardern discovered it among some documents and sent a copy to the BBC.[19]

Stewart, a perspicacious engineer, was able to describe and give details of the equipment and impressions which were not mentioned by *The Times* reporter.

> The subject, which in the demonstration was limited to a size about 10 inch × 8 inch is brightly illuminated, about 500 candle power being used at one foot distance, and placed before an optical device of revolving lenses which continuously explores the whole surface in 32 vertical bands, each $\frac{1}{4}$ inch width is thus treated as being uniform . . . at the demonstration the received image was one ninth the area of the subject being $3\frac{1}{4}$ inch × $2\frac{1}{2}$ inch before magnification.

Stewart went on to describe the quality of reproduction:

> I found it possible to distinguish between two human faces I had previously seen in the life whilst opening and closing of the mouth, protusion of the tongue, orientation of the head and passing of the hand over the face could clearly be followed.
>
> At the same time it would be very difficult to recognise an individual previously unknown from the television representation . . . The inventor agreed however that the image was distorted and attributed it to, one, inferior optical equipment and, two, to insufficient sub-division of the pictures. He assured me that his lenses now were only lantern condensers and cycle lamp bull's eyes. This would certainly not add clarity to the picture and it would be interesting to see the effect of properly ground and treated lenses. With regard to sub-division of the picture the inventor has planned to divide the existing picture into four squares. This will quadruple the number of transmitted impulses and inevitably add detail. The amount however of detail added will then only bring the image to the order of 1/30th of the detail for perfect perception.

So far as is known Baird did not pursue this latter possibility, which not only would have very considerably complicated his apparatus but in addition would have required a frequency band width four times as great as that which he was using. Stewart's report showed that Baird was now again using a scanner having 32 bull's-eye lenses, a number which was indicated in his first provisional patent with Day but which was changed to 18 in the complete specification and further altered to 8 in the case of the lensed discs of a later patent. Baird used discs having 30 holes for his broadcast transmission as this was effectively the maximum number which could be used at a picture frequency of 12.5 picture per second when transmitting on the medium waveband.

As mentioned previously, Baird was concerned that his ideas should not be exploited by any large films and so had his apparatus 'entirely enclosed except for the input lens' when giving his early demonstrations. Stewart also wrote '. . . he has definitely decided to give a minimum of information upon the details of construction and operation to anyone'. In particular the light-sensitive cell which Baird used was 'a closely guarded secret of the inventor and he told me only sufficient of its construction to demonstrate that it was entirely different from existing cells on the market'.

At about this time Baird was having considerable difficulty in obtaining adequate photocell sensitivity, and as suitable cells were not available Baird experimented with different types of photo-sensitive devices.

> I made a number of efforts to increase the sensitivity of the photo-electric cell and to find other materials which would give greater reactions. to light. The light sensitivity of the human eye, according to Ederidge Green, and certain others, resides in a purple fluid in the retina of the eye and is called the visual purple.
>
> I decided to make an experimental cell using this substance and called at the Charing Cross ophthalmic hospital. Asking to see the Chief Surgeon, I told him I wanted an eye for some research work I was doing on visual purple. He thought I was a doctor and was very helpful.
>
> 'You've come at an appropriate time', he said, 'I am just thinking of taking out an eye and shall let you have it if you will take a seat until the operation is over'.
>
> I was handed an eye wrapped in cotton wool, a gruesome object. I made a crude effort to dissect this with a razor but gave it up and threw the whole mess into the canal.[20]

On the future of television, Stewart felt from his observations and discussions with Baird that, if satisfactory assurances could be given by the inventor on his ability to produce better images, the invention would be worthy of financial encouragement and development – assuming of course that successful television was a needed and valuable invention.

He also considered that it would be an error of judgement if the inventor placed his application on the market in its present state — which, of course, was what Hutchinson wanted.

> The apparatus as now developed gives a crude image which is not even physically pleasant to view. Again distortion is present and only com-

paratively small fields of view (e.g. the face) can be presented. While the existing type of apparatus would undoubtedly achieve a temporary market the public would heartily tire of the results and would either expect a rapid improvement or failing such improvements leave the idea in disgust resulting in a severe check to a desirable invention both in regard to sales and to capital. Those well known personalities whom the public would most desire to see would be scared off television by the present reproductions so that deserving developments later on would be hampered in securing support. The development of broadcasting by the gramophone lends colour to this view.

Stewart's carefully written report is probably the most important document available on the state of Baird's system during the early part of 1926. He did not explain why he went to Frith Street, except that he was following instructions. It is possible that his employer was interested in financing Baird's invention but like Odham required assurances from an expert that the system was worth supporting. Stewart made it clear that the time was not appropriate for the sale of televisors to the public even if permission was given for television transmissions.

Hutchinson, from whom Baird later became estranged, was not the kind of business man who calculated about long term results. Baird too was keen that no attempt to steal his thunder should pass unchallenged. In later years he wrote:

> Even after the lapse of time, at an age when such things should not matter, I feel my anger rise again against those who sought to brush my work aside. They did not, however, find me easy to crush. Years of fierce struggling to keep alive had certainly taught me to fight and I kept kicking and shouting, sometimes, I am afraid, paying little attention to dignity or reticence in the publicity methods I found it necessary to employ.

As Moseley said: 'He made no bones about it, in this world one had to shout one's achievements from the housetops'.[8]

Baird was in a dilemma: he had successfully demonstrated the transmission of crude images and urgently needed capital to develop his ideas. As Hutchinson and Baird saw the position this could only come by stimulating the public's interest in television with glowing accounts of its possibilities: they continued to give demonstrations and interviews to the press.

Following one press demonstration on 27th April 1926,[21] when the image 'was far brighter, clearer and larger than it was at the demonstration three months ago' – Baird was obviously making progress – the reporter mentioned, presumably following discussions with Hutchinson and Baird: 'Comparatively imperfect as the apparatus may be applications have been received for no fewer than 50 000 receivers . . .' at a cost of approximately £30 each. This number was vastly in excess of the number which the Baird Television Development Company was able to report to the BBC as having been sold after the BBC's low definition service had commenced operations in 1929.[8] It is not surprising that such gross exaggerations tended to increase the opposition of certain notable persons, including, in addition to Eckersley, Campbell Swinton, who made no attempt to disguise his great dislike for mechanical methods of television.

Hutchinson continued his correspondence with the Post Office for a licence and when this had not arrived by June 1926 wrote again stating that 'we are completely at a standstill until we obtain this permit and the delay is causing us great inconvenience and financial loss.[22] Meanwhile the Post Office had referred the matter to the Head of the Wireless Telegraphy Board,[23] and on the 18th June[24] he wrote to the Secretary of the Wireless Sub-committee of the Imperial Communications Committee saying: 'the W.T. Board concur in granting authority under the conditions recommended by the PMG . . . except as regards the employment of a wave of 200 m. The Board proposes a wavelength of 275 m and feel that it would be preferable that the wavelength employed should not be in the band in regular use by the services for training . . .'

The choice of 275 m proposed by the Wireless Telegraphy Board probably arose because Col. F. W. Home (Head of the Board), felt television was bound to be added to the BBC's services and consequently considered the equipment should be designed, *ab initio*, to operate as near as possible to the broadcasting band.[25]

The Post Office pointed out that the wavelength of 275 m was very near to some of the wavelengths recently agreed for certain broadcasting stations; in particular those for Sheffield at 275.2 m, Leeds at 277.2 m and Nottingham at 277.8 m.[26] They mentioned that the reason for suggesting 200 m was because the proposed stations were for experimental purposes and because that wavelength was in the band allocated for amateur experimental stations. They went on to ask whether the Services would agree to the use of a wavelength of 200 m on the understanding that it was for experimental work only.

By this time Hutchinson was very concerned about the delay – he had first approached the Post Office regarding a licence six months previously – and again communicated with the Secretary. He now altered his approach in an attempt to force a decision and wrote (29th June 1926)

> The above information is urgently required, not only in view of the financial loss now being incurred by this Company . . . but to the fact that a tempting offer has been made for the world's rights in our invention by a powerful American syndicate. Acceptance of this offer on our part would necessarily mean the transference of our apparatus to the USA. It is for many reasons our desire to keep control of television in this country but if the necessary facilities for advancement are withheld we shall have no alternative but to make the best deal we can and allow the invention to go elsewhere.[27]

Unfortunately, no information is available concerning the American offer and it may well have been a ruse to speed things up. It certainly produced a rapid response from the Post Office, for a few days later the GPO wrote to Television Limited saying:

> We are now in a position to grant you permission to install wireless experimental transmitting stations at Motorgraph House, and at the Green Gables, subject to the power and wavelength being fixed at 100 W max. and 270 m respectively.[28] It would also be a condition that transmission should not take place during broadcasting hours of the London Station.[29]

But of course, these conditions were not those which Hutchinson had originally requested, namely a power of at least 250 W and a wavelength between 150 and 200 m. He wrote '. . . we had our plant designed at considerable expense to work on 200 m wavelength and to attempt alteration at this juncture would put us in a very backward position apart from the financial loss entailed'.[30] He again appealed to the 'national point of view' and on 15th July 1926 heard that the Postmaster General had given his permission, for the use, for experimental purposes, of a wavelength of 200 m and a power of 250 W.[31] He was later to say that he was unable to state what wavelength could be allocated in the event that a licence for a television broadcasting service was granted.[32] The experimental transmission licences for 2 TV and 2 TW were dated 5th August 1926.

Meanwhile, through the long period of procrastination, Baird had continued his experiments[33] and demonstrations, including one which was reported in *Nature* by Dr Alexander Russell, FRS, the Principal of Faraday House.[34] He was agreeably surprised by the great progress made in solving the television problem and stated:

> We saw the transmission by television of living human faces, the proper gradation of light and shade, and all movements of the head, of the lips and mouth and of a cigarette and its smoke faithfully portrayed on a screen in the theatre, the transmitter being in a room in the top of the building. Naturally the results are far from perfect. The image cannot be compared with that produced by a good kinematograph film. The likeness, however, was unmistakable and all the motions are reproduced with absolute fidelity . . . This is the first time we have seen real television, and, so far as we know, Mr Baird is the first to have accomplished this marvellous feat.

Confirmation of Baird's achievement came from a number of sources. The *New York Times*,[35] 6th March 1926, gave a whole page to the subject and said 'no one but this Scottish Minister's son has ever transmitted and received a recognisable image with its gradations of light and shade . . . Baird was the first to achieve television'. Later, in September 1926, the *Radio News* (of USA) sent a reporter to investigate Baird's claims: 'Mr Baird has definitely and indisputably given a demonstration of real television . . . It is the first time in history that this has been done in any part of the world.'[36]

In his early experiments with television Baird had great difficulty in reducing the intensity of the light (see Fig. 2.1) used to illuminate his subjects without impairing the results achieved by the apparatus. The photocells which were

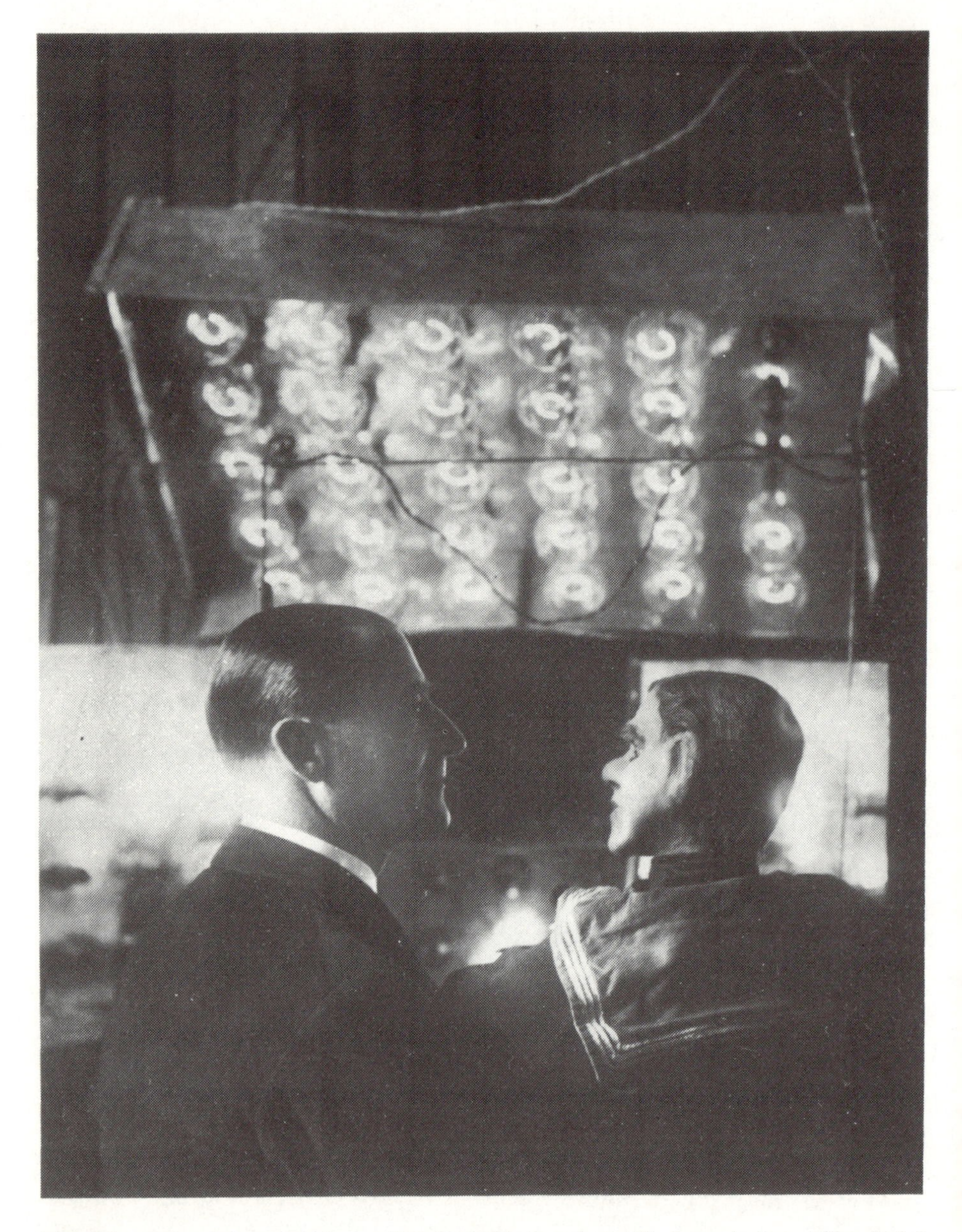

Fig. 2.1 *Arthur Prince and his ventriloquist doll being televised in the early days of 'floodlight' television*

available in 1926 were small gas filled cells which had a high ambient noise level. This, together with the parasitics introduced by the then dull emitter valves which were used in the amplifiers, caused the received picture to have a very poor signal/noise ratio.[37] Baird accordingly set out to try and select cells whose colour response matched the luminosity–wavelength characteristic of his flood-lights and this led him to experiment with various coloured lamps and filters. According to J. D. Percy it was during these trials that the lamps were masked as an experiment with wafer thin ebonite sheet so that all visible light was cut off and only infra-red light played on the subject. 'Much to Baird's surprise, the picture was not only visible at the receiver, but the signal/noise ratio, since he was using red sensitive cells, was surprisingly good'.

Baird had earlier tried ultra-violet rays but these had proved to be very objectionable for the subject and had the added disadvantage that the lenses used attenuated the radiation very considerably. Infra-red rays, however, did not cause any discomfort and so 'Noctovision', as Baird christened his latest discovery, came into being and occupied much of his attention for the next five years (Fig. 2.2).

These experiments have been graphically described by the Scottish inventor in the following words:

> At this time my only assistant was the office boy imported from Hutchinson's soap works. He was ignorant but amiable. The ultra-violet rays hurt his eyes, but he did not complain, but I got a fright and tried the infra-red. At first I used electric fires to produce these infra-red rays which are practically heat rays. I could not get a result and added more fires until Wally was practically roasted alive. Then I put in a dummy's head and added more fires until the head went up in flames.
>
> I decided to try another tack and used the shorter infra-red waves. To get this I used ordinary electric light bulbs covered with thin ebonite which cut off all light but allowed the infra-red to pass. Wally sat under this without discomfort, and after one or two adjustments I saw him on the screen although he was in total darkness. That again was a thrill, new and strange. I was actually seeing the person without light.[20]

Baird gave a demonstration of the use of infra-red rays for television to Dr A. Russell, Mr R. W. Paul and Mr Creed on 23rd November 1926 and followed this by extending an open invitation to members of the Royal Institution to witness the results on 30th December 1926. Approximately 40 members of the Institution were given demonstrations at Baird's laboratories.[38]

Russell later sent a letter to *Nature* dated 28th January 1927[39] describing the experiments following some criticism of Baird in the editorial of *Nature* for 15th January 1927. An extract from this letter reads:

> One of us stayed in the sending room with a laboratory assistant in apparently complete darkness. In the receiving room, on another floor, the image of the assistant's head was shown brilliantly illuminated on a screen,

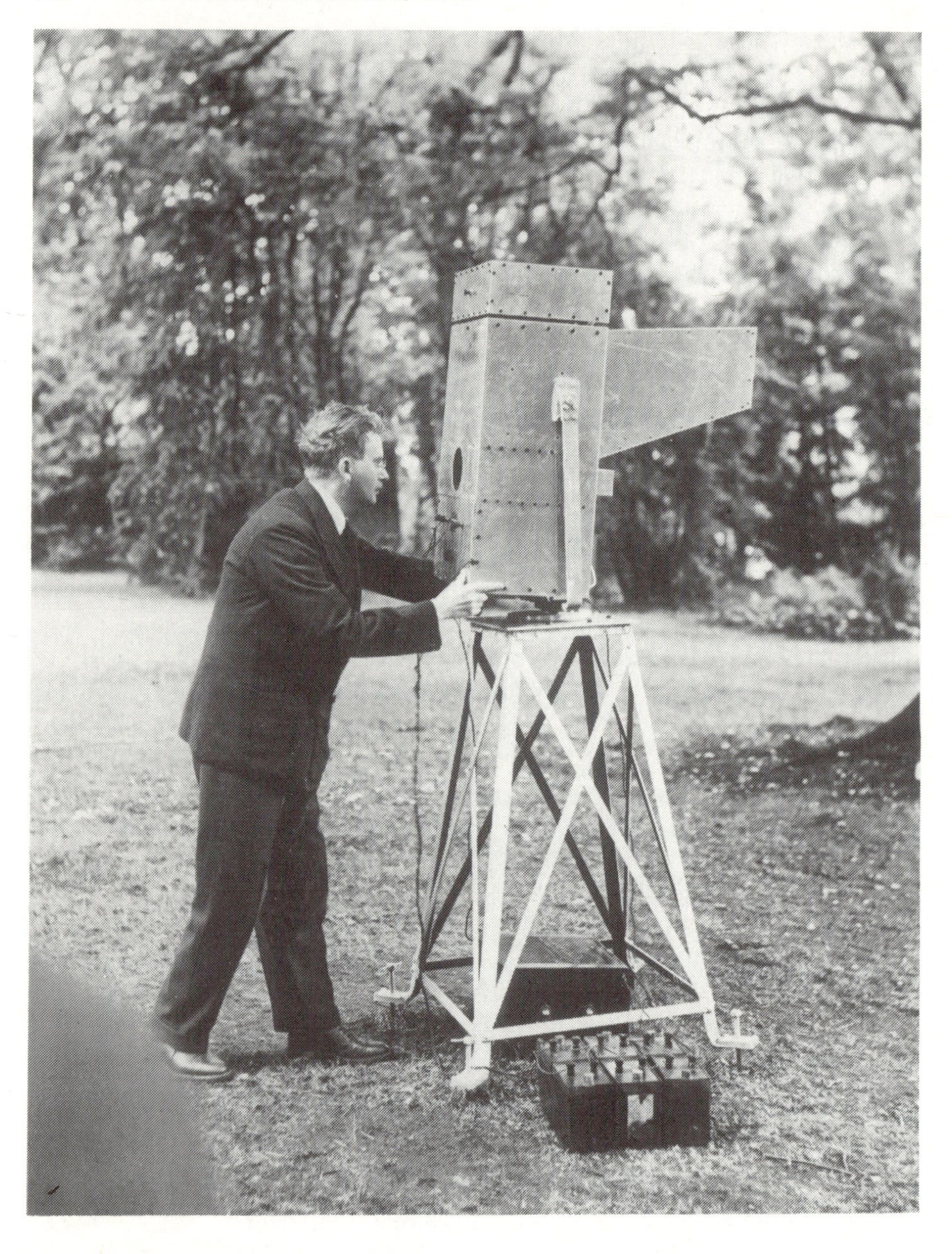

Fig. 2.2 *J. L. Baird with his 'Noctovisor'*

> and all the motions he made could be readily followed. These images were not outlines or shadowgraphs but real images by diffusely reflected rays. The application of these rays to television enables us to see what is going on in a room which is apparently in complete darkness. So far as I know, this achievement has never been done before.
>
> We had the impression that the image on the screen was not quite so clearly defined as when visible rays were used, but we easily recognised the figures we saw, and made out their actions. The direct application of Mr Baird's invention in warfare to locating objects apparently in the dark seems highly probable, but I hope that useful peace applications will soon be found for it.

Baird certainly thought his 'Noctovisor' had great potential,[37] particularly to penetrate fog, and years later, when he lived in a house on Box Hill, experiments were still going on. The very limited range of the apparatus proved a formidable obstacle[40] and it was not until the Second World War that infra-red rays were used for the penetration of darkness.

For normal television the use of infra-red radiation had certain disadvantages. First, it was generally inconvenient for the subject to sit in total darkness, and secondly, with the early type of photoelectric cells the correct colour tones were difficult to achieve; red appeared as white while blue did not appear at all. The effect, therefore, was to give a rather ghostly appearance to the image of the person being transmitted. An added difficulty, aptly put by Sir Oliver Lodge, who attended the December demonstration with his daughter, was that it was very hot.

Hence, although when noctovision was demonstrated at the British Association meeting in Leeds in 1927 it proved such a popular scientific exhibit that the police had to be called in to regulate the queues, it really represented no advance for television.

The ideas contained in the above invention formed the basis of Baird's patent 288,882,[41] dated 15th August 1927. This was formed from three provisional patents, the first of which was applied for on 26th January 1927. Taken together the patents show that Baird was still using a light interrupter, in conjunction with a Nipkow disc, and glow discharge lamp. Some of the diagrams contained in the complete specification do not show the interrupter, but in the text Baird stressed the need for its continued use: 'It is important that the radiation should be interrupted, as the intermittent impulses thereof which impinge on the sensitive receiving device are most readily controllable and rendered operative, by amplification, in that arrangement'.

One of the more important concepts contained in the above patent was the use of a single Nipkow disc for both transmission and reception purposes. Baird had in mind the utilisation of the device for the location of ships in fog, as well as its use in warfare. In this case the object was to be illuminated by natural light, or a searchlight, rich in infra-red emission, and the object viewed by means of a glow discharge tube positioned at the receiving device. Consequently the apparatus could be made mobile and mounted on a stand, allowing both azimuthal and elevating motions.

An alternative approach to the use of infra-red radiation to shield a subject from the intense light of the floodlights was the use of the 'spotlight' method which Baird patented in January 1926.[42] Figs. 2.3*a* and *b* show the 'floodlight' and 'spotlight' systems, respectively.

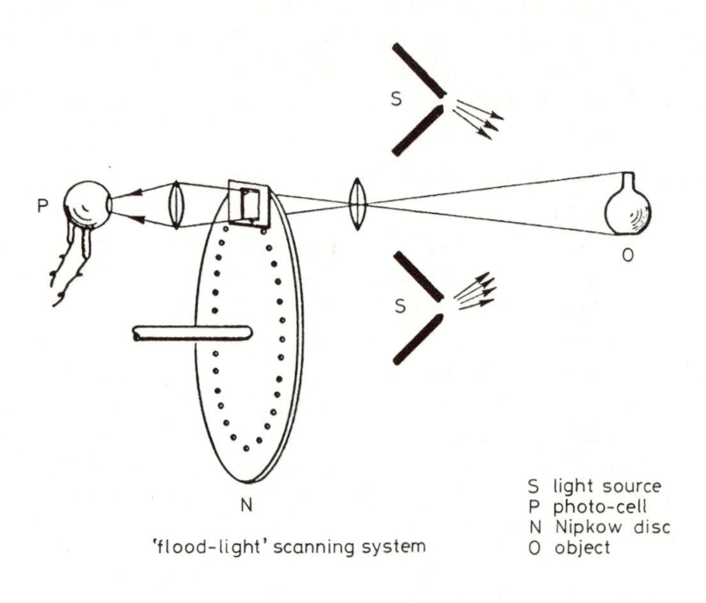

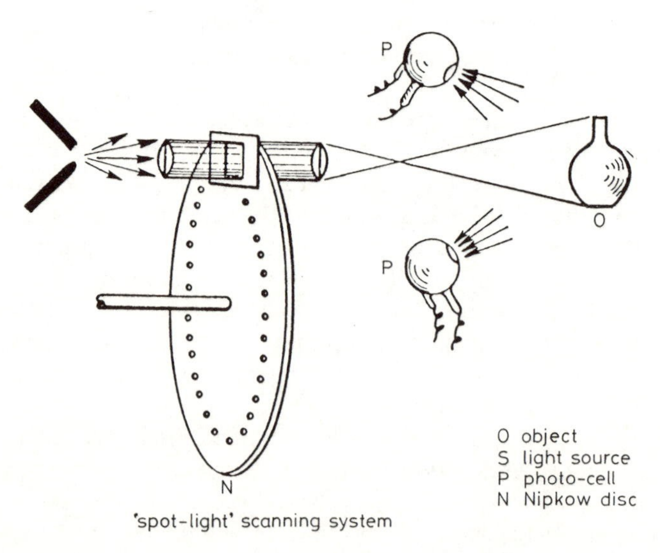

Fig. 2.3 *Diagrams showing 'floodlight' and 'spotlight' television*

In any method, such as that shown in Fig. 2.3*a*, which depends upon scanning an image, as formed by a lens, of the object, the efficiency of the system is ultimately limited, for any given size of image that can be scanned, by the ratio of the aperture to the focal length of the lens.

Experiments showed that with the best lens then available to form a one-inch square image, it would be necessary to illuminate a subject with a 16 000 candle power arc at a distance of about four feet in order to secure an image bright enough for the photoelectric cell to give an output current above the noise level of the amplifier.[43]

Baird reversed this process as follows: 'At the transmitter the scene or object to be transmitted is traversed by a spot of light, a light-sensitive cell being so placed that light reflected back from the spot of light traversing the object falls on the cell'.[42]

This method of scanning permitted two very large gains to be made in the amount of light flux available for producing the photoelectric current. First, the transient nature of the light allowed a very intense illumination to be used without inconvenience to the subject, and secondly, the optical efficiency of the system was not limited by the apertures of the available lenses but could be increased by using large photoelectric cells and more than one cell, all connected in parallel.

Baird applied for provisional patent protection for this very important principle on 20th January 1926 and submitted the complete specification for patent 269,658 on 18th November 1926. His patent was accepted on 20th April 1927. These dates are of some interest, for on 6th April 1927, Electrical Research Products Inc. (assignees of F. Gray), applied for patent protection for the same principle in the United States and made an application for a similar patent, 288, 238 in the United Kingdom, on 18th January 1928 (accepted 18th April 1929). Baird's application was thus made before that of the American company, but notwithstanding the similarity of the first two claims of Baird's patent with the first claim of the ERP patent, the latter was not invalidated by the Comptroller General of the Patent Office. The relevant claims are as follows:

Baird

1 In a television or like system using a high intensity of illumination of the object whereof an image is to be transmitted, the method of illuminating the object with an intensity so high *that if the illumination were continuous on any one part the object would be damaged* (*i.e. burned*), which method consists in traversing over the object a spot of light of the desired high intensity.

2 In a a television or like system the combination with means for illuminating the objects set forth in Clam 1, of a light-sensitive cell so placed that light reflected back from the spot of light traversing the object falls on the cell.

ERP

1 A television or like system in which the object, or subject is scanned by means of a beam of intense light which is moved rapidly so as to cause a cyclic

> point-by-point illumination of the object – the light reflected from the object being received directly on a photoelectric cell or cells of wide aperture, there being no obstruction between the cell and the object or subject.

The italicised section of Baird's claim introduced a weakness in its validity which presumably allowed ERP's claim to stand. Shoenberg of EMI was to use this point in discussions with the Television Committee in 1934.[44]

Apart from the inherent value of the 'spotlight' principle, the idea is important as it probably stemmed from the practice which obtained in the field of picture telegraphy and therefore represents another application of the techniques of this branch of electric telegraphy to television. Several proposals had previously been made by inventors for scanning a transparency by a spot of light so that the transmitted light influenced a photosensitive cell. Korn had put forward this method for phototelegraphy purposes in 1904[45] and in 1923 the Westinghouse Electric and Manufacturing Company included in its patent 225,553[46] an arrangement whereby the spot of light produced on the screen of a cathode-ray tube by fluorescence was used to scan a moving film and thus cause a varying signal to be produced by a photocell placed behind the film.

Actually, although both Baird and ERP were probably inspired to put forward their 'spotlight' patents by events which had taken place in picture telegraphy, the original claim for the principle was made by Ekström in a Swedish patent 32,220 dated 24th January 1910. Neither Baird nor ERP made reference to this little known (in 1926–27) patent, and it is unlikely that they knew about its existence.

One of the difficulties associated with spotlight scanning lay in the random reflection and scattering which occurred from an opaque object. Both Baird and ERP referred to the solution to this problem, namely, the use of several light cells, in their patents. This solution was necessary so that as much of the reflected and scattered light as possible could be utilised. For their 1927 demonstration of television, the American Telephone and Telegraph Company used photosensitive cells 15 inches long and 3 inches in diameter – then the largest in the world. Even so, very special precautions had to be taken to ensure a satisfactory output from the cells.[47]

The spotlight system was to be used for the next ten years and was an essential feature of the Baird installation at Alexandra Palace in 1936–37.

Baird desparately needed money and facilities for experimental transmissions to develop television. 'Luckily for him', Moseley wrote, 'television was born at a time when company promoting was running rather wild. Speculators, greedy for fat profits and quick returns, were ready to gamble on very slim chances'.[8]

One speculator who was persuaded to take an interest in television was Colonel Anderson, then a partner in the firm of Vowler and Company. He has told[8] how he first met Baird:

> I originally met Mr Baird about the beginning of 1927 on the introductin of Captain O. G. Hutchinson. Hutchinson told me about Mr Baird's ideas

> and asked if I would go and see his small laboratory on the top floor of a building in Long Acre. Hutchinson told me that he had known Mr Baird and/or his family in Glasgow and that he had met him walking in the Strand and asked him where he was going; he said he was walking home to some address, which I forget, in the East End of London. Hutchinson asked him why he did not take a bus and said he had no money and could not take buses. Hutchinson asked him what he was doing and he said he was working on television. Hutchinson decided to finance him in a very small way, as he could afford nothing more, and had taken the little laboratory for him in Long Acre and helped him to get some very primitive equipment.
>
> I went to Long Acre with one of my partners and met Mr Baird, who gave us a demonstration of television, using both electric light and infrared; he televised me on a screen in another room connected by wire, where my partner could watch, and then he televised my partner so that I could see him. The whole of his equipment was a most Heath Robinson affair, tied together with bits of string, bits of wire, old bicycle lamps etc. I was tremendously impressed with what I saw and felt certain there were enormous potentialities in the invention.
>
> There and then, having heard from Hutchinson that the amount required to buy some reasonable equipment would be £50 000, I decided to raise £100 000 provided he (Hutchinson) could raise one-third of the sum, i.e. about £33 000. Hutchinson went up to Glasgow to see various friends of his and within a week came back with promises to put up £33 000. On 27th April 1927 we made the issue, which raised for the Company approximately £100 000. Sir Edward Manville, the then Chairman of the Daimler Company, was the Chairman of the first Company.

The other directors were Sir James Percy, Francis A. Shortis, J. L. Baird, and O. G. Hutchinson,[48] and their remuneration was fixed at £300 per annum after the deduction of tax.

On the following day, 28th April, the national newspapers contained reports of the successful television transmission, from Washington to New York, by the American Telephone and Telegraph Company. Baird's monopoly was thus broken and Vowlers felt that they had been cheated.

The AT & T Company seriously began a study of the problems involved in implementing a system of television two years prior to their public demonstration on 7th April 1927, when 'it began to be evident that scientific knowledge was advancing to the point where television was shortly to be within the realm of the possible'.

The company were of the opinion that it would have a real place in the world's need of distant communication and that it would be closely associated with telephony. They were certainly in a very strong position to advance development in the field of television, not only because of the vast expertise and facilities available in the Bell Laboratories, but also because of the tremendous experience which had been acquired by the research and development work which had made possible transcontinental and transoceanic telephony and telephotography.

In January 1925, development work under the direction of Dr H. E. Ives had been completed on a system for sending pictures over telephone lines and so research facilities and expertise existed for a new scientific venture. Ives and Arnold, the Director of Research, agreed that the next problem to be undertaken was television. Wrote Ives:[49]

> At Arnold's request I prepared and submitted to him on 23rd January 1925, a memorandum surveying the problem and proposing a programme of research. The survey discussed the characteristic difficulties of securing the requisite sensitiveness of pick-up apparatus; the wide frequency bandwidths which, from our experience with picture transmission, were indicated as necessary for television; the problem of producing enough modulated light in the received image to make it satisfactorily visible; and the problem of synchronising apparatus at separated sending and receiving ends. It concluded with the proposal of a very modest attack capable, however, of material expansion as new developments and inventions materialised.

Ives's suggestion was for a mechanically linked transmitter and receiver, each utilising a Nipkow disc, operating on a 50 lines/picture, 15 pictures/second standard. A photographic transparency, later to be superseded by a motion picture film, was to be used at the sending end, together with a photoelectric cell and a carbon arc lamp; and at the receiving end Ives proposed the use of a crater type gaseous glow lamp. His scheme was thus based on the transmission of light through an 'object' rather than on the reflection of light from an opaque body: the latter problem was of an altogether different order of difficulty.

By May the aparatus was completed and in operation (Fig. 2.4). A memorandum of 12th May 1925, by J. G. Roberts, General Patent Attorney, records:

> I witnessed today a demonstration of Mr Ives's system of television. He has constructed and put into operation substantially the system he described in his memorandum of 23rd January 1925, to Mr Arnold. In viewing the picture at the receiving end, I could distinguish with fair definition the features of a man's face like that of a picture at the transmitting end and also observed that, when the picture at the transmitting end was moved forward or backward, or up or down, the picture at the receiving end followed these motions exactly.[49]

With this success behind them Ives's group next tackled the problem of transmitter–receiver synchronisation, when the two Nipkow discs were mechanically uncoupled. H. M. Stoller was given the responsibility for this particular phase of the project, and by December 1925 the group was able to show motion pictures from a projector driven in synchronism with the discs.

A further landmark in Bell Telephone Laboratories' research programme was passed on 10th March 1926, when, at the conclusion of the ceremonies commemorating the 50th anniversary of the telephone, F. B. Jewett, President, and E. B. Croft, Executive Vice-President, were invited to visit the telephone laboratory. There they talked over the telephone, with the expressions and movements of the face of the speaker being clearly seen by the distant listener.

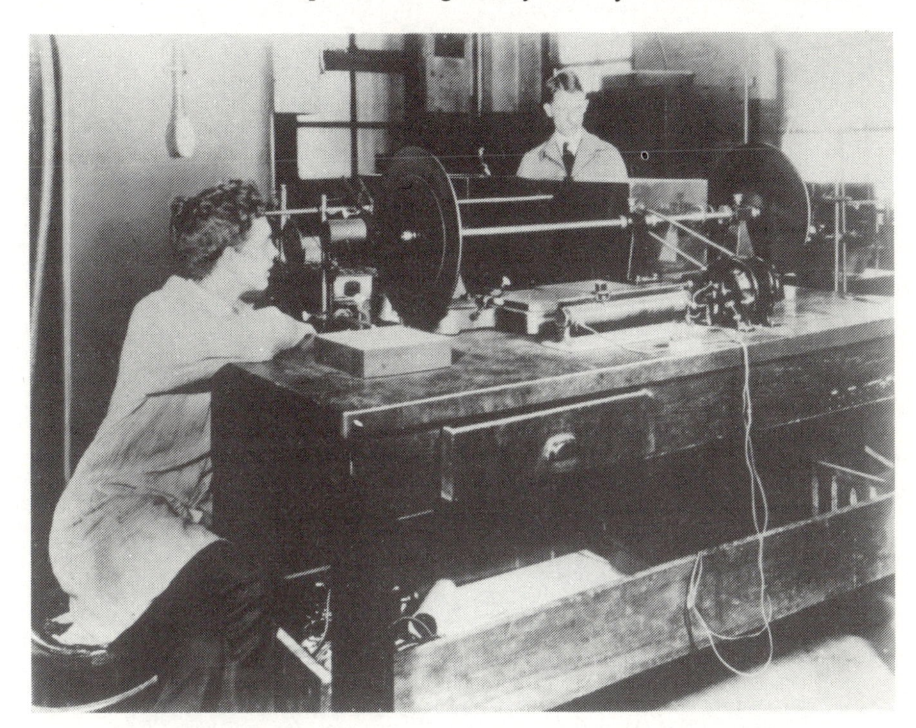

Fig. 2.4 *Bell Laboratories' apparatus used in the development of television. The photograph shows J. R. Hefele (left) and Dr F. Gray. An illuminated transparency is being used as the object*

According to Ives, the group forbore to announce their achievement because, from the beginning of their investigation, it had been considered that only when vision signals could be sent over large distances – to parallel what had been done for voice signals – would their apparatus be worthy of the appellation television system. 'It would be television when the laboratory experiment was expanded to cover distances beyond any the eye could reach'.

By 7th April 1927 the system was ready to be demonstrated (Fig. 2.5). It has been estimated that 200 engineers, scientists and technicians contributed to the success of the project, although some reports quoted a figure of 1000.

The AT & T demonstration is particularly interesting in that it probably represented the best that could be achieved with the known state of the art at that time. The company sought to show not only small scale domestic reception of television but reception suitable for large audiences: they wanted to transmit signals not only by wire but also by radio; they were obviously keen to demonstrate the best which could then be obtained.

Fig. 2.5 *Close-up of television transmitting apparatus installed in a laboratory in Washington, USA. Mr E. F. Kingsbury of Bell Telephone Laboratories is being televised. The photoelectric cells are situated behind the three screens of the panel immediately in front of him (April 1927)*

The demonstration[50] using a wire link consisted of the transmission of images from Washington, DC, to the auditorium of the Bell Telephone Laboratories in New York, a distance of over 250 miles. In the radio demonstration, images were transmitted from the Bell Laboratories experimental station at Whippany, New Jersey, to New York City, a distance of 22 miles. Reception was by two forms of apparatus. In one, a small image approximately 2.0 inch by 2.5 inch was produced, suitable for viewing by a single person, and in the other a large image approximately 2.0 ft by 2.5 ft was formed for viewing by an audience of considerable size.

Ives, who worked on the photoelectric cells used in the system and who co-ordinated the entire research and development programme, stated in his paper: 'the smaller form of apparatus was primarily intended as an adjunct to the telephone, and by its means individuals in New York were enabled to see their friends in Washington with whom they carried on telephone conversations', while 'the larger form of receiving apparatus was designed to serve as a visual adjunct to a public address system'.

The first demonstration consisted of the transmission of an image and an address by Herbert Hoover, Secretary of Commerce, from Washington to new York over telephone wires. Hoover's speech was short and non-technical but contained the following notable remarks:

> Scientists for many years, in many countries, have struggled to solve the problems of television. We may all take pride in the fact that its actual accomplishment is brought about by American genius and its first demonstration is staged in our own country.[50]

There is no doubt that Bell Laboratories had on its staff some of the cleaverest and most brilliant telephone and telegraph engineers and scientists in the world: such persons as Nyquist, Nichols, Zobel, Colpitts and others are now known to electrical engineering students throughout the world. Their scheme was no string, sealing wax and glue affair with hat boxes and darning needles, but a very carefully thought out and superbly engineered system which represented the zenith of achievement possible at the time. This is not to say that it introduced any radically new scientific discoveries or inventions – it did not.

Indeed Dr Dauvillier, the eminent French physicist, who saw the wireless transmission, as an independent witness, commented later in the *Revue Generale de L'Electricite*: 'Finally, the Bell Telephone Company recently succeeded in transmitting to a considerable distance the human face, using (without acknowledgment) the Baird system'.[51]

The system was seen by Colonel Angwin (the deputy Chief Engineer of the GPO), some time after the public demonstration, and in his report he mentioned:

> This system reproduces a clear and undistorted picture and results obtained are undoubtedly far in advance of those claimed for by the Baird system. The American system is a very costly and elaborate piece of mechanism and requires a special circuit for line transmission and exceptionally stable conditions for wireless transmission.

Angwin's visit occurred approximately one year after the first announcement of the AT & T Company's work in the television field, but during that time nothing had been done by the company to exploit the system commercially and the Secretary of the GPO was given to understand that they had no intention of proceeding further with it at that time.[52]

Ives and his colleagues used a Nipkow disc, for scanning purposes, having 50 apertures. They arrived at this figure by taking as a criterion of acceptable quality, reproduction by the half-tone engraving process in which it was known that the human face could be satisfactorily reproduced by a 50-line screen. Assuming equal definition in both directions, 2500 elements or 40 000 elements per picture had to be transmitted for a rate of picture transmission of 16 pictures per second. The frequency range necessary to transmit this number of elements per second was calculated to be 20 kHz.

A spotlight scanning method was adopted to illuminate the subject, the beam of light being obtained from a 40A Sperry arc. Three photoelectric cells of the potassium hydride, gas-filled type were specially constructed and used to receive the reflected light from the subject. They were probably the largest cells that had ever been made at the time and presented an aperture of 120 $inch^2$ (Fig. 2.6).

For reception purposes a disc similar to that used at the transmitter was employed together with a neon glow lamp. The largest disc used, three feet in diameter, gave a 2.0 inch by 2.5 inch rectangular image. The other type of receiver used for presenting television to a large audience consisted of a single, long, neon-filled tube bent back and forth to give a series of 50 parallel sections of tubing. The tube had one interior electrode and 2500 exterior electrodes cemented along the back side of the glass tubing. A high frequency voltage applied to the interior electrode and one of the exterior electrodes caused the tube to glow in front of that particular electrode. The high frequency voltage was commutated to the electrodes in succession from 2500 bars on a distributor with a brush rotating synchronously with the disc at a transmitting station. Consequently, a spot of light moved rapidly and repeatedly across the grid in a series of parallel lines one after the other and in synchronism with the scanning beam at the transmitting station. With a constant exciting voltage the grid appeared as a uniformly illuminated screen, but when the high frequency voltage was modulated by the received picture current, an image of the distant subject was produced on the screen.

The use of a high frequency current enabled the glow to be dependent upon the amplitude of the high frequency current which was amplitude modulated by the received picture current. Provision was made to control the maximum and minimum peak values of the modulated high frequency waves in order to produce the correct voltage for the lamp and to control the range between the lightest and darkest portions and the average tone value of the produced image.

Although the success of the American Telephone and Telegraph Company's demonstration upset Vowlers, Baird convinced them that he was quite unaware that his monopoly in television would end so soon: he himself was taken aback but the development was only to be expected. He wrote:

> Our demonstrations and the whole system, with drawings and details of our apparatus, had been given the widest publicity. Every electrical firm

> had been aroused and their experts had been put to work. It was surprising that our monopoly had lasted so long.

Baird endeavoured to emulate the success of the AT & T Company, albeit in a rudimentary way and on a smaller scale on the 24th and 26th May 1927, when he transmitted television signals, by telephone wire, between London and Glasgow. Baird used two operators, one at the transmitting end and one at the receiver.[53]

Fig. 2.6 *Photograph of one of the photoelectric cells employed in the April 1927 demonstration of television by Bell Laboratories*

His apparatus was placed in the sitting room of the Central Station Hotel, Glasgow, and was seen by leading scientists and public figures of Glasgow. Professor E. Taylor Jones, who held the Chair of Physics at Glasgow University and who was received by Hutchinson, described the demonstration in an article[54] published in *Nature* (18th June 1927):

> . . . the receiving apparatus was set up in a semi-darkened room, the lamp and shutter being enclosed in a case provided with an aperture. The observer looking into the aperture saw at first a vertical band of light in which the luminosity appeared to travel rapidly side-ways, disappearing at one side and then reappearing at the other. When any object having contrast was placed in the light at the sending end, the band broke up into light and dark portions forming a number of images of the object. The impression of side-ways movement of the light was then almost entirely lost, and the whole of the image appeared to be formed simultaneously. The image was perfectly steady in position, was remarkably free from distortion and showed no sign of the streakiness which was, I believe, in evidence in the earlier experiments.
>
> The size of the image was small, not more than about 2 inches across when the object was a person's face, and it could be seen by only a few people at a time. The image was sufficiently bright to be seen vividly even when the electric light in the room was switched on, and I understand that there is no difficulty in enlarging the image to full size. I was told that arrangements will soon be made for transmitting larger objects, and for increasing the number of appearances of the image per second.
>
> The amount of light and shade shown in the image was amply sufficient to secure recognisability of the person being televised, and movements of the face or features were clearly seen. At the second demonstration some of those present had the experience of seeing the image of Mr Baird transmitted from London while conversing with him (over a separate line) by telephone.
>
> My impression after witnessing these demonstrations is that the chief difficulties connected with television have been overcome by Mr Baird and that the improvements still to be effected are mainly matters of detail. We shall doubtless all join in wishing Mr Baird every success in his future experiments.

Possibly sentimental considerations dictated Baird's choice of Glasgow as the setting for the greatest trial to which his apparatus had been subjected. He was hailed by the *Glasgow Herald* as the brilliant young inventor and accorded a long report on his work.[53]

References

1 Anon.: a report, *The Times*, 26th January 1926
2 Anon.: a report, *Daily Express*, 8th January 1926
3 WILSON, J. C.: 'Television Engineering' (Pitman, 1937)
4 Press Photonachtrichtendienst, letter to the GPO, 28th January 1926, Minute 51/1929, file 1
5 MOSELEY, S. A.: 'Broadcasting in my time' (Rich and Cowan, 1935)
6 Anon.: a report, *Evening Standard*, 8th January 1926
7 CAMPBELL SWINTON, A. A.: a letter to the BBC, T16/42, file 1, c. 1928

8 MOSELEY, S. A.: 'John Baird' (Odhams, London, 1952)
9 MOSELEY, S. A.: letter to Sir Kingsley Wood, 28th January 1933, Minute 4004/33
10 HUTCHINSON, O. G.: letter to the Postmaster General, 4th January 1926, Minute 4004/33, file 1
11 Anon.: a report, *Daily Telegraph*, 11 January 1926
12 Secretary of the GPO: memorandum to the Solicitor of the GPO, 7th January 1926, Minute 4004/33, file 1
13 Solicitor of the GPO: opinion, Minute 4004/33, file 1
14 HUTCHINSON, O. G.: letter to the Secretary of the GPO, 11th January 1926, Minute 4004/33, file 1
15 Anon.: a report 'Baird television receiver tested', *Wireless World*, 12th March 1930, p. 277
16 Secretary of the GPO: letter to O. G. Hutchinson, January 1926, Minute 4004/33, file 1
17 Secretary of the GPO: a memorandum to the Engineer-in-Chief of the GPO, 13th January 1926, Minute 4004/33, file 1
18 STEWART, E. G.: a report, April 1926, BBC file T16/214
19 HARDEN, L.; letter to N. Collins (Director of Television, BBC), 15th April 1948, BBC file T16/214
20 BAIRD, M.: 'Television Baird' (HAUM, South Africa, 1974)
21 Anon.: a report 'Television problem. Listening-in and seeing-in', *Daily Telegraph*, 28th April 1926
22 HUTCHINSON, O. G.: letter to the Secretary of the GPO, 21st June 1926, Minute 4004/33, file 1
23 Secretary of the GPO: letter to the Head of the Wireless Telegraphy Board, 7th June 1926, Minute 4004/33, file 1
24 Head of the Wireless Telegraphy Board.: letter to the Secretary of the GPO, 18th June 1926, Minute 4004/33, file 1
25 PHILLIPS, F. W.: internal memorandum, Minute 4004/33, file 2
26 WISSENDEN, J. W.: letter to the Head of the Wireless Telegraphy Board, Minute 4044/33, file 2
27 HUTCHINSON, O. G.: letter to the Secretary of the GPO, 29th June 1926, Minute 4004/33, file 2
28 LEE, A. G.: memorandum to the Secretary of the GPO, 27th May 1926. Also see F. L.: memorandum to the Secretary of the GPO, 31st May 1926, Minute 4004/33
29 WISSENDEN, J. W.: letter to Television Ltd., 7th July 1926, Minute 4004/33, file 2
30 HUTCHINSON, O. G.: letter to the Secretary of the GPO, 8th July 1926, Minute 4004/33, file 2
31 PHILLIPS, F. W.: letter to Television Ltd., 15th July 1926, Minute 4004/33, file2
32 WISSENDEN, J. W.: letter to Television Ltd., 5th August 1926, Minute 4004/33, file 2
33 MOSELEY, S. A., and BARTON CHAPPLE, H. J.: 'Television today and tomorrow' (Pitman, London, 1931)
34 A. R.: 'Television', *Nature*, **118**, 3rd July 1926, pp. 18–19
35 Anon.: a report, *New York Times*, 6th March 1927
36 Anon.: a report, *Radio News*, September 1926, p. 283
37 PERCY, J. D.: 'The founding of British Television', *Journal of the Television Society*, 1950, pp. 3–16
38 BAIRD, J. L.: 'Television' (a letter), *Nature*, 1927, **119**, 5th February, pp. 161–162
39 RUSSELL, A.: 'Television' (a letter), *Nature*, 1927, **119**, 5th February, pp. 198–199
40 See, for example: JONES, R. V.: 'Infra-red detection in British Air Defence, 1935–38', *Infrared Physics*, **1**, pp. 153–162
41 BAIRD, J. L.: 'An improved method of and means for effecting remote control', British patent no. 288,882, application dates, 15th October 1926, 26th January 1927 and 10th March 1927
42 BAIRD, J. L. and Television Ltd.: 'Apparatus for the transmission of views, scenes or images to a distance', British patent no. 269,658, 20th January 1926

43 GRAY, F.: 'The use of a moving beam of light to scan a scene for television', *JOSA*, 1928, **16**, March, pp. 177–190

44 SHOENBERG, I.: evidence to the Television Committee, 17th June 1934, Post Office minute 33/4682

45 KORN, A.: 'Uber Gebe und Empfangsapparate zur elektrischen Fernubertragung von Photographien', *Physikalische Zeitsschrift*, 1904, **5**, no. 4, pp. 113–118

46 GARDNER, J. E., and HINELINE, H. D.: 'Improvements in and relating to television systems', British patent no. 225 553, application date (UK) 25th November 1924

47 GRAY, F.: 'Production of television signals', *Bell Syst. Tech. J.*, 1927, **6**, October, pp. 560–603

48 Anon.: a report, *Financial Times*, 4th April 1927

49 IVES, H.E.: 'Television: 30th anniversary', *Bell Laboratory Record*, May 1947, pp. 190–193

50 IVES, H. E.: 'Television', *Bell Syst. Tech. J.*, 1927, **6**, October, pp. 551–559

51 DAUVILLIER, A.: 'La television electrique', *Revue Generale de l'Electricite*, 7th January 1928, pp. 5–23

52 ANGWIN, A. S.: memorandum to the Secretary of the GPO, 7th June 1928, Minute 51/1929, file 9

53 Anon.: 'Television marvels. Glasgow looks in on London. Young Scottish inventor's great achievement', *Glasgow Herald*, 27th May 1927

54 TAYLOR JONES, E.: 'Television', *Nature*, 1927, **119**, 18th June, p. 896

Chapter 3

Company formation and progress, 1926–1928

Baird Television Development Company was established in April 1927. The object of the new company was essentially 'to develop commercially the Baird television and other inventions' and to achieve this the company acquired from Television Ltd. the sole right to exploit the inventions in the United Kingdom and a 66.66% interest in the net proceeds arising from and payable to Television Ltd. in respect of the sale and exploitation of the foreign and colonial rights. The purchase price was £20 000 and this was paid and satisfied by the allotment to Television Ltd. of 400 000 fully paid up deferred ordinary shares of 1 shilling each.

A nominal capitalisation of £125 000 was set for the size of the company divided into:

100 000 preferred participating ordinary shares of £1 each	£100 000
500 000 deferred ordinary shares of 1 shilling each	£25 000

Each of the two classes of shares carried the same voting rights, and as 400 000 of the deferred ordinary shares had to be allotted to Television Ltd., the latter company had an important stake in the new enterprise. When the shares of the new company, and in 1928 Baird International Television Ltd., became available for purchase there was no shortage of buyers: Hutchinson had certainly succeeded in acquainting the general public of the accomplishments of Baird.

The statutory meeting of the Baird Television Development Company was held on 18th July 1927 with Sir Edward Manville in the chair, the Secretary being Mr T. W. Bartlett. The Chairman announced that it was the Board's intention to implement the following initial programme:[1]

1 to acquire more extensive premises in a suitable locality,

2 to have manufactured for them certain component parts required for new and improved television apparatus and experiments in connection therewith;

3 pending the acquisition of new premises and the construction of new aparatus to endeavour by means of telephone lines to demonstrate television over a much longer distance than had heretofore been possible.

At the meeting Manville commented on both the American Telephone and Telegraph Company's demonstration and that of Baird. He referred to the former as a crude form of television and noted the number of Baird features which it contained and the patent position concerning some of Baird's patents – including the spotlight scanner and large screen. On the latter he said:

> This, however, was also patented by Mr Baird in July 1923, nearly four years ago, and is protected by patents the property of this Company, though not in the United States, but we have no intention of using this system since it is, in our opinion, far too complicated a device and our present system for effecting the same results is, we believe, greatly superior.

Notwithstanding this comment, Baird was to use the ideas contained in the above patent for a demonstration of large screen television at the London Coliseum on the 28th July 1930.

The new premises which were obtained were in Long Acre, in the West Central district of London. These were needed to extend the existing laboratory facilities so that more powerful and accurate transmitting machinery and apparatus than the original experimental apparatus could be constructed so as to accomplish the broadcasting of vision to a greater distance and with greater perfection.

It is interesting at this stage to consider Ives's views of the applications and future developments of television transmission. He foresaw three kinds of services:[2]

1 a service from individual to individual, parallel in character to the telephone service, and as an adjunct to it:
2 a public address service, by which the face of a speaker at a distant point could be viewed by an audience while his voice was transmitted by a loudspeaker;
3 the broadcasting of events of public interest, such as athletic contests, theatrical performances and the like.

Although Ives did not mention whether the above order represented the order of importance of the applications of television in his view, he considered the state of the art was not sufficiently developed for a practical television broadcasting system.

> The third type, because of the uncontrolled conditions of illumination and the much finer picture structure which would be necessary for satisfactory results, will require a very considerable advance in the sensitiveness and the efficiency of the apparatus, to say nothing of the greatly increased transmission facilities. For all three types of service, wire or radio transmission channels could be utilised . . . However, the very serious degradation of image quality produced by the fading phenomena characteristic of radio indicates the practical restriction of radio television to fields where the much more reliable wire facilities are not available.

With the leader of the research and development team holding such views, the way ahead for the American Telephone and Telegraph Company was clear – the development of two-way television, as an adjunct to the telephone service, using wire transmission.

Baird's requirement for substantial capital to develop television may be appreciated by noting that the AT. & T Company approved the expenditure of \$308 100 on low definition television developments from 1925 to 1930 (inclusive), and a further \$592 400 on other aspects of television from 1931 to 1935 (inclusive).[3]

Having achieved the successful production of rudimentary television images in the laboratory and the transmission of these along 400 miles of telephone wire, Baird wanted to experiment in broadcasting vision signals if the objective of the new company was to be achieved. He therefore communicated with H. L. Kirke, an engineer of the BBC, who was later to become its Senior Research Engineer. 'Kirke was very interested and helpful', wrote Baird. Several transmissions were arranged, the television picture being sent from Baird's laboratory at Frith Street to a BBC studio by telephone line. Kirke then put it on the air through the BBC radio transmitter and Baird received it again by wireless at his laboratory.[4] 'Complete success was achieved by this method.' No details of the equipment used for this demonstration appear to be available and in particular the method of synchronisation adopted is not known but was probably the same as that used several months later for the transatlantic transmission.

Meanwhile, questions on television and related matters had begun to be asked in the House of Commons. On the 12th April 1927, Sir H. Brittain asked the Postmaster General whether he had had his attention called to the results of experiments in television carried out in the United States and 'whether any research of a similar nature was being undertaken in this country'.[5] The reason for Sir. H. Brittain's question seems a little obscure but it was answered by Viscount Walmer, the Assistant Postmaster General. 'The answer to both questions is in the affirmative. As I informed the Honorable Member for the Central Division of Southwark on the 8th March the matter is not yet out of the experimental stage.' Perhaps this was the answer Brittain wanted to hear. Certainly there were some persons, notably Campbell Swinton, who were much opposed to Baird's mechanical methods and his desire to broadcast his rather crude images. Brittain's supplementary question appeared to show his own position: 'Will the noble Lord be very careful about encouraging what may prove to be a very tiresome invention?' The Assistant Postmaster General expressed the view that 'the Post Office (was) bound to accept all inventions which may be for the service of the community whether they are tiresome or not'.

The Post Office's attitude seemed to be one of cautious co-operation but this view was to change a few weeks later when they objected to the proposed tests which had been planned by Baird and the BBC with Kirke acting as their advisor.[6] Only three experimental transmissions were given but although Baird recorded that ' . . . the picture came through the BBC practically unaltered', the BBC's view was that the three experimental transmission were unsatisfactory.[7]

Hutchinson persevered in his task and in August wrote to the Secretary of the GPO mentioning:

> We are negotiating for the formation of a French Company called Baird Television Development Company (France) Ltd. and application to the French government for a 3 kW power licence was made by us to facilitate the speedy development of our inventions as we were rather doubtful about getting such facilities in this country without considerable delay. Since that time however we have been afforded the help of the BBC who were kind enough to broadcast our television image and sounds and has helped considerably in the development of our apparatus. By an order received by the Post Office authorities, the BBC were forced to cease working in conjunction with us and in subsequent interviews with your Mr F. W. Phillips and Mr Wissenden it was suggested we should apply for a licence to use a more powerful transmitting set. Formal request is now made. Owing to the success obtained in our co-operation with the BBC it is our intention, in the event of the Post Office authorities upon reconsideration still withholding from the BBC permission to co-operate with us, to erect a transmitter as nearly a duplicate as possible to the transmitter now used by the BBC, but we sincerely trust that such permision will not be withheld. We confirm our verbal application for facilities to broadcast television to the meetings of the British Association at Leeds through 2LO at Daventry.[8]

Thus Hutchinson was hoping that a repetition of his earlier (1926) tactics would force the Post Office's position. He was not disappointed, for the next day Phillips wrote[9] to the Baird Company and indicated the PMG would be glad to consider their application but that it was essential that his technical officers should have the opportunity of 'ascertaining the present stage of development of television apparatus'. The letter went on to say that the PMG proposed to discuss further with the BBC the question of whether one of their stations might be used in connection with the company's experiments during a limited period, under suitable conditions, and that further provision could not be given for the use of a broadcasting station for the demonstration at the meeting of the British Association.

Hutchinson's letter of 17th August 1927, however, was not a bluff, for he had written on 27th June to the Minister of Commerce and of Industry, Paris,[10] and mentioned that his company wished to make an application for a permit to establish a television transmitting and receiving aparatus in or near Paris. The object in view was the experimental intercommunication of television between Paris and London on a wavelength of 45 m using a transmitting power of 3 kW.

At this time the Baird Company wanted to achieve a great deal of publicity for their system in order to establish a broadcasting system in this country and in addition Baird himself wished to satisfy his desires and pride by achieving as many 'firsts' as possible: he had been the first person to demonstrate the live television of subjects, and the first to televise objects in total darkness using infra-red radiation; he was later to be the first to show colour television, to record television signals on to discs, to transmit television signals across the Atlantic and so on. However, the Americans had been the first to propagate signals over 200 miles by telephone line and over 200 miles by wireless. Baird had transmitted experimental television signals by wireless and by telephone lines and now felt that television broadcasting in this country was possible on medium wavelengths. Apart from satisfying his pride, of course, he now had a public company with shareholders who would want some return from the company for the money they had invested. The most likely source of any profits would be from the sale of television receivers and this demanded a service.

The French Minister of Commerce and of Industry did not reply to the Baird Company's letter immediately but took the precaution of enquiring from the General Post Office whether they had details of any objections to the establishment of television between Great Britain and France.[11] The Post Office was now alerted to Baird's plans and on 19th July 1927 Wissenden wrote to the company pointing out:[12] 'the use of your licensed stations for this purpose will be contrary to the conditions on which the licences were granted', and went on to ask if additional facilities were required they would be so good as to furnish full particulars.

Baird's transmission between London and Glasgow had not satisfied his desire to surpass the Americans and all through 1927 he worked in secret for a transmission across the Atlantic. Clapp,[13] who worked for Baird and was a radio amateur, had a transmitter at his home in Coulsdon, Surrey, and was in touch with another amateur, a Mr Hart, who had a transmitter at Hartsdale, near New York. Baird was to use this link for the transmission of television across the Atlantic.[4]

The first inkling of these experiments appeared in a number of newspaper reports in April 1927. The *Daily News* carried a column headed 'Faces across the sea'[14] and went on to describe the propagation of television signals via an amateur shortwave station in Coulsdon, which were being picked up regularly in New York. The *Morning Post* in a short paragraph on 22nd April[15] described the station in slightly more detail: 'A wireless telephone transmitter capable of delivering an electrical power of one kilowatt[16] to the aerial, which has a vertical wire about 35 feet long, was used.' The wavelength mentioned was 45 m and the installation took up the whole of a bedroom (Figs. 3.1).

In August 1927 Clapp left for America and Hartsdale to prepare for the reception of television there. Meanwhile, Baird and Denton, one of his assistants, spent many nights at Motorgraph House working on the project, for this was the only time the work could be carried out. Problems were encountered with the smallness of the power transmitted and with the synchronisation process.[4]

Fig. 3.1 *Antennas for the transatlantic shortwave transmitter erected at B. Clapp's house (1928)*

During these experiments the small company moved premises again, this time to 133 Long Acre, from which the BBC were eventually to make their first experimental transmissions. Two impressive masts were erected on the roof of the laboratories, an aerial was installed and the experiments continued.

By February 1928 Baird felt that he was in a position to announce his experiment. The transmission commenced at midnight on the 8th, London time (7 p.m. on the 8th, New York time). The first night was a failure. Baird had asked the actress Elisa Landi to come to the studio and be televised, but nothing was received in New York. The following night the signals were strong and Elisa Landi's place was taken by a Mr Howe, who was clearly seen.[4]

Immediately afterwards, the Television Society, recently formed with Lord Haldane as its President, issued a special report on the event:

> . . . there assembled at the offices of the Baird Company in Long Acre a small party of Press representatives and privileged guests. The transmission commenced at midnight, London time, or 7 p.m. New York time. In order to give watchers at the New York end an opportunity to adjust the receiving apparatus, the image of a ventriloquist's doll was first transmitted. The image sound produced by this doll . . . was sent over a telephone line to the Company's private experimental wireless station at Coulsdon. From this station the image sound was then flashed across the Atlantic on a wavelength of 45 metres.
>
> On the American side, the signal was picked up by an amateur receiving station at Hartsdale, a suburb of New York. After amplification the signal was then applied to the receiving televisor upon the ground-glass screen of which the image appeared. This screen measured about two inches by three inches.
>
> Four watchers were anxiously gathered round the apparatus. They were Captain O. G. Hutchinson, the Joint Managing Director of the Baird Company, who had gone to New York specially to conduct the experiments, Mr Clapp, one of the Company's engineers, Mr Hart, the owner of the amateur station at Hartsdale, and Reuter's Press representative.
>
> When the image of the doll's head had been satisfactorily tuned in, Mr Hart started up his transmitter, called a receiving station operator at Purley, near London, and asked that Mr Baird should take his place before the transmitter instead of the doll. This message was telephoned from the receiving station to the laboratories at Long Acre.
>
> For half an hour Mr Baird sat before the transmitter, moving his head this way and that until the message came through from New York that his image had come through clearly.
>
> . . . The demonstration proved quite conclusively that if a much higher powered transmitter had been employed the image would have been received in New York entirely free from atmospheric and other disturbances. An important feature is that only two operators were required to attend to the television transmission, one at each end of the circuit.[17]

Tribute soon came from the *New York Times*, which on 11th February 1928 said:[18]

> Baird was the first to achieve television at all, over any distance. Now he must be credited with having been the first to disembody the human from optically and electrically, flash it piecemeal at incredible speed across the ocean, and then re-assemble it for American eyes.
>
> His success deserves to rank with Marconi's sending of the letter 'S' across the Atlantic — the first intelligible signal ever transmitted from shore to shore in the development of transoceanic radio telegraphy. As a communication, Marconi's 'S' was negligible; as a milestone in the onward sweep of radio, of epochal importance. And so it is with Baird's first successful effort in transatlantic television . . . All the more remarkable is Baird's achievement because he matches his inventive wits against the pooled ability and vast resources of the great corporation physicists and engineers, thus far with dramatic success . . .

The quality of the images received was naturally not perfect. 'It could be made out', said *The Times* New York correspondent, 'that the man turned his head from side to side and opened his mouth, and that the woman showed first her full face, then her profile. The features were too blurred and dim to be recognisable as those of particular persons.'

The imperfections were caused by atmospherics and fading and by interference from other shortwave stations in Paris and Mexico City. A report[19] in the *New York Herald Tribune* drew a pointed comparison between the resources of American and British television pioneers. 'It is said that probably one thousand engineers and laboratory men were involved in the American (Washington–New Yorks) tests', the report mentioned; then it noted 'only a dozen worked with Baird'.

The possibility of transmitting television signals across the Atlantic was put forward by Ives of the Bell Laboratories in an important memorandum,[20] dated 4th May 1927, to H. D. Arnold, the Director of Research. In this Ives listed certain specific problems which might be taken up, namely

1 the establishment of a two-way appointment service between New York and Washington
2 the development of a public address system using television
3 the transmission of television signals across the Atlantic
4 the development of natural light scanning
5 the refinement of scanning means
6 the use of multichannel transmission
7 the elimination of the synchronising channel.

On the third of these prospects Ives wrote:

> The supreme achievement in television would be, beyond doubt, two-way transmission across the Atlantic. The conquest of time and space which would be exemplified by providing for the King of England and the

> President of the United States to talk with and see each other from their own official residences would have an appeal to the imagination of all ranks of humanity which would be unsurpassable.

Ives thought that this was within the realm of possibility and could be carried out by utilising two shortwave channels if sufficient money and talent were put on the task. He felt that the greatest uncertainty in realising the project would stem from the nature and effects of fading associated with the wavelengths which would be used. Ives ended his brief note by stating: 'It is probable that this project, if undertaken, would have to be carried out on the grounds of the acquisition of information as to radio transmission, and publicity value.' The latter aspect was probably the one which motivated Baird and it is likely that he wished to emulate (*mutatis mutandis*) Marconi's 1901 transatlantic experiment.

Shortly after his success Baird opened a new television section in Selfridge's wireless department. At a luncheon Baird said the new section would sell parts which would enable the amateur constructor to build his own television sets, adding that these would only allow very crude silhouettes to be transmitted. He further mentioned that the amateur would be able to use the elementary televisor as a basis for building a more technical machine for receiving the television transmissions sent out on a wavelength of 45 m. The cost of the sets was £6.10.1d.[21]

However, while Baird was keen to establish 'firsts' and to give demonstrations which would produce good publicity and a favourable response from the press, he was very diffident about showing his equipment to experts of the Post Office. They had requested a demonstration several months[22] before the transatlantic experiment and now, 1st March 1928,[23] Angwin was writing again on behalf of the GPO and asking 'if the company is now in a position to fix a date'. Baird's reply was unusual: 'Further contingencies have arisen which to some extent alter the situation. However I shall put the matter to our Board and will communicate with you further in due course' (3rd March 1928).[24]

Still, there was no reason why an engineer of the GPO's staff should not attend one of Baird's public demonstrations of television as a member of the general public, either with or without Baird's knowledge. Although the Engineer-in-Chief of the GPO had suggested this to Television Ltd. at the time of the Leeds experiments in September 1927, it would appear that the company did not reply to his letter. Nevertheless, a member of the headquarters technical staff attended a demonstration and submitted a long report on his observations.

Baird obviously did not want to demonstrate his apparatus to experts and yet he knew he could not broadcast in this country unless the Postmaster General gave his permission, which would be based on his engineers' reports. The most likely explanation for this odd situation is that Baird felt his images were not yet good enough for a satisfactory test before a critical audience but were sufficiently good to bring in much needed publicity from a duly impressed lay public.

Confirmation of this view is given by Baird's refusal to accept the £1000 challenge made by *Popular Wireless*.[25] The conditions of the challenge were that he should transmit over a distance of 25 yards three recognisable faces, various simple objects in motion, such as a tray of dice and marbles, simple geometrical figures and the moving hands of a clock. The test was to be supervised by a committee of competent investigators appointed by the editor of the magazine, Mr Norman Edwards.

Baird declined the invitation and said, at a luncheon in the Aldwych Club, that he proposed to say nothing about it except that he thought recent demonstrations had been a sufficient answer without going further.[26]

The reason for the challenge was given by Edwards:

> The technical problem of television is extremely complicated and not easy for the lay mind to understand, but in the opinion of scientists it will undoubtedly be a practical proposition one day. But it is also the opinion of scientists – certainly of the many I have consulted – that known television systems are not capable of sufficient development to warrant great optimism with regard to their early utility for television in the home.
>
> Our challenge to Mr Baird was based on a desire to clarify the position. If he could carry out the conditions, he would at least have demonstrated that his system had merits which would justify his optimism.[27]

Although the offer by *Popular Wireless* was renewed at the expiration of the seven days allowed for its acceptance Baird failed to take it up.[28] Yet, had he accepted and been successful, he would not only have had a good press in the technical journals (which would possibly have silenced Eckersley, the Chief Engineer of the BBC, who was a contributor to *Popular Wireless* and a strong opponent to the broadcasting of Baird's images), but in addition would have had a worthwhile sum of money for further technical development.

The plain fact was that Baird's images were poor and could not withstand scrutiny by professional scientists and engineers: Eckersley at that time was right. Even Moseley, Baird's most staunch supporter and friend, conceded this point years later when he wrote:

> Actually, *Popular Wireless* was showing no enthusiasm for our television because it thought – honestly, no doubt – that the invention had not reached anything like a practicable stage (nor had it for that matter: but it should have been encouraged).[7]

What undoubtedly caused some scientists to have mixed, if not entirely hostile, feelings about television were the somewhat extravagant claims made by Hutchinson. Even at the time of the £1000 challenge, while recognising the need for more detail, he could not resist making further statements about the prospects of an imminent service: 'Television', said Hutchinson on the 10th March 1928,[29] 'has now very nearly reached the commercial stage – a few more experiments and we shall be there. The next move will be to get more detail and to increase the picture to life size. We intend to start a two-way service between London and New York, and the machine has already been built in the latter city.' This service never started.

Baird's successes were now following each other rapidly. Rather than concentrate on improving detail he was determined to achieve 'firsts' in all aspects of television transmission. A few days after the transatlantic experiment, pictures were transmitted to the Cunarder, *Berengaria*, in mid-Atlantic. The experiment took place during the night of 5th March 1928[30] using the same wavelength as before, namely, 45 m. The images were seen by a small group of people, including passengers and ship's officers, crowded together in a small reception room. Mr W. Sutcliffe, staff chief engineer of the *Berengaria*, described what they saw in the May issue of the *Television* magazine:

> On looking at the screen of the televisor I saw rapidly moving dots and lines of orange light which gradually formed themselves into a definitely recognised face. This image varied from time to time in clarity, but movements could be clearly seen, and the image, when clear, was unmistakable.

Another witness of the experiments was Mr S. W. Brown, chief wireless operator of the liner, who recognised on the screen the image of his fiancee, Miss I. Selvey, who had been specially invited by the Baird Company to sit before the transmitter in London. Mr Brown recognised Miss Selvey first by her characteristic style of hair dressing, but was later convinced of her identity when she turned to show her profile.

Notwithstanding the crude nature of the images which could be produced at this time, certain American businessmen were sufficiently impressed by what could be achieved to enter into negotiations for patent rights with the Baird Company.

Two such businessmen were Mr N. Feldstern and Mr H. Z. Pokness who represented a group of American financiers. The announcement that they had acquired the American rights for certain patents of the Baird television system was made at a dinner at the Savoy Hotel on 19th April 1928 by Sir James Percy, one of the Directors of the Baird Television Development Company.[31]

In responding to Sir James's proposal for their health Mr Pokness said they came, saw, and were conquered, but did not get 'it' as cheaply as they wanted. 'There had been a great deal of publicity devoted to television in the United States,' he declared, 'but we found you folks owned the rock bottom foundation of it, it was with that knowledge we negotiated and made a deal.' They were now perfectly satisfied, he said, and were in a position to do a hundred per cent job. The people that were with them were people who really controlled the radio situation in America.

Vowlers, the stockbroking firm, were delighted with Baird's achievements and successes and arranged to underwrite a new company, Baird International Television. This was launched on 25th June 1928 and the subscription list closed on 26th June. The company was formed with the object of acquiring from Television Ltd. the vendors, the rights and interest of that undertaking in the Baird inventions and patents. The intention was to form or collaborate in the formation of overseas manufacturing and marketing companies and it was stated that negotiations 'with interests of first importance were in hand for the commercial exploitation of the inventions relating facsimile telegraphy and phono-vision'.[32]

The purchase price payable to Television Ltd. was satisfied by an allotment of all the 1 400 000 'B' shares of 5 shillings each credited as fully paid. As a consequence only the 1 400 000 'A' shares of 5 shillings each were available for issue and of these the vendors were entitled to the call of 400 000 until the 31st December 1928, at a price of 6 shillings per share. Thus only 1 000 000 'A' shares were on issue to members of the public.[32]

Lord Ampthill was Chairman of the new company and the other directors were Sir Edward Manville, Lt. Col. George B. Winch, J. L. Baird and O. G. Hutchinson. The share capital of the company was £700 000 divided into two classes of shares of 5 shillings each, and when the subscription list was closed it was heavily oversubscribed, with the result that the shares were allocated according to the number originally requested.

In many respects Baird's early commercial activities and difficulties mirrored those which had been experienced by Marconi. Both inventors were keen to exploit their initial successes, and, as their work could be furthered by obtaining funds from the public, companies were formed. The Wireless Telegraph and Signal Company was established in 1897, the year following Marconi's visit to England. Television Ltd. was registered in 1925, the year of Baird's first demonstration of rudimentary television. When further companies were floated these incorporated the name of the inventor in each case. None of the public companies was initially rewarding for their shareholders: Marconi's Wireless Telegraph Co. Ltd. did not pay any dividends from 1897 to 1910. Both inventors felt that the creation of overseas interests were necessary for the advancement of their plans. In 1899 the Marconi Wireless Telegraph Company of America was set up and was followed by the Marconi International Marine Communication Company Ltd. in 1900.

The original capital of the Wireless Telegraph and Signal Company was £100 000, that of the Baird Television Development Company was £125 000, and was subscribed largely by wealthy individuals. Presumably, in each case the subscribers wanted a speculative investment in the new communication systems. Marconi was handsomely rewarded for his enterprise and received £15 000 in cash and 60% of the original stock in exchange for nearly all his patent rights. He was only 23 years of age at the time. Similarly, Baird Television Development Company acquired the sole rights to exploit the inventions of Baird in the United Kingdom by paying a purchase price of £20 000 to Television Ltd. which had Baird as one of its founder members.

Moreover, both Marconi and Baird encounted difficulties in advancing their commercial interests. Aggressive tactics were adopted by the two companies to further their objectives, and much goodwill towards Marconi and Baird was dissipated later by this approach. Sir William Preece, who initially championed Marconi, said in 1907: 'I have formed the opinion that the Marconi Company is the worst managed company I have ever had anything to do with . . . Its organisation is chiefly indicated by the fact that they quarrel with everybody.'

Following the establishment of Baird International Television, short term speculators were richly rewarded: the 5 shillings shares were quoted at 12 shillings or exactly double their issue price by 9th July 1928.[33] It appears that this rise was induced mainly by knowledge of the signing, on the previous Friday, of two contracts for the flotation of a United States subsidiary and a Canadian subsidiary (with the parent company retaining a half interest in both). An additional reason for the advance was the progress which Baird was making on the technical side; he had succeeded in televising objects in daylight and had demonstrated colour television.

The daylight demonstrations took place towards the end of June 1928.[30] They were brought about because Baird had been able to improve the sensitivity of his photoelectric cells. Sir Ambrose Fleming, a staunch supporter of Baird and President of the Television Society, wrote an account of this achievement in the *Television Magazine* for July 1928.

> The writer has had the opportunity of seeing in practical operation in Mr Baird's laboratory a very striking advance in the apparatus for television which has been recently made by Mr Baird.[34]
>
> In this vast improvement it is not necessary for the face or object, the image of which is to be transmitted for television, or 'televised' (if one may venture to coin such a word), to be scanned by a brilliant beam of light traversing it, or to be flooded by powerful infra-red rays, as explained below. The object whose image is to be transmitted can be simply placed in diffused daylight, just as if the ordinary photograph of it had to be taken. The transmitting apparatus is then placed near to the object, and the image of it appears on the screen at a distance when proper synchronism is secured. The advantage of this important advance will be clear. It means that the face of a singer or speaker can be transmitted by television at the same time that the voice is being picked up by a microphone for ordinary wireless broadcasting. It means a great step forward in the possibility of transmitting to a distance.

For several years following his early (1923) experiments in Hastings, Baird was very secretive about the constructional features of his light-sensitive cells. Even in January 1927 the inventor wrote: 'The exact nature of my light sensitive device is being kept private.'[35] It is significant that no cell exists with the televisor in the Science Museum, London.

The year 1928 was one of great activity for the Baird organisation. Not only was Baird achieving an impressive number of firsts but the company was growing. Commander W. W. Jacomb was appointed Chief Engineer of the Baird Television Development Company[36] and under his direction the first home receivers were designed and built and the transmitting equipment rationalised for the first time. Among the technical assistants were B. Clapp, T. H. Bridgewater, J. Denton, D. R. Campbell, J. D. Percy, A. F. Birch, J. Collier and J. C. Wilson, a theoretician who did much work on colour television before he died in 1941, (Fig. 3.2).

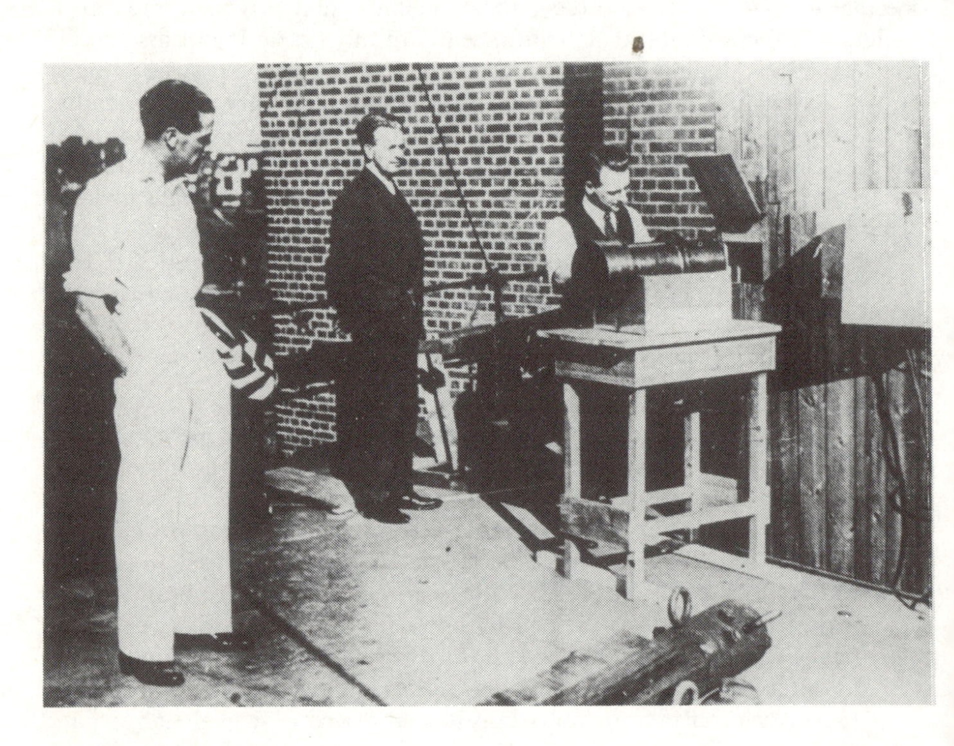

Fig. 3.2 *J. L. Baird with Jack Buchanan on the roof at Long Acre. Mr Collier is on the right (1928)*

Percy has described Jacomb in the following words:

> He probably did more than any other single individual for the rationalisation and development of the Baird principle. This remarkable man, whose association with Baird lasted over seven years, displayed a brilliance and foresight, which, had mechanical television not been eventually eclipsed by electronic methods, would have placed him in the forefront of television history. As it is, his niche is assured by the surprising progress which was made in mechanical television under his guidance during his term of office as Baird's Chief Engineer.

Life in the laboratories was hectic: no sooner was one of Baird's ideas past the experimental stage than he was on to the next leaving Jacomb and the staff to perfect earlier ideas as best they could. Hours and holidays meant nothing to Baird and time was always too short for the implementation of the many thoughts which occurred to him. Baird was the kingpin of the organisation and after the formation of the public company the directors were much worried about this: if anything happened to Baird, who did not enjoy good health, the company would collapse.

The solution was to insure Baird for the sum of £150 000 for one year at a premium of £2000.[7] As he put it:

> My technical staff consisted of half-a-dozen new men who had not yet attained a mastery over the many weird devices which I then used. There were no television engineers in those days, and to the wireless experts much of my apparatus contained features entirely strange . . . So it was decided to insure me for a hundred and fifty thousand pounds. I was extremely unwilling and nervous about this. I remembered my wartime card, 'Unfit for service'. If the insurance company turned me down, the Directors might refuse to carry on. The Chairman was adamant, so I was prodded about by two doctors who whispered together in a corner of the room. They obviously did not like the proposition. They were reluctant to turn down such a magnificent piece of business. Finally, the insurance company decided to take the risk for twelve months at a whacking premium, two thousand pounds, I think it was. The situation was saved.[4]

Baird and the Directors of the Baird companies made a number of errors of judgement in pursuing their objectives.[37] First, they failed initially to appoint a sufficient number of high-calibre research scientists and engineers. The importance of this point is well illustrated by the successes which the highly qualified research teams of RCA and EMI, led by Drs Zworykin and Engstrom and by Mr Shoenberg, respectively, achieved in the 1930s. Although the staff recruited by two Baird organisations were keen and eager to advance the development of television, they lacked the formal university research training which many of Shoenberg's and Engstrom's staffs possessed. In this respect, Baird's approach to television development differed from the plan of action of Marconi to the advancement of radio communication. Marconi surrounded himself from an early stage with a group of very able engineers and technicians, including Dr W. H. Eccles, Dr Erskine Murray, W. W. Bradfield, A. Gray, C. S. Franklin, H. J. Round, *et al.* By 1900, there were 17 professional engineers in the Marconi company in the UK. Electric and Musical Industries, had the brilliant A. D. Blumlein as one of its first research engineers, and they found it necessary to engage many university trained research workers: their research department included Dr L. Klatzow, Dr B. M. Miller, Dr B. M. Crowther, Dr J. D. McGee, Dr H. G. Lubszynski, Dr W. Stewart Brown, Dr L. F. Broadway, Dr E. L. C. White, C. O. Browne, F. Blythen, G. E. Condliffe, and many junior technical staff.

Secondly, insufficient importance was attached to the value of consultants. Marconi fully appreciated their worth and employed them for many years. He sought out some of the most promising university scientists who were interested in wireless and appointed them as consultants. Professor Ambrose Fleming and Professor M. Pupin were two of the notable consultants of the Marconi organisation. Marconi was not hesitant about engaging staff with a greater intellectual and technical competence than his own. Scientists of the proven ability of, say, Watson-Watt or Appleton, could have had a most beneficial effect on Baird's work, particularly in the use of the shortwave bands for television. Marconi's policy certainly brought benefits to his company. One of his most important early demonstrations of wireless telegraphy, the transatlantic venture, owed much to Fleming's work. It was Fleming who designed the transmitters, each of which incorporated an ingenious circuit having two spark-gaps operating in cascade at different frequencies.

With Jacomb in charge of the development of the original Baird system of television, Baird was able to devote time to the achievement of more firsts. After the demonstration of daylight television in June 1928 he went on to show colour television, for the first time anywhere in the world, on 3rd July 1928. The *Morning Post* for 7th July 1928[38] contained an account of the progress which Baird had made:

> One of the party went onto the roof of the building where a transmitting televisor had been set up, and the rest of the party went with Mr Baird into a room where there was a receiving apparatus. The receiver gave an image, about half as large again as an average cigarette card, but the detail was perfect.
>
> When the sitter opened his mouth his teeth were clearly visible, and so were his eyelids and the white of eyes and other small details about his face. He was a dark-eyed, dark-haired man, and apeared in his natural colours against a dark background. He picked up a deep red coloured cloth and wound it round his head, winked and put out his tongue. The red of the cloth stood out vividly against the pink of his face, while his tongue showed as a lighter pink.
>
> He changed the red cloth for a blue one, and then, dropping that, put on a policeman's helmet, the badge in the centre standing out clearly against the dark blue background.
>
> The colour television proved so attractive that the sitter was kept for a long time doing various things at the request of the spectators. A cigarette showed up white with a pink spot on the end when it was lit. The finger nails on a hand held out were just visible and the glitter of a ring showed on one of the fingers.

Although Baird's first demonstration of colour television did not take place until July 1928, the inventor had been reflecting on the problem for a number of years. His patents, 266,564 and 267,378 were applied for on 1st September

1925 and they both contained references to the transmission of images in their natural colours.

At that time Baird was more concerned with the reproduction of monochrome images having half-tones than with coloured images, and his patent 267,378[39] was an attempt to provide a solution to his problem. However, as with other patents, the ideas it contained were adaptable to related situations and so Baird was able, in the same patent and in patent 266,564,[40] to refer to the possibility of achieving images in colour.

For the 1928 demonstration Baird used transmitting and receiving Nipkow discs with three spirals and associated filters, one for each of the primary colours, a single transmission channel, and effectively three receiver light sources (giving red, blue and green outputs). These sources were switched sequentially by a computer so that only one lamp was excited at a time.

Another person who saw the display was Baird's supporter, Dr Alexander Russell. He contributed an article on the progress of television to the 18th August 1928 issue of *Nature*.[41] His report was naturally more technical than the *Morning Post* article:

> The process consisted of first exploring the object, the image of which is to be transmitted, with a spot of red light, next with a spot of green light, and finally with a spot of blue light. At the receiving station a similar process is employed, red, blue and green images being presented in rapid succession to the eye. The apparatus used at the transmitter consists of a disc perforated with three successive spiral curves of holes. The holes in the first spiral are covered with red filters, in the second with green filters and in the third with blue. Light is projected through these holes and an image of the moving holes is projected onto the object. The disc revolves at ten revolutions per second, and so thirty complete images are transmitted every second – ten blue, ten red, and ten green.
>
> At the receiving station a similar disc revolves synchronously with the transmitting disc, and behind this disc, in line with the eye of the observer, are two glow discharge lamps. One of these lamps is a neon tube and the other a tube containing mercury vapour and helium. By means of a commutator the mercury vapour and helium tube is placed in circuit for two thirds of a revolution and the neon tube for the remaining third. The red light from the neon tube is accentuated by placing red filters over the view holes for the red image. Similarly, the view holes corresponding to the blue and green images are covered by suitable filters. The blue and green lights both come from the mercury helium tube, which emits rays rich in both colours.
>
> The coloured images we saw which were obtained in this way were quite vivid. Delphiniums and carnations appeared in their natural colours and a basket of strawberries showed the red fruit very clearly.

Marconi and Baird had the same policy with regard to their technical contributions. They applied for patents on every method and apparatus which they devised. In this respect they were inventors and innovators rather than scientists. The tradition of pure science did not allow university scientists, for example, to seek commercial gain for their work.

Baird had to devote a great deal of time and labour to acquire a patent holding which would place his companies in a favourable position commercially. Until about 1930–1931, he engaged in this task almost single-handedly. From the start of his work in 1923 to the end of 1930 Baird applied for 88 patents; the number of patents originating from other members of the Baird companies in the same period totalled four.

Fig. 3.3 *Stereoscopic television apparatus which was used to give demonstrations at the Glasgow British Association meeting in 1928, and to the Prince Consort of Holland in Rotterdam in 1928. Mr Collier is seated on the right*

Not surprisingly, Baird had little time for writing scientific papers and engaging in field trials of the type demonstrated by the AT & T Co. in April 1927. He tried to anticipate every likely development and application of the new art. Daylight television, noctovision, colour television, news by television, stereoscopic television (Fig. 3.3), long-distance television, phonovision, 2-way television, zone television and large screen television were all demonstrated by Baird during a hectic period of activity from 1927 to 1931. A former Baird engineer has said: 'Too much time was spent on adventurous sidelines and in exploiting the 30-line system – largely in pursuit of publicity, which, rightly or wrongly, was considered necessary for the attraction of public inerest and capital; too little time on essential technical improvements'. However, when an inventor wishes to further the commercial prospects of his ideas and there are competitors about it is essential that he should protect his basic methods and apparatuses by patents as quickly as possible: once these have been safeguarded, technical enhancement can take place at a more leisurely pace.

Fig. 3.4 *Bell Laboratories' apparatus for colour television. With the exception of the photoelectric cabinet on the left, the apparatus is identical with that used for the original demonstration of monochromatic television (July 1929)*

Certainly there were many industrial organisations interested in television by the late 1920s. The Western Electric Company, Westinghouse Electric and Manufacturing Company, the Radio Corporation of America, General Electric, the American Telephone and Telegraph Company, the Jenkins Television Corporation and others in the USA, Telefunken, Fernseh AG and Telehor AG in Germany; while in the UK several companies were considering pursuing investigations in this field, including both HMV and MWT Co. (Fig. 3.4).

Baird's strategy in the 1920s was clearly to secure as many firsts as possible, not only for publicity purposes, but to give his companies commercial bargaining power. He was not pursuing adventurous sidelines wholly for good public relations, but as a means to an end. It is not possible always to foresee the likely outcome of an invention or the manifold applications of a basic idea.

Notwithstanding Baird's successes and the good publicity which they achieved, there was a cloud on the horizon; the British Broadcasting Corporation was not impressed by Baird's work.

References

1 Anon.: 'Baird Television Development Company, Statutory Meeting', *The Times*, 18th July 1927
2 IVES, H. E.: 'Television', *Bell Syst. Tech. J.*, 1927, **6**, October, pp. 551–559
3 Case 33089, AT and T Co., Warren Record Centre, USA
4 BAIRD, M.: 'Television Baird' (HAUM, South Africa, 1974)
5 BRITTAIN, Sir H.: Question in the House, 12th April 1927
6 Memorandum to the Postmaster General, November 1928, Minute 16806
7 MOSELEY, S. A.: 'John Baird' (Odhams, London, 1952)
8 HUTCHINSON, O. G.: letter to the Secretary (GPO), 17th August1927, Minute 4004/33, file 3
9 PHILLIPS, F. W.: letter to Baird Television Development Company, 18th August 1927, Minute 4004/33, file 3
10 Baird Television Development Company, letter to Minister of Commerce and Industry, Paris, 27th June 1927, Minute 4004/33, file 4
11 Secretary General of PTT: letter to GPO, 12th July 1927, Minute 4004/33, file 4
12 WISSENDEN, J. W.: letter to Baird Television Development Company, 19th July 1927, Minute 4004/33, file 4
13 GPO: letter to Television Ltd., 26th August 1926, Minute 4004/33, file 3
14 Anon.: a report 'Faces across the sea', *Daily News*, April 1927
15 Anon.: a report, *Morning Post*, 22nd April 1927
16 WEAVER, R. C.: memorandum to Engineer-in-Chief (GPO), 3rd May 1927, Minute 4004/33, file 3
17 Anon.: a report, Television Society, quoted by M. Baird in Reference 4
18 Anon.: a report, *New York Times*, 11th February 1928
19 Anon.: a report, *New York Herald Tribune*, 11th February 1928
20 IVES, H. E.: memorandum on 'Development program for television', 4th May 1927, Case 33089, AT and T Co., Warren Record Centre, USA
21 Anon.: a report 'Television now on sale', *Daily Chronicle*, 21st February 1928
22 PHILLIPS, F. W.: memorandum to W. T. Leech, Minute 4004/33, file 3
23 ANGWIN, A. S.: letter to Baird Television Development Company, 1st March 1928, Minute 4004/33, file 4
24 BAIRD, J. L.: letter to the Engineer-in-Chief (GPO), 3rd March 1928, Minute 4004/33, file 4
25 Anon.: a report, *Popular Wireless*, March 1928
26 Anon.: 'Seeing by gramophone. Future of television', *Daily Telegraph*, 9th March 1928
27 Anon.: 'Television challenged. Simple facts demanded', *Daily Telegraph*, 10th March 1928
28 Anon.: 'Challenge withdrawn', *Daily Telegraph*, 20th April 1928
29 Anon.: a report, *Morning Post*, 10th March 1928
30 MOSELEY, S. A., and BARTON CHAPPLE, H. J.: 'Television today and tomorrow' (Pitman, 1933)

31 Anon.: 'American rights sold', *Daily Telegraph*, 21st April 1928
32 Anon.: a report, *Evening Standard*, 25th June 1928
33 Anon.: a report, *Financial Times*, 9th July 1928
34 FLEMING, Sir Ambrose.: a report, *Television Magazine*, July 1928
35 Anon.: a report 'Triumph of television', *Evening Telegraph and Post*, 10th January 1927
36 PERCY, J. D.: 'The founding of British television', *Journal of the Television Society*, 1950, pp. 3–16
37 BURNS, R. W.: 'Success and failure', *Proc. IEE*, 1979, **126**, no. 9, pp. 921–928
38 Anon.: a report, *Morning Post*, 7th July 1928
39 BAIRD, J. L.: 'Improvements in or relating to transmission and/or reproduction of views, scenes or images by wires or wirelessly', British patent no. 267,378, application date 1st September 1925
40 BAIRD, J. L.: 'Improvements in or relating to the transmission and/or reproduction of views, scenes or images by wires or wirelessly', British patent no. 266,564, application date 1st Setember 1925
41 Anon.: a report, *Nature*, 1928, *122*, 18th August, pp. 283–234

Chapter 4

The BBC view, 1928

The BBC was most unhappy about the sporadic attacks which were being made on it for doing nothing about television, and considered at a Control Board meeting on 15th May 1928[1] how best to counter them. Their Chief Engineer, Eckersley, suggested 'some form of cover for the public eye, such as the appointment of a commission to investigate the matter' (he already being thoroughly in touch, and being satisfied that there was nothing to be given to the public at this moment). It was decided that the issue should be covered by a series of articles in the *Radio Times* by experts whose names were to be agreed between Eckersley and Gladstone Murray, the BBC's Assistant Controller (Information). The experts who were invited to contribute were Professor L. B. Turner, Dr W. H. Eccles, and Professor E. V. Appleton.

Eckersley's letter to Turner set the scene: 'The world is apt to go mad and it is the duty of experts to restrain enthusiasm which is based on insecure foundations', he wrote.[2] He then went on to give his views on the difficulties associated with the practical realisation of television and invited Turner to write an article for the *Radio Tiumes*.

Turner readily agreed[3] and after Eckersley had asked him whether he could do it at once[4] produced a contribution headed 'The problem of television' within a few days.[5] The task had not been an easy one: 'I have found it very difficult to say anything satisfactorily on so intricate a subject which will be both true and intelligible to non-expert readers', he said in a letter to the Chief Engineer.[6]

Ashbridge, the Assistant Chief Engineer, was not happy about its publication in the *Radio Times*: 'Personally I think it is of very little use for publication in the *Radio Times*. It is too technical I think for the average reader and although the last page sums up the position fairly well I am afraid very few *Radio Times* readers will get as far as that,' he told Eckersley in an internal memorandum.[7]

Meantime, Eccles had declined to write, as an article from him on a contentious matter would not be 'consistent with his position as President of the Institution of Electrical Engineers', and Appleton had not replied.[7]

Turner was to share Ashbridge's view, for on the same day that the Assistant Chief Engineer gave his views Turner wrote: '. . . I have come to realise that only part of my article which the incorrigible lay reader will swallow at all is the last paragraph, and I think this certainly ought to be deconcentrated a bit.' He enclosed another sheet to replace the last paragraph but the article was never published.[8] After receiving a letter from Turner, dated 21st September 1928,[9] asking '. . . have you dealt with my typescript', Eckersley wrote back saying[10] 'We are keeping your article for the time being in cold storage as the psychological moment has not yet come.' Although he did not explain what he meant by this, Eckersley, of course, would have been conscious of the considerable successes which Baird had achieved during 1928 – spotlight scanning, the transatlantic experiment, daylight television, colour television and later stereoscopic television – and probably felt that to publish an article antagonistic to the possibilities of Baird's systems would be to typecast the BBC and cause alienation which would lead to further attacks on it. Additionally Appleton had failed to produce an article for the *Radio Times* although he had done so for *The Times*,[11] and Eckersley probably felt that he personally was in a weak position as he had not been inside the television laboratories for two years.

By the middle of 1928 the BBC were becoming distinctly concerned at the Baird Company's plans and pretensions. At that time Bairds had no licence for public broadcasting but it was known to the BBC that they had applied to the PMG for broadcasting rights similar to those of the BBC and that it was examining the legal question of whether a licence was necessary for television transmissions. Baird was promoting public interest in his system by his considerable achievements and on 22nd June 1928 inserted an advertisement in *The Times* headed: 'Television for all – a Baird Seeing-in-Set for £25.'[12]

According to this the Baird Televisor would be purchasable either as a separate instrument or in combination with a listening-in-set; and so 'at one and the same moment' the owner would be able 'both to hear and see' a performer at the broadcasting station. Reith was certainly concerned by this announcement and sent an appropriate letter[13] to the GPO, for the implication of the advertisement was clearly that the public could be led to expect an early inauguration of a television service in co-operation with the BBC. The Director General thought the advertisement appeared to contain such 'palpable misrepresentations as to create a situation' which the Corporation should not ignore.[14] His anxiety was not allayed by the GPO's reply,[15] which referred to their view which had been given in an answer[16] to a question[17] put by Mr Maloney, in the House of Commons, to the Postmaster General on 22nd May 1928. This was that the Post Office still considered television to be in the experimental stage and that the time had not yet arrived when the PMG should make provision for a public service. Reith felt that the issue, by the GPO, of an authorative statement containing a definite warning to the public on the matter was necessary, but Phillips[18] of the GPO disagreed, although he had no objection to the BBC issuing its own press statement.

The advertisement was also seen by Wissenden who, in a memorandum to the Engineer-in-Chief, wrote:

> It seems advisable that the Television Company should again be asked when they are prepared to give the promised demonstration to the Post Office. It seems also desirable to remind Captain Eckersley of the arrangement that the broadcasting stations are not to be used for experiments with this apparatus until a demonstration to the Post Office has been given.[19]

The Engineer-in-Chief agreed and three days later letters were despatched to both the BBC[20] and Bairds.[21]

The Secretary to the Baird Company did not reply at once but wrote on 16th July 1928 and mentioned that Hutchinson was in America and would not be returning to the United Kingdom until towards the end of August. 'On his return arrangements will be made for a private demonstration to you,' Bartlett wrote.[22] The company was again procrastinating and obviously did not want to give a demonstration to the Post Office's experts, although it had no reluctance to give some indication of its successes to the press and its friends.

Following the GPO's letter of 25th June to Bairds and the BBC there appeared in the *Evening Standard* for 26th June an article with the title 'Television export, what will listeners do then'.[23] Phillips had his attention drawn to it and immediately wrote to Television Ltd. and referred them to the following extract from the reports: '. . . broadcast television, (with) a licence from the Post Office, is to begin this autumn, the Secretary of this far sighted broadcasting concern stated officially today'. Phillips went on: 'If there is any foundation in this report the PMG would be glad to have the Company's observations on it.'[24]

Hutchinson replied[25] stating that there was no foundation whatever for the report and enclosed a copy of a letter[26] which Bartlett had sent to the newspaper in question when the announcement had been brought to the company's notice.

Rumours of this nature and other mysterious leaks of information which usually aided Baird's publicity and purpose were to be a feature of the company's relationship with the BBC and the GPO. Despite assurances of intensive investigations by the company, the rumours were never tracked down, but it would seem that someone with access to policy decisions also had easy access to the press.

Bartlett's letter to the *Evening Standard* stated:

> I wish to place on record that no statement has been made by me and the paragraph in question so far as it is attributed to me is entirely inaccurate. I would be glad if you would enquire into the matter and let me know how it was the statement appeared.

The reports did nothing to ease the BBC's misgivings about the whole situation. With the mediumwave bands as crowded as they were, the BBC thought that the granting of facilities to Baird might interfere with its programmes, and it was much concerned about rumours that the BBC would take an active part in the development of television. In order to prevent the public from being misled, the BBC issued a press statement[27] in July 1928 which read:

> In order that its listeners may not suffer disappointment by anticipating the possibility of seeing as well as hearing its performances, the BBC wishes to make it plain that it has not so far been approached with apparatus of so practical a nature as, in the opinion of the Corporation, to make television possible on a service basis.
>
> It should be noted that the Postmaster General, in replying to questions in the House of Commons, has indicated that in the opinion of his officers, television is still in the experimental stage, and that the time has not yet come to make arrangements for the provision of a public service.
>
> When the development of the service has reached the stage where some form of service will benefit listeners may be guaranteed, the BBC will be prepared, subject to the approval of the Postmaster General, to co-operate in the matter.[27]

In spite of this the Baird Company took newspaper space in *The Times* on 4th August 1928 to make an important announcement: 'Practical Television is Here'.[28] On the previous day the *Daily Mail* carried a report headed, 'television for all – broadcasting to begin in the autumn'.[29]

The advertisement stated that wihin a very short time the company would begin broadcasting their own programmes for those of the public who owned the new Baird combined wireless receiver and television set. After stating that the new dual set would be on show at the Radio Exhibition at Olympia in September, the advertisement continued:

> These services, in the first instance, may naturally be somewhat restricted: but owing to the adaptability of the new Baird Dual Set, you will be capable of receiving the ordinary broadcast programmes.
>
> The Baird Televisor will show, on your screen, the head and shoulders of the person being transmitted and give a living picture with perfect synchronism of movement and sound.

The price of the television sets was expected to be between £25 and £30, and of the combined sets between £50 and £60. However, in view of Baird's statement more than a year later, it seems doubtful whether there was any large-scale manufacture of the televisor in progress. The above advertisement was soon followed by another.[30]

Phillips wrote to the Baird Television Development Company on 7th August referring them to *The Times* advertisement and in particular the extract: 'Our new transmission station at 133 Long Acre is nearing completion. Within a very short time we propose broadcasting our own programmes for those of the public who own the Baird combined receivers and televisor sets . . .' Phillips went on to mention that the licences which had been issued were for experimental purposes only and that the use of the apparatus for general calls or for sending news, advertisements, or communications of a business or non-experimental character or a similar matter was expressly forbidden and that 'having regard to these limitations I am to ask under what circumstances your Company made these public announcements.'[31]

Fig. 4.1 *Captain O. G. Hutchinson, J. L. Baird and S. A. Moseley photographed in Berlin when the German company Fernseh A. G. was formed*

Bairds were certainly forcing the pace and probably hoping that public opinion and J. L. Baird's achievements would encourage the Post Office to grant a licence for broadcasting whether the BBC liked it or not. The company was fortunate in having a powerful ally at this time in the form of Sydney A. Moseley, the journalist (Fig. 4.1) Moseley had been interested in broadcasting for a number of years, had written letters to the BBC, had broadcast often enough for Broadbent of the BBC to write to him and say 'I expect you feel by now that you have thoroghly mastered the peculiar but by no means easy art of broadcasting',[32] he had produced criticisms of radio plays and considered himself to be Britain's first radio critic. He knew everybody in Fleet Street and was on cordial terms with the various BBC chiefs. In addition he had a shrewd knowledge of business and finance and had written a book called 'Money making in stocks and shares'.

Moseley was introduced to Baird shortly after the magazine *Popular Wireless* had published some adverse comments on Baird's system. The meeting occurred in the boardroom one day when Hutchinson was talking to Moseley and another person whose name Moseley could not later recall.[33] Baird referred to the two strangers in some notes which he wrote towards the end of his life:

> One of these, a stout and jovial man, with a merry and wicked twinkle in his eye immediately attracted me. It was Sydney A. Moseley, who was to play a very prominent part in our future activities. Both Hutchinson's and my own knowledge of high finance were infantile and this was also true of our knowledge of journalism. We were both much upset because a certain wireless paper had seen fit to publish unfavourable criticisms of us which we considered utterly unfair and prejudiced.
>
> We talked to Moseley about this and he said he would fix it and he did. The next issue of the paper contained an article by Moseley refuting all the previous attacks and hailing the invention as a great achievement. From that day on his position was established. He became one of the family and I welcomed him with open arms. What a relief to have a man about the house who took a real interest in our affairs.[33]

Moseley made it his immediate mission in life to put Baird television on the map. 'Realising that television was struggling against odds, ignorance, scepticism and hostility,' he has written, 'I decided to take up the cudgels on its behalf and determined to carry the fight through to the finish.'[32] He became Baird's self-appointed champion. 'One more into the fray' he exclaimed with his usual zest.[34]

After refuting the attack in *Popular Wireless* he began to attend meetings of the Board, where he was a great help to Baird who was not at his best in the boardroom. Baird found that the meetings of the various boards were long rigmaroles to be slept through. He had no patience with the reading of minutes and the propositions common in business, but Moseley was not the sort of man to fall asleep at board meetings or anywhere else where money was concerned.[35]

Moseley's association with the Baird companies was of great importance to the furtherance of low definition television systems in the UK. With his knowledge of finance, journalism and publicity and his contacts in Fleet Street, Parliament and elsewhere, Moseley brought much valuable expertise and experience to bear on the side of the television pioneer. He played a central and crucial role in the early development of television in Britain.

Marconi was one of the first to foresee the commercial possibilities arising from the work of himself and the early radio pioneers. He worked energetically throughout his life to progress and extend wireless communications. His primary interest, however, was in the technical development of the subject and his knowledge of business methods was such that, when he managed both the commercial and innovative aspects of his company, considerable difficulties arose. However, when Godfrey Isaacs, a born businessman, was appointed managing director, thereby allowing Marconi more time to further his research and development work, the company prospered. Marconi, like Baird, had a well balanced personality and worked easily with others: Isaacs and he formed an effective team, each respecting the other's skills to the advantage of their common interest.

A similar parallel can be inferred from the relationship between Baird and Moseley: Baird was quite content to allow Moseley to look after his business and financial interests and a bond of friendship was established between them shortly after Moseley's introduction to Baird in 1928 which lasted until the latter's death in 1946.

Unfortunately for Baird, Moseley resigned his directorship in 1933. The part played by Moseley in advancing Baird's interests was stressed in the BBC's letter of regret to him (on hearing of his resignation):

> Although there has not always been agreement either in policy or method, it should be recognised that your consistently active advocacy has been an important, perhaps the decisive, factor in the progress that Baird Television has made to date.[36]

On the other hand, Marconi's principal rivals in the USA, de Forest and Fessenden, were not so well adjusted personally as either Baird or Marconi and did not team up with such skillful operators as Moseley or Isaacs. Marconi, de Forest and Fessenden were associated with the three most important early American communication firms, namely Marconi Wireless Telegraph Company, de Forest Wireless, and National Electric Signalling. Only the Marconi Companies survived, although each of the three communication concerns was organised around one outstanding inventor.

Hutchinson replied[37] to the GPO's letter of 7th August on 22nd and gave an assurance that there was no intention on the part of the company to do otherwise than use the existing licences for experimental purposes and that an application would be made to the Department for the necessary broadcasting facilities in due course. He mentioned that the company felt it had reached a position where Baird Television's commercial and public use could only be established by extensive demonstration and experiment, and went on to say that for this purpose the co-operation of the public was essential. The company had in mind the issue of a limited number of sets to members of the public competent to assist in the experiments so that reports could be obtained from different localities and under varying conditions.

Bairds naturally were anxious to establish a new industry for the benefit of its shareholders by issuing licences to radio manufacturing companies for the purpose of manufacturing and selling their television apparatus, but these licences would be of little use unless coincidentally some part of broadcasting at stated intervals could be assured to purchasers of their sets. Bairds consequently adopted the ploy of enlisting some prominent radio manufacturing firms to their aid by giving them a demonstration of their combined television and radio reception apparatus and persuading them it was in a fit state to be adopted by the public. 'There was unanimity of opinion amongst those who witnessed the demonstrations that the invention was now more advanced than radio reception was when a public service was first instituted in this country.'[38]

This view, however, did not quite accord with the hesitancy Bairds had adopted towards the demonstration of their equipment to experts of the General Post Office, who, as late as 8th September 1928, had not been given an opportunity to see the capabilities of it, although they had requested to view a demonstration months earlier and had followed this request by sending a number of reminders to the firm. 'We regret,' said Bairds, 'that so far a recent demonstration has not been accorded to your representative, the reason for this being that rapid progress was being made in the development of the system and we were desirous when demonstrating it to your representative that he should see the very latest improvements which have been made.' Bartlett ended his letter by offering a demonstration to the GPO at our 'mutual convenience'.

While the Baird Companies received great support from a number of notable scientists, including Sir Ambrose Fleming, Dr Alexander Russell, Professor Taylor Jones and others and from a wide section of the press, owing in no small measure to the strenuous efforts of Hutchinson and now Moseley, the companies had their detractors. Captain P. P. Eckersley, the BBC's Chief Engineer, was in a powerful position to influence the adoption or rejection of television broadcasting and his view was that '. . . a radical discovery is necessary before television will be practicable, just as the valve made broadcasting possible'.[39] He was not opposed to television, *per se*, but opposed to the adoption of a system which might not fulfil the hopes which had been made for it. 'Now if television were perfected,' he commented, 'that would be a different proposition. There would be, I believe, a very popular demand for the BBC to take it up. But in its present form it would be useless for us to do anything. We might just as well have inaugurated a broadcasting system twenty years ago with the Poulsen arc as the nucleus of our transmitting equipment.' He considered that television was still definitely in the experimental stage of development.

Eckersley had his supporters, among whom Campbell Swinton was to be particularly vociferous, both privately and publicly.[40] In a letter to *The Times* dated 20th July 1928[41] he felt 'some comments should be made on the many, and in some cases very absurd, prognostications that have appeared during the last few weeks in the daily Press on this important subject'. His arguments were based on the very large bandwidth required for successful public television compared with that adopted by Baird and other low definition television workers. As an illustration he estimated that a 10 inch × 16 inch photograph of the Eton and Harrow match in an issue of *The Times* contained a quarter of a million dots and that if this picture was transmitted by television at the rate of 16 per second, following cinematograph film pactice, a rate of transmission of 4 000 000 dots per second would need to be realised. 'Such achievements,' he wrote, 'are obviously entirely beyond the possible capacity of any mechanism with material moving parts, and this view, which I have personally been inculcating in scientific circles for many years, has recently been thoroughly endorsed by no less an authority than Sir Oliver Lodge, himself a notable pioneer inventor in wireless telegraphy.'

Swinton very definitely wished to see the adoption of a system which utilised beams of electrons in both the transmitting and receiving apparatus and completely discouraged the advancement of any method of mechanical scanning. Not content to air his views publicly he wrote[42] to Eckersley in a more pungent and trenchant tone: '. . . I fear that my view is simply that Baird and Hutchinson are rogues, clever rogues and quite unscrupulous, who are fleecing the ignorant public, and should be shown up'. The difficulty was, he said, that papers like fat adverts. He went on to state: 'I have saved the Institution of Electrical Engineers having a paper on television by Baird . . . One of the past presidents was anxious for this, but I hope I have squashed the idea with the new President.'

Actually, as no televisors had at that time been placed on the market, Swinton was not correct in stating that the ignorant public were being fleeced. They had been invited to purchase shares in a public company which, for most companies, involved a degree of risk. His action in squashing a paper by Baird seems to have been quite unworthy of him. Had this been adopted serious discussion amongst scientists and engineers could have taken place on the merits of the various systems which had been proposed.

Swinton's action was particularly unfortunate for the state of the art, as Baird had been criticised in an editorial[43] in *Nature* the previous year (1927) for a discourse which he had given at an exhibition of the Physical Society. 'It is to be regretted that, possibly on account of patent considerations, Mr Baird had hitherto been unable to submit to a proper authentication of his claims by a learned society', the editorial noted. 'The policy of withholding publication of an essential item does not commend itself to modern inventors. It savours too much of medieval practice and usually defeats its own object of securing to the inventor the fruits of his invention', it went on.[44]

Campbell Swinton's letter to *The Times* was replied to by J. Robinson.[45] Dr Robinson was well known as a former member of the Imperial Communications Committee and as the Head of the Department of Wireless Telegraphy and Photography at the Royal Aircraft Establishment. Robinson set out to show, through letters and articles in the technical press and in *The Times* that all those critics who were calculating the wide range of frequencies for high-quality television – basing their arguments on a comparison with the dots in photographic blocks – were completely wrong. Actually, of course, he was wrong, but his views on this matter (of bandwidth) did much to support the low definition television methods in use in the 1920s. He asserted that '100 000 cycles per second is more nearly correct for a good picture than 3 000 000' and even ventured to add that 'the transmission of small scenes will be possible with a much smaller frequency band than that required for a telephony service.'

Robinson's views were contained in the above reply:

> I understand that the image is not divided into dots, but rather into strips, and a spot of light is made to move continuously along each of the strips in turn. Thus our conception turns immediately from the huge numbers of millions of dots to the comparatively small number of strips, the image being made up, not of dots but of continuous adjacent shaded areas – it is really surprising how few of them are required for reasonable definition. I witnessed a demonstration recently at the Baird Television Laboratories and saw the visual image of a man transmitted, the teeth when smiling, and such like details, being visible, the images also being in colour. I was informed that only 15 strips were employed in this apparatus. I also witnessed a transmission in monochrome, in which the detail was remarkable, though in this case 30 strips were employed. There thus appears nothing to prevent us having television of a serviceable nature in the near future.

The flaw in Robinson's argument is easy to appreciate; nevertheless, he found supporters for his views 'but they did no service to Baird who, as a result, was lulled into a rather obscure outlook on this very important dimensiuon of the art'. Bridgewater, who worked with Baird, mentioned in a paper:[46] 'It is only fair to remark that not many of the protagonists, even in those controversial days, fell into quite such depths of error.' (Fig. 4.2).

By the middle of 1928 the BBC knew that Bairds were examining the legal question as to whether a licence was necessary for television transmissions and had applied to the Postmaster General for broadcasting rights similar to those which they, the BBC had.[47]

In August 1928 the Baird company bought newspaper space[48] to announce that at the Radio Exhibition Olympia (22nd–29th September) orders for the new dual sets would be taken and that 'as soon as possible afterwards, our special broadcasting services from Long Acre, will commence'. The advertisement stated that the service, in the first instance, might

> naturally be somewhat restricted, but owing to the adaptability of the new Baird Dual Set, you will be capable of receiving the ordinary broadcast programmes. The Baird Televisor will show on your screen the head and shoulders of the person being transmitted and give a living picture with perfect synchronism of movement and sound.

The price of the televisors was expected to be about £25 and the set would be like an ordinary suitcase, 24 inch square and 18 inch in depth. The picture would be seen on a 'little glass screen about 8 inch in diameter' and the combined sets would be connected to the aerial in the ordinary way.

The BBC knew nothing of these proposals and told inquiring reporters that no scheme of co-operation with the BBC was in existence.[49] In fact Bairds were not proposing to proceed with plans to broadcast television through the BBC's stations at this time. Bartlett, the Secretary of the Baird Television Development Company, stated[50] that the company intended relying on expert legal advice

SEPTEMBER 1928 TELEVISION

OLYMPIA — 1928

Injured one: "I WILL see that Televisor, even if they kill me."

4

Fig. 4.2 *Cartoon published in the magazine* Television *in September 1928* which was that no licence for the transmission of a television service was required.

Meanwhile, Baird continued his quest for firsts, and possibly publicity, by demonstrating stereoscopic reception of television[51] (August 1928). This was also demonstrated together with colour television at the British Association meeting at Glasgow in September.

Another invention of this year was the phonovisor, an apparatus for recording television signals (Fig. 4.3). Baird has written:

> In testing the amplifiers I used headphones and listened to the noise which the signals made. I became expert at this and could even tell roughly what was being televised by the sound it made. I knew whether it was the dummy's head or a human face. I could tell when the person moved. I could distinguish a hand from a pair of scissors or a matchbox, and when two or three people had widely different appearances I could tell one from the other by the sound of their faces, each having its own sound.[52]

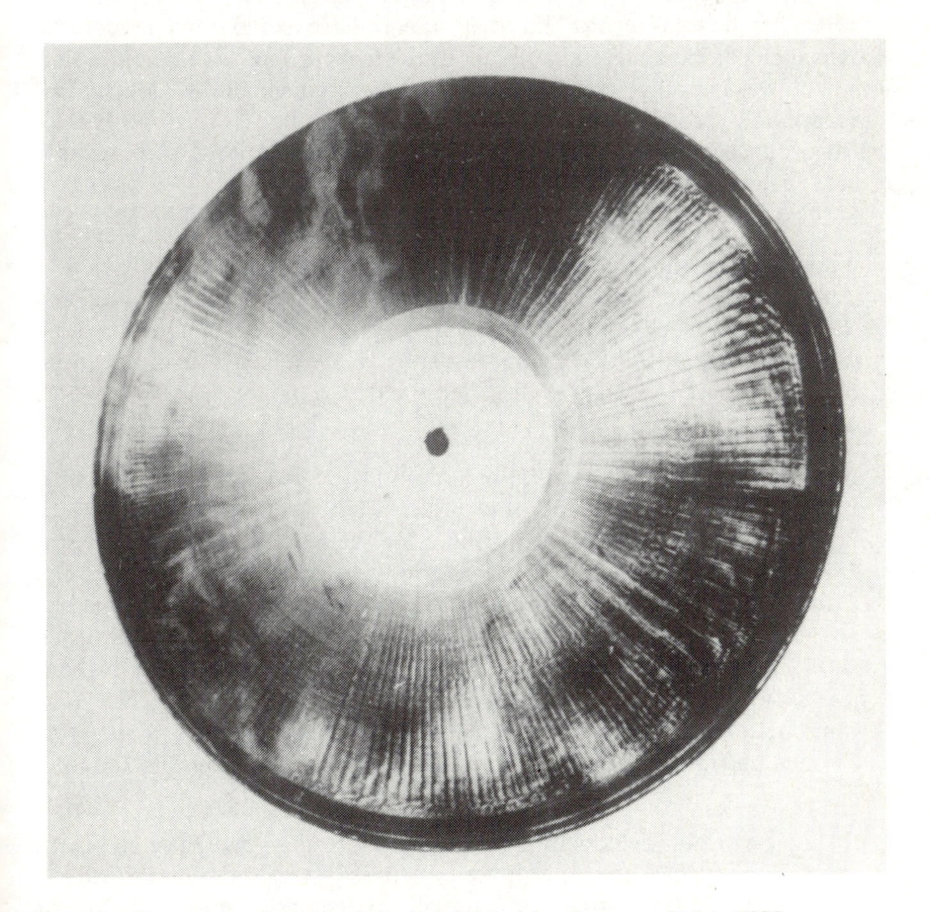

Fig. 4.3 *Phonovision record made by Baird Television Company Ltd. in 1928*

As usual Baird patented his idea (patent 289,104, dated 16th April 1929). His provisional patent[53] was taken out on 15th October 1926, less than two months after a Mr R. Hall had made an application for a patent[54] to protect his ideas. The complete specification of this (280,630 dated 17th November 1927) made reference to the recording of the received fluctuations in current from the transmitter in the form of sound impulses on a gramophone recording machine. Hall gave no details of his idea and his patent was essentially concerned with a novel form of scanning disc, but the closeness of the above dates shows that other people were thinking along similar lines to Baird, although in the case of most patent applications from private persons there seemed to be a lack of perseverence in developing the ideas contained in them.

Baird's recording system made provision for both vision and sound signals to be recorded on a gramophone disc by means of two spiral tracks. The novel aspect of his system concerns the reduction of the two recorded grooves to one: 'It also lies within the scope of this invention to combine the two records in a single groove, using the side walls for, say, the sound record and the bottom of the groove for the sight record.' Baird appears to have had in mind a recording process which combined lateral and hill-and-dale recording techniques. A similar method was later used and described by A. D. Blumlein in his classic patent on stereophonic recording and reproduction (394,325 dated 14th June 1931).[55]

Baird produced several patents on phonovision but nothing of immediate usefulness came out of them. He later wrote:

> I had a gramophone record made of these sounds and found that by playing this with an electric pick-up and then feeding the signal back to a television receiver I could reproduce the original scene. A number of such records were made but the quality was so poor that there seemed no hope of competing with the cinematograph. If the cinema had not been invented, the phonovisor, as I called the device, might have been worth developing.[52]

His ideas and patents on magnetic recording also came to nought. Possibly Baird gave too much thought and time to such adventurous sidelines, as one of his former colleagues later described them.[46] Had Baird concentrated on reproducing a good television image his early relations with the BBC might have been better and possibly the subsequent fortunes of the Baird Companies thereby changed. The British Broadcasting Corporation had been singularly disgusted with Baird's demonstrations and by Hutchinson's rash pronouncements about the fitness of the Baird television system for public broadcasting, but now, following a letter from Bartlett[56] to the Post Office stating that the company was ready to demonstrate its system to the GPO, Sir Evelyn Murray told Sir John Reith:

> In the event of a demonstration being given to the Post Office engineers providing satisfactory it is proposed if you see no objection to inform the Company that the Postmaster General is prepared to agree to the use of

> one of the Company's stations for further experiments subject to suitable conditions and to suggest the Company should communicate with the Corporation with a view to making the necessary arrangement.[57]

Reith considered that before the Corporation decided what to say to Baird the 'Corporation would be interested to receive the Post Office's engineers comments on the experiments which are about to be conducted between them and the Company'.[58]

References

1 BBC Control Board Minute, 15th May 1928
2 ECKERSLEY, P. P.: letter to Professor L. B. Turner, 11th June 1928, BBC file T16/214
3 TURNER, Professor L. B.: letter to P. P. Eckersley, 12th June 1928, BBC file T16/214
4 ECKERSLEY, P. P.: letter to Professor L. B. Turner, 18th June 1928, BBC file T16/214
5 TURNER, Professor L. B.: 'The problem of television', BBC file T16/214
6 TURNER, Professor L. B.: letter to P. P. Eckersley, 16th July 1928, BBC file T16/214
7 Assistant Chief Engineer: memorandum to the Chief Engineer (BBC), 17th July 1928, BBC file T16/214
8 TURNER, Professor L. B.: letter to P. P. Eckersley, 17th July 1928, BBC file T16/214
9 TURNER, Professor L. B.: letter to P. P. Eckersley, 21st September 1928, BBC file T16/214
10 ECKERSLEY, P. P.: letter to Professor L. B. Turner, 25th September 1928, BBC file T16/214
11 Chief Engineer: memorandum to the Assistant Controller (Information), 26th September 1928, BBC file T16/214
12 Anon.: a report, *The Times*, 22nd June 1928, Minute 51/1929, file 13
13 REITH, Sir J. F. W.: letter to the Secretary (GPO), 25th June 1928, Minute 51/1929, file 13
14 REITH, Sir J. F. W.: letter to the Secretary (GPO), 30th June 1928, Minute 51/1929, file 13
15 PHILLIPS, F. W.: letter to Sir J. F. W. Reith, 26th June 1928, Minute 51/1929, file 13
16 MALONEY, C.: Question in the House, 22nd May 1928, Minute 51/1929, file 13
17 Postmaster General: reply to question in the House (reference 16), Minute 51/1929, file 13
18 PHILLIPS, F. W.: letter to Sir J. F. W. Reith, 4th July 1928, Minute 51/1929, file 13
19 WISSENDEN, J. W.: memorandum to the Engineer-in-Chief (GPO), 22nd June 1928, Minute 4004/33
20 Engineer-in-Chief's Office: letter to Baird Television Development Company, 21st June 1928, Minute 4004/33
21 Engineer-in-Chief's Office: letter to the BBC, 25th June 1928, Minute 4004/33
22 Secretary, Baird Television Development Company: letter to Engineer-in-Chief (GPO), 16th July 1928, Minute 4004/33
23 Anon.: 'Television export — and what will listeners do then', *Evening Standard*, 26th June 1928
24 PHILLIPS, F. W.: letter to Television Ltd., 28th June 1928, Minute 4004/33
25 HUTCHINSON, O. G.: letter to the Secretary of the GPO, 3rd July 1928, Minute 4004/33
26 Secretary, Baird Television Development Company, letter to the editor of the *Evening Standard*, 2nd July 1928, Minute 4004/33
27 BBC press announcement, July 1928
28 Anon.: 'Practical television is here', *The Times*, 4th August 1928
29 Anon.: 'Television for all — broadcasting to begin in the autumn', *Daily Mail*, 3rd August 1928
30 Anon.: a report, *The Times*, 4th August 1928
31 PHILLIPS, F. W.: letter to Baird Television Development Company, 7th August 1928, Minute 4004/33
32 MOSELEY, S. A.: 'Broadcasting in my time' (Rich and Cowan, 1935)

33 MOSELEY, S. A.: 'John Baird' (Odhams, London, 1952)
34 MOSELEY, S. A.: 'The private diaries of S. A. Moseley' (Max Parrish, 1960)
35 BAIRD, J. L.: autobiographical notes, quoted by BAIRD, M. *in* 'Television Baird' (HAUM, South Africa, 1974)
36 MOSELEY, S. A.: 'The private diaries of S. A. Moseley' (Max Parrish, 1960), pp. 320–321
37 HUTCHINSON, O. G.: letter to the Secretary (GPO), 22nd August 1928, Minute 4004/33
38 BARTLETT, T. W.: letter to the Secretary (GPO), 8th September 1928, Minute 4004/33
39 ECKERSLEY, P. P.: 'BBC and television', *Popular Wireless*, 14th July 1928
40 CAMPBELL SWINTON, A. A.: letters to the editor, *The Times*, 20th July 1928, 4th August 1928, 20th September 1928
41 CAMPBELL SWINTON, A. A.: 'Television methods of reproducing pictures' (letter to the editor), *The Times*, 20th July 1928
42 CAMPBELL SWINTON, A. A.: letter to P. P. Eckersley, BBC file T16/42, sub-file 1
43 Editorial, *Nature*, 1927, **119**, 15th January, pp. 73–74
44 BAIRD, J. L.: letter to the editor, *Nature*, 1927 **119**, 29th January, pp. 161–162
45 ROBINSON, J.: 'Television' (a letter) *The Times*, 24th July 1928
46 BRIDGEWATER, T. H.: 'Baird and television', *Journal of the Royal Television Society*, **9**, No. 2, pp. 60–68
47 REITH, J. F. W.: letter to the Secretary (GPO), 20th September 1928, Minute 4004/33
48 Copy of advertisement in Post Office file, Minute 4004/33
49 PHILLIPS, F. W.: letter to E. H. Robinson of the *Observer*, 7th August 1928, Minute 51/1929, file 17
50 Anon.: 'Television plans', *Daily Mail*, 4th August 1928
51 Anon.: 'Colour television', *Glasgow News,* 5th September 1928
52 BAIRD, M.: 'Television Baird' (HAUM, South Africa, 1974)
53 BAIRD, J. L.: 'Improvements in or relating to the recording of views of objects or scenes, or optical images or the like', British patent 289,104, application date 15th October 1926
54 HALL, R.: 'Transmitting and receiving apparatus for a television system', British patent 280,630, application date 17th August 1926
55 BLUMLEIN, A. D.: 'Improvements in and relating to sound transmission, sound recording and sound reproducing system', British patent 394,325, application date 14th December 1931

Chapter 5

An important demonstration, 1928

At long last, a demonstration[1] took place on 18th September 1928 at Long Acre and at the Engineers Club about 600 yards away. The Post Office had waited since August 1927 for this.

Two tests were carried out, one in which a line transmission circuit was used for a short transmitter–receiver distance in the company's offices at Long Acre, and another which involved a radio link, on a wavelength of 200 m between the same building and the Engineers Club in Coventry Street. The demonstrations showed the facial images of several individuals while they carried on a conversation or sang and these were reproduced by the receiver to give an image size about $3\frac{1}{2}$ inch x 2 inch. This was viewed by a lens (approximately 5 inch in diameter) to enable a magnification of nearly two to be achieved.

Of the two tests, that which utilised the line circuit gave the best results. Angwin observed:[1] '. . . the faces were shown with features outlined as clearly as they would have been seen from reflection in a metal reflector'. Twelve images per second were shown and hence some flicker was produced but this did not seem to be objectionable. The synchronisation method and control were 'definitely superior in efficiency and simplicity to that used by other systems' and only required two external handles for speed and phase adjustment.

The wireless television transmission experiment was less satisfactory because of some apparent interference in the radio channel from local sources. The result was a tendency to swinging of the image in the place of observation and the flicker effect was more pronounced.

Angwin concluded his report by stating:

> The system merits consideration from the simplicity of the receiving apparatus and the possibilities of improvements if further developed . . . An experimental trial from one of the BBC stations with observations on a set of the model it is proposed to sell to the public would, I consider, be desirable to test out the quality of the reproduction that might be expected under normal broadcasting conditions.

Another observer had a different view and described the demonstration in the following words:[2]

> The Baird machine may be said to give a recognisable human head. It is curiously unlike any particular face. I suspect that the eyebrows were heavily made up. Only slow movements are possible, anything of even normal speed producing a wild blur. The impression is of a curiously ape like head, decapitated at the chin, swaying up and down in a streaky stream of yellow light. I was reminded of those shrunken heads favoured by such persons as Mr Hedges. Not even the collar or tie were visible, the effect being more grotesque than impressive. The faces of those leaving the show showed neither excitement nor interest. Rather like a fair crowd who had sported 6d to see if the fat lady was really as fat as she was made out to be. The image was held for $4\frac{1}{2}$ minutes before receiving adjustment which seemed a simple operation.

The BBC did not have a representative at the demonstration but it did receive a report[3] two days later. In acknowledging the report Reith said he assumed that the Baird Television Development Company would in due course make an application to the Corporation for facilities to broadcast television.[4] He was clearly not impressed by the Post Office report on the system and enquired of Murray: 'Do you not feel that this Corporation should insist on a demonstration to its technical experts before being committed to anything in the nature of a public test for a trial period of any duration.'

Reith was not only concerned about the quality of any proposed transmission but also the effect such transmissions might have in producing serious jamming internationally in view of the inadequate wavebands available for broadcasting generally. He thought that the sidebands required, of the order of 15–16 kc/s, might prohibit the use of television internationally and that short waves would appear to be essential.[5] The Post Office too recognised this problem but did not at this stage raise it with the BBC.[6]

The BBC was in a rather embarrassing position at this time in relation to the use of its transmitters for experimental television broadcasting as a precedent had been set for the use of one of these by Wireless Pictures (1928) Ltd.[7] to make experimental transmissions of still pictures by the Fultograph process. This process, however, did not require the wide bandwidths necessary for good television transmission.

Notwithstanding Reith's comments, the Post Office wrote to him on 25th September 1928 stating[8] 'that in view of the favourable report' the Postmaster General was of the opinion 'that the stage of development which is now reached is such as to justify him in agreeing to the use of a broadcasting station for further experiments . . .'. The Postmaster General recognised the difficulties which Reith had mentioned concerning the bandwidth requirements of television transmission but felt that in view of the facilities afforded to the proprietors of the Fultograph apparatus it would be difficult to refuse similar facilities to the Baird Company.

Reith was not happy and two days later telephoned the GPO. They[9] informed him of the terms of the proposed letter to Bairds but Reith did not like it as it threw the entire reponsibility on to the BBC if they did not for, say, technical reasons, see their way clear to use one of their stations. He wanted the Post Office to add a paragraph to the effect that the Post Office's authority was contingent on the BBC experts being given a demonstration and being satisfied that the use of a BBC station was feasible.[10] The Post Office acceded to this request and notified the Baird Television Development Company on the 28th September.[11,12]

Bairds immediately wrote to the BBC asking for facilities for television transmission, but Reith considered that the BBC might have to appeal to the Postmaster General to decline to permit their system to be used for television if the demonstration mentioned in the Post Office's letter were found 'to be really bad'.[13]

This demonstration was arranged for 9th October 1928. It was attended by J. F. W. Reith, P. P. Eckersley, N. Ashbridge, W. Gladstone Murray, R. Eckersley and Major C. F. Atkinson.[14]

The day before the demonstration the BBC's Chief Engineer had written a long memorandum[15] on 'Suggested attitudes towards television' which was seen by the members of the visiting party prior to the demonstration. By this action he rather unfairly endeavoured to pre-empt an adverse decision against Bairds. He mentioned that the Control Board would see a demonstration of the head and shoulders of a man and that it would be extremely interesting and quite likely better than what they had been led to expect, but, he added brusquely:

> if Control Board feel that this would justify a service then let us go ahead, but I warn everyone, that in my opinion, it is the end of their development, not the beginning and that we shall be forever sending heads and shoulders. Are heads and shoulders a service? Has it any artistic value? Is it not in fact simply a stunt?

Eckersley had clearly made up his mind before the demonstration. As the BBC's Chief Engineer he was not simply concerned about the quality of the transmission but also had to consider the implications of the service if Control Board thought the system had merit, and here there were difficulties. He wrote:

> It must be remembered that in effect an extra wavelength must be sacrificed for television. It has been pointed out that an extra wavelength is almost impossible as we have so few as it is, and I think to take up so much ether in the broadcasting band simply to give the small picture of the head and shoulders of a man to people who can afford sets, is rather ridiculuous.

Eckersley was not sanguine about the likely attitude of the Post Office and felt that if they 'let us down again' the BBC should say that a special wavelength should be given to Bairds for use out in the country 'with which they can do experimental work, but that it is quite unecessary that they should sell apparatus: selling apparatus to the public is not experimental work'. For Eckersley the Baird system was not capable of development, and as Chief Engineer he did not, presumably, wish to be responsible for sending out a poor transmission. Later events showed in fact that considerbale development work could be done, even with 30-line working.

The Chief Engineer thought that it was unfair to say that wireless broadcasting was as undeveloped as television when it started: '. . . the spoken word was perfectly intelligible, music was rough and this has been improved, but on the basis of the spoken word alone, the Writtle transmissions for instance were entirely successful'.

For him the nature of a television service was clear; it was not to consist only of 'heads and shoulders', but of two men standing talking together, of a lot of men playing football together, of a liner arriving at Plymouth, of topical events and so on.

> If Baird can show us the interior of a room with the people in it fairly clearly that would have a different service aspect, we could do plays, but if his plays are going to be silhouettes exchanging places with one another as they speak, I doubt if that has service value.

Eckersley's memorandum had the hoped for result. 'From the angle of service, yesterday's demonstration would be merely ludicrous if its financial implications did not make it sinister', thought Gladstone Murray, the BBC's Assistant Controller (Information).[16] 'The demonstration considered in terms of service might well be considered an insult to the intelligence of those invited to be present. The sinister implication is formed by the attached cutting from a financial article published last week', he wrote.

Murray formed the impression the Baird system provided an interesting laboratory experiment but that from the point of view of successful television being an adjunct to a service of broadcasting the Baird method was 'either an intentional fraud or a hopeless mechanical failure'. Harsh words indeed for a system which had been praised by many leading scientists, had satisfied Post Office officials that it would merit further development and was being utilised in its essential features by experimenters and companies in Germany and the United States of America.

As the BBC's publicity spokesman, Murray thought the BBC had a primary duty to the listening public 'to promulgate the truth and prevent the excitement of fake hopes in the crash of which we are bound to suffer by a remote implication'. He thought the BBC should issue a carefully considered statement in which they expressed their benevolent interest in all new inventions concerned with wireless and say quite frankly what they thought of the Baird device on the scientific side.

Murray was in no doubt what the Post Office attitude would be:

> It is obviously the intention of the Post Office to dump the blame for the further obstruction on our shoulders. This does not worry me in the

> slightest. We can carry well-informed and disinterested opinion with us. This is all that matters.

The policy to be adopted was clear in Murray's view:

> Keeping in mind the fundamental fact that the intrusion of Baird transmissions into the broadcasting band will gravely disturb our normal service and prejudice the regional scheme I think it our duty to resist or delay the suggestion in every reasonable and possible way.'

Eckersley had won the day.

However, while Murray and the Chief Engineer were firm and harsh in their opinions, Leech, of the Post Office, who saw a demonstration of the Baird system in Selfridge's as a member of the public two days later, thought[17] 'the image of the person and facial expression were very good – apart from the flickering effect – in fact much better than I had expected'.

Reith communicated with Phillips, on the outcome of the tests, on 15th October[18] and enclosed a draft statement for the Post Office's comments. Phillips, [19] while accepting that the responsibility for the statement rested with the BBC, considered it was 'rather premature to make a statement of so uncompromising a character'. He felt that if it was made the Television Company would almost certainly demand increased facilities for the development of their system from their own station, and further, that if the use of a BBC station for tests was now definite it was doubtful whether the Post Office could permanently refuse to grant permission for Bairds to establish a more powerful station.

Nevertheless, while the BBC Governors considered[20,21] the risks raised by Phillips, they decided to issue a notice to the press on the lines of the draft contained in Reith's letter of 15th October 'but slightly toned down'. They did however add a line at the end to the effect that the Corporation's statement could be revised when further development merited it.

The statement read:[22]

> The opinion of the BBC representatives was that while the demonstration was interesting as an experiment it failed to fulfil the conditions which would jusitfy trial through a BBC station. The Board of the Corporation has decided that an experimental transmission through a BBC station shall not be undertaken at present. The Corporation would be ready to review this decision if and when development justified it.[23,24]

The immediate effect for Bairds was a slump in the value of their shares which fell back from a recent high of 11 s 3 d, for the 5 shilling shares, to 6 s 3 d on the day of the statement.[25] On the Stock Exchange television shares had had a very good run. The share capital of both the Baird Television Development Company and the Baird International Television Company had gone to a heavy premium.[26]

In September 1928 there had been a sudden rise in the belief that the Baird system was going to be put into use, while earlier in the year, in April, the Development Company's shares had risen in value on rumours of a sale of foreign rights. The September rise is shown in the following:

	20th September	26th September	1st October	17th October
BTD Co. 10% Pr. Ord.	42/6	47/6	48/9	50/–
BTD Co. 1/– Deferred	16/–	19/6	17/–	17/3
BIT Co. 'A' 5/– Ord.	7/9	11/–	10/–	9/3

Baird took the BBC's decision with a certain degree of stoicism:[27]

> I regard the decision of the BBC to grant no facilities to television as a challenge which I mean to take up. The attitude of the Corporation, though not unexpected is inexplicable, because it is a direct contradiction to that adopted by the Post Office and many leading scientists. Whatever may be behind the decision I intend to go forward independently. In a short time I shall be broadcasting vision on my own.

Baird had faced many challenges previously and he had succeeded; he now had the support of the Post Office, scientists of the calibre of Fleming, Moseley and his many contacts in Fleet Street, and he was not, in his view, going to fail. For him television was inevitable.

The *Daily Herald's* wireless correspondent stated:

> It is not likely that the Company will rest satisfied with the BBC refusal, and we are likely to see a renewal of the campaign against the monopoly of broadcasting as a semi-state perquisite.[28]

The battle of words was about to begin and on 22nd October the Baird Television Development Company issued a press statement describing the events which led up to the BBC's rejection of their system.[29] They laid the onus for this on Eckersley:

> Certain adverse criticisms had previously appeared from time to time in the technical press under the hand of the chief technical officer of the BBC. The Baird Television Company refrained from answering these criticisms, in view of the fact that at that time no demonstration of the Baird system had been given to the BBC. It is unfortunate that the officer in question should have committed himself to such statements before he had had an opportunity of examining the results of the system.

The BBC was under attack. They emphatically denied that the engineers who had witnessed the demonstration were in any way prejudiced, or that their decision was due to any considerations except those of the merit and practicability of the invention.[30] However, Eckersley had written a memorandum on the subject of television shortly before the test which had possibly influenced some of the witnesses. Baird had stated that when he looked for friendly co-operation similar to that which had been extended to him abroad he had found definite hostilities.[27]

Moseley's attitude was true to character: 'Anyhow I feel that the television people have got a real grievance and I'm fighting on this side', he wrote to Murray.[31]

Moseley was not alone in 'knowing everone in Fleet Street'; Murray too was a very capable and efficient publicity and public relations officer. He also had friends in Fleet Street. Following a letter from Eckersley[32] in which he had suggested that a letter to *The Times* should be sent giving the BBC's views, Murray wrote back to say that this was not necessary as he had already written privately to 'pivotal people' of the press medium. These letters had been sent to the editors of the *Daily News, The Times, the Spectator, the Daily Express*, the Press Association and other newspaper organisations and contained 'a private tip to instruct your people to avoid accepting rumours of the favourable attitude of the BBC towards the Baird Television scheme'.[33] (This was before the BBC's statement of 22nd October had been issued).

The BBC continued to prepare for the expected onslaught and at a Control Board meeting held on 24th October 1928[34] 'it was considered desirable for the Director General to see the Postmaster General to put him personally au fait with our views'. In addition it was 'decided that A. C.(I) should himself write an article to be ready for publication should an attack necessitate it'. The Chief Engineer felt that a statement should be prepared by Murray and that it should be based on his (Eckersley's) article in *Popular Wireless* and Murray's impressions of the demonstrations.[35]

Meanwhile the Baird Television Development Company had written to the PMG for a request to 'grant a licence to this company to use a station of their own' for broadcasting television.[36,37] Manville followed this up the next day with a long letter[38] to the Postmaster General in which he outlined the events which had occurred during the past few months. He explained that the BBC, without any previous intimation to his company, had sent to the press 'an announcement to the effect that they were not prepared to grant facilities for broadcasting television', and further that the press statement made reference to certain conditions not having been fulfilled although no conditions whatsoever 'were communicated to us by the BBC'.

Manville was concerned: 'The attitude of the BBC amounts to an arrogation of a function at no time designated to them by you', he wrote, 'and if their refusal to co-operate is to result in the prevention of public broadcasting of television altogether in this country, we must inevitably fall very far behind other countries in this venture, the importance of which is universally recognised. Such a result would I venture to think be a national blunder, the ultimate effects of which it seems impossible to foresee'.

The position was one of regret for Sir Edward Manville:

> Television was first shown in this country and my Company has now been the first to show a commercial apparatus. Broadcasting facilities are being freely given by the governments of foreign countries on the Continent and in America. It will be infinitely regrettable if this country is to lose the lead so hard earned owing to difficulties such as those now under discussion.

In the USA television broadcasting[30,40] began in 1927 when the Federal Radio Commission issued the first television licence to C. F. Jenkins, authorising him to broadcast television transmissions from a station, W3XK, situated in the suburbs of Washington, DC. The station operated on a radiated power of 250 W initially, but by 1931 the Federal Radio Commission had approved an increase to 5000 W.

During 1928 seven further television licences were issued in the USA, and, in 1929, 22 additional stations were granted broadcasting rights. The early station planners had a wide latitude in their choice of frequency and almost any frequency above 1500 kHz could be used if no interference were caused to other services. Many of the experimenters were inexperienced, however, and caused some damage to the development of television by their use of scanners, having large numbers of holes, which were rotated at high speed to reduce flicker. Consequently, interference was produced because of the wide freqency bands employed.[41] The belief was held that listeners would be antagonised by the irritating signals radiated by their loudspeakers to the exclusion of the wanted, regular, broadcast programmes, and so, in January 1929, the Federal Radio Commission issued a preemptory order banning television broadcasting in the sound broadcasting waveband. In the same month a North American Conference[42] held in Ottawa decided that radiations of television signals were to be limited to a bandwidth of 100 kHz, within the frequency bands 2000–2100, 2100–2200, 2200–2500, 2750–2850, and 2850–2950 kHz. The powers employed varied from 10 W to 20 kW, with the majority of stations operating at 5 kW (see Table 5.1).

One of the consequences of the FRC's order was the restriction it imposed on definition standards. Weinberger, Smith and Rodwin of the Radio Corporation of America showed that a standard based on 60 lines per picture, 20 pictures per second, and on an aspect ratio of 5:6 (height to width) would be realistic and apposite for the television bandwidth specified by the FRC. These suggested television parameters became the norm for American television transmissions, although in 1928 a committee appointed by the Radio Manufacturers Association of the United States had recommended that all radiovision pictures then being broadcast should be standardised on the Jenkins system, namely 48 lines per picture, 15 pictures per second. Presumably the RCA proposals prevailed because they enabled a better definition to be realised: the RCA and RMA

Table 5.1 *List of stations licensed to transmit television experimentally in the United States (1931)*

Call letters	Company and location	Power (watts)
	2000–2100 Kc	
W3Xk	Jenkins Laboratories, Wheaton, Md	5000
W2XCR	Jenkins Television Corporation, Jersey City, N.J.	5000
W2XAP	Jenkins Television Corp. Portable	250
W2XCD	De Forest Radio Co., Passaic, N.J.	5000
W9XAO	Western Television Corp. Chicago, Ill.	500
W2XBU	Harold E. Smith, North Beacon, New York (School)	100
W1OxU	Jenkins Laboratories, aboard cabin monoplane	10
W1XY	Pilot Electrical & Manufacturing Co., Springfield, Mass.	250
W1XAE	Westinghouse Elec. & Manufacturing Co. Springfield, Mass.	20 000
	2100–2200 Kc	
W3XAK	National Broadcasting Co. Bound Brook, N.J.	5000
W3XAD	R.C.A.–Victor Co., Camden, N.J.	500
W2XBS	National Broadcasting Co., New York	5000
W2XCW	General Electric Company, South Schenectady, N.Y.	20 000
W9XAP	Chicago Daily News, Chicago, Ill.	1000
W2XR	Radio Pictures Inc. Long Island City, N.Y.	500
	2750–2850 Kc	
W2XBO	United Research Corporation, Long Island City, N.Y.	20 000
W8XAA	Chicago Federation of Labour, Chicago, Ill.	1000
W9XG	Purdue University, West Lafayette, Ind.	1500
W2XBA	W.A.A.M. Inc., Newark, N.J. (broadcasting station)	500
W2XAB	Columbia Broadcasting System, New York	500
	2850–2950 Kc	
W1XAV	Shortwave and Television Labs., Boston, Mass.	500
W9XR	Great Lakes Broadcasting Co., Downers Grove, Ill.	5000

picture standards were based on 2500 and 2304 picture elements, respectively (Baird used a 2100 element image).

Another result which stemmed from the FRC's ruling was that, as none of the commercial receivers then on the market could receive signals in the 100–150 m wavelength range, an entirely new and distinct receiver was required for television reception. The existing broadcast receivers which covered the band of 200–600 m had been developed to a considerable degree of excellence, but now manufacturers were faced with the prospect of designing and constructing sets for a frequency band which had not previously been fully investigated for commercial use. Somewhat naturally, a number of firms felt uneasy about embarking on large scale manufacturing programmes for television sets when the art was still in the experimental state.[42]

The Radio Corporation of America,[43] which did more than any other company in the USA to advance high definition television, commenced its activities in 1927 when the Technical and Test Department's laboratory group at Van Cortland Park carried out some experimental investigations on mechanically scanned television systems. Their approach and work was similar to that which had been undertaken by E. F. W. Alexanderson at General Electric in Schenectady, and, by April 1928, the group had advanced their developments to the trial stage. They applied for and were granted in April 1928 a permit for the first RCA television station, known as W2XBS, and the initial broadcasts took place from the Van Cortland site.

T. J. Buzalski, of the NBC's commercial television outlet in New York, has described[44] these early transmissions.

> The programmes material usually consisted of (a) posters with black images upon a white background, such as W2XBS-New York- USA . . ., (b) photographs, (c) moving objects such as Felix the cat . . .revolving on a phono turntable, and (d) human talent. Communications from observers indicated that the service area of the station was quite large and, under favourable conditions, good television reception was reported at distances of several miles, the farthest point being Kansas. There are letters on file from some 200 observers and numerous telephone calls have been received from time to time reporting local reception. Many interesting tests were conducted between W2XBS at Times Square and the RCA laboratory at 711 Fifth Avenue.

The system used the 60 lines per picture, 20 pictures per second standard and the received image was approximately $1\frac{1}{2}$ inch x 2 inch in size. Regular transmissions took place from 7.00 to 11.00 p.m. daily but these were for experimental purposes only, as it was thought that the state of development of television was not sufficiently advanced to entertain and maintain the interest of the general public.

The company was at this time working on mechanically scanned television systems, as were several other firms, but apart from some work carried out by

Bell Laboratories, no organisation had systematically investigated the many parameters which define a 'good' television picture. RCA devoted much effort to this issue and published a series of papers on their findings in the *Proceedings of the Institute of Radio Engineers* from 1929 to 1936. The corporation's conclusions very considerably helped to rationalise the standards of the television industry for both low and high definition systems.

The work of Weinberger *et al.*[45] did not represent a definitive study on picture standards (Engstrom later investigated the problem much more fully), but nevertheless their results provided a basis on which low definition systems could be satisfactorily designed and they allowed manufacturers to produce sets which were able to receive radiations from many different television transmitters using these standards. The RCA group particularly had in mind criteria for a commercial television service. Fortunately a definition of what constituted commercial television had been given in 1929 by a special committee of the National Electrical Manufacturers Association appointed to deal with the subject and this is quoted in the following:[45]

> Commercial television is the radio transmission and reception of visual images of moving subjects comprising a sufficient proportion of the field of view of the human eyes to include large and small objects, persons and groups of persons, the reproduction of which at the receiving point is of such size and fidelity as to possess genuine educational and entertainment value and accomplished so as to give the impression of smooth motion, by an instrument requiring no special skill in operation. having simple means of locating the received image and automatic means of maintaining its framing.

So far as is known this was the first formal, explicit definition of television ever advanced. It posed the question of the meaning to be given to the phrase 'genuine entertainment and educational value', and in an effort to clarify it, Weinberger's group consulted various persons engaged in the motion picture and theatrical fields and came to the conclusion that genuine entertainment value would be achieved if a clear reproduction of a semi-close-up of two persons was obtainable. From the group's discussions it appeared that the majority of dramatically interesting situations reduce to two, or at most three persons, and in motion picture practice it is customary to photograph 80 or 90% of a picture in close-up or semi-close-ups. Consequently, the RCA research team felt that, if it chose a degree of picture detail that would acceptably render two persons in a semi-close-up view, the entertainment requirement would be met, at least in the early stages of the art.

Baird's images in 1928 were restricted to head and shoulders and so did not satisfy the above criterion. Eckersley was correct in his view about what constituted entertainment television. However, patronage and encouragement are important in the development of any new invention.

Murray's article 'The truth about television', which he had been asked to write by Control Board, was published as a press statement[46] on 3rd November 1929.

In it Murray went over ground which had already been well trodden by Campbell Swinton in his letters to *The Times* regarding bandwidth requirements for a system of television which could give 'satisfactory television'. He mentioned that television's needs for twinwave stations (one to transmit vision and another for sound) was contrary to the Corporation's policy of reorganisation, which was to provide a few stations of higher power for the comparatively numerous low power single-wave stations. 'The demands of other countries impose a severe limitation on wavelengths available for this country' he wrote . 'It follows therefore that if television were to be incorporated with broadcasting the scheme for providing alternative programmes of speech and music through the five twin stations would have to be abandoned and what would be given to the millions of listeners to compensate them for the loss of the alternative programmes for which they have been waiting for years and for which the ten channels are not actually sufficient. The answer is not in doubt', he said. 'In view of the expense of the apparatus and the difficulty of its adjustment even under conditions which cannot be fulfilled a negligible minority of well-to-do listeners would be provided with a mechanical toy'. There was another point to consider: 'Take the problem from the programme end', he continued. 'Under the Baird system for instance it is necessary for an artist to sit or stand nearly rigidly at one place for microphone work. To apply this to broadcasting would seriously prejudice the programmes'.

For Murray the conclusion was obvious: 'Television would seriously impair the broadcasting transmission in this country and abroad – threatening the interests of the majority for the dubious advantage of the curious few. Such a move would upset the democratic foundation of British broadcasting and disturb its basic tradition of the greatest good for the greatest number. To protect the efficiency and democratic character of the broadcasting service from what might be in practice a serious menace is regarded as part of the primary duties of the Corporation', he ended.

All this was not meant to imply that the BBC did not wish to abandon its interests in television; on the contrary, the BBC considered that 'should the Television Company be given an experimental station for itself, the BBC monopoly in regard to broadcast telephony should be strenuously upheld'.[47] Essentially the BBC was not interested in the Baird television system as it had developed by September 1928, and did not wish to actively participate in its progression to a high grade public service. The Corporation was really demanding a system which would give good public entertainment from the outset: for them the interim period which involved the testing and improvement of the equipment as well as the development of the art of television production was one they did not wish to know about. Neither Reith nor any of his colleagues proposed any suggestions for the implementation of a viable television broadcasting system (from a financial point of view); they gave Baird no encouragement but criticised him and his associates when they formed companies to raise money for their ideas. The Corporation did not explain how the inventor should proceed with the development of his inventions although it must have been clear to them that Baird could not finance his activities unaided.

Later in November Murray had lunch with both Moseley and Hutchinson at Long Acre to ascertain what the Baird people had in mind.[48] He was 'anxious to discover without saying anything' himself, 'whether and how much Hutchinson knew of pending proposals'. 'It became obvious', Murray noted, 'that he was aware of everything that is in hand and also of the extreme importance of secrecy'.

Following this lunch Murray prepared a long report[48] on the conversations and sent a copy to Moseley.[49] Moseley was taken aback by this and wrote that he should have been warned of the formal significance that Murray placed on the meeting.[50]

During their talks Hutchinson mentioned that the Post Office engineers had spent about two hours examining transmissions both by land line and by wireless during the demonstration held on 18th September, and that at the end of this time seemed 'enthusiastically disposed'. By contrast, he complained, the BBC inspection had been 'most perfunctory, casual and frivolous'. Most of the advisors of the Baird Company had been against any demonstration but chiefly on the advice of Moseley, which Hutchinson supported, the company had decided to accept the challenge to Captain Eckersley being present at the official demonstration 'in view of his repeated public committment against them'. Another point which Murray recorded was the view of the Baird Company that the issue by the BBC to the press of a unfavourable verdict late at night before there could be an opportunity of consultation was 'a gross breach of faith and manners'. 'Their theory was that some financial motive must have been behind it – certainly an endeavour to depreciate the value of Bairds shares on the market'. For his part Murray stressed there were 'no extraneous or hidden motives' behind any action of the BBC or any of its officials in the matter of television. The policy of the Corporation, he told Moseley and Hutchinson, was 'to resist television while it was still premature and to welcome it and give it every encouragment once it had reached a stage where it could be fairly regarded as being available for a general service as a useful adjunct to broadcasting'.[48]

The Baird Company, Murray wrote, believed that the Chief Engineer's opposition was due to the influence of a merger of the Marconi Company, British Thomson Houston, Metropolitan Vickers and the General Electrical Company – a merger under which the GE method of television would be exploited in this country.

Murray concluded his memorandum by noting that the BBC's attitude to publicity had been one of restraint and that only such steps had been taken as were necessary to repel the definite attacks of the protagonists of the Baird case.

Bairds were understandably apprehensive about their own position at this time as reports were being published in the press of the 'successes' of other individuals and firms. The American Telephone and Telegraph had broken Bairds monopoly to demonstrate true television in 1927 and it was known that both the Radio Corporation of America and the General Electric Company were conducting experiments in television transmission. In Germany Mihaly had announced in July[51] that he was ready to manufacture sets which would bring 'into the drawing room a horse race or a boxing match so perfectly that the jockeys' caps will be distinctly visible and you will be able to recognise the faces of the boxers'. He told a reporter of the *Daily News* that he intended to start a television company in London in a few weeks 'with a system which would be different from Baird's and far more effectual'. The sets would have one valve and would sell for £20 each.

Against this background of increasing activity to broadcast television, Baird and his friends had formed Baird International Television with the policy to 'form or collaborate in the formation of manufacturing and marketing companies' in various parts of the world in order to commercially exploit television.[52] Since it was authorised to commence business on 26th June, the company had made some progress with regard to the exploitation of the American, Canadian and Mexican rights in the Baird inventions; it had prepared heads of agreement for the formation in the Argentine of a company to exploit the inventions in that country and other South American states and was pursuing every opportunity to secure the establishment of the Baird system in various European countries.

Naturally the Directors of Baird International Television Ltd. and the Baird Television Company had to consider the interests of the shareholders of these companies but without facilities for television broadcasting the Directors were not able to complete their licence arrangements, with members of the radio manufacturing industry, to make and market Baird televisors.

The BBC, too, were aware of developments in television other than those put forward by Baird. Eckersley reported to Control Board on 27th Novermber 1928[53] that there were rumours that the Marconi Company and the British Thomson–Houston Company had developed television schemes. Actually the Marconi Company did not participate in low definition television development until 1930, but the rumour was sufficient for Reith to suggest that it ought to be mentioned to the Postmaster General 'at the next opportunity in order that the latter's attitude in regard to Baird might be strengthened'.

Baird, however, had prepared certain contingency plans following a successful demonstration of their stereoscopic television at the Rotterdam Exhibition. 'Principally owing to the enthusiasm with which our exhibits were received at the Rotterdam Exhibition', Lord Ampthill said in his speech[54] to members of Baird International Television on 25th September, 'we have been successful in obtaining from the Dutch Government the use, on very favourable terms, of one of the broadcasting stations controlled by the Dutch Government in Holland'. He was to add that with these facilities 'television in Holland will make rapid progress'. Approximately one month later Baird announced that he had reached an agreement with the Dutch Wireless authorities to transmit television images from their station at Scheveningen.[55] A state transmitter had been placed at the disposal of the Dutch company 'holding the rights of Mr Baird's apparatus' and transmission would be on a wavelength of 1470 m.[56]

The Manchester Guardian's reference to a wavelength of 1470 m was strange, for the Scheveningen Station was shown in the list of broadcasting stations as having a radiated power of $2\frac{1}{2}$ kW and an operating wavelength of 1950 m. However, even with a wavelength of 1470 m, the frequency separation between this station and the BBC's Daventry 5XX station (which worked on a wavelength of 1562.5 m and a radiated power of 25 kW) would have been 12 kHz and this was more than the normal separation allowed in order to clear two broadcasting stations from mutual interference.[57]

The chief financial backer of this scheme was a Mr Gordon Sherry, a theatrical producer and composer.[58] 'We are renting from Continental stations in existence and giving our programmes every night an hour at a time' he said. 'This hour will be occupied by English artists in our employment who will either sing, play or act'. He went on to state that it would be possible to 'pick up' the signals 'by all people in Great Britain possessing an ordinary two-valve wireless set and that the first programme would be ready by about February. 'We are going to give the BBC a surprise' commented Sherry. Baird himself did not hide the fact that it would be a challenge to the Corporation's monopoly.[59] 'The attitude adopted by the BBC towards my system of television does not make one feel any too well disposed towards them' he said.

Newspaper reporters were quick to note the rather piquant situation which had arisen as a result of the BBC's refusal to allow Baird access to facilities for televising from a home station. 'In the eyes of his supporters he is unconsciously assuming the aspect of a man honoured everywhere except in his own country' wrote an observer in the *Glasgow Herald*.[60]

However, Sherry's proposals were not going to be cheap: 'This will be a tremendously costly undertaking' suggested the *Daily Herald* reporter to Sherry. 'We shall get our money back one way and another' Sherry told him.

Actually the offer to Bairds to use the Scheveningen station was not made because of any benevolence on the part of the Dutch authorities but represented an arrangement open to any commercial undertaking. The Dutch stations were open to hire and 'time on the air' could be bought. There was nothing to prevent an organisation, under the guise of broadcasting music, from broadcasting publicity for their manufactured goods or other saleable articles. Indeed one British manufacturer of wireless sets was already broadcasting a musical programme from Hilversum on alternate Sunday evenings.[58]

Bairds had in mind the 'American principle of making the advertiser pay for the programmes'.[61] This was hinted at in the preliminary announcement which described the programme to be broadcast as 'a programme in English – musical, dramatic and recitatory'. Also Bairds were not going to limit their broadcasting to one continental station, for newspaper reports indicated that six continental wireless stations would shortly transmit programmes in English, these being in Germany, Belgium and Holland.

All of this was no doubt intended to put the BBC on the spot and enable Baird to further advance his image of the lone inventor up against the powerful monopoly. The BBC was highly suspicious of these events and in an internal memorandum dated 11th December 1928 the observation was made that repeated efforts to get from France the truth about the reputed Tour Eiffel arrangements for television and English hours for next February had drawn no response or *démenti*.[62]

Lord Angus Kennedy, the Vice-President of The Television Society, thought the BBC's rejection of Baird's request had a slightly sinister tone:

> Is this the era when we are not allowed to use our own intelligence, but rely on officials to think for us? Must the man in the street rely on foreign countries to provide him with the means of receiving enjoyment from the latest great inventor in spite of the fact – or is it because of the fact – that it emanated from a Scotsman?[63]

The opposition of the BBC to the use of its stations by Baird had its parallel in the history of the Marconi organisation. Although Preece had given Marconi much needed support in the early stages of his work, Austen Chamberlain, as Postmaster General, took quite a different attitude. He saw the Marconi company as a potential competitor of the government-controlled telegraph industry, and at first stubbornly refused to allow the Marconi overseas service to utilise the Post Office's telegraph lines. Later, an agreement between the Post Office and the Marconi company was signed on 11th August 1904 and facilities for wireless telegraphic traffic were granted to it.

Criticism of the BBC was growing, and at the Control Board meeting held on 9th January 1929, Murray mentioned[64] that 'the press were massing a serious attack on us in support of television' and had asked for a statement from the BBC. This the Corporation provided but it did nothing to appease the mounting condemnation of their refusal to allow Baird the use of much needed facilities for broadcasting. 'Baird Television Company', it was stated, 'has not yet intimated to the BBC any claim of improvement. Any such claim would be examined by the BBC with a view to determining whether the above decision should be modified'.[65] Captain Hutchinson added a little intrigue to the situation by observing:

> In view of the real facts, which are within the knowledge of the BBC as well as ourselves, regarding the broadcasting of television in conjunction with music, singing and speech in this country – which facts we are under a pledge of secrecy not to reveal, and have not revealed – we are extremely surprised that such a statement should have been repeated and issued.[66]

The most efficacious attack, however, came not from the press but from a former Postmaster General, the Right Honorable Sir Herbert Samuel, GCB, CBE. On the 14th January 1929 he wrote[67] to the Postmaster General and enclosed a copy of a letter which he proposed to make public if the Post Office did not take some initiative in the matter. 'It is possible that the Post Office may be desirous of bringing about a change in attitude of the BBC', he hinted, 'and that it may succeed in doing so'. Samuel's letter had an almost immediate result. The PMG replied on the 29th January[68] that he had had discussions both with the Baird Company and the Broadcasting Corporation with a view to achieving 'some *modus vivendi*'.

> As a result they have both agreed in principle that a strictly private demonstration of an attempted television broadcast should take place simultaneously at the Post Office and the BBC headquarters to a party of witnesses at each place, who will include myself, the Governors of the Corporation and some Members of Parliament of all parties whom I propose to invite

the PMG replied. He went on to say that some of the Baird Company's advertisements and publicity had not been conspicuous for their accuracy and candour and that the public, whether in its capacity as potential shareholders or as potential purchasers of the Baird apparatus should not be exploited.

The PMG ended by saying he thought a letter from Samuel to the press 'might be rather embarrassing, in particular it might possibly have the effect of stiffening the objections of the BBC which I hope the demonstration, if it is in fact successful, may help to remove'.

There seems little doubt that the Post Office's attempt to establish a rapport between the BBC and Bairds was given greater urgency by Samuel's letter, aided, of course, by the hostile view of the BBC by sections of the press media. Samuel stressed that he had no interest of any kind in the matter except the general interest of every citizen that useful inventions should be encouraged, and that the development of what might become an industry of world-wide importance should not, after having been initiated here 'pass to other countries'.

He had, moreover, the special interest as a former Postmaster General to see that the Post Office used its influence to help the advancement of a new 'and perhaps a most useful' method of human communication.

Samuel's proposed letter to the press commented on the facilities given by broadcasting stations in America and France but 'in England there are none'. He was critical of Eckersley's approach to the development of television:

> It is understood that the Chief Engineer of the BBC holds the view that progress cannot be made along the lines so far pursued. He may be right. On the other hand he may be wrong. In any case the road to further experiment ought not to be closed. If the apparatus proves to be valueless, it will find no patrons and the question will solve itself.

The BBC, too, did not escape his pungent comments:

> But that the dead hand of monopoly should be laid upon a most remarkable invention, that the public should be denied access to it; that a brilliant inventor should be met, not with the gratitude and the active encourage-

> ment which are his due, but with an attitude of indifference and even of obstruction, this would indeed be in accordance with many precedents in the past, but it would not be in accordance with an enlightened view of the public need

he wrote.

Samuel acknowledged the PMG's letter on the 30th January[69] and said he would take no further action 'at all events for the the time being' and expressed the hope that the BBC would 'prove more helpful'.

While Samuel's action considerably influenced the PMG's January decision, he had, unkown to the general public, discussed the matter with the Chairman of the Corporation and Reith and Murray several weeks beforehand and after the BBC's statement of 17th October.[70] On this occasion the PMG had asked the BBC to outline the conditions for a demonstration which would be acceptable to them.

These were communicated to the Post Office on 16th November.[71] In the meantime Sir Evelyn Murray had written to the Director General to say that the Postmaster General had seen Sir Edward Manville and that he had agreed 'in principle to a demonstration on the lines we discussed'.

Bairds, in agreeing to give this, stipulated that, before the test was arranged for the official party, the Postmaster General should personally witness a laboratory demonstration. The company's strategy was clear: many notable scientists and public figures had seen television in operation and had been impressed; the BBC, on the other hand, was scathing in its criticism after the September 1928 test. Thus, if Sir William Mitchell-Thomson saw a demonstration before the main party and reported favourably upon it, Bairds were well placed to capitalise on the result. Their strategy paid off, for the PMG 'was distinctly impressed'.[72]

The BBC asked that aboslute secrecy should be a fundamental condition of any test and that if this condition were violated by Bairds 'the demonstration should be automatically cancelled'.[73] Manville accepted the BBC's conditions and on 26th November Murray told Reith the PMG agreed with the BBC views as to the television demonstration.[74] The BBC further agreed to the use of the London station for this.

Sir William Mitchell-Thomson exercised commendable initiative by his personal intervention in the television issue: he was, of course, anxious that the public, whether in its capacity as potential shareholders or as potential purchasers of the Baird apparatus, should not be exploited but at the same time he felt that it would be unfortunate if television were 'throttled in its infancy' by the refusal of adequate facilities for demonstrating its possibilities.[72]

Notwithstanding the BBC's request that secrecy should be observed, so as to prevent stock market manipulations of the Baird Companies shares, the proposed test was in fact the subject of general gossip in the lobby of the House of Commons as early as the first week in December. By New Year's Day the 'leak' was public knowledge for the *Daily Telegraph* carried the following paragraph:

> Further tests of the Baird television apparatus will take place, it is understood, in February, and one or more of the BBC stations will be used for this purpose. It is part of the agreement, however, between the BBC and the Baird Company that these tests take place secretly and that no publicity shall be given to them until the BBC has thoroughly considered the results of the trials.[75]

Popular Wireless also mentioned the tests a few days later,[76] but Bairds categorically denied that they had participated in the inspiration or suggestion of the paragraphs. Indeed their view was that the publicity was designed to do them harm by imperilling the proposed test.[74]

Nevertheless, newspaper reports continued to be produced which referred to the secret negotiations. The *People* in particular on 6th January 1929,[77] made a strong, if not sensational, attack on the BBC under the heading of 'Great Wireless Mystery' remarking that '. . . the attitude of the BBC in regard to this amazing British invention is absolutely incomprehensible . . .'. The *Manchester Guardian* on 9th January 1929[78] stated: 'The British Broadcasting Corporation has consented after all to carry out experiments with the Baird system of Television and these will begin next month', but a BBC press announcement on the same day mentioned the October 9th demonstration and said it failed to fulfil the conditions which would justify a trial through a BBC station.[79]

The effect of these announcements was to create active trading in Baird shares and at one time on 10th January the Baird Television 'A' shares were valued at over 10 shillings.[80] Bairds, however, denied that the statements about the test had any effect on the market value of their shares and ascribed the rise to the news of important contracts and developments in other countries. These were certainly taking place, for on 9th January,[81] Hutchinson told a reporter of the *Morning Post*:

> Important negotiations have been successfully concluded with seven important Continental broadcasting stations, including Radio Toulouse and Radio Lyons, for the regular broadcasting of the Baird system in conjunction with music, singing and speech. It is significant that these negotiations were concluded after the experts representing these stations had been to London and investigated the system.

He further added:

> During the last two weeks or so the German Government sent to London an official deputation, consisting of Dr Bredow, German Secretary of State for Wireless, Dr Bareth and Dr Reisser to investigate the system. As a result we are installing a television transmitter in their Berlin studios for the purpose of public demonstrations with a view to their adopting the system generally.

The sequel to this visit was that Bairds received a letter from the Minister of the German Government in charge of broadcasting services asking for an installation of the Baird system in his country. This letter arrived at an opportune time when two Members of Parliament, C. G. Amon and E. Maloney, were being given a demonstration in the television company's studio and seems to have created a favourable effect. They thought 'if there was no catch in this business' the BBC would be well advised to extend its 'active co-operation or assistance' as they felt it would be a 'great pity if in this as in other inventions in the past Germany stepped in and reaped the advantage that might fall to this country'.[82]

The activity in some of the Baird Companies shares was to persist for most of January. On Wednesday, 23rd January, for example, the International Television shares stood at 8/9d while two days later they had risen to 14/1½d.[83] The Television Development Company's shares did not reflect the appreciation of the others. Bairds gave as the reason for the share movement the fact that they had received an enquiry from America for 50 000 International Shares. Murray's information was that this explanation, even if genuine was not the sole factor and that E. M. Vowler, of 28 Throgmorton, who was apparently the 'brains' behind the finance of Bairds, was engaged in some stock market scheming. Vowler, who corresponded to Jackson of Wireless Pictures (1928) Ltd. had earned and maintained 'the reputation of being perhaps the most astute market manipulator in the city' and appeared 'to be an expert in handling rumours'. Murray had discussed the position relating to the shares of the Baird Companies and Wireless Pictures with an experienced financial observer who had given it as his opinion 'that the finance of Baird and Fultograph have been interrelated from the beginning, and that it has been the common policy of the manipulators to make money both ways, by bearing the Fultograph shares and bulling Bairds'. Murray continued: 'It is now pretty common knowledge in the city that Jackson and Vowler are discussing terms for the absorption of Wireless Pictures by Baird'.

This explanation would account for the buoyancy and strength of the Baird position financially compared with the steady decline and weakness of the Fultograph position financially, which had been puzzling, Murray observed.

Reith raised the issue of the financial state of the Baird Companies with Hutchinson at their meeting on 30th January 1929, but the latter assured Reith that Bairds were in no way 'market jugglers', that neither he nor Baird had at any time sold their shares – although they had bought more – and that the Board was not responsible for the recent market trends.[84] Hutchinson did tell Reith, however, that Bairds were on the point of bringing off, at the time of the BBC press announcement, a very advantageous deal in France, which would have given them 51% of the holding, which was going to exploit television in practically all European countries except Germany, but that the BBC's decision on the demonstration had severely 'prejudiced them'.

Reith also raised the question of the demonstration with Hutchinson and told him the time for this would be 11.00 a.m. on 27th February subject to certain conditions which the BBC wished to impose. These were that absolute secrecy was to be observed and that the instruments which the company proposed to sell to the public should be used both at the Post Office and the BBC with their prices revealed and that access to the transmitting end of the system should be permitted.

The condition of secrecy was unusual in that, following a reply to a question in the House of Commons, the test could not be carried out under such a restriction. For on 29th January Colonel L'Estrange Malone, socialist MP for Northampton, had asked the Postmaster General whether his Department had at any time suggested to the British Broadcasting Corporation that facilities for experimental transmission should be afforded to the Baird television system as had been arranged with the Fultograph apparatus.[85,86]

Actually the demonstration took place a little later than the date Reith had given Hutchinson. The invitations were sent out by the Postmaster General[87] on 22nd February 1929 to C. G. Amon, MP, Lt. Commander J. M. Kenworth, MP, J. I. McPherson, MP, R. Hamilton, MP,and W. Smithers, MP, all of whom, with one exception, replied that they would aceept. The above members were to see the demonstration in the PMG's room at 10.30 a.m. (later changed to 11.15 a.m.).[88]

References

1 ANGWIN, A. S.: 'Baird television demonstration', a memorandum, 19th September 1928, Minute 4004/33.
2 WHITEHOUSE, J. H.: memorandum to the Assistant Controller (Information), BBC, 1st October 1928, BBC file T16/42.
3 LEECH, W. T.: letter to Sir J. F. W. Reith, 20th September 1928, Minute 4004/33.
4 REITH, Sir J. F. W.: letter to Sir E. Murray, 21st September 1928, Minute 4004/33.
5 REITH, Sir J. F. W.l: letter to the Secretary, GPO, 20th September 1928, Minute 4004/33, file 6.
6 LEECH, W. T.: memorandum to T. Daish, 21st September 1928, Minute 4004/33. file 6.
7 BBC announcement: 'The broadcasting of pictures, BBC experiments', 8th September 1928.
8 LEECH, W. T.: letter to Sir J. F. W. Reith, 25th September 1928, Minute 4004/33.
9 LEECH, W. T.: memorandum, 27th September 1928, Minute 4004/33.
10 REITH, Sir J. F. W.: letter to Sir E. Murray, 27th September 1928, Minute 4004/33.
11 LEECH, W. T.: letter to Baird Television Development Company, 28th September 1928, Minute 4004/33.
12 LEECH, W. T.: letter to Sir J. F. W. Reith, 28th September 1928, Minute 4004/33.
13 Control Board Minute, extract, 2nd October 1928, BBC file T16/42.
14 Control Board Minute, extract, 9th October 1928, BBC file T16/42.
15 Chief Engineer (BBC): memorandum, 8th October 1928, BBC file T16/42.
16 MURRAY, G.: memorandum to the Controller, 10th October 1928, BBC file T16/42.
17 LEECH, W. T.: memorandum to F. W. Phillips, 11th October 1928, Minute 4004/33.
18 REITH, Sir J. F. W.: letter to F. W. Phillips, 15th October 1928, Minute 4004/33.
19 PHILLIPS, F. W.: letter to Sir J. F. W. Reith, 16th October 1928, Minute 4004/33.
20 PHILLIPS, F. W.: memorandum to the Secretary (GPO), 17th October 1928, Minute 4004/33.
21 REITH, J. F. W.: letter to F. W. Phillips, 17th October 1928, Minute 4004/33.
22 BBC statement, 17th October 1928, Minute 4004/33.
23 Anon.: 'BBC and television. No transmission at present', *The Times*, 19th October 1928.

24 Anon.: 'No broadcasting of television. BBC's adverse report. Result of recent tests', *Manchester Guardian*, 19th October 1928.
25 Anon.: 'BBC and television. Baird shares slump', *Glasgow News*, 18th October 1928.
26 Anon.: 'Share movements', *Daily News*, 18th October 1928.
27 Anon.: 'Mr Baird and the BBC. "Inexplicable" attitude to television And the reply', *Evening Standard*, 18th October 1928.
28 Anon.: 'Television surprise', *Daily Herald*, 18th October.
29 Anon.: 'BBC and television. Statement by the Baird Company', *The Times*, 22nd October 1928.
30 MURRAY, G.: memorandum to the Director General (BBC), 19th October 1928, BBC file T16/42.
31 MOSELEY, S. A.: letter to G. Murray, 20th September 1928, BBC file T16/42.
32 SHEPHERD SMITH, Miss.: memorandum to A. C.(I), 17th October 1928, BBC file T16/42.
33 MURRAY, G.: letters to the editors of the *Daily News, The Times*, the *Spectator*, the *Daily Express*, the Press Association, 15th October 1928, BBC file T16/42.
34 Control Board Minute, extract, 24th October 1928, BBC file T16/42.
35 Chief Engineer (BBC): memorandum to A. C.(I), 24th October 1928, BBC file T16/42.
36 Baird Television Development Company: letter to the GPO, 25th October 1928, Minute 4004/33.
37 Anon.: memorandum to the Postmaster General, November 1928, Minute 16806/1928.
38 MANVILLE, Sir E.: letter to the Postmaster General, 26th October 1928, Minute 4004/33.
39 BURNS, R. W.: 'The history of British television, with special reference to the work of J. L. Baird', PhD thesis, University of Leicester, 1976.
40 FINK, D. G.: 'Television broadcasting practice in America- 1927 to 1944', JIEE, 1945, ***92***, part III, pp. 145–160.
41 Anon.: 'Some impressions of the New York Radio Show, Sept. 23–29, 1929', *Television*, November 1929, pp. 360–463.
42 DINSDALE, A.: 'Television in America today', *Journal of the Television Society*, 1932, pp. 137–149.
43 BITTINGS, R. C.: 'Creating an industry', *J. S. M. P. T. E.*, 1965, **74**, November, pp. 1015–1023.
44 BUZALSKI, T. J.: 'Experimental television station W2XBS', Development Group Engineering Report, No. 95, NBC, 1st March 1933.
45 WEINBERGER, J., SMITH, T. A., and RODWIN, G.: 'The selection of standards for commercial radio television', *Proc. IRE*, 1929, **17**, No. 9, September, pp. 1584–1594.
46 Anon.: 'The truth about television', 3rd November 1928, BBC file T16/42.
47 Control Board Minutes, extract, 6th November 1928, BBC file T16/42.
48 A. C.(I): memorandum to the Director General (BBC), 19th November 1928, BBC file T16/42.
49 MURRAY, G.: letter to S. A. Moseley, 20th November 1928, BBC file T16/42.
50 MOSELEY, S. A.: letter to G. Murray, 21st November 1928, BBC file T16/42.
51 Anon.: 'Television. Big claim from Germany. 1-valve sets. Demonstration in London soon', *Daily News*, 29th July 1928.
52 Anon.: a report, *Evening Standard*, 25th June 1928.
53 Control Board Minute, extract, 27th November 1928, BBC file T16/42.
54 Anon.: 'Baird International Television. Statutory Meeting', *The Times*, 1928.
55 Anon.: 'A television development', *Manchester Guardian*, 1st November 1928.
56 Anon.: 'Mr Baird's announcement', *Manchester Guardian*, 1st November 1928.
57 PHILLIPS, F. W.: memorandum to private secretary, 6th November 1928, Minute 51/1929, file 20.
58 Anon.: 'Broadcast rival. Peer said to be backer of BBC opposition. From Abroad. Prospect of advertising by radio', *Daily Herald*, 5th November 1928.
59 Anon.: 'English radio hour from the continent. Challenge to BBC monopoly. Television 700', *Daily Chronicle*, 5th November 1928.

60 Anon.: a report, *Glasgow Herald*, 1st November 1928.
61 Anon.: 'Six rivals to the BBC', *Daily Mirror*, 5th November 1928.
62 ATKINSON, C. F.: memorandum to A. C.(I), 11th December 1928.
63 KENNEDY, Lord Angus: letter to the editor, *Daily Telegraph*, 12th November 1928.
64 Control Board Minutes, extract, 9th January 1929, BBC file T16/42.
65 Anon.: 'BBC and television', *Manchester Guardian*, 10th January 1929.
66 Baird Television Development Company: a statement, 9th January, 1929, Minute 51/1929, file 26.
67 SAMUEL, Rt. Hon. Sir H.: letter to the Postmaster General, 14th January 1929, Minute 4004/33.
68 MITCHELL-THOMPSON, Sir W.: letter to Sir H. Samuel, 29th January 1929, Minute 4004/33.
69 SAMUEL, Rt. Hon. Sir. H.: letter to the Postmaster General, 30th January 1929, Minute 4004/33.
70 Control Board Minutes, extract 20th November 1928, BBC file T16/42.
71 Anon.: 'Statement of Baird Television Company', *Daily Herald*, 10th January 1929.
72 Postmaster General: draft letter to Sir H. Samuel, 29th January 1929, Minute 4004/33.
73 MURRAY, Sir E.: memorandum, 21st February 1929, Minute 4004/33.
74 Report on television, for Control Board, c. January 1929, BBC file T16/42.
75 Anon.: a report, *Daily Telegraph*, 1st January 1929.
76 Editorial, *Popular Wireless*, 5th January 1929.
77 Anon.: a report, the *People*, 6th January 1929.
78 Anon.: 'Television next month. BBC experimental broadcasts. Public reception', *Manchester Guardian*, 9th January 1929.
79 BBC announcement, 9th January 1929, Minute 51/1929, file 26.
80 Anon.: 'Television activity', *Daily Sketch*, 10th January 1929.
81 Anon.: a report, *Morning Post*, 9th January 1929.
82 AMON, C. G.: letter to the Postmaster General, 16th January 1929, Minute 51/1929, file 27.
83 A. C.(I): memorandum to the Director General, 28th January 1929, BBC file T16/42.
84 Director General: memorandum to A. C.(I), et al, 31st January 1929, BBC file T16/42.
85 MALONE, C.: question in the House to the Postmaster General, 29th January 1929, Minute 51/1929.
86 Postmaster General: answer to the House (reference 85), Minute 51/1929.
87 Postmaster General: letters to C. Amon *et al.*, 22nd February 1929, Minute 4004/33.
88 Private Secretary: letter to Sir E. Murray, 4th March 1929, Minute 4004/33.

Chapter 6

The start of the experimental service, 1929

A meeting to discuss the details of the test was held on 14th February.[1] It was attended by Jacomb and Hutchinson of Bairds, and by Murray, Ashbridge and Gambier Parry of the BBC. While the general conditions of the demonstration had been laid down by Reith on the 16th November, the form this was to take had not been agreed. Bairds said they wished to demonstrate television and music combined and it was agreed that 2LO should be used for the vision part of the experiment. There was some difficulty in selecting the music channel as 5XX had no convenient free period, but finally it was decided that Bairds own transmitter should be used for this purpose. Their television transmitter was to be erected in the BBC's building so that it would be open for inspection and the Baird Company was to provide two receivers each for the GPO and BBC. The BBC undertook to provide a piano to accompany the singers and Bairds were to be allowed to transmit one trial programme.

At this time the BBC was not only making preparations for the demonstration but considering how events should be controlled after the demonstration. The general opinion of the lobby was that the BBC was being forced by political pressure to change its policy towards television, which was considered to have been unfairly reactionary in the technical sense. Murray was of the view that the verdict on the test would almost certainly be cautiously friendly to television with a recommendation to some further action on the part of the BBC.[2]

He thought the panel should encourage the Baird people to claim sufficient improvement over the position from October to merit a fresh trial by the BBC. Murray felt this would remove any suggestion that the BBC should take the initiative. 'Of course,' he said, 'if we know definitely that public tests of television through the BBC station would be a sequel of this demonstration on 27th February, it might be worthwhile to short circuit the whole thing by arranging for the Baird people to apply in the ordinary way for a test by us under the last clause of our statement of policy. If this line were taken I think it would be difficult to sustain the case that we were wrongly reactionary before, and that the cause of science were served by the timely and effective intervention of an enlightened PMG.'

Following the demonstration, the date of which had been changed from February to 5th March, the Post Office issued a brief press statement. This coincided with questions in the House of Commons by a Mr Thurtle who raised the issue of the Baird demonstration and in a supplementary question asked:

> Will the Right Honourable Gentleman bear in mind that there is a very wide-spread opinion that the whole system will not stand examination, and that there is great speculation going on in the Stock Exchange in regard to this Company, and will he see that his Department does not lend itself to any of these marketing operations?

Sir W. Mitchell-Thomson replied non-committally that there were many factors to be taken into consideration.[3]

On 28th March 1929 the Postmaster General published in *The Times* the letter he had sent to the Secretary of the Baird Television Development Company.[4] He confirmed his earlier opinion that the Baird system was capable of producing, with sufficient clearness to be recognised, the features and movements of persons posed for the purpose at the transmitting point, although he added that it was not yet practicable to transmit a scene or performance which required a space of more than a few feet in front of the transmitting apparatus. In the PMG's view the system represented 'a note-worthy scientific achievement' but had not reached a stage of development which would merit the inclusion of television programmes within the broadcasting hours. He was anxious that facilities for further development should be granted and mentioned that he would assent to a station of the BBC being used for this purpose outside broadcasting hours. The PMG thought it was probably essential that television should be accompanied by speech and that two transmitters would be required, but pointed out that a second transmitter would not be available until the new station at Brookmans Park was completed, possibly in July. In the meantime the company was to open negotiations with the Corporation on the financial and other arrangements which might be necessary.

The Postmaster General made two further points; first, the need to press on with experiments on a much lower band, as the bands being used for speech broadcasting were already highly congested, and second, that neither the PMG nor the BBC could accept any responsibility for the quality of the transmission or for the results obtained. 'The purchaser must understand that he buys at his own risk at a time when the system has not reached a sufficiently advanced stage to warrant its occupying a place in the broadcasting programmes.'

Lee-Smith's advice to Bairds to press on with experiments on a much lower band of wavelength was to prove highly significant in the later history of television. Unfortunately for the Television Company, Baird did not immediately initiate an urgent programme of research and development work in this region of the electromagnetic spectrum, but concentrated instead on achieving some success with the low definition system. In retrospect, the position of Bairds companies *vis-à-vis* EMI Ltd. in the 1930s might have been much improved had he appreciated earlier than he did the limitations of the 30-line system and the need for a television service operating on a higher definition standard. If Baird had produced a 150 or 180-line picture system (say), working on short waves, by 1931, the future of his three companies probably would have been more secure. Events later were to show that Baird made the change to the higher standard too late for his companies to surpass the impressive EMI and MEMI television systems, and yet the writing was on the wall for the 30-line system, even in 1929. Eckersley and others had given their views on 'head and shoulders' television broadcasts and the medium waveband did not allow, owing to bandwidth limitations, any extension to a better definition system, and hence an improved and more varied picture. In addition the Post Office had indicated to the Baird Television Company in 1929 that they had no objection to them experimenting for demonstration purposes on 75 m or 50 m.[5] The 75 m band was reserved, by the Washington Convention for Broadcasting, for experimental work, while the 50 m band was allowed for broadcasting (Baird had, of course, used a wavelength of 45 m for his transatlantic experiments but only for a 30-line picture).

There were certain unknowns in working in the above wavebands and these may have influenced Baird's determination to press on with his imperfect system. In 1929 there was insufficient information available to show how far either of the 50 m or 75 m wavebands would be effective inside the London area, although it was known that the presence of steel-framed buildings had a deleterious influence on the reception of shortwaves.

Any pioneer company clearly must keep ahead of its competitors if it is to remain operationally viable. Baird realised this point in so far as it related to low definition television, for he expended much thought and effort in modifying his basic system so that it could be utilised for colour television and noctovision *inter alia.*

He failed, however, to appreciate at a sufficiently early stage that 30-line television could never give rise to an all-embracing television broadcasting service. While Baird's low definition system was capable of improvement, as evidenced by the successive demonstrations given by the company and later the BBC, the system was effectively restricted to head and shoulders type shots. Baird persisted in this view until he was almost overtaken by EMI and RCA. He believed that television broadcasts should be transmitted in the medium waveband to ensure a wide coverage of the population, even though the bandwidth available severely restricted the definition of images which could be sent out.

This view was shared by the directors of the Baird companies until approximately 1932, and it seems that they overrated the commercial possibilities of 30-line television, failing to recognise that television broadcasting would only become widespread in popular appeal when the Derby, sporting and athletic events, motor racing and so on could be televised to give adequate image detail, as Eckersley had pointed out.

It is possible that, if the company had retained a number of distinguished and eminent radio engineering consultants, it may have heeded the Postmaster General's advise at an earlier stage. It is significant to note that it was Appleton, the radio scientist, who urged Baird to investigate the use of the HF and VHF bands.

Rather interestingly, Marconi too had a 'blind spot'. His dominant urge was to extend his system of wireless communications, but he failed initially to appreciate the importance of continuous wave operation in transalantic working, and thereby failed to visualise the benefits of radio telephony. Several of Marconi's contemporary inventors took a different view, but Marconi did not share their optimism. He considered the Morse code to be quite suitable for ship communications and for transoceanic signalling, and saw no real need for a wireless telephone. Like Baird, his approach to his work was pragmatic. He was not interested in pursuing scientific investigations in fields which had a doubtful commercial viability, and, in this respect, Baird and Marconi lacked some foresight, although both inventors possessed an entreprenurial outlook. This blind spot was unfortunate initially for the Marconi Company and the furtherance of communications. Luckily for Marconi some of the important early work on radio telephony was undertaken by two of his rivals, de Forest and Fessenden, and neither inventor had access to financial or engineering resources or skills comparable to those of the Marconi companies.

Perhaps unfortunately, the 5th March demonstration using the 2LO transmitter showed that television caused no undue interference outside the normal band of the station: fate might have been kinder to Baird had interference effects condemned the use of this band for television broadcasting. The lack of interference was probably due to the characteristic of the 2LO antenna which cut off fairly sharply outside certain well-defined transmission limits.

For Baird the demonstration had been a most anxious experience. He wrote[6] many years later:

> It was a nerve-racking ordeal as we were to stand or fall by the result of one crucial demonstration. If a wire were to slip or a valve to burn out at a critical moment, the demonstration would fail and we should be faced by a devastating fiasco. The night before the test transmission passed like a nightmare on the top floor of Savoy Hill, trying to make sure that our transmitter would not fail.
>
> By morning all seemed to be well and I set out for St Martin Le Grand accompanied by Sir Ambrose Fleming. Here in a large hall, on the first floor, four receivers had been installed and soon, to my relief, they were running properly and receiving images of the artists we had assembled at the studio at Savoy Hill.
>
> Mr F. W. Phillips, who was in charge of the Post Office's arrangements, told the Committee that we were ready, and they trooped in, headed by the impressive figure of Lord Clarendon. They took their places in front of the receivers and watched the little programme, simply consisting of

> head and shoulder views of singers and comedians. Captain Eckersley himself provided an unexpected turn when he went before the transmitter and was seen by the Committee.

The immediate periods prior to and succeeding the all-important demonstration were not without rumours and innuendos. Moseley, Baird's champion, felt so strongly about two of these that four days before the crucial test he sent a personal letter[7] to the Prime Minister, the Right Honourable Stanley Baldwin, and mentioned the speech which the Postmaster General proposed to make to the members of the panel. Moseley considered the speech would have the effect of prejudicing a jury which had not yet seen the evidence, for certain alleged technical defects were to be remarked upon and the PMG was to refer to the high cost of the televisor which it was proposed to place on the market. 'To me the whole tenor of this memorandum is one-sided, and so obviously defeats the purpose of the demonstration, that I have taken the risk of addressing myself to you direct'. Moseley told the PM.

There was another point: he had learnt from certain members of Wireless Pictures (the Fultograph Co. Ltd.) that they had been informed by members inside the BBC 'that in no circumstances would the BBC agree to the transmission of television, and that the demonstration on Tuesday was foredoomed to failure'. Moseley was prepared to call upon the Prime Minister to tell him frankly what the position behind the scenes really was, but regrettably the Post Office files do not contain any follow up correspondence to Moseley's letter. The latter did, however, give some indication of the thinking behind the Baird Companies' business pretensions, for Moseley concluded: 'It is vital that British control of television throughout the world should be procured. This would indeed be super-salesmanship'.

Rumours continued to upset the Baird Television Company after the test and Hutchinson had occasion to write to Sir William Mitchell-Thomson about them. One concerned the contention that the test television transmission could not have been broadcast within a 10 kHz bandwidth,[8] and another held that each one of the guests, at the demonstration, had been given a pamphlet which, *Popular Wireless*[9] stated, had the approval of the President of the Board of Trade. The pamphlet, a copy of which had been obtained by the magazine, (which had Eckersley as its chief radio consultant) contained the clause '. . . no intelligent guest at the television demonstration could possibly fail to understand why the BBC turned down the television scheme on technical grounds'.

Actually no pamphlet of any character was issued to the guests[10] and Hutchinson was able to confirm that the demonstrations given at Olympia in September 1928 were made within a 7 kHz bandwidth and that results similar to those obtained at the official demonstration on the 5th March 1929 could be duplicated inside a 10 kHz bandwidth.

These rumours and innuendos show, perhaps, something of the feeling which existed in some quarters against Baird and his business associates at this time. While no libel actions were ever initiated by Baird against his detractors, nevertheless harsh things were said in private about the Television Company's publicity and business methods. Possibly these private communications were the breeding ground for many of the rumours which circulated during the 1920s and 1930s. Hutchinson and Moseley – fortunately for the company – were both skilled in negating the effect of these and had no hesitation in writing to the Postmaster General, the Prime Minister, the Prince of Wales or any other high ranking individual to achieve the aims and objectives of the Baird Companies. They worked tirelessly for these companies and never ceased to take action when any event occurred which tended to put the Baird television system in a poor light. Naturally Baird International Television Ltd. was anxious to advance the superiority of the British system and exploit the rights of the Baird Companies overseas, in for example France, Italy, Spain, Portugal, Czechoslovakia, Poland and the Scandinavian countries as well as several others: the publication of rumours and criticisms did nothing to help the Baird cause. The vigilance of Moseley and Huchinson was therefore well-based. A particular example of their interests being 'seriously prejudiced' concerned the 'existing unsatisfactory position regarding the demonstration' on the 5th March. Following this the Publicity Department of the Post Office had on the 5th March issued a very brief notice:[11]

> The Postmaster General and others were present at a demonstration of the Baird system today.The Post Office does not propose to make a statement on the subject. There are many aspects which the Postmaster General will have to consider and it is likely to be some considerable time before a statement can be made.

This terse wording could possibly be interpreted, particularly by overseas interests, as an indication of the lack of interest evinced by the Post Office towards television broadcasting, and so Hutchinson wrote to the PMG[11] to say that he thought it 'only fair to the Baird Companies that an announcement should be published mentioning the successful demonstration of the instantaneous reception of living images and sound synchronism'. Sir William Mitchell-Thomson, however, did not think it would be expedient to make such a preliminary or provisional statement[12] – much to Hutchinson's regret.[13]

Following the welcome news from the PMG on the 28th March about the tests, Hutchinson wrote to Eckersley and requested an appointment.[14] He was disappointed, for the Chief Engineer was in Prague,[15] but on 18th April the two antagonists arranged to meet.[16] The BBC was in no hurry to provide facilities for television and by 2nd May Hutchinson was complaining about the delay in making definite arrangements.[17] He told Eckersley the delay might 'have the effect of losing for the British companies and British interests control of television in foreign and colonial countries'. Eckersley countered [18] by referring to the Postmaster General's letter which 'says that broadcasting of television cannot take place until Brookmans Park is opened for service' and this would not occur before October 'at the earliest'. He went on to say: 'It would appear therefore inevitable that we should have to wait until, say, July before discussing the matter further'.

Hutchinson thought this prospect was quite unacceptable.[19] He said to Eckersley:

> The present position as outlined by you is so grave indicating as it does a seemingly incomprehensible attitude to British national interests that unless some satisfactory solution is immediately found I fear it will be necessary to reopen the whole matter.

This hint of renewed hostilities between Bairds and the BBC followed Hutchinson's visit to Berlin to arrange television tests, for the German Secretary of State for Broadcasting, Dr Bredow, of the Baird system.[20] It had been arranged that the Berlin broadcast transmitter would be used, but before the preliminary work had been carried out Baird had returned to London and informed Dr Bredow that he had to prepare for an important demonstration in London and that because of a lack of personnel he was unable simultaneously to carry out any tests in Germany. The Press Service of the Reich-Rundfunk Gesellschaft had an alternative explanation: it noted that the BBC appeared to have given up its absolute reluctance to adopt a television service and observed pointedly: 'The fact that the German Secretary of State for Broadcasting has in London made successful tests of the Baird system and has suggested that these tests be continued from German broadcasting stations should have played a certain role in this connection'.

Baird had thus delayed his proposed tests from a German station so as not to interfere with his demonstrations in England, only to find, after these had been pronounced successful, that the BBC were still procrastinating.

Hutchinson was furious and wrote a long letter to Eckersley instancing the 'immense harm which may result from this present position'.[21] He recalled the meeting which Baird and he had had with Eckersley and how it was 'definitely arranged' so far as the technical side was concerned that Bairds shouuld be permitted to transmit daily from 2LO between the hours of 11.00–12.00 in the morning and on three nights of the week after the station had closed down for regular broadcasting. 'In fact,' said Hutchinson, 'we were so pleased at this prospect of such whole hearted co-operation we felt we could not avail ourselves of your valuable services without at least offering some tangible form of appreciation . . .'

Hutchinson went on to say that at the 18th April meeting

> 'it was quite obvious that you also understood that we wished to transmit television only, as apart from music and speech. Also if it could be arranged tele-cinema pending the completion of the Brookmans Park Stations when music and speech could also be transmitted making the transmission a better entertainment.'

The Brookmans Park Transmitter was one of a number of transmitters which together were to provide the broadcast transmissions of the Regional Scheme. This was conceived by Eckersley in 1924[22] and had as its objective the provision of a dual-programme service throughout the United Kingdom. Prior to the opening of the new service the Corporation had nine main stations and a number of relay stations, and these, with the high power longwave transmitter at Daventry, enabled 80% of the population to receive their transmissions, even if simple cheap receivers were used.

One programme was available to listeners in any particular area except that those who possessed a suitable receiver could receive the longwave station as well as the local mediumwave station and thus had a choice of programme at those times when the local station was originating its own broadcasting material.

Eckersley's idea of the Regional Scheme was, specifically, to cater for national and regional programmes: the former being of interest to the whole country and the latter being of local appeal.

The Regional Scheme had to be designed around an allocation of only ten medium wavelengths and one long wavelength under the Prague Plan of 1929 and all the relay stations had to operate on a UK Common Wave of 288.5 m, with the exception of Leeds, which used an International Common Wave.

In the scheme the transmitters were to be of a much higher output power, (up to 100 kW) than had previously been used so as to allow the stations to be more economically run than a larger number of low power stations giving the same coverage, especially as each station had to be staffed. Capital costs would also be lower and the use of increased power would enable the best use to be made of the limited number of frequency channels available.

The development plan of the Regional Scheme made provision for five twin-wave stations to be built to cover the following areas:

1 London and the Home Counties
2 Manchester and the industrial north of England
3 Glasgow, Edinburgh and the Scottish Lowlands
4 South Wales and the West of England
5 Birmingham and the Midlands

Each of these stations was provisionally estimated to cost £115 000. Because of the large sums of money which would be involved in implementing the scheme, Reith thought it wise to have an independent technical committee to examine and report upon the BBC's Chief Engineer's proposals. The Committee, known as the Eccles Committee, comprised:

Dr W. H. Eccles	Chairman
Professor E. V. Appleton	University of London
Dr L. B. Turner	University of Cambridge
Mr R. T. B. Wynn	Technical Secretary (BBC)

(It was to the first three members that Eckersley turned when Control Board decided in May 1928 that the 'sporadic attacks made on us for doing nothing about television' should be countered by a series of articles in the *Radio Times*).[23]

The Eccles Committee endorsed Eckersley's Regional Scheme without suggesting any modifications, but, notwithstanding this, the Post Office withheld its approval as it felt that the permanent allocation of wavelengths for broadcasting purposes was the responsibility of the Telecommunications Administrations and as there was to be a World Wireless Conference in 1927 the decision regarding the Regional Scheme had to be delayed until the Conference reported. However, on 20th April 1928 the GPO authorised the BBC to go ahead and the construction of the first permanent-twin-wave transmitting station was commenced.

London and the South-Eastern Counties was the region which it had been decided should have the first of the regional transmitters, and much thought was given to the location of the site. There were a number of conditions to be satisfied. The site had to be:

1 within about 15 miles of Oxford Street, so that the signal would be strong enough in Central London to permit the continued use of existing sets;
2 far enough from populated areas to avoid blanketing a significant number of receivers;
3 far enough from the coast to avoid wasting a large part of the radiated energy over the sea;
4 in an open position to avoid absorption of energy by neighbouring buildings;
5 accessible to a Post Office cable route.

Brookmans Park, 36 acres in extent, 16 miles north from Charing Cross and 414 ft above sea level, was chosen for the first site and work started there in July in 1928. It was anticipated that the construction work would take about twelve months, but owing to a severe frost in the early part of 1929 the first transmitter (356 m, 45 kW), was not put into service until 21st October 1929, while the second transmitter (261 m, 67·5 kW), was not completed and brought into service until four and a half months later.

This then was the background to Eckersley's statement to Hutchinson on 18th April 1929 that television could not take place until Brookmans Park was opened for service. From Bairds point of view the delay was unsatisfactory.

Hutchinson felt the reference Eckersley made to the Postmaster General's letter was quite incomprehensible, but the Chief Engineer saw the position differently. After receiving a second letter on this subject from Hutchinson, asking why he had not received a reply,[24] Eckersley wrote back: 'I am amazed by your letters of the 15th and 24th of May. I should not have thought it possible without the facts that any one should have gained so complete a misunderstanding on practically every point we discussed.[25] A further meeting was clearly desirable and this was arranged for the 5th June 1929.

Hutchinson and Bairds were not alone in misinterpreting the other side's thoughts. Reith told Sir Evelyn Murray on 29th April[26] that the Television

Company had a system which only required one wavelength and was really the projection of a cinema into people's homes, communicating vision without speech. This was a new suggestion and represented an extension to the Fultograph system which only allowed a single picture to be broadcast at a time. It seemed that Bairds were anxious to keep themselves before the public and did not want to have to wait until the late summer or autumn before broadcasting speech and vision from the BBC's system.

The Director General was not averse to the supposed temporary scheme and he informed Murray: 'We could do this after midnight on Tuesday, Wednesday and Thursday when the Fultograph people are not on, and there is also time in the morning between 11 and 12'. Reith felt a demonstration of the cinematograph would be necessary but 'if it is good enough we will then let them go on for a few half hours weekly out of programme times . . . making them pay on the same rate as the Fultograph Company pays'. Sir John's unexpected display of generosity towards Bairds on 29th April contrasted markedly with the tone of a letter he sent to Sir Evelyn the following day in which he mentioned that the Fultograph people had been bought out by the Baird people and hence, as the company had ceased to exist, the BBC was not bound to continue transmitting Fultograph pictures. 'By stopping them there would cease to be a precedent for agreeing to the cinematograph idea of the Baird people.[27] This reversal of feeling for the Television Company compared unfavourably with the helpful spirit of the previous day's letter.

From the Post Office's point of view there were no reasons why either still or moving pictures should not be transmitted (subject to certain conditions),[28] and accordingly Murray wrote to Reith and acquainted him of the GPO's lack of objection to the scheme.

The interesting aspect of the tele-cinema episode is contained in a letter[29] (dated 14th June 1929), which Reith sent to the Secretary of the General Post Office:

> . . . we have recently had a visit from representatives of the Company from which it appears that there was a misunderstanding with regard to the proposal to broadcast the tele-cinema . . .

What Bairds actually desired was to commence television broadcasting, without sound (as only one wavelength was likely to be available), from the London station immediately. Notwithstanding the 18th April meeting between the BBC and the company, the position regarding the latter's wishes was not resolved until nearly two months later. Certainly the Corporation was in no hurry to provide facilities for television and Reith's letter to Murray was tinged with a note of regret concerning the PMG's decision of 28th March.

> The corporation would have been glad had the late Postmaster General, after the definitely unconvincing demonstration which he and others witnessed, been able to deal with the Baird Company in such a way as would not have involved an early promise of co-operation by this Cor-

> poration, and without giving the Baird Company such increased facilities as might either technically or otherwise, prejudice the Corporation.

The Board was still most 'unwilling' to permit experiments from the London station as their attitude was that public trials through a BBC station were not justified partly because erroneous deductions were likely to be made. Another meeting between the GPO and BBC was necessary and this was arranged for 24th June 1929.[30]

Whether Bairds knew about this arrangement is not known and their request to the GPO to have a discussion three days beforehand,[31] which tended to pre-empt a favourable decision from the Post Office, may have been fortuitous. When Hutchinson saw Phillips and Wissenden he was able to tell them of the favourable attitude of the German authorities towards television and also that in deference to their desire that television in Germany should be conducted by a German company, a company had been formed consisting of the Robert Bosch, Zeiss, Löewe Radio and the Baird Television Companies to develop television in that country. The Managing Director of the Baird Television Company was able to contrast this support with the hostile attitude of the BBC and stated that they had already frustrated schemes of development in both foreign and colonial countries and in the dominions by describing the televisor as a crude apparatus being used mainly for the purpose of Stock Exchange manipulation. On the other hand, in Germany television development seemed to be encouraged; no charge was made for the experimental transmissions and Hutchinson anticipated that in view of the close association between the new company and the government no charges would be made when a regular programme service was afforded. The eagerness of the Baird Companies to start television broadcasting was no doubt associated with the responsibilities of the companies towards their shareholders. Approximately £150 000 had been spent by Bairds on the development and exploitation of their patents and it was perhaps natural that the shareholders should look for some return. Hutchinson asked for the support of the Post Office to ensure television transmissions being given at once from the London Station (2LO) between 11.00 a.m. and noon on weekdays and on three nights a week after midnight, and for some guarantee of security of tenure so as to enable the company to make arrangements with manufacturers for the bulk production of televisors.

In the face of Hutchinson's pleadings, the strong BBC team (Reith, Carpendale, Ashbridge and Atkinson), which met Phillips and Leech on the 24th June had little success.[32]

Subsequently, on 26th June the Controller wrote to the Secretary of the Baird Television Development Company and informed them that the transmitter of the London Station (2LO), would be placed at the disposal of the company for three periods of about fifteen minutes each per week, at 11.00 a.m.[33] The company was to reimburse the BBC at the rate of £3 for each period of fifteen minutes or less and such capital cost as would be involved, although the latter was not likely to involve a sum greater than £250 so far as the Oxford Street transmitter was concerned. In addition the company was to indemnify the BBC against all claims that might be made upon the BBC because of infringement of patent rights and defray all costs due on account of copyright charges.

Carpendale further stated that the BBC had been instructed by the Postmaster General to emphasise four more points, first, 'that no broadcasting monopoly would be created for any particular system', second, 'that we must reserve the right, subject to Post Office approval, to curtail or discontinue any transmission as might be required', third, 'that you should press on with experiments on a much lower band of wavelengths than those hitherto used', and fourth, 'that the Post Office and the BBC dissociate themselves from responsibility for the quality of the transmission or the results obtained and will give the public to understand that the purchaser of receivers apparatus buys at his own risk'.

Carpendale's stern conditions illustrated the BBC's general lack of enthusiasm for the new form of broadcasting and contrasted markedly with the excitement which Hutchinson had displayed after his meeting with Eckersley. Not surprisingly Bairds regarded the offer as inadequate and as being a withdrawal of the proposals which the Chief Engineer had made at the above meeting. They wrote to the BBC and said it would be necessary for them to re-open the matter with the PMG and added darkly that they would 'take such other steps as they may be advised forthwith'.[34] Once again friction had been caused between the BBC and the Baird Television Development Company and sections of the press rallied to their aid. The *People* accused the BBC of 'holding up television'[35] and the *Electrician*[36] failed to see why the Corporation could not provide better facilities for experiments in a science which must eventually become as public a service as that of broadcasting. There was a rumour that the government had decided to examine the whole problem of television service in this country but this was squashed when the PMG, in answer to a question from Mr Malone in the House of Commons said he had not received such a request from the Baird Company.[37]

The BBC was to some people, behaving in an unreasonable and unpatriotic way during the three months of negotiations following the PMG's letter of 28th March. Their television policy seemed to be one of procrastination and obstructionism and the whole business of television broadcasting appeared to be rather tiresome for the Corporation.

> We desire to say with a full sense of responsibility that throughout our negotiations with the British Broadcasting Corporation we have been met with constant obstruction. The attitude of the officials throughout has led us to infer that they do not desire to be associated in any way with television . . . We wish to point out in all seriousness that the privileged position conferred on the BBC by licence of the PMG has been used by the officials of that Company, for reasons unknown to us, to hinder and delay the progress of a system which is of national importance . . .[38]

So wrote the respected Lord Ampthill and Sir Edward Manville to Mr H. B.

Lees Smith – the new PMG – on 9th July 1929. The BBC was adopting a most ungenerous stance, in its relations with Bairds, and its offer to the Company of broadcasting time for three fifteen minute periods per week seems to have bordered on naïvety in its likely acceptance by Bairds, particularly as Eckersley himself had agreed to the use of the London Station (2LO) for one hour each day between 11.00 a.m. and noon and on three nights a week for an unspecified length of time. However, Eckersley had been obliged to resign his position as Chief Engineer to the Corporation (following a divorce action), and his departure presumably left the BBC free to amend any verbal offers made by him to the Television Company.

Ampthill and Manville considered

> the minimum fair arrangement would be:
> 1 An assurance of a service over a period of years comparable with that allowed to the BBC on its inception.
> 2 A daily service beginning with not less than six hours per week during broadcasting hours to be increased as and when the growth of the industry, as evidenced by the sale of receiving sets, demonstrates increasing public interest.
> 3 Facilities for an extended service outside broadcasting hours.

Points 2 and 3 were similar to those agreed by the former Chief Engineer and point 1 was necessary to give some security to manufacturers of television receiving equipment: the latter were hardly likely to embark upon a programme of receiver production until some guarantee of security of tenure was given. Hutchinson thought three years was the minimum which would be satisfactory. Coupled with this aspect of the problem there was also the need to have television broadcasts during the weekdays so that retailers could demonstrate their televisors to prospective customers.

All of these negotiations and representations compared adversely with the situation in Germany where the Minister, corresponding to the Postmaster General and his staff had not only afforded Bairds every assistance in inaugurating the new service but had gone to the length of encouraging important manufacturers to initiate the production of sets 'and of removing difficulties caused by private interests which stood in (their) way'. In addition, President Kruckow of the German Central Post Office had taken such a personal pleasure in developing television that he found time to attend a conference[39] (held in Berlin on 12th July 1928), to decide on the type of disc to be used for the transmission and reception of broadcast television and also on the method of synchronising to be adopted. These decisions were required by the German manufacturers so that they could go into immediate production of a set for a television exhibition which was scheduled to take place at the end of August in Berlin.

The Postmaster General replied to Lord Ampthill's and Sir Edward Manville's letter of 9th July on 14th August, although Sir Evelyn Murray had produced a draft reply for the PMG's approval on 15th July.[40] The reason for the delay in the despatch of this letter is not evident from an examination of the appropriate Post Office files but a possible explanation is that Lees-Smith was not satisfied with Murray's draft. This view is supported by a comparison between the draft and actual letters. Whereas the former was non-concessionary in tone and supported the status quo, the latter was more generous. Leech told Hutchinson the Corporation had agreed[41] to extend its offer of three fifteen minute periods of broadcasting per week to five thirty minute periods per week – subject to the same conditions as stated in the BBC's letter of 26th June[42] (this extension of time had been suggested by the GPO to the BBC).

In acknowledging Leech's letter, Hutchinson mentioned that his Board felt the offer was 'inadequate but had to accept it as a temporary measure as they had no alternative'.[43] The next day, 5th September 1929, Leech wrote to the Managing Director of Bairds and somewhat darkly pointed out: '. . . the PMG does not anticipate that you will experience any difficulty or delay in making the necessary arrangements with the Corporation regarding the agreed experimental transmissions.[44]

This was certainly a strong hint that the procrastination and obstruction of the BBC were at an end – albeit temporarily: Lees-Smith seems to have exerted his influence and authority both with the BBC and his own department. He was also keen that his officers should follow the experiments and asked that a television receiving set should be installed for the purpose, on loan from Bairds.[45] The company readily agreed to this request and a set was despatched to the Aldersgate address of Lt. Col. A. S. Angwin – the deputy Chief Engineer of the Post Office Engineering Department.[46]

Captain Hutchinson and Mr Baird discussed the arrangements for the experimental transmissions with officials of the BBC on 11th September 1929 and the following points were agreed:[47]

1 transmissions should commence on Monday, 30th September from the Oxford Street transmitter, and later from the Brookmans Park transmitter;
2 the transmission times would normally be from 11.00 a.m. to 11.30 a.m. on Monday to Friday inclusive, but after 31st October, when it was anticipated that additional periods would become available, the morning transmissions might be replaced occasionally by transmission after midnight, or at other times outside programme hours as mutually agreed;
3 the extent of any interference would not be greater than that given by a music transmission as normally radiated from programmes;
4 the transmission would take place from television studios in Long Acre and the Baird Company would be responsible for renting all lines;
5 a BBC engineer would be allowed reasonable access to the television transmitter;
6 The company would instal and maintain one televisor at Savoy Hill, one at the BBC receiving station and one at the GPO;
7 the BBC would answer technical queries from home constructors;

8 the BBC reserved the right to curtail or discontinue any particular transmission should it conflict with the BBC's own programme, on a particular day;
9 transmission would not be curtailed, discontinued or at any time altered by the BBC except with three months' notice given in writing.

For this concession the Baird Company was to pay £5 per half hour and reimburse the BBC for any capital cost involved.

Thus, although the September 1928 demonstration of the Baird television system was considered to be sufficiently satisfactory for the Post Office to agree, as far as it was concerned, to the use of a BBC station for further experiments, the start of the experimental service had been delayed by a full year.

References

1 Assistant Chief Engineer (BBC): memorandum to the Chief Engineer *et al.*, 14th February 1929, BBC file T16/42.
2 A.C.(I): memorandum to the Director General (BBC), 7th February 1929, BBC file T16141.
3 MITCHELL-THOMSON, Sir W.: answer to a question in the House, 5th March 1929, Minute 51/1929.
4 Postmaster General; letter to Baird Television Development Company, *The Times*, 6th March 1929.
5 Assistant Engineer-in-Chief: memorandum to the Secretary (GPO), 6th March 1929, Minute 4004/33, file 12.
6 BAIRD, M.: 'Television Baird' (HAUM, South Africa, 1974).
7 MOSELEY, S. A.: letter to the Prime Minister, 1st March 1929, Minute 4004/33, file 12.
8 HUTCHINSON, O. G.: letter to the Postmaster General, 8th March 1929, Minute 4004/33 file 12.
9 Anon.: a report, *Popular Wireless*, 16th March 1929, pp. 37–38.
10 Private Secretary to the Postmaster General: letter to Baird Television Development Company, 15th March 1929, Minute 4004/33, file 12.
11 HUTCHINSON, O. G.: letter to the Postmaster General, 7th March 1929, Minute 4004/33, file 12.
12 General Post Office: letter to Baird Television Development Company, 11th March 1929, Minute 4004/33, file 12.
13 HUTCHINSON, O. G.: letter to the Postmaster General, 14th March 1929, Minute 4004/33, file 12.
14 HUTCHINSON, O. G.: letter to P. P. Eckersley, 4th April 1929, BBC file T16/42.
15 Assistant Chief Engineer (BBC): letter to O. G. Hutchinson, 5th April 1929, BBC file T16/42.
16 ECKERSLEY, P. P.: letter to O. G. Hutchinson, 15th April 1929, BBC file T16/42.
17 HUTCHINSON, O. G.: letter to P. P. Eckersley, 2nd May 1929, BBC file T16/42.
18 ECKERSLEY, P. P.: letter to O. G. Hutchinson, 6th May 1929, BBC file T16/42.
19 HUTCHINSON, O. G.: letter to P. P. Eckersley, 15th May 1929, BBC file T16/42.
20 Press statement, no. 129, Reich-Rundfunk Gesellschaft.
21 HUTCHINSON, O. G.: letter to P. P. Eckersley, 15th May 1929, BBC file T16/42.
22 PAWLEY, E.: 'BBC engineering 1922-1972' (BBC Publcations, 1972), pp. 83–101.
23 Control Board Minutes, extract 15th May 1928, BBC file T16/214.
24 HUTCHINSON, O. G.: letter to P. P. Eckersley, 24th May 1929, BBC file T16/42.
25 ECKERSLEY, P. P.: letter to O. G.: Hutchinson, 24th May 1929, BBC file T16/42.

26 REITH, Sir J. F. W.: letter to Secretary (GPO), 29th April 1929, Minute 4004/33, file 14.
27 REITH, Sir J. F. W.: letter to Sir E. Murray, 30th April 1929, Minute 4004/33, file 14.
28 PHILLIPS, F. W.: memorandum to W. T. Leech, 2nd May 1929, Minute 4004/33, file 14.
29 REITH, Sir J. F. W.: letter to Sir E. Murray, 14th June 1929, minute 4004/33.
30 LEECH, W. T.: memorandum, 25th June 1929, Minute 4004/33, file 15.
31 PHILLIPS, F. W.: memorandum, 21st June 1929, Minute 4004/33, file 15.
32 LEECH, W. T.: memorandum, 24th June 1929, Minute 4004/33, file 15.
33 Controller (BBC): letter to Baird Television Development Company, 26th May 1929, BBC file T16/42.
34 BARTLETT, T. W.: letter to the Controller (BBC), 9th July 1929, BBC file T16/42.
35 Anon.: 'BBC holding up television. Foreign countries lead the way with British invention', *People*, 28th July 1929.
36 Anon.: 'Television and the BBC', *The Electrician*, 26th July 1929.
37 MALONE, C.: question in the House to the Postmaster General, July 1929.
38 AMPTHILL, Lord and MANVILLE, Sir E.: letter to the Postmaster General, 9th July 1929, Minute 4004/33, file 16.
39 HUTCHINSON, O. G.: letter to H. B. Lees-Smith, 12th July 1929, Minute 4004/33, file 16.
40 MURRAY, Sir J. F. W.: memorandum to the Postmaster General, 15th July 1929, Minute 4004/33, file 16.
41 REITH, Sir J. F. W: letter to Sir E. Murray, 12th August 1929, Minute 4004/33, file 16.
42 LEECH, W. T.: letter to O. G. Hutchinson, 14th August 1929, Minute 4004/33, file 16.
43 HUTCHINSON, O. G.: letter to the Postmaster General, 4th September 1929, Minute 4004/33, file 16.
44 LEECH, W. T.: letter to O. G. Hutchinson, 5th September 1929, Minute 4004/33, file 16.
45 PHILLIPS, F. W.: letter to O. G. Hutchinson, 24th September 1929, Minute 4004/33, file 16.
46 HUTCHINSON, O. G.: letter to F. W. Phillips, 25th September 1929, Minute 4004/33, file 16.
47 CARPENDALE, Adimiral Sir C.: letter to Baird Television Ltd., 11th September 1929, BBC file T16/42

Chapter 7

The low definition experimental service, 1929–1931

On 30th September 1929 the BBC transmitted its first experimental television broadcast. 'A great day for Baird and all of us', wrote Moseley.[1] The honour of opening the proceedings at 11.04 a.m. fell to Moseley who announced: 'Ladies and Gentlemen: You are about to witness the first official test of television in this country from the studio of the Baird Television Development Company and transmitted from 2LO, the London Station of the British Broadcasting Corporation'. Then followed a number of short messages from the Rt. Hon. William Graham, PC, MA, LLD, MP, Sir Ambrose Fleming and Professor E. N. da C. Andrade. Graham took as his theme one which had been raised on a number of occasions previously by Hutchinson – the establishment of a new industry which would provide employment for large numbers of our people, and would prove the prestige of British creative energy. This point was further mentioned by Fleming who acknowledged that the creation of the new industry owed so much to the genius of Mr Baird. Andrade compared the occasion to that

> 'on which the records of the early phonograph were publicly tried. The voices that then issued from the horn were not of the clarity which we now expect, and the faces that you will see today, by Mr Baird's ingenious aid, are pioneer faces, which will no doubt be surpassed in beauty and sharpness of outline as the technique of television is developed. One face, however, is as good as another for the purpose of today's demonstration and I offer mine for public experiment in this first television broadcast.'[2]

The second half of the proceedings consisted of very short solo performances:

11.16 a.m. Sydney Howard: televised for two minutes
11.18 a.m. Sydney Howard: gave a comedy monologue
11.20 a.m. Miss Lulu Stanley: televised for two minutes
11.22 a.m. Miss Lulu Stanley: sang 'He's tall, and dark, and handsome' and 'Grandma's proverbs'
11.24 a.m. Miss C. King: televised for two minutes
11.26 a.m. Miss C. King: sang 'Mighty like a rose'

Only one transmitter had been allotted for the transmission and so each artist had to be televised for two minutes and then repeat the act before a microphone.

The programme was not without its difficulties – the most serious of which was the reproduction of a negative image instead of a positive image. However this was rapidly corrected and fairly clear images received, said *The Times*.[3] The *Daily Herald* thought that Sydney Howard and a woman artist were 'quite recognisable – looking like the earliest photographs in the daily paper' but 'the image jerked up and down like a film when the operator is having serious trouble with his machine'.[4]

Baird told a *Manchester Guardian* reporter that 'he was satisfied with the demonstration but he hoped to obtain much better results as the experiments continued. He pointed out that there had been very little time indeed for tuning-in a most important operation'.[5] *Amateur Wireless* also referred to the hunting but mentioned that the results were of good quality.

> One sees the image through a wide lens about eight inches in diameter and the general effect is similar to that of looking into an automatic picture-machine as installed in amusement halls. The image appears as a 'soft tone' photograph illuminated by a reddish-orange light.

The general impression, the writer stated, was that the present televisor had reached the stage of development of the early flickering cinematograph. He thought there was

> much, very much, yet to be done, but the present stage is highly creditable and the fact that public broadcasts are now being given will undoubtedly hasten progress.[6]

Unfortunately, very few people witnessed this historic event. Asked by an *Evening News* reporter immediately after the broadcast how many people he thought had been able to receive the transmissions, Baird himself put the total at under 30.

> There is one receiving set at my home at Box Hill, and I believe the BBC and the Post Office each have one. That makes three and I should say there are half a dozen other sets in the country. Add to these the receivers which clever amateurs have built for themselves from our directions and you might count another twenty. That makes twenty-nine in all.[1]

Shortly after the inauguration of the first public experimental service the Fultograph transmissions were withdrawn and Bartlett seized the opportunity to write to Reith to enquire whether Bairds could have the periods released by their suspension.[7] He told the Director General that amateurs found the present television transmission times unsuitable and that manufacturers looked upon an extension of the present facilities as being of vital importance to this new industry. The BBC considered this request at a Control Board meeting the following day and agreed to let the company have some extra time on two nights a week after programme hours.[8] They had to wait for a definite assurance on this point, however, as the Director General wanted to be supplied with 'any definite particulars indicative of public interest generally, or manufacturing interest in particular, as you will agree that this must have some bearing on the matter' before agreeing to implement Bartlett's request.[9] The latter replied that at the last Radio Exhibition Bairds had received 'about 35 applications from manufacturers for licences, including some of the leading concerns'.[10] Referring to the public interest, the Secretary wrote that until the second circuit was put at the disposal of the company so that both sound and vision could be transmitted simultaneously, the public was not in a position to judge on the entertainment value of the system. He hoped that the BBC would soon be able to provide this facility. Ashbridge thought this latest request might tend to mask the issue as to whether television in its present state of development has any practical application to the broadcast service.[11] Apart from that objection he felt the BBC ought to allow the company a certain number of tests using both sound and vision.

Reith replied to Bartlett's letter of 22nd November five days later and indicated that while he considered it a vague reply mentioned that the company could have the extra two nights a week transmission time but that if and when they got a second frequency care would have to be taken to obviate undue concentration on the sound.[12,13]

The BBC could have been more generous, for the weekly programmes of the experimental transmissions of still pictures totalled 2 h 20 m.

The Television Company's request was entirely reasonable as it was at a disadvantage with respect to the public generally in view of the fact that the company had no convenient transmission times. Whatever the Corporation's reasons for limiting the extra time allocation to Bairds (Reith's letter of 13th November 1929 to the Secretary of the GPO on the extension of time to Bairds did not give any explanation), nevertheless the BBC considered it might be good policy to grant some additional viewing facilities for the public after normal working hours.

This rather half-hearted, unenthusiastic approach by the higher powers in the BBC – Reith, Carpendale, Ashbridge and Murray – towards Baird television was to continue for a number of years, including the period when the Corporation took over reponsibility for programme production. There seems to be little doubt that the fortunes of the Baird Company would have been altogether different had the BBC in 1928 given some encouragement and expertise to the new broadcasting system on the lines of the assistance which the Post Office gave to Marconi shortly after he arrived in this country.

The additional evenings subsequently chosen were Tuesday and Friday,[14] although the company would have preferred Wednesday and Saturday, and the first of the new transmissions was sent out on Friday, 3rd January 1930. An essential requirement which the BBC imposed was that all programme material had to be censored and to enable this to be done the BBC required two weeks notice. Bairds offered to go further but the BBC did not want this as it was anxious to avoid the inclusion of their programmes in the *Radio Times*.

The next big step forward was the granting by the BBC of an additional wavelength so that sound and vision could be synchronised. This demanded the use of both the Brookmans Park transmitters, one operating on 261 m for vision and the other on 356 m for sound:[15] the first combined transmission was sent out on 31st March 1930.[16]

For the wireless correspondent of the *Evening Standard* the reception of the transmission was remarkable: 'It was, so to speak', he wrote, 'a "talkie" by wireless, but a "talkie" that consisted of close-ups'.[17]

The programme, which started at 11.00 a.m., consisted of introductory remarks by Sidney A. Moseley followed by short speeches by Lord Ampthill and that staunch supporter of John Baird, Sir Ambrose Fleming. Songs by Annie Croft and Gracie Fields then followed. R. C. Sherriff, who was to have spoken, was unavoidably prevented from performing.[18]

The Times commented on the remarkable detail of the images transmitted considering the relative simplicty of the receiving apparatus. This apparatus, the televisor receiving set, was on the market at 25 guineas although it was possible for amateurs to buy a kit of parts for home construction from a number of London stores for 16 guineas, They were designed to work in conjunction with a wireless receiver capable of delivering a good quality output of the order of $1\frac{1}{2}$ watts.

Head and shoulders only were shown on this occasion but it was Baird's eventual aim to be allowed to introduce a portable television transmitter into the BBC's studios so that no limitation would be imposed on artists movements and so that it would not be necessary to make use of the dentist's chair arrangement.[19]

Following the inauguration of the new service, Bairds installed a television set at Number 10 Downing Street. Ramsay MacDonald wrote back in glowing terms:

> I must thank you very warmly for the television instrument you have put into Downing Street. What a marvellous discovery you have made! When I look at the transmission I feel that the most wonderful miracle is being done under my eye. I congratulate you most heartily and send you my sincere hopes for your future success. You have put something into my room which will never let me forget how strange is the world – and how unknown.[20]

With praise like that it seemed Bairds could never look back.

The first television play was transmitted on 14th July 1930 and was produced by Lance Sieveking and Sidney Moseley.[1] It was Pirandello's 'The man with a flower in his mouth' and was chosen because it had three characters only. The cast was Gladys Young, Earle Gray and Lionel Millard (Val Gielgud was to have participated but fell ill at the last moment). 'They came to Long Acre and were made up in yellow, with navy blue shading around the eyes and nose', wrote Margaret Baird, 'these colours on the face improving the picture'.[20] Again head and shoulders were all that could be seen and not much of C. R. W.

Nevinson's scenery could be observed. Only one face at a time was shown and between each sequence a checked curtain was drawn across the screen, incidental music filling in the pauses (Fig. 7.1).

Fig. 7.1 *The first television play produced in the UK was 'The man with a flower in his mouth' by Pirandello. S. A. Moseley and L. Sieveking were the producers. Left to right: L. Sieveking, C. D. Freeman, M. Eversley and L. Millard*

'Allowing for such things as the televisor going out of synchronism every now and then and the poor quality of the sound transmission, it must be recorded that the broadcast as a scientific achievement was a success', wrote *The Times*.[21] 'As a play – well, it left a certain amount to the imagination, but it was a good entertainment and certainly an advance on the mere reception by sound', it continued.

The *Daily Mail* found the pauses tended to minimise the dramatic interest of the performance 'and constitute one of the problems which it is hoped to solve when bigger receiving sets are cheaper. But it was certainly startling, as well as helpful to the dialogue, to be able to see their every expression – even to the lifting of the eyebrows. We even saw the gestures of their hands – although we had to sacrifice their faces for the time being'.[22]

A select audience in a canvas 'theatre' on the roof at Long Acre also saw the play on Baird's big screen. Senatore Marconi, who had been invited by Colonel Winch, a Director of Baird Television, was a member of the select audience but unfortunately his views on the play were not recorded. Baird, however, was much impressed by Marconi's 'aloof politeness and almost regal manner' and described him in words which could equally have been applied to himself:

> Although the invention of no single device of fundamental importance can be attributed to Marconi, it was he who ventured forth like Christopher Columbus and forced upon the attention of the world the existence of a new means of communication'.[20]

All was not well though. Notwithstanding the fact that true television had been demonstrated four years previously and that Baird himself had given hundreds of demonstrations, the quality of reproduction left much to be desired. The characters of the play for example 'moved and had their being in a heavy and persistent shower of rain'.[1]

Bridgwater has described Baird's aims at this time:

> Baird was hastening to anticipate every possible application of television. He wanted to be the first all along. In those early years he did not seem in the least worried about the difficulties of improving the quality of his crude flickering images. Promise and potential came before polish and perfection: improvements could follow later. Enthusiastically, perhaps almost naïvely, he felt his machines were sufficiently 'marvellous', to use an epithet already bestowed by a fellow of the Royal Society, to be commercially exploitable'.[23]

Technical improvements were certainly required. Val Gielgud, who found the experience a most interesting one, wrote to Moseley two days after the transmission to suggest a meeting, at which the results could be frankly discussed 'because, as you hint, a certain number of inevitable difficulties in the past experiment not only could be, but definitely should be, eliminated in the event of further developments'.

Lance Sieveking, the co-producer of the play, who was 'exceptionally imaginative', made a report for the BBC after the second experimental rehearsal on 24th June. He described how Gielgud, Baird, Moseley and himself conducted exhaustive experiments with all sorts of make-up:

> As I suspected would prove to be true an approximation between normal flesh tints and dead white is most suitable for a general face background, the raised parts of the face being brought out with lines of pure black . . . The other conclusion concerns the fading board. At present it is painted a dark brown. Every time the fading board came down, when it was removed the picture became desynchronised and started to whirl about, the reason for this being that while the fading board alone was

> before the transmitter it was giving the photo-cells nothing to occupy their energy. To obviate this, I am having a broad white line painted all around the board. This will ensure that the synchronisation of the transmitter continues uninterrupted. There were a certain number of minor points of interest, one only of which is of sufficient importance to mention here. That is, that a very satisfying effect of perspective is obtained in a picture in which the back of the nearest speaker's head is seen, while beyond it, smaller, the face of his *vis-à-vis*.[1]

The above relatively minor points are mentioned here to illustrate the fact that improvements could be made not only to the technical quality of the transmission but also to the programme quality. Both Val Gielgud and Lance Sieveking were quick to realise the latent possibilities of television in the earliest stages. As Birkinshaw and Robb[24] were to show a few years later when the BBC changed its attitude to a more wholehearted co-operation with Bairds, much could be done to improve 30-line television. However, the BBC was neither excited nor enthusiastic about Baird's television in 1930. They had been forced to accept the position as it was in that year only after much pressure had been put onto them and then only after some prevarication. There were, however, people inside the BBC who wished to co-operate on the programme side, who could see that improvements could be made and who were excited about the possiblities television offered for broadcasting. Such persons were Sieveking and Gielgud.

Sieveking, in particular, gave the matter very considerable thought and produced, in July, a long memorandum[25] headed: 'Television – Final Report on Present Situation', which he forwarded to Mr Goldsmith for the attention of Reith. So convinced was Sieveking for the need to put his point of view that he was prepared to risk a rebuke or a snub should there be factors in the problem of which he knew nothing, and he fully realised that he might be plunging in where more wisely he might fear to tread. Sieveking listed eleven points in his report and after mentioning the recent production of Pirandello's play observed (point four)

> The collaboration proved that given the necessary inventiveness and enthusiasm the medium of radiovision could be exploited to a very much greater degree than had hitherto been thought possible. Mr Moseley and his assistant who are responsible for making up the Baird programmes seem to have been content to do the same thing over and over again every day since they began to transmit on our wavelengths. During the period of our collaboration research was carried out in the greatest possible exploitation of the Baird method of television at its present state of development. The ground work of a technique was evolved and conclusions were come to on such matters as fading from one scene to another, the use of make-up, music effects and the handling of composite programmes generally. The result of a use of this technique was appre-

> ciated by those whose interest prompted them to be spectators. Since the transmission, the Baird people seem to have returned quite happily to their old methods, which perhaps suggests a mental attitude to be deprecated.

Sieveking thought the policy for the BBC was clear: 'Namely the furtherance of radiovision whether by Baird, by ourselves or by anyone else'. He was pressing for the necessity of regarding radiovision in the abstract, quite apart from persons and from its previous history, and was keen that the Corporation should later, looking back, be seen to have acted wisely, energetically and rightly in relation to all the problems which from time to time faced it.

For him the prospect for the future was readily apparent – assuming that matters were left exactly as they were:

> They (Bairds), will continue to spend until all is spent, and then there will be nothing left except the income from the patents. A very negligible amount. It then becomes a political and a business question as to who will take any steps to acquire control or reinstate research into the possibilities of radiovision. The regrettable picture rises before us of the BBC having to hire from the German Government the right to transmit radiovision programmes with the apparatus invented and developed in German laboratories. Though in your judgement, Sir, this possibility may seem merely ridiculous, I would urge that though this hypothetical possibility would be much to be deplored, any other eventuality except the one which I am proposing would be in a certain degree almost as regrettable.

Sieveking considered that two main factors had militated against closer co-operation between Bairds and the BBC: first, the share market operations of certain individuals, and second, the opinion that a new invention was required before radiovision could be successful (these were Eckersley's and Swinton's views). However, notwithstanding these points Sieveking thought the BBC was the obvious body to see that in some form radiovision should be perfected. He referred to the position concerning 'talking films' and how the entire control of the talking film throughout the world was in the hands of a few internationals apparently unhampered by any moral or aesthetic scruples of any kind, who had met a week ago in order to divide the countries between them. The implication of this was that in regard to radiovision it could not be too early to act and to act strongly.

Sieveking reinforced his views by referring to some conversations he had had with different people, the most important of which was with Kirk and Howe. He appended to his report a brief summary of those talks.

Both Kirk and Howe were inclined to agree with Baird's view that if a sufficiently wide waveband was allotted to television he could produce a screen with a much larger area of vision, capable of showing a considerable number of people at once, and with a very much better definition than had been shown. They were inclined to the view that a new discovery might be necessary before perfection could be achieved but without all the facts before them they could not commit themselves. Moreover, both Kirk and Howe felt that if they had had charge of research into television a year ago, probably greater developments would have been seen.

Sieveking, along with Kirk and Howe, therefore suggested to the Director General that research should be instituted under Kirk in the near future. He 'visualised something in the nature of two or three first class men together with whatever plant was thought necessary. These men to be considered as concentrating solely on the problem of radiovision for five years, with a budgeted expenditure of something in the neighbourhood of five thousand a year'.

There was a feeling among the three that there was a slight suspicion of complacency in Bairds laboratories and that the laboratory staff were possibly not exactly working at high pressure against time. They considered that some strategic step should be taken to enable the Corporation to goad Bairds into the fullest exertions of which they were capable.

Finally Sieveking suggested that the whole question of radiovision be looked into afresh and that the technicians under the Chief Engineer should thoroughly investigate the laboratory workings at Bairds.

Noel Ashbridge, who was now the BBC's Chief Engineer, gave his views[26] on Sieveking's report in a memorandum dated 5th September 1930. Ashbridge considered there were four points to stress. First, that although television had got very little nearer to a commercial standard during the past two years, nevertheless it was highly probable that such a standard would be achieved during the next ten years and that therefore the BBC ought to remain in 'close touch' with any research being carried out on the subject. Second, a successful television broadcast service was likely to begin on shorter waves and consequently the BBC should keep in touch with developments likely to lead to a suitable service. Third, it would not be justifiable . . . to spend vast sums of money acquiring the rights to any one existing system at the present time. Ashbridge suggested the BBC should wait until the best system had established itself, although then it might be faced with the prospect of paying heavy licence fees to someone in the same way as they do now for the Marconi patents. Also Ashbridge said it would seem to be a wrong thing to spend a large amount of money in developing a service from which a very large majority of the public would not receive any benefit owing to the cost of the receiving apparatus. Fourth, the policy for the future should be to keep close liaison with the Baird system and any other system which came along on the technical side only. He did not think that television was sufficiently developed to admit of serious co-operation on the programme side although he understood Sieveking's interest. The most that Ashbridge thought the BBC should concede was the rendering of technical assistance where possible in connection with the existing experimental transmissions.

Both Sieveking's and Ashbridge's memoranda were read at a Control Board meeting[27] held on 9th September, and Ashbridge's views prevailed; namely that the Baird television transmission should continue for the present, that there should be no programme co-operation with Baird and that the technical liaison should be closer. Naturally Sieveking was unhappy with the Chief Engineer's views and further expanded his own in another memorandum to Gielgud.[28] He stressed the importance intuition had played in his findings but mentioned 'the direction of commerce on a large scale, and politics is governed to a greater degree by instinct and intuition than by logic'. Gielgud sympathised with Sieveking's keenness and his belief in the power of his intuition in the matter but thought that once it 'reaches the sphere of economic and general policy it must naturally pass from our hands to Control Board and finally the Director General'. Gielgud felt the policy of encouraging closer liaison on the engineering side, but not on the programme side, was inconsistent and thought it was desirable that the programme branch should not be entirely caught napping when the time arrived when television would be an actuality.

All of this was, of course, unknown at the time to Moseley, who during this period had been to Berlin.[29] When he came back he wrote to Murray and referred to the great amount of interest manifested by the German Post Office and the authorities in Germany generally and suggested that the BBC should give the television transmissions free of cost, and additionally give an extension or alteration of the times of transmissions, so as to give a greater number the opportunity of looking in. In support of this request he mentioned that the German Post Office and Rundfunk transmitted television free of cost. 'Their idea', he said, 'is to develop television so that they shall lead the world'.

Murray was not impressed[30] and requested a meeting with Moseley to discuss whether and to what extent the experimental broadcasting of television by the Baird process was to be continued by the BBC next year. For Moseley the hint of a possible discontinuation of the experimental broadcasts must have sounded slightly ominous and Murray's tone seemed to be in conflict with Control Board's minute of the 9th September which referred to closer technical co-operation and the continuation of the Baird television transmissions. The meeting, however, which had been originally arranged for the 17th November[31] was postponed *sine die* – owing to developments.[32]

The German authorities' approach to television was quite different to that adopted by Reith and his colleagues. In Germany the transatlantic transmission of 1928 had aroused intense interest, and following this Dr Bredow, Managing Director of the German Broadcasting Corporation, had come to England accompanied by his two chief technical experts, Dr Banneitz and Dr Reisser.[20] (It is interesting to compare this situation with the fact that Reith turned down a number of requests to see Baird's apparatus although only a relatively short journey was involved). According to Baird's notes, Dr Bredow, after seeing the laboratory experiments, invited the company to send representatives over to Germany and install a transmitter in the Berlin broadcasting station.[20]

> I found myself arriving at the Adlon Hotel in Berlin, complete with Hutchinson, a great load of apparatus, a team of technicians and last, but dominating the whole picture, Sydney A. Moseley. Hutchinson had done

> the preparatory work and done it well. Our apparatus was erected at German Broadcasting House, the Reichsrundfunk, and a meeting had been arranged, to be presided over by the State Secretary of Posts and Telegraphs and to include Dr Bredow, the Postmaster General of the Reich, Dr Growirow, and all those interested in television. [Baird]

After this and other meetings a company was formed to develop television in Germany, under the auspices of the General Post Office. The formation of this company was referred to by Sir Edward Manville at an Extraordinary General Meeting of the Baird Television and Development Company in April 1930.[33] The firms involved were the Bosch Magneto Company, the Zeiss-Ikon Company and the Loewe Radio Company. The Baird Company was to supply the television, the Zeiss Company to supply the television parts, the Loewe Company to supply the wireless parts and the Bosch Company to supply the electric motors then used. Dr Banneitz, the Chief Technician of the German Post Office, was appointed consultant. This company was duly incorporated, the four concerns (including the Baird Company) having equal shares, and Fernseh AG came into being. 'For some time', continued Baird, 'we kept a number of engineers permanently in Berlin. I spent a lot of time there, and these visits were very happy until Hitler stepped in'.[20]

The Chairman of the Baird Television and Development Company, Sir Edward Manville, believed that with the help of the German Companies, not only would progress be made in the practical applications of television in Germany, but also that his company would receive technical assistance from them in the general development of the system by virtue of their agreement for a reciprocal exchange of patents and improvements.[33]

Later in the year, Lord Ampthill, Chairman of Baird Television Ltd., told the Annual General Meeting that their company Fernseh AG had supplied a spotlight television transmitter to the German Post Office and that the company was making rapid and substantial progress in development work.[34]

Previous to the establishment of the new company, Fernseh AG, Bairds in 1929 had rented ground at Couldsdon in order to set up large aerials to receive television signals from the continent.[20] These were first received in July 1929 and the fact is recorded on a memorial stone, given by the Wembley County Council, which was unveiled on 30th July 1953 by David Gammans, MP, Assistant Postmaster General (the stone was erected under the auspices of the Wembley Historical Society).[20] (Fig. 7.2)

During the financial and political perambulations of the companies which Hutchinson and Moseley either engineered or dealt with, Baird was devising new ways of using television. In July 1930 he introduced a large screen receiver to the public.[35]

The first demonstration of this took place on the roof of the Baird company in Long Acre at 10.00 a.m. on 1st July 1930 to a number of press representatives. The size of the screen was 5 feet by 2 feet and consisted of 2100 very small metal filament lamps – each lamp being placed in a cubicle so that the screen had the appearance of a large honeycomb.

The front of the screen was covered by a ground glass sheet – presumably to soften the harsh light intensity of the lamps. An alternative earlier proposal,

Fig. 7.2 *The Baird Company's television transmitting station at Kingsbury Manor, (1929)*

described in patent 326,192,[36] made use of a lens to project an image of the lamp array on to a viewing screen. The advantage of this arrangement according to Baird (1929) was that the quality of the viewed image was to a certain degree under control, in that, for instance, a 'soft image' could be obtained by throwing the projected image slightly out of focus. He found that a projected image gave a more pleasing effect than one obtained by viewing the lamps directly, Also the image of the array could be made larger than the array itself by means of an optical system – although the coarseness of reproduction necessarily would be more obvious. An added disadvantage would have resulted from the need for a stage having a depth of approximately 20 feet to accommodate the projection equipment.

Each lamp was connected to a separate bar of a commutator which switched on one lamp at a time – the various lamps being excited in succession. As the speed of the selector arm was 750 rev/min each lamp was energised every one-twelfth of a second.[2]

At the press demonstration again only head and shoulder views were projected, the performers being Miss Lulu Stanley, Miss Pearl Greene and Mr Ben Lawes. For the *Daily Mail*[37] the demonstration was interesting, while the *Daily Worker*[38] found it was remarkable.

The first public demonstration of Baird's latest invention occurred at the London Coliseum, where Sir Oswald Stoll had arranged with Bairds to allocate part of each performance, for a period of two weeks, to show the transmission of television to a large audience (Fig. 7.3). Moseley, as the Director of Baird Television Programmes, was responsible for the arrangements, and, anxious for success, booked many celebrities for the two week period.[20] Sir Francis Goodenough, Lt. Commander Kenworthy, Robert Young, MP, Lord Baden Powell, Rt. Hon. George Lansbury, MP, Lord Marley, Frederick Montague (Under-Secretary of State for Air), Col. L'Estrange Malone, MP, Young Stribbling, Sir Nigel Playfair, H. W. Austin (tennis champion), Herbert Morrison (Minister of Transport), Miss Isabel MacDonald, Miss Ellen Wilkinson, MP, Sir Oswald Mosley, MP, Bombardier Billy Wells, A. V. Alexander, MP, The Rt. Hon. Lord Mayor of London, and Miss Irene Vanbrugh were all televised during the two week period.[2] The list is interesting as it perhaps indicates some of the influential people who were prepared to support Bairds – including a number of Members of Parliament.

Public interest in the demonstrations was certainly great and on many occasions the house full notice had to be displayed. Bairds made a profit of £1500.[20]

The *Sphere*[39] described the performance as follows:

> There is no doubt as to the thrill which one receives on seeing for the first time a television demonstration of this kind. When the television number was announced, the theatre was plunged into darkness. Then, in a setting of black curtains, we suddenly saw the screen upon which the features of various people, taking their place before the transmitter in Long Acre, were shortly to appear before us. A demonstrator, telephone in hand, informed us of the nature of the invention. Then a human face – Sydney Moseley's – appeared upon the screen and began to speak to us. The image wavered up and down, but gradually steadied . . . There was a kind of rustling effect all over the screen, but through it one could distinctly make out the features of one well known personage after another. One could not only hear them speak, but see their lips moving. The face on the screen turned from right to left. 'Would any member of the audience like to ask a question of the speaker?' said the demonstrator. 'Tell him to put his hand up' cried a voice from the darkness of the auditorium. The request was

telephoned to Long Acre and immediately the speaker put his hand up to his chin in reponse to it. The movement was perfectly visible to every part of the house, and in spite of some surface defects, the screen rendered all the various tones of the face before us.

THE WORLD'S FIRST
PUBLIC PERFORMANCE OF
TELEVISION IN A THEATRE

BAIRD
TELEVISION
at the
LONDON
COLISEUM
commencing
JULY 28th, 1930

LIVING CELEBRITIES AND ARTISTES
TELEVISED THREE TIMES DAILY
BY THIS MARVELLOUS INVENTION

Fig. 7.3 *Poster advertising the first public performance of television in a theatre in the UK*

There were some reservations however. The *Daily Express*,[40] while praising the enterprise, commented: ‘Television obviously has a long way to go, but it was proved last night that its young Scottish inventor has something which will soon arrest the attention of the world’.

Following the success of the screen at the Coliseum, the television equipment was taken first to the Scala Theatre, Berlin, where performances were given from 18th September to 30th September, then to the Olympia Cinema, Paris and finally to the Roda Kvarn Cinema in Stockholm. According to Moseley the demonstrations were a great success and amazing enthusiasm prevailed.[2]

Interest in television was growing rapidly. Mihaly, Fernseh AG and Telefunken (under the direction of Dr Karolus) in Germany; the American Telephone and Telegraph Company, the Radio Corporation of America, the General Electric Company, the work of Jenkins and of Farnsworth in America and other companies in France and elsewhere were stimulating people with their demonstrations of television and the opportunities which could follow. Bairds were finding it increasingly difficult to be first in the race to show something new. While Baird's large screen system was based on a patent dated 26th July 1923,[41] nevertheless the first demonstration of television to a large audience was given by AT & T in 1927.[42] The first play broadcast by television was produced in America by the National Broadcasting Company (NBC) in April 1930,[43] three months before the production of Pirandello's 'The man with a flower in his mouth'. Also, television companies abroad either were not hampered by a hesitant broadcasting monopoly or were encouraged in their activities.

However, while Baird could plead with some justification that the BBC had not been excited by his particular demonstrations until the early part of 1930, the general public, whatever its views on television, could not have purchased a Baird television receiver for its own use. 'At last', said *Wireless World* on the 12th March 1930,[44] 'a Baird receiver built for sale to the public has arrived'. Hutchinson's remark of 1926 had come true. 'High-class workmanship, but with a none too pleasing external appearance, owing to the use of light metal cabinet and poorly devised controls, are one's first observations on acquaintance with the instrument', said the writer, but it did 'give reception of images with sufficient definition to be readily intelligible', he conceded.

The receiver (see Figs. 7.4, 7.5 and 7.6) consisted essentially of a thin, 20 inch diameter aluminium scanning disc having 30 apertures, the majority of which were square (about six near the ends were rectangular), an electirc motor which carried the disc and synchronising mechanism, a neon tube and a voltage regulating resistance. A width/length ratio of the image area of 1:2.5 was produced by the spiral design so that the resulting picture was suited to accommodate the head and shoulders image of a person speaking. The universal motor could be operated from either an a.c. or d.c. supply and synchronising was effected by means of a toothed wheel running between the poles of an electromagnet. According to the company, a power output, from the last stage of an amplifier, of 1.5 W was sufficient to actuate both the neon lamp and the synchronising gear (which were in series), but the *Wireless World* reviewer found that the process of synchronising became easier if the L55A output valve used in the amplifier was changed to a L56A as this had a rated output of 5 W. 'When once correctly set up and little practice gained in the operation of the speed regulating control, reception became reliable', he said. 'For quite long intervals the picture remained steady though in the case of head and shoulders images the lighting effect was far from perfect, and it was not possible to glean the significance of the movements, though if accompanied by speech the effect might have been different'.

Fig. 7.4 *Internal view of the Baird 'Televisor', back view*

The television receiving set was produced to give am image only and not sound. It was assumed that the purchaser of the set already possessed a suitable receiver and aerial installation for the reception of both the vision and sound signals.

A consequence of the above arrangement of having separate vision and sound reproducers, when both visual and aural information were transmitted, was the lack of coincidence of the two stimuli affecting the viewers senses, thereby causing an unnaturalness in the reproduced scene.

Baird gave some thought to this problem and in patent no. 318,278[45] suggested a means whereby the above defect could be overcome. His idea was to use the receiver's Nipkow scanning disc as an element of a loudspeaker in addition to its function as an image synthesiser. The disc was to be capable of moving both rotationally and translationary according to the motions of the shaft of the electric motor and the armatures of the electromagnets, respectively. In this way vibrations would be established in the disc according to the speech signals fed to the electromagnets and sound waves would be radiated from it.

There is no evidence to show that Baird constructed a model or prototype of this reproducer: *prima facie* the difficulties of design to give a reasonably flat and smooth frequency response over the audio frequency range would seem to be formidable. In any case, coincident sound and vision fields can be obtained (for suitably placed observers), by employing a pair of loudspeakers symmetrically situated about the viewed scene.

Fig. 7.5 *Internal view of the Baird 'Televisor', front view*

The first dual transmission – of aural and visual signals simultaneously – took place on 31st March 1930.[46] Prior to this date the test transmission announcements were therefore given in the form of wording running across the aperture. According to the author of the aforementioned review 'the capital letters forming the words were clearly defined and easy to read, while a clock face could be read to the nearest half minute'.

Finally the reviewer said: 'Such results will interest the enthusiast, and these have become possible since the adoption of the signal controlled toothed-wheel method of synchronising first introduced towards the end of last year'. It was thought that the price of the set would be about £20.

The introduction of the television sets had not been without some delay – indeed, some mystery. Bairds had announced towards the end of 1929 that their receivers would be on the market by 1st February, but when a reporter from the *Birmingham Daily Mail* asked the company why they had not been able to follow their plans he was told that delivery was being held up by the difficulties and expense attendant on procuring the right kind of glass for the lenses.[47] It seems that when tenders were sought for the supply of lenses required in the sets

Fig. 7.6 *Plan view of the Baird 'Televisor'*

the lowest British quotation was £500 for 1000 6 inch lenses, and £200 for the same number of 4 inch lenses, One of each size was required in the construction of the set (the focal lengths of the larger and smaller lenses were 18 inches and 12 inches, respectively).

On the other hand, a French quotation, which was accepted, mentioned £87.10d for 1000 of the 4 inch lenses; and these could be supplied at that price with the 50% import duty included and with a shorter delivery time than the three months given by British manufacturers. The explanation offered for the long delivery time was that the special kind of sand needed for the grinding of the glass used in the lenses was not obtainable but was plentifiul in parts of France.[48]

These comments immediately led to an exchange of correspondence in the columns of the *Birmingham Post* and elsewhere and a British Optical Lens Company representative wrote and stated that his company had quoted £225 and £75, respectively, for 1000 each of the 6 inch and 4 inch lenses and that the promised delivery was six weeks.[49] Chance Brothers mentioned that sand was not used in grinding lenses and that the British optical industry was particularly busy making lenses for cinematograph and talkie apparatus for the USA in very large quantities. 'It seems a pity that the tendency to run down British products should be so widespread', the spokesman for the company said.[50] 'The general effect of the paragraph in Saturday's *Post* is to give the impression that British lens manufacturers are out of date', wrote another correspondent.[51] 'There is much that is mysterious . . .' commented the *Engineer*, but unfortunately Bairds were not able to clarify the position.[52]

There was another mystery in 1930 involving Bairds: 'The Olympia television mystery'. This arose following an interview between the official organiser of the Radio Exhibition and the Baird Television Company for space for the company at Olympia. Although agreement had been reached on the amount of floor area required and the price to be paid for it, when the organiser put the matter before the Exhibition Board for ratification it was rejected without any explanation and no further amended arrangement invited. The Board, however, offered the company a tiny space in which to exhibit their literature and an official of the Radio Manufacturers Association later said: 'In the view of the Exhibition Committee demonstrations of television, as well as of wireless sets of individual manufacturers, are technically impracticable. For this reason we banned both kinds of demonstration.[53]

Naturally Bairds were quick to point out the position abroad and were supported by the wireless correspondent of the *Daily Mail* who was in Berlin at that time.[54] This cold-shouldering of a British invention seems in strange contrast to what I have seen here' he reported:

> The organisers of this exhibition have given generous facilities to television. Here television has its own hall. Affixed to the entrance is the legend: 'Reichspost Fernseh'. The German Government has thus far placed the seal of its approval on the exhibition by allowing it to have a postal department. Help has been given by the Reichspost, and British and German television authorities, working together, have made the demonstrations successful. Many thousands of visitors have crowded to see this practical demonstration of a new wonder of an age of wonders.

Baird woefully summed up the position: 'The system is receiving every support abroad while in the land of its birth it is receiving every discouragement'.

Towards the end of 1930, Gladstone Murray had hinted to Moseley that there might be a prospect that the television experimental broadcast would not continue.[55] If this had happened the future for sales for televisors would have been bleak indeed. Bairds were concerned and a few days later Lord Ampthill met Lord Gainford, the Vice-Chairman of the Board of Governors of the BBC, at Brook's Club and had a half hour's chat.[56]

Ampthill argued that if more facilities for television broadcasts were given by the BBC more televisors would be purchased, and pointed to the attitude which had been adopted in America where the process was being warmly taken up. Gainford replied that the Corporation's first duty was to its 3 200 000 listeners, paying 10 shillings to listen to programmes, and for the BBC to deprive these people of one alternative programme for longer intervals than was now conceded was unreasonable. The position seemed one of deadlock. There were only about a thousand television sets in use and the BBC was not prepared to give up more time from that which was available to its huge listening audience for the sake of such a small minority. On the other hand the television broadcasts consisted of a half-hour transmission from 11.00 to 11.30 a.m. each morning from Monday to Fridays inclusive and from midnight to 12.30 a.m. on two nights each week and these were unsuitable viewing times for most people, occurring as they did during normal working hours or normal sleeping hours. For Bairds the sale of televisors represented the only reasonable way to recover some of the very heavy expenditure of the past and the heavy expenditure still to be faced, but if the company was denied satisfactory viewing times the sale of televisors would not increase.

Ampthill[57] later pointed out the generous facilities granted in Germany where television was broadcast from 9.00 to 10.00 a.m. on four days each week, between 1.00 p.m. to 1.30 p.m. on five days each week and an hour at midnight every night except Saturday and Sundays, all without a charge being made, whereas Bairds had to pay £2000 per annum.

He also appealed to Gainford for consideration on national and imperial grounds – a tactic which Bairds used on many occasions and one which they considered of great importance. 'The inventions we are developing are purely British and will if successfully developed provide increased trade and work in this country and anything which will help to do this should receive special consideration at the present juncture'. Bairds were alarmed at the interest being taken in America by companies with enormous resources of capital at their disposal, companies such as the Radio Corporation of America which were obtaining a stranglehold in the United States and which would without doubt be striving to obtain a similar position in this and other countries. Bairds' prognostications for the future of British television development were gloomy – unless the BBC granted better facilities. The company, argued Ampthill, was countering this by its activities in Germany and France – and elsewhere.

Another point to be borne in mind was the imperial aspect. Television was inevitable and would become international if only from the point of view of standardisation of transmitting and receiving apparatus, to the country whose system was adopted would fall the manufacture of the apparatus, said Ampthill. He illustrated his point by mentioning that America might impose its standards on the world and achieve a position in the manufacture of television apparatus such as it has already achieved in the radio business and the film industry.

However, the future could be different. Bairds were working on apparatus in their laboratories which was capable of showing scenes of several people, wrote Ampthill and intended to pursue experiments using a broader waveband for this purpose from their shortwave station at Hendon. Also they were working on a portable transmitter for the transmission of television from the BBC so that there would be no interference with speech and music programmes. 'Naturally', he said, 'these developments will take time and it may not be out of place for me to refer to the heavy financial burden which has fallen on this Company in respect of all this development work involving apart from the maintenance of research laboratories, the payment of broadcast transmissions, the upkeep of studios and the necessary staff and artists'.

Bairds suggested that half-an-hour's transmission from 10.00 to 10.30 a.m. should be granted each evening and they hoped the Corporation would take a closer interest in the company's affairs and assist in some of the many ways in which they could assist it.

Gainford could only give some assurances[58] regarding reasonable facilities and stressed the need for the BBC to protect its major interests. 'Anything which even temporarily cripples or reduces this service must be avoided if we would observe our tradition and practice'. The correspondence was brought to a close when Ampthill thanked Gainford for his assurances and observed 'Let me, however, say that it is our confident hope that television will in due course become an important part of the practice of the BBC and add to the lustre of the traditions that you are rightly observing'.[59]

For Bairds the matter of greater BBC assistance was not over: John Baird wrote to Reith on the 2nd January 1931 and requested a meeting.[60] Murray thought that Baird would likely ask if he could have an allocation out of their licence revenue to pay artists used in his broadcast television transmissions.[61] The Assistant Controller added that the existing charges made to Bairds did not cover expenses and other overheads and considered it probable that the BBC would ask for a further sum of approximately £25 per month.

Undoubtedly the company was experiencing financial difficulties. In May the Baird Television Development Company Ltd. and the Baird International Television Ltd. amalgamated to form a new concern known as the Baird Television Ltd.[62] Manville, Chairman of the Development Company had told shareholders that it had been evident to the Directors of the two companies that it was practically impossible to separate the interests in the working of the two companies, each of them necessarily doing their best to further the interests which were common to both.[63] A lawyer, however, told Mr Justice Eve that the original company had now exhausted a large part of its capital, but the International Company had still a large cash capital left, £170 000 in fact.[62]

Bairds had been pouring money out on travelling, publicity, transmissions through the BBC, staff and equipment. Both Baird and Hutchinson drew large salaries from the companies. John Logie's salary from Baird International was £3000 a year.[20] Margaret Baird has related how 'Hutchinson, carried a long way on the tide of success, had ideas which bordered on the extravagant, seeking to take John's place at meetings on the Board and dreaming of financial deals that had more connection with a dream world than with the real world'. In 1930 Hutchinson visited the United States in the hope of founding a vast company there – he was always longing for the big deal – but unfortunately for him and the company he could not pin any American down to a definite undertaking. When he eventually came back all he had with him were some vague statements and a heavy bill for expenses. The Board was annoyed and Hutchinson resigned.

After a short interregnum Moseley was appointed business manager and made some economies. Baird had his salary halved.[20] The number of television receivers sold passed a thousand but the outgoings were a severe drain on the company's resources and when the BBC gave their permission for television broadcasts from a new studio, additional expenditure was required to equip the studio and pay the artists fees; the studio cost money to maintain and the company was paying the BBC for the use of the radio transmitter. As Baird put it 'We were behaving more like a philanthropic institution for the benefit of the television-minded public than a business concern'. The position was precarious and Moseley, the author of 'Money making in stocks and shares' and a realist with money 'told John that funds were once again dangerously low'.[20]

A partial way out of these difficulties would have been for the company to have received a percentage of the 10 shilling licence fee, even though the time occupied by television broadcasts amounted only to a very small fraction of the BBC's total programme time and the number of viewers was small indeed. Of the licence fee the Post Office was entitled to $12\frac{1}{2}$% and required its portion to cover administrative expenses, the Treasury was entitled to a further percentage and the BBC entitled to the balance. The PMG told Baird[64] it would be quite useless for him to suggest to the Treasury that they should give up any portion of their receipts and Lees Smith suggested that the proper course for the company to take was to make whatever representations they though fit to the BBC concerning the terms under which television was transmitted and that, if they considered the BBC's reply was unreasonable and they were being treated unfairly, the company could make further representations to the PMG.

It was against this background that Baird wrote to Reith and requested a meeting. Reith replied: 'There is really no case from our point of view for subsidising your activities or for bearing part of the cost of the experimental transmissions'.[65] He did say to Baird 'There is here the maximum goodwill towards you personally and I am asking those concerned to continue to explore with you and your colleagues the various avenues of constructive co-operation'. In replying in this tone Reith was following the recommendations of his Assistant Controller, Murray, who in a memorandum[66] listed four points for the Director General to consider. First: 'Nothing has happened in the way of development or improvement which would justify any extension of facilities or change of attitude towards Baird Television'. Second: 'it is essential to avoid the admission of Baird Television to programme time'. Third: 'there was no case for relieving the Baird Company from the burden of the expenditure involved' in the experimental transmissions, and fourth, there were various directions in which the BBC could help the Baird Company to make more effective and more economical use of existing facilities within the limits of the present policy.

Naturally Baird was disappointed[67] with Reith's letter, but the BBC soon found support for their policy from an article in the *Berliner Tageblatt* headed 'Television scandal'.[68] 'Not so long ago one heard almost every day of a new completely revolutionary television invention the introduction of which was to be only a question of weeks. However weeks became months and months years and nothing was heard'. The writer went on to note that a well known big German firm had given up its television experiments as completely useless at the moment and that the Bell Telephone Company had similarly ceased working on television after spending on them a monstrous sum said to be $5 000 000. 'An interesting point is that one of the pioneers of television sees the reason for the unpopularity of television in the fact that the public is not used to it and must first learn to see with the televisor'. But he added: 'No one should put a bad thing before the public label it good enough and reply to complaints that it is good enough only the public cannot see it properly'. The article then mentioned that firms were trying to obtain money from the Post Office and the RRG (which had already given many thousands of marks), to continue their experiments and hence the matter was not a private but a public issue. There was another point: 'Those interested in order to mislead the public into wasting good money on such imperfect apparatus are now trying to get the Post Office to agree to transmit the experiments at the most popular listening hours . . . This must be fought in the interests also of listeners who cannot afford television receivers and do not desire to have to listen patiently to the buzzing of the television transmission. As long as films only can be transmitted, and films which flicker badly are very poorly lighted and show no details there is no occasion whatever to hasten the introduction of television as an official part of the programme'.

The article in fact supported what had always been the BBC policy and the BBC were quick to draw the attention of Bairds to it.[69] Baird considered the report was full of the most flagrant mis-statements of fact carefully calculated to damage the progress of television in Germany and showed the writer to be either deliberately malicious or completely misinformed.[70] Moseley did not think 'we ought to be deterred from our endeavours by these foolish and unsubstantiated reports from disappointed foreign sources'.[71]

The German television company referred to in the *Berliner Tageblatt* was the

Table 7.1 *Comparison of Baird's contributions in the 1920s to those of Bell Laboratories*

Use of/demonstration of	Baird	Bell Laboratories
1 Nipkow discs	From 1923	From 1925
2 Means to reduce time lag of photocells	From c. 1925. Used derivative of photocell current. Patent 270 222, 21st Oct. 1925	From c. 1925/26. Used C–R coupling circuit to enhance high-frequency gain. Internal memorandum 27th Feb. 1926
3 Coloured filters on lamps to reduce discomfiture of persons being televised	Various experiments in 1925. Demonstration of television using infra-red radiation, 23rd Nov. 1926	Various experiments in 1925. Mentioned in internal memorandum, 26th August 1925
4 Large Nipkow discs	Utilised discs up to 8 ft (2.44 m) in diameter sometime during the period 1923–25	Advantage of using discs up to 10 ft (3.05 m) in diameter mentioned in an internal memorandum, 27th July 1925
5 Spotlight scanning	Employed from 1926 to 1936. Patent 269 658, 20th January 1926	Employed from c. 1925–26. US Patent applied for on 6th April 1927. UK Patent 288 238, 18th January 1928
6 Two-way television	Patent 309 965, 19th October 1927	UK Patent 297,152, 17th June 1927. Demonstrated from 9th April 1930 to 31st December 1932

Table 7.1 (continued)

Use of/demonstration of	Baird	Bell Laboratories
7 Transatlantic television	Demonstrated 9th February 1928	Suggested as a publicity event in an internal memorandum 4th May 1927
8 Intercalated images to improve resolution	Patent 253 957, 1st January 1925. Various experiments c. 1924	Mentioned in internal memorandum 9th September 1927
9 Colour television	Demonstrated 3rd July 1928	Demonstrated July 1929
10 Large-screen television	Demonstrated 28th July 1930	Demonstrated 7th April 1927
11 Daylight television	Demonstrated June 1928	Need to work on natural light scanning mentioned in an internal memorandum 4th May 1927
12 Zone television	Demonstrated 2nd January 1931. Patent 360 942, 6th August 1930	Described in a paper by H. E. Ives, 'A multi-channel television apparatus', *J. Opt. Soc. Am.*, 1931, **21**
13 Commutated lamp bank/display	Patent 222 604, 26th July 1923.	Demonstrated 7th April 1927

Telefunken Company, according to Moseley, which had always regarded television as one of their sidelines and had never intended to go out big on it. When the German Post Office took up the Baird system Telefunken were naturally disappointed and rather chagrined. In addition the Chief Engineer of the Reichrundfunk had been attached to Telefunken and to the Karolus system and it did not surprise Moseley that he was not friendly towards television, particularly Baird television. Both Moseley and Baird thought it was absurd to suggest that any American company had stopped its experiments and Baird pointed out that the HMV Company had given, a short time ago, a demonstration of television using most complex equipment, and that in the USA very large sums of money were being spent by the Radio Corporation of America on the development of television.[70]

Bairds were particularly anxious at this time about the developments of the HMV Company and for some time in the early 1930s tried very hard to get their work on television delayed. The Baird Company's diffculties were certainly compounded when the HMV Company entered the television field and Bairds never made any secret of their dislike for the developments which were taking place in this country by what they felt was a subsidiary of RCA.

However Baird's approach to Reith in January 1931 had a desirable effect in that the extra £25 per month payment from the company was waived – while maintaining the principle that it was due.[72]

Baird had now (the end of 1930) been working on television for rather more than eight years. He had demonstrated a rudimentary form of television in 1925 and had during the succeeding years applied his basic concepts to schemes for showing colour television, stereoscopic television, daylight television, *inter alia*. Nevertheless he had been criticised: even today (1986) his work is the subject of some controversy. Baird's contribution, in the 1920s, to the art of television, albeit low definition television, may be compared with that of Bell Laboratories (Table 7.1). The table is not comprehensive and does not serve to determine whether Baird or a staff member at Bell Laboratories was the first to patent or demonstrate a particular aspect or method of television. Rather, it is a basis to indicate Baird's sound appreciation of the low definition television problem and to show that, in the 1920s, his thoughts and his implementation of those thoughts were vindicated by the activities and ideas of a well endowed and well staffed research organisation.

Baird produced more than 170 patents. A careful reading of some of these patents and of those of his contemporaries in the same fields, and a comparison of the efforts and successes of Baird and of his competitors in the 1920s, again in the same fields, clearly shows that his understanding and knowledge of television systems using mechanical scanning was considerable and appropriate for the 1920–1930 decade.

References

1 MOSELEY, S. A.: 'John Baird' (Odhams, London, 1952)
2 MOSELEY, S. A., and BARTON CHAPPLE, H. J.: 'Television today and tomorrow' (Pitman, London, 1933)
3 Anon.: 'Television. First experimental broadcasts', *The Times*, 1st October 1929
4 Anon.: 'Television humour. Machine breaks down in first day broadcast. Mugs Monday', *Daily Herald*, 1st October 1929
5 Anon.: 'Broadcast of television begins', *Manchester Guardian*, 1st October 1929
6 Editorial, *Amateur Wireless*, October 1929
7 BARTLETT, T. W.: letter to Sir J. F. W. Reith, 11th November 1929, BBC file T16/42
8 Control Board Minutes, extract, 12th Novmber 1929, BBC file T16/42
9 Director General (BBC): letter to T. W. Bartlett, 13th November 1929, BBC file T16/42
10 BARTLETT, T. W.: letter to the Director General (BBC), 22nd November 1929, BBC file T16/42
11 Chief Engineer (BBC): memorandum to the Director General, 25th November 1929, BBC file T16/42
12 Control Board Minutes, extract, 26th November 1929, BBC file T16/42
13 Director General (BBC): letter to T. W. Bartlett, 27th November 1929, BBC file T16/42
14 MURRAY, G.: letter to S. A. Moseley, 1st January 1930, BBC file T16/42
15 Lines engineer: memorandum, 13th March 1930, BBC file T16/42
16 MURRAY, G.: letter to S. A. Moseley, 12th February 1930, BBC file T16/42
17 Anon.: 'First sound television Broadcast. Singers seen 10 miles away. Dual transmission by BBC twins', *Evening Standard*, 31st March 1930
18 Anon.: 'Television broadcast with sound. Success of the Baird experiments', *The Times*, 1st April 1930
19 Information executive: memorandum to A. C.(I), 18th December 1929, BBC file T16/42
20 BAIRD, M.: 'Television Baird' (HAUM, South Africa, 1974)
21 Anon.: 'First play by television', *Daily Herald*, 15th July 1930
22 Anon.: a report, *Daily Mail*, 15th July 1930
23 BRIDGEWATER, T. H.: 'Baird and television', *Journal of the Royal Television Society*, **9**, no 2, pp. 60–68
24 Vide Chapter 11
25 SIEVEKING, L.: 'Television– final report on present situation', July 1930, personal collection
26 Chief Engineer: internal memorandum to the Director General (BBC), 5th September 1930, BBC file T16/214
27 Control Board Minutes: extract, 9th September 1930, BBC file T16/42
28 SIEVEKING, L.: memorandum to V. Gielgud, 17th September 1930, BBC file T16/214
29 MOSELEY, S. A.: letter to G. Murray, 10th September 1930, BBC file T16/42
30 MURRAY, G.: letter to S. A. Moseley, 31st October 1930, BBC file T16/42
31 MURRAY, G.: letter to S. A. Moseley, 4th November 1930, BBC file T16/42
32 MURRAY, G.: letter to S. A. Moseley, 17th November 1930, BBC file T16/42
33 Report: 'Company Meetings', *The Times*, 8th April 1930
34 Report: 'Baird Television Ltd. Annual General Meeting', *The Times*, 23rd December 1930
35 Anon.: a report, *Today's Cinema*, 1st July 1930
36 BAIRD, J. L.: 'Improvements in or relating to television and like systems', British patent 326,192, application date 5th November 1928
37 Anon.: 'New television advance', *Daily Mail*, 2nd July 1930
38 Anon.: 'Television. A remarkable display of Baird's process', *Daily Worker*, 3rd July 1930
39 Anon.: a report, the *Sphere*, quoted in Moseley, S. A., reference 1
40 Anon.: a report, *Daily Express*, quoted in reference 1
41 BAIRD, J. L., and DAY, W. E. L.: 'A system of transmitting views, portraits and scenes by telegraphy or wireless telegraphy', British patent 222,604, 26th July 1923

42 Anon.: a report, Bell Laboratories Record, May 1927
43 Anon.: 'First play by television', *Sunday Chronicle*, 6th April 1930
44 A report, *Wireless World*, 12th March 1930, p. 277
45 BAIRD, J. L.: 'Improvements in or relating to loudspeaking telephones applicable for use with television apparatus and the like', British patent 318,278, application date 31st May 1928
46 Anon.: 'First sound television broadcast', *Evening Standard*, 31st March 1930
47 Anon.: a report, *Birmingham Daily Mail*, 7th February 1930
48 Anon.: 'Television lenses order goes to France', *Birmingham Post*, 8th February 1930
49 ELLIOTT, E.: a letter, *Birmingham Post*, 11th February 1930
50 CHANCE, W. H. S.: a letter,*Birmingham Daily Mail*, 20th February 1930
51 WATTS, M.: a letter, *Birmingham Post*, 12th February 1930
52 Anon.: 'British lenses and television', *Engineer*, 26th February 1930
53 Anon.: 'Television ban. Declared impracticable at an exhibition', *Daily Mail*, 26th August 1930
54 Anon.: 'Television surprise. Wireless show ban', *Daily Mail*, 27th August 1930
55 MURRAY, G.: letter to S. A. Moseley, 31st October 1930, BBC file T16/42
56 GAINFORD, Lord.: letter to the Director General (BBC), 8th November 1930, BBC file T16/42
57 AMPTHILL, Lord.: letter to Lord Gainford, 17th November 1930, BBC file T16/42
58 GAINFORD, Lord.: letter to Lord Ampthill, 9th December 1930, BBC file T16/42
59 AMPTHILL, Lord.: letter to Lord Gainford, 15th December 1930, BBC file T16/42
60 BAIRD, J. L.: letter to Sir J. F. W. Reith, 2nd January 1931, BBC file T16/42
61 MURRAY, G.: memorandum, 6th January 1931, BBC file T16/42
62 Anon.: 'Law reports. Chancery Division. Baird Television Amalgamation', *Daily Telegraph*, 20th May 1930
63 Anon.: 'Company Meetings, *The Times*, 8th April 1930
64 PHILLIPS, F. W.: memorandum. 3rd January 1931, Minute 4004/33, file 18
65 Director General (BBC): letter to J. L. Baird, 15th January 1931, BBC file T16/42
66 A. C.(I): memorandum to the Director General, January 1931, BBC file T16/42
67 BAIRD, J. L.: letter to Sir J. F. W. Reith, 27th January 1931. BBC file T16/42
68 Anon.: 'Television scandal', *Berlin Tageblatt*, 28th January 1931
69 Control Board Minutes, extract, 17th February 1931, BBC file T16/42
70 BAIRD, J. L.: a statement, 28th February 1931, BBC file T16/42
71 MOSELEY, S. A.: a statement, 28th February 1931, BBC file T16/42
72 Control Board Minutes, extract, 26th March 1931, BBC T16/42

Chapter 8

A possible way forward, 1931

In 1930 the Gramophone Co. (HMV) which later merged with the Columbia Graphophone Company to form Electric and Musical Industries Ltd.,[1] began to take a serious interest in television. The research staff, under G. E. Condliffe, decided to produce an image of greatly improved definition by effectively combining five separate television channels, each using mechanical scanning.[2] C. O. Browne engineered equipment on these lines (Fig. 8.1) and gave a number of demonstrations at the Physical and Optical Societies Exhibition in January 1931.[3] The demonstrations created great interest and queues of people waited to see the new system outside the small theatre which the Gramophone (HMV) Company had set up for their show. G. A. Atkinson, the film critic of the *Daily Telegraph*, referred to the great advance in the television tests and wrote of the new system: 'It marks a considerable technical advance on any system yet demonstrated, especially in the direction of bringing television rapidly into use for entertainment purposes.[4] The *Evening Standard* noted that television seemed now to be within sight of realisation.[5]

During the exhibition the Gramophone Company showed images from ordinary cinematograph films projected by means of their apparatus on to a screen measuring 24 inch by 20 inch. No longer were heads and shoulders shown, but instead the audience saw images of public buildings, soldiers marching, cricketers walking on and off the field and so on. 'Everything was easily recognisable. An LCC tramcar showed up so clearly that its number on the front was decipherable without difficulty. The pictures were steady on the screen. They were in good focus at very short range. They would probably stand enlargement up to four or five times the size of the screen actually used . . . The general effect was that of looking at a performance of miniature films lacking full illumination.'[4]

HMV's success was due to the operation of a system working not on 30 lines but on 150 lines per picture. This standard allowed much greater detail to be shown and clearly pointed the way to be successful adoption of a public television broadcasting system. Bairds now had a rival in this country and considerable animosity was to be engendered by this fact. Baird wrote about the gigantic Radio Trust of America spending vast sums of money on the perfection of television 'not only in America but also in England where their subsidiary the HMV Company (had) recently (given) a demonstration of most elaborate apparatus.'[6] This reference to HMV and later EMI being a subsidiary of RCA with the implication that a great deal of the know-how of television was imported from the US was to be a feature of the Baird Company's writings to the BBC and the Post Office for the next few years. Bairds were clearly anxious at this time about the emergence of a competitor in this country who could compound the difficulties the company was experiencing with financial matters and the introduction of a satisfactory low definition service.

Fig. 8.1 *Early 1930s EMI television equipment based on the use of five channels*

Unfortunately for the Baird company, but fortunately for the prestige of Great Britain the work on television carried out in the laboratories of the HMV and EMI Companies was undertaken by British technicians and engineers and all their equipment was designed and constructed in British research laboratories and workshops. The multichannel television apparatus shown at the exhibition was the work of Browne – a British engineer *par excellence*. Unlike Baird, who never published a technical paper in the UK suitable for a professional engineering audience of a learned society, Browne wrote and delivered a paper on his work to the Institution of Electrical Engineers in January 1932. This lecture clearly showed Browne's considerable grasp of the problems which had to be overcome in the implementation of a practical television system for the reproduction of scenes involving moving people.

Browne referred in his paper to the principles which had to be observed in any new system of television:

> Either we must realise the practical limitations imposed by the very limited frequency band it is possible to transmit along one channel from the transmitter to the receiver and design apparatus to accommodate only this range of frequencies, or we must disregard this practical limitation and attempt to produce correspondingly better results at the expense of a number of transmission channels.

Browne adopted the latter alternative and described the advantages which would be realised by its adoption.

> The total frequency band necessary to transmit a given picture may, in the case of a multichannel television system, be divided into a number of channels, each of which accommodates a frequency band given by the total frequency range divided by the number of channels. On this account the difficulties of design, not only of the apparatus situated at the transmitter and receiver but also of the transmission line between the two stations, are considerably reduced. Apart from this advantage of the multichannel system, the amount of light available for illuminating the receiver screen is increased in proportion to the number of channels used. Further the velocity with which the scanning spots travel over the surface of the picture to be transmitted is decreased so that the accuracy necessary for synchronising is reduced as the number of channels is increased.

Five channels were chosen and each picture was scanned at a rate of $12\frac{1}{2}$ times per second (this figure being adopted because then the 50 Hz mains could be used without the inclusion of awkward gear ratios between the synchronous driving motors and scanning devices at the transmitter and receiver).

The aspect ratio of each picture was 3 (width) by 2 (height) and the amount of detail which it was decided to transmit was that corresponding to 15 000 picture points. From these factors the frequency band could be calculated to be 117 000 Hz.

The HMV television apparatus did not represent true television, in which reflected light is received from an object and allowed to fall on a photoelectric cell, but was more in keeping with the early systems of television of Jenkins and Baird in which a powerful light source was situated behind the object or scene to be propagated. There was an important difference, however: whereas the early workers had used opaque objects Browne employed cinematograph film bearing the subject to be transmitted. Half-tones were thus taken into account as in true television. Film was used mainly because it was plentiful and it enabled the conditions existing for any particular transmission to be repeated with comparative accuracy on any subsequent occasion. Browne, like the engineers of the Bell Telephone Laboratories who were responsible for the 1927 demonstration of television between Washington and New York, was particularly interested in investigating the problems associated with the transmitter and receiver and examinimg the electrical and optical conditions necessary for the picture channels in order to secure good results. Repeatability was thus an important point to be considered and the employment of film allowed this to be achieved.

The film was scanned, while stationary (using the intermittent motion provided by a Maltese cross shutter), by a revolving lens drum containing 38 lenses, and the light passed over five equidistant photoelectric cell apertures which distributed the light simultaneously into five vertical sections. Although 30 lenses were used in the lens drum for scanning purposes, actually 38 were included – eight of these being operative during picture changing. Each photoelectric cell, and associated amplifier, was required to handle frequencies up to 23 400 Hz.

At the receiver the modulated image signals were further amplified by a bank of five amplifiers and the outputs impressed on a bank of five Kerr cells, complete with crossed Nicol prisms. These cells caused the light flux from a powerful arc lamp to be modulated. Reconstruction of the picture from the five fluctuating pencils of light was accomplished by a revolving drum fitted with polished steel mirrors corresponding to the arrangement of the lenses at the transmitter.

Synchronisation was achieved by electrically coupling a 1200 Hz generator of the phonic wheel type, mounted on the lens drum spindle of the transmitter, to a motor of similar design on the receiver mirror drum. Phase adjustment was made by viewing a predetermined mark on the mirror drum in the light of a neon lamp, which was excited once every revolution of the transmitter lens drum.

Great attention was paid by Browne to a consideration of the various kinds of distortion which could be produced in the received picture and to ways of overcoming or minimising their effects. Effort was also expended in achieving a high standard of workmanship and technique, both optical and electrical, in the apparatus, although the Gramophone Company had no intention, at the time of the Physical and Optical Society Exhibition, of manufacturing television apparatus on a commercial basis. Even so, the reproduction of the images was not without some defects:

> The effect was, however, marred by a series of five wavebands which kept travelling constantly across the picture, cutting it with travelling light and dark areas. The general effect can best be described as a small poorly lit photograph seen through a disc of concentric circles constantly travelling to the centre. Nevertheless the picture was in great, but poorly lit, detail and if the scanning circles could have been eliminated might have passed for a good film inadequately projected.[3]

For a public used to 30-line head and shoulders images the HMV Company's display of medium definition (150-line) television created quite a stir. Newspaper reporters were quick to realise the potentialities of such a system and for some it opened up the near prospect of a new entertainment era, apart from the other almost infinite uses of television. 'For the first time in history a stage or film producer will be able to take his show to the public instead of waiting for the public to come to his show.'[4]

The demonstration of telecinema at the Imperial College of Science and Technology, London, was interesting not only from the technical point of view but also from a personality point of view. In the front row at one of the shows sat John Baird. He was recognised by His Master's Voice demonstrators and later was allowed to examine the apparatus. When asked for his opinion he replied 'Wait a while.'[7]

Subsequently, Baird Television Ltd. brought an action against the Gramophone Company and alleged that 'it had manufactured, exhibited and used at the Imperial College of Science and Technology, Kensington, a certain apparatus in January 1931' – which presumably infringed a patent of the plaintiff company. However, in a hearing before Mr Justice Clauson, Chancery Division, on 15th March 1932 – by which date Bairds were out of time in prosecuting the action – the defence advocate stated that the defence was that the apparatus was experimental, of purely scientific interest and of no commercial interest and that no such apparatus had ever been offered for sale. Bairds did not proceed with the case and costs were awarded against them.[8]

The HMV Company had not been alone in working on multichannel television in 1930. In the same month that they had given a demonstration of their apparatus to audiences at the Physical and Optical Societies Exhibition, Bairds had shown their zone television system to newspaper reporters.

Todays Cinema referred to the new equipment which had been displayed to the press in the Baird laboratories on 2nd January as 'an amazing new development in the Baird television process, which makes it possible to project pictures on to an ordinary full sized cinema screen, televise people and objects illuminated only by arc lighting or daylight instead of an intensive exploring beam and show an unlimited amount of detail in the picture'.[9] The technical editor, in a special editorial note, wrote:

> It is my firm belief that Baird has at last hit on the very method which will bring television into the cinema. It is a bold statement, but I make it in all seriousness and when I saw yesterday's demonstration I could see beyond the tiny screen shown to a visitor to a new revolution in our industry.

In pursuing development work in this field Bairds probably had in mind two important points: first, the restriction of their 30-line television system to head and shoulders images, and secondly the enthusiasm of the London Coliseum screen television shows. These had been very successful both here and on the continent, the press had reported favourably on them and there had been suggestions that a new form of entertainment had been founded – telecinema.

However, the system of lamps was somewhat crude and cumbersome and did not lend itself to higher definition images owing to the complexity of the commutator. Also the use of the 30-line system was completely unsuited for showing outdoor scenes such as cricket matches, processions, and, in general, any scene having a number of artists or performers.

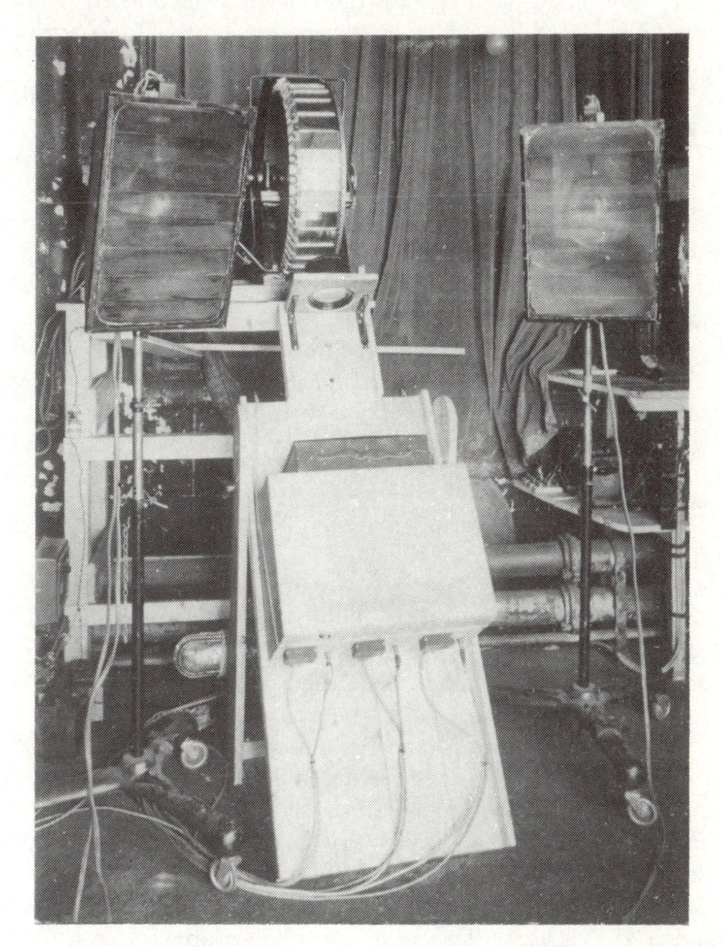

Fig. 8.2 *The triple zone receiving equipment employed at the Metropole cinema to back project a 3-zone television image on to a screen for public viewing. Three arc lamps had their outputs modulated by three Kerr cells (1932)*

Baird's solution was to use three channels in a similar way to that adopted by the HMV Co., but whereas Browne had restricted his equipment to show films only, Baird's apparatus had the advantage that it could be used to televise objects or subjects both indoors and outside. Unfortunately John Baird never emulated Browne's openness in giving exhaustive technical details of his equipment and such information about it has to be gleaned from newspaper and other reports (Fig. 8.2).

The Times of 5th January 1931[10] devoted several paragraphs to the new system and stated:

> The chief difficulty in broadcasting large images by television is that of the scarcity of available wavelengths. When the communication is by means of telephone lines, however, the difficulty is overcome by using several pairs of lines, each line being, as it were, responsible for a portion of the picture.
>
> A demonstration of such zone television was given by the Baird Company when Mr H. Strudwick, the England and Surrey cricketer, was 'televised' in action as a batsman. His movements and those of the wicket-keeper could be clearly seen.
>
> In this latest apparatus the scene to be transmitted is not scanned by a rapidly moving spot of light, but is illuminated by ordinary floodlighting, such as is used in theatres, and ordinary daylight is equally suitable. The picture was shown, made up of three sections, transmitted side by side. The transmitter consisted of a large mirror drum with 30 mirrors, which revolving rapidly, caused a succession of images to be moved over three different appertures admitting light to three photo-electric cells. These cells controlled the light emitted by three neon lamps, which produced the final image on a ground glass screen.

Baird's patent 360,942[11] describes several different embodiments of this basic configuration.

There was also demonstrated a variation of this scheme in which for the neon lamp and its accessories was substituted a powerful arc lamp, the illumination from which was controlled by the simple process of injecting the television signals into the arc supply circuit. This scheme produced a picture of great brilliancy.

Baird's three-zone system was later to be used to televise the Derby in 1932, but in the meantime in America work had also been progressing on the development of a multichannel television apparatus by engineers of the Bell Telephone Laboratories. This work was described, as were the HMV and Baird Company's systems, in January 1931. Ives, who had been the engineering director of the AT & T's earlier, 1927, demonstrations, submitted a paper on the above subject to the *Journal of the Optical Society of America* towards the end of 1930.[12]

That three quite independent companies should publish descriptions of their multichannel apparatus in the same month indicates the state of the art at that time. Cathode-ray television systems had not been perfected by 1931 and the techniques of electronics had not been developed sufficiently to enable a single-channel high definition system to be evolved.

Nevertheless, the statement of the problem as given by Ives was simple:

> An electrically transmitted photograph 5 inch × 7 inch in size, having 100 scanning strips per inch, has a field of view and a degree of definition of detail, which, experience shows, are adequate (although with little margin)

> for the majority of news events pictures. It is undoubtedly a picture of this sort that the television enthusiast has in the back of his mind when he predicts carrying the stage and the motion picture screen into the house over electrical communications channels.

The difficulty about achieving this desirable result was readily apparent. In the aforementioned photograph the number of picture elements is 350 000, and at a repetition speed of 20 per second (24 per second had now become standard with sound films) this meant the transmission of 7 000 000 picture elements per second and a bandwidth of 3·5 MHz for the system on a single sideband basis. Ives compared the criteria for high definition television and the results which had been achieved in America and observed: 'All parts of the television system are already having serious difficulty in handling the 4000-element image.' (This was the number of image elements used in the 72-line picture of the two-way television/telephone installation of the American Telephone and Telegraph Company in New York.)

The difficulties which had to be overcome before a high definition system could be demonstrated concerned the use of the scanning discs at the transmitter and receiver, the photoelectric cells, the amplifying systems, the transmission channels and the receiving lamps.

Assuming the utilisation of discs in a television system arranged for direct scanning, an increase in the image detail transmitted required that either a loss of light resulted or an increase in the size of the disc had to be achieved, and in either case the factor of change involved was large. Thus, if the number of scanning holes is doubled in a disc of a given size, providing four times the number of image elements, the holes must be spaced at half the angular distance apart and for the same aspect ratio the diameter of the holes must be one quarter the diameter or 1/16th the area of the holes in the original disc. The light flux falling on the photoelectric cell is therefore reduced by a factor of 16 for the disc having half the number of holes, or in general the light transmitted by the disc to the cell decreases as the square of the number of image elements. However, if the disc size is enlarged to give the same light transmission then its diameter increases directly as the number of image elements.

An added disadvantage in using many small holes is that they are susceptible to blockage by dust particles, leading to streakiness in the reproduced picture. Baird and his rivals later used discs containing 45 holes and in order that a 180-line scan could be produced the discs were run at 6000 rev/min *in vacuo*. Discs up to 30 inch in diameter were being utilised by 1934.[13] Ives noted that the disc, while quite the simplest means for scanning images of few elements, was entirely impractical when really large numbers of image elements were in question and wrote: 'As yet however, no practical substitute for the disc of essentially different character has appeared.'

Turning next to photocells there were, in 1930, two types of cell which could be utilised for television; a gas-filled cell which had a good sensitivity but poor frequency response, and the vacuum cell, which was much less sensitive than the gas-filled cell, although it was free from its failing. The self-capacity of the cells and the associated wiring and amplifier valves caused the high frequencies to be attenuated relative to the lower frequencies and consequently equalising circuits with their attendent problems of phase adjustment, together with more amplification were required, but amplifiers, capable of handling frequency bands extending from low frequencies up to 100 000 Hz or more gave serious problems, observed Ives.

The communication channels, either radio or wire, used for television also posed grave difficulties for high definition television and its associated wide bandwidth specification.

> In radio, fading, different at different frequencies, and various forms of interference stand in the way of securing a wide frequency channel of uniform efficiency. In wire, progressive attenuation at higher frequencies, shift of phase, and cross-induction between circuits offer serious obstacles. Transformers and intermediate amplifiers or repeaters capable of handling the wide frequency bands here in question also present serious problems.
>
> Finally at the receiving end of the system the neon glow lamp could not follow satisfactorily television signals well below 40 000 Hz, and, in the case of the 4000-element image the neon had to be assisted by a frequently renewed admixture of hydrogen, which again could not be expected to increase the frequency range indefinitely. With the receiver disc, as at the sending end, increasing the number of image elements, rapidly reduced the amount of light in the image and with a plate glow lamp of given brightness, the apparent brightness of the image is inversely as the number of image elements.

Ives's rather gloomy survey of the problems facing research workers in this field led him to one clear and definite conclusion: 'The existing situation is that if a many-element television image is called for today, it is not available, and one of the chief obstacles is the difficulty of generating, transmitting, and receiving signals extending over wide frequency bands.' This statement, made in 1931, by a leading engineer of Bell Telephone Laboratories (which provided many hundreds of scientists, technicians and other workers for the 1927 demonstration of television), must be regarded as definitive.

For Ives, the alternative, which prompted the experimental work described in his January 1931 paper, was the use of multiple scanning and multiple channel transmission.

His experimental apparatus used scanning discs over whose holes were placed prisms, as illustrated in Fig. 8.3, so that at the sending end, the beams of light from the successive holes were thus diverted to different photoelectric cells. At the receiving end, the prisms enabled beams of light from three lamps to be diverted to a common direction.

In Fig. 8.3a the disc holes are shown disposed in a spiral, at such angular distances apart that three holes are always included in the frame f. Over the first

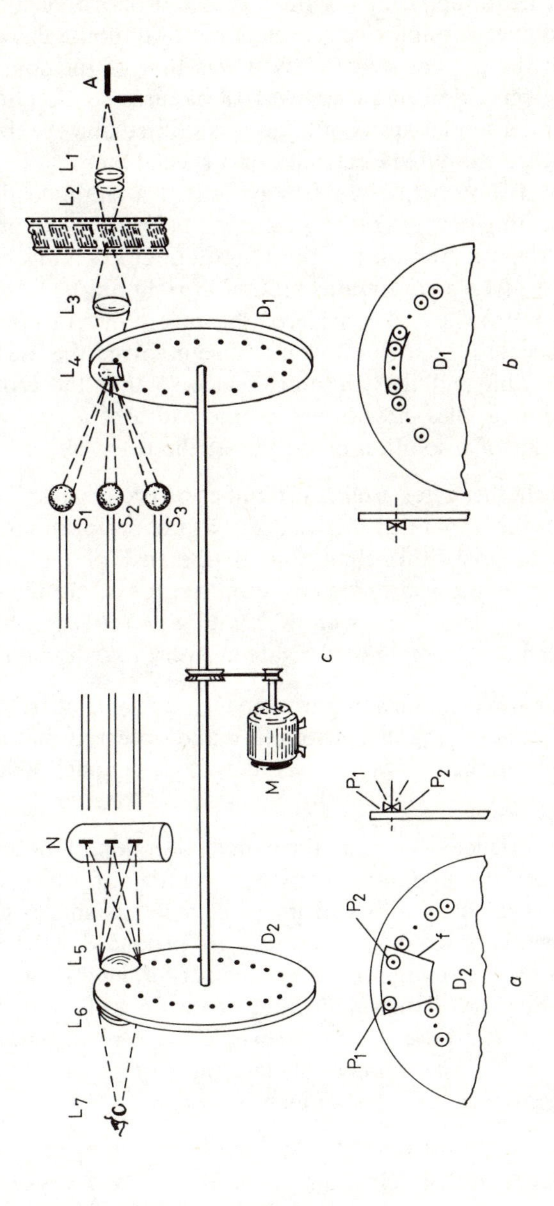

Fig. 8.3 *Three-channel television apparatus of the Bell Laboratories*

hole of a set of three is placed a prism P_1 which diverts the normally incident light upward: the second hole is left clear; the third is covered by a prism P_2 turned to divert the light downward.

This arrangement led to apparatus of manageable proportions and allowed the signals generated in each photoelectric cell to be continuous. The number of holes used was 108; the picture shape chosen was that of the sound motion picture, 7:6, and the repetition frequency was 18 pictures per second.

At the receiving end a similar spiral of holes was utilised and the three sets of signals were supplied to the three electrodes of a special lamp.

The employment of motion picture film permitted a simplification in the design of the transmitting disc in that the scanning holes could be arranged in a circle instead of a spiral by utilising the longitudinal scanning of the film obtained when it was given a continuous motion at right angles to the motion of the scanning holes. Browne had considered this possibility for his apparatus but had rejected it because the method would have involved the use of perfect gearing between the film and the scanning device so that the errors in the positions of the scanning lines did not exceed the width of a line, which was hypothetically the limit of resolution of the eye at the receiver.

> It is doubtful whether after a limited number of traversals of one length of film through the apparatus the accuracy of the perforations would be good enough to satisfy this condition. In the case of the intermittent motion, however, which is provided by a Maltese cross, the film is stationary during scanning and, moreover, even with a defective mechanism the film obtains identical framing in the gate at every fourth picture.[2]

Ives found that his three-channel apparatus yielded results strictly in agreement with the theory underlying its construction and observed that the 13 000 element image was a marked advance over the single-channel 4000 element image.

> Even so, the experience of running through a collection of motion picture films of all types is disappointing, in that the number of subjects rendered adequately by even this number of image elements is small. 'Close-ups' and scenes showing a great deal of action, are reproduced with considerble satisfaction, but scenes containing a number of full length figures, where the nature of the story is such that facial expressions should be watched are very far from satisfactory. On the whole the general opinion . . . is that an enormously greater number of elements is required for a television image for general news or entertainment purposes.

This, of course, was the BBC's view.

As a result of his work at Bell Telephone Laboratories, Ives was in a position to make a general comparison of the principal means available, at that time, for achieving a television image of extreme fineness of grain; namely, by an extension of the frequency band, and by the use of several relatively narrow frequency bands. Both these methods demanded enhanced sensitivity of the photosensitive elements at the sending end, and increased efficiency of the light sources at the receiving end, but Ives felt the multichannel scheme he described had some advantages in compactness over the equivalent single-channel apparatus. The overall efficiency of light utilisation in the two cases was not essentially different. However, in comparing the demands made upon the electrical systems the differences between the two methods were clear cut. The first method required an extension of the frequency range of all parts of the apparatus 'the attainment of which,' wrote Ives, 'depended upon physical properties and technical devices whose mastery lay in the indefinite future'; while the second method required a multiplication of apparatus parts and a careful design and construction of those parts so as to ensure accurately similar operations of a considerable number of electrical circuits and terminal elements. 'The attainment of the necessary uniformity of performance of the several electrical circuits and terminal elements while involving no fundamental problems, must present increasing difficulty with the number of channels used.'

Ives's prognosis that a successful single-channel high definition television system depended on physical properties and technical devices whose mastery lay in the indefinite future was the view which the BBC's former Chief Engineer, Eckersley, and Campbell Swinton had propounded on numerous occasions. Ives's and Browne's papers in effect sounded a warning about the pursuit of perfection by mechanical means. The HMV Company (and later EMI), quickly accepted the limitations imposed by non-electronic scanning and directed their research accordingly. Bairds did not, or were not able to, change the direction of their development work, although the limitations of mechanical scanning must have been just as apparent to them as they were to others. They persevered with a fairly straightforward scaling-up of the mechanical processes until it was too late to change.

Shortly after the display of the Gramophone Company's multichannel system at the Physical and Optical Societies Exhibition in January 1931, the company and the Columbia Graphophone Company united to form Electric and Musical Industries Ltd.[14] The research teams of the two companies were also merged and Isaac Shoenberg who had been with the Columbia Company became Director of Research and Head of the Patents Department.[15] The new company decided to extend the work which had been started by Browne and Tedham working under the management of Condliffe at HMV Co. at Hayes, but there was a dilemma which Shoenberg later described in 1952 at an IEE meeting.[16] 'In deciding the basic features of our system we frequently had to make a choice between a comparatively easy path leading to a mediocre result and a more difficult one which, if successful, held the promise of better things.' One of the examples Shoenberg gave to illustrate his point concerned mechanical scanning.

> When we started our work in 1931, the mechanically scanned receiver was the only type available and was under intensive development. Believing that this development could never lead to a standard of definition which

would be accepted for a satisfactory public service, we decided to turn our backs on the mechanical receiver and to put our effort into electronic scanning.

Meantime Bairds continued with the development of their optico-mechanical system. One reason for this was probably that Baird did not possess a sufficient knowledge of electronics to guide a research and development programme on a purely electronic television system. J. D. Percy, who worked for the Baird Companies from 1928, has written:[17] 'Baird's main interest always lay in the mechanical and optical problems of television. An amplifier was just a necessary and rather unimportant box to him, and one amplifier was very much like another.' Percy, during all the years that he knew Baird, could never recall being able to detect any gleam of enthusiasm from Baird when amplifiers or circuits were being discussed. 'Optics and mechanics, however, were his very life blood' he wrote.

During the month, January 1931, that Bairds, HMV, and Bell Telephone Laboratories had released details of their multichannel television systems, Baird disclosed for the first time the use of direct arc modulation as a means of increasing the brilliance of the image received on a screen. The demonstration was given before representatives of the scientific press, including *Nature*, and the technical representative of *The Times*. In this demonstration the television signals from the Baird transmitter were applied after suitable processing to a specially adapted arc. 'The detail and definition of the received image was comparable to that received on the standard commercial "Televisor" receiver, and the brilliance of illumination was remarkable. This demonstration of the successful modulation of the arc with television signals appears to open up considerable possibilities, and the television arc would appear to have a useful future.' commented *Nature* (10th January 1931).[18]

The first public demonstration of the modulated arc did not take place until the meeting of the British Association in September 1931 at the French Institute, Cromwell Gardens, when it was shown in the section devoted to Mechanical Aids to Learning. Here the new form of light source for television signals was used to project a picture on to a screen approximately 4 feet by 2½ feet.

The most noticeable effect for a reporter of the *Manchester Guardian* was that 'the light was no longer orangy (as with the neon tube), but white. This put it on a level more comparable with photography and the cinema, whereas the black and orange picture received on the ordinary televisors reminds one of the early flickering films and so emphasises the distance television has yet to travel. A black and white picture seems to mark a definite milestone.'[19]

Although the directly modulated high intensity arc had not previously been used for television, its properties and susceptibility to modulation had been known for some time.

The discovery of the speaking arc was made by Bell and Hayes in the United States of America and independently by Simon in Germany in 1897 and numerous circuit arrangements were devised to control the arc by speech signals.[20]

Monasch,[21] in his book 'Alternating and direct current electric arcs' (1904) refers to the speaking arc:

> With a ten ampere direct current arc of between three and five millimetres, between either solid or cored carbons, a clearly audible sound was produced even when an alternating current of one milliampere was superimposed on the direct current, and having a periodicity of 50 to 5000 Hz. The sound became inaudible only when the frequency was raised to 30 000 Hz.

Ernest Ruhmer in 1901 seems to have been the first person to have used the luminous properties of the speaking arc in a system for transmitting speech over an optical path.[22] In his apparatus the fluctuating light beam was rendered parallel by a parabolic mirror at the transmitter and after propagation was allowed to fall on a selenium cell placed at the focus of a similar mirror at the receiver. The consequential changes in the cell's resistance, owing to the action of the variable light intensity incident on it, produced corresponding changes in the receiver circuit containing the telephone, so that the whole apparatus comprised a speech communication system, rather similar to Bell and Tainter's photophone.

Many experiments were carried out in photophony in the first quarter of the 20th century, by Ruhmer, Schuckert, Blake and others. Ruhmer and Schuckert were particularly successful in their work on photophony on the Wansee and the Havel, near Berlin, and in 1902 two permanent stations were erected, one at the works of Siemens, Schuckert and Co., in Berlin, and the other at the parish school in the Baumschulweg, 2½ kilometres away. Their results showed that speech could be transmitted by day or by night with good clarity, and even in wet weather transmission was possible though with poorer fidelity of reproduction.[20]

Blake, in 1925, conducted some experiments in photophony using the speaking arc, in which he used rods of glass and silica – with optically polished ends – at the receiver to filter out the infrared radiation and thereby prevent damage to the selenium cell.[23]

Another application of the directly modulated arc was made by Bernouchi, who sent phototelegraphic signals over a beam of light,[24] while the first utilisation of the speaking arc for television was proposed by E. W. Whiston[25] in British patent 185,463 dated 4th May 1921. The objective of Whiston's invention was to provide an improved system in which all mechanical inertia was entirely avoided. There seems to be no evidence to show that Whiston actually tried to implement his invention, which was basically sound, and which used a circuit essentially similar to one advanced by Duddell. Whiston belonged to a fairly large group of inventors who produced a single patent sometimes containing useful ideas which they never implemented probably because of a lack of suitable resources.

Baird's first patent[26] (269,219) on the utilisation of arcs was published on 21st April 1927, although the initial application was made on 21st October 1925 –

the month in which he first privately demonstrated true television. The six year delay between this initial application and the actual demonstration of the use of the arc in television may have arisen because the convential arcs available in the 1920s when modulated gave a light output which consisted of an alternating component superimposed on a direct component. Consequently, in the application of the device for television purposes such a steady component would decrease the contrast range of a reproduced picture, whereas in photophonic uses the d.c. component would be unresponsive in producing an audible effect.

The Baird company carried out several investigations on speaking arcs. These showed that light variations were dependent upon the luminosity of the gaseous envelope partly surrounding and partly within the crater of the negative electrode. As a consequence later arc electrodes were made of metal, which permitted the heat generated at the electrode tip to be rapidly conducted away, and allow the crater gases to contribute to the major portion of the light from the arc. This discovery greatly increased the contrast of the reproduced television picture.[27] An additional improvement was the coring of the negative electrode with a refractory material such as cerium oxide, mixed with certain other salts of the metals of the alkaline earths, to increase the brightness and improve the colour of the light given out by the flame of the arc.

In order to fully modulate the arc, considerable power was necessary, and for an arc capable of passing a current of 10 A and dissipating 300 W an amplifier having an output of over 100 W was required.[27]

Experiments by Baird showed that an arc, such as that described above, had a frequency response which was essentially flat from 10 to 35 000 Hz. As the 30-line transmitter bandwidth did not have to exceed 10 000 Hz the arc was entirely suitable for low definition telecinema type applications and Bairds successfully demonstrated television on screens measuring 3 feet by 7 feet by its use.[28] A slight disadvantage was that small lateral movements of the flame of the arc (due to inhomogeneities in the cerium coring of the negative electrode), caused variations in screen brightness. However, Banks of Baird Television Ltd. successfully designed an optical system comprising a pair of separated lenses (patent 380,234)[29] to overcome this drawback.

During the period when Banks was working on the modulated arrc, in an endeavour to increase screen brilliance, others members of the company were investigating the use of the Weiller mirror drum scanner as a means of increasing the optical efficiency of the system and thereby lead to an improvement in the signal to noise ratio at the transmitter (and hence also at the receiver).

Baird first demonstrated his mirror drum transmitter on May 8th 1931. This was to be a prelude to the televising of the Derby the following month and for this purpose Bairds had equipped a van with a transmitter of the drum type. Natural light was used during the demonstration, but although there were considerable variations in the quality of the reception owing to the varying cloud cover reporters saw in a room in Long Acre the images of people passing along the street outside.[30]

The *Daily Telegraph* noted: 'The televising of great national events, Mr Baird considers, is now well within the bounds of possibility, though so far no definite arrangements for such broadcasting have yet been made.[31] Moseley, the radio critic of the *Daily Herald*, observed on the same day however: 'Mr Baird said that the fact that one was able to pick up the street scene showed that the idea of televising the Derby or the cricketers at Lords was not so fantastic as some imagined.'[32]

Following this latest success, Moseley[33] wrote to his friend Gladstone Murray and enquired whether the BBC could give Bairds five or ten minutes on Derby Day for a television broadcast during the race. Moseley pointed out that his company had willingly given up its half-hour transmission when asked to do so and consequently such a request would more than make amends.

Murray's letter [34] stated that while Baird could not have the London Regional wavelength for their vision transmitter in connection with the Derby it might be possible 'to arrange for the London National wavelength, i.e. 261 m, to be placed at (their) disposal for the television signals from approximately 2.45 to 3.15 p.m. on Wednesday, 3rd June (provided) the following conditions were fulfilled;

1 that the speech accompaniment, which of course would not be broadcast, would be on a telephone line quite separate from any telephone rented by the BBC,
2 that the BBC engineers would be satisfied in a preliminary test that nothing involved in this television transmission should in any way interfere with the normal service transmission of the running commentary'.

Thus on Derby Day, 3rd June, Bairds' outside television van was taken to Epsom and parked opposite the winning post. It was connected by telephone line direct to Bairds' London studio, from whence the signals were sent to the Brookmans Park transmitter.[34]

The parade of the horses was seen, although the horses were not individually recognisable, together with shadowy images of moving people. 'As the moment of the finish of the race approached interference became worse and the screen, seen through its enlarging lens, at times dissolved into a blurred mass of flickering lines. Notwithstanding this, however, the horses were plainly seen as they flashed past the post.'[35] The *Daily Telegraph* was more charitable: 'Fifteen miles from the course, in the company's studio at Long Acre, all the Derby scenes were easily discernible – the parade of the horses, the enormous crowd, and the dramatic flash past at the winning post.'[36]

After the transmission Baird said he was quite satisfied with the experiment. 'This marks the entry of television into the outdoor field', he noted, 'and should be the prelude to televising outdoor topical events.'

Moseley wrote[37] many years later: 'The first Derby broadcast captured the imagination of the general public so strongly and so fired the enthusiasm of amateur experimenters as to impress even the BBC.' He hinted that because of this Bairds were offered the use in August of Studio 10 near Waterloo Bridge for their portable transmitter.

Mention has been made of the growing interest in television by 1930 by major companies in the USA, Germany and the UK. 1930 was also the year in which the Marconi Wireless Telegraph Company considered it prudent to enquire whether the new art could have an influence on its commercial activities.

References

1 CLARK, A., *et al.*: evidence to the Television Committee, 27th June 1934.
2 BROWNE, C. O.: 'Multi-channel television', JIEE 1932, **70,** pp.340–349.
3 Anon.: 'Television nears the cinema. Great advances by H. M. V.. New system demonstrated', *The Bioscope*, 7th January 1931.
4 Anon.: 'Great advance in television tests. Nearing practical success. Broadcast of films and plays. When all may see the Derby', *Daily Telegraph*, 6th January 1931.
5 Anon.: 'Success of new tests today. Film of everyday life projected. Ride on a bus. Everything clearly recognisable', *Evening Standard*, 6th January 1931.
6 BAIRD, J. L.: a statement, 28th February 1931, BBC file T16/42.
7 Anon.: 'Rival inventors', *Daily Herald*, 9th January 1931.
8 Anon.: 'Baird television v. gramophone', *Financial Times*, 16th March 1931.
9 Anon.: a report, *Today's Cinema*, 2nd January 1931.
10 Anon.: a report, *The Times*, 5th January 1931.
11 BAIRD, J. L.: 'Improvements in or relating to television and like apparatus', British patent 360, 942, application dates 6th August and 8th December 1930.
12 Ives, H. E.: 'A multi-channel television apparatus', *J. O. S. A.*, 1931, **21**, Jan–June, pp. 9–19.
13 GRADENWITZ, A.: 'The Ferneseh A. G. television systems', *Television Today*, **1**, pp. 252–261.
14 Notes of a meeting of the Television Committee held on 17th June 1934, evidence of Messrs. Clark and Shoenberg on behalf of EMI and the Marconi – EMI Television Col. Ltd., p.12.
15 McGEE, J. D.: 'The life and work of Sir Isaac Shoenberg, 1880–1963', *The Royal Television Journal*, 1971, **13**, No. 9, May/June.
16 SHOENBERG, Sir I.: discusion on a paper by Messrs. Garratt and Mumford, JIEE, 1952, **99**, pp. 41–42.
17 PERCY, J. D.: 'The founding of British television', *Journal of the Television Society*, 1950, pp. 3–16.
18 Anon.: a report, *Nature*, 10th January 1931.
19 Anon.: 'A stage nearer cinema television', *Manchester Guardian*, 26th September 1931.
20 BARNARD, G. P.: 'The selenium cell: its properties and applications' (Constable, London, 1930).
21 MONASCH, A.: 'Direct and alternating current arcs' (Julius Springer, 1904).
22 RUHMER, E.: 'Neue Sende und Empfangsanordnung fur drahtlose Telephonie', *Phys. Z.*, 1901, **2**, pp. 339–340.
23 BLAKE, G. G. 'Communications on wavelengths other than those in general use', *Experimental Wireless*, 1925, **2**, pp. 561–572.
24 THORNE BAKER, T.: 'Wireless pictures and television', (Pitman, 1926).
25 WHISTON, E. W.: 'A mode of and /or means for transmitting photographs, messages, views and like devices by wire or wireless telegraphy', British patent 185, 463, 4th May 1921.
26 BAIRD, J. L.: 'Improvements in or relating to television and to apparatus for use in transmitting views, scenes, images, pictures and other objects of an animated or inanimated nature to a distance', British patent 269, 219, 21st October 1925.
27 BANKS, G. B., and WILSON, J. C.: 'Modulating the arc', *Television*, September 1934, pp. 397–398.
28 Anon.: 'A stage nearer cinema television', *Manchester Guardian*, 26th September 1931.

29 BANKS, G. B.: 'Improvements in or relating to optical systems for use in electro-optical transmission systems', British patent 380, 324, application date 25th August 1931.

30 Anon.: 'Street television. Test of new apparatus in London', *Daily Mail*, 9th May 1931.

31 Anon.: 'Television by daylight. Watching the man in the street. National events to be brought near. First open air apparatus', *Daily Telegraph*, 9th May 1931.

32 Anon.: 'Televising the Derby', *Daily Herald*, 9th May 1931.

33 MOSELEY, S. A.: letter to G. Murray, 19th May 1931, quoted in 'John Baird', *op. cit.*, p.150.

34 MURRAY, G.: letter to S. A. Moseley, May 1931, quoted in 'John Baird', *op. cit.*, p.151.

35 Anon.: 'Television scenes. Horses seen flashing past post', *Daily Mail*, 4th June 1931

36 Anon.: 'Race at home. Television success', *Daily Telegraph*, 4th June 1931.

37 MOSELEY, S. A.: 'John Baird' (Odhams, London, 1952), p.153.

Chapter 9

A commercial prospect, 1930–1932

The Marconi Wireless Telegraph Company pioneered the development of wireless telegraphy in this country and so it was obvious that sooner or later they would turn their attention to television. They were cautious in their approach though, and asked one of their consultants, Sir Ambrose Fleming, to let them have a confidential report on the Baird television system. Fleming, who had given Baird much moral support and was President of the Television Society, submitted his report to the Company in February 1930.[1] He stressed that he was able to take a perfectly disinterested view of the system's possible future and present results, that he was in no way whatever connected with Baird's company, that he was not a shareholder and had never received any consideration for any professional report or retainer.

Fleming had been asked by H. W. White, Managing Director of the Marconi Company, to consider a number of questions: the first two concerned the prospect of employing the Baird television apparatus for the commercial transmission and reception of messages. Marconi's consultant reported that, based on his experience, the only way this could be achieved would be by printing the message in rather bold letters about half an inch high on a strip of paper and then passing the strip through the Baird televisor. At the receiving end a clerk would be required to take down in long hand or short hand a word or two at a time, but Fleming thought it would be doubtful if anyone could read a printed message by eye even at the rate of 50 words per minute whereas in the case of a beam wireless message received by an undulator and recorded on Morse tape a speed of between 100 and 200 words a minute was possible. It was practicable, however, to transmit script like Chinese, Arabic or Hebrew letters which could not be telegraphed.

On the question as to whether the results achieved by the Baird system warranted the belief that it was now suitable for telegraphy over long distances, Fleming observed that it was not suitable. White's third question referred to the necessity of further improvements in television equipment before such equipment could be applied to telegraphy. Here again Fleming was not definite in his response for he reported that if facilities could be given by the company for experimentation on the transmission of television images of living and moving objects by the shortwave beam system it might be possible to achieve good results over extended distances.

Finally Fleming was asked to consider whether there was any system of television now in the course of development which it would be prudent for the Marconi Company to acquire the control of or rights in. After listing seven points relating to his views on television and the prospect of the company investigating the patent and financial position of Bairds and progress in America, Fleming stated his broad general conclusion. He could not recommend the Marconi Company acquiring the rights in or control of any particular television system although he felt that enough had been accomplished to suggest the possibility of a great more being done. Fleming urged that a careful watch should be kept on its progress and realisations.

Fig. 9.1 *A television picture was broadcast by Marconi Wireless Telegraph Company across the world as early as 1932. The head and shoulders of A. Longstaff, then London Manger of AWA, were clearly recognisable at Sydney, Australia, in a picture transmitted by a 30-line mechanical scanner (the black box with aperture) and photoelectric cell pick-up (slung from the top brackets)*

The Marconi Company was sufficiently impressed by the opportunities likely to be made available by a good television broadcasting system that it very quickly initiated work on a low definition system using mechanical scanning, so that by August the following year they were able to demonstrate their equipment at a meeting of the British Association in Leicester (Figs. 9.1 and 9.2). Possibly they had in mind not only the application of television to telegraphic

Fig. 9.2 *Marconi low definition television apparatus of the early 1930s, using the Nipkow aperture disc and optical system to produce the 'flying spot'*

systems but also the need to ensure that their expertise on transmitters was fully updated. In addition it is interesting to note that the patent for which Zworykin applied on the 17th July 1929 in the USA, covering a television transmitter system using an iconoscope type of electron camera, was also taken out in Britain as patent 369, 832,[2] and was granted to the Marconi Wireless Telegraph Company in 1931.

Fleming's report followed a long letter[3] dated 28th June 1929 from a Mr Aisenstein (of 8 Zgoda, Warsaw), to H. A. White of the Marconi Wireless Telegraph Company Ltd. Probably the conclusions reached by Aisenstein precipitated Marconi's commission to Fleming, for Aistenstein was firmly of the opinion that television would be a growth industry and because of the development of German broadcasting there was an expectation of a large market for television instruments in Germany.

Aisenstein's important letter, which gives some very interesting information concerning the Mihaly, Telefunken and Baird systems, only came to light in 1963 when a book seller, Mr Graham Scott, presented it to the BBC. This book seller had actually typed a letter to one of his customers on the reverse side of Aisenstein's letter before realising its historical significance.

The opening sentence of the letter makes it clear that MWT Co. Ltd. had asked Aisenstein to undertake a survey for them: 'According to your instructions, I carefully investigated this question (television ?) in Berlin.' Unfortunately the excellent historical archives possessed by the Marconi Company do not provide any information of Aisenstein's precise terms of reference, but it would seem that he was required to report on the Mihaly system. Aisenstein wrote that after a careful examination of the situation he agreed an arrangement with Mihaly would be of the greatest interest for the Company.

Mihaly, whose contribution to television has been briefly mentioned, worked on the problem of seeing by electricity from 1918. His experiments had utilised oscillating mirrors and had largely been unsuccessful, but, when Baird showed the way forward to successful television, in January 1926, using Nipkow discs, Mihaly abandoned mirror scanning and adopted the Nipkow disc and the neon glow discharge lamp as the two essential components of his receiver. In 1928 he had shown the transmission of crude shadowgraphs at the Berlin Radio Exhibition and in 1929 had persuaded the German Post Office to allow the Witzleben transmitter to be used for experimental transmissions using his system. These had consisted of cinematographic picture broadcasts without sound, for Mihaly was not able to demonstrate direct television at this time, although Baird had achieved this feat more than three years earlier.

Mihaly seems to have created a very favourable impression, for Aisenstein drew attention to the following points made by the inventor:

1 Mihaly possessed a good, strong German patent which would result 'in all other companies' having to pay royalties, or to exchange their patent rights.
2 The inventor had ready available designs and constructions for television which could be put on the market in a short time.
3 He was one of the most experienced men in the field of television.

Aisenstein stressed the first of these points and mentioned his great surprise on finding out that the patent included features used in the Karolus, the Alexanderson and the Baird systems. It seemed, he wrote, that no inventor could proceed with television without infringing some part of the patent.

Moseley, who was a shrewd judge of people and business matters, was not impressed by Mihaly, whom he later met in Berlin. Writing in the journal *Television* in June 1929 (the same month that Aisenstein sent his letter to White), Moseley observed:[4]

> Mihaly is clever enough however to surround himself even after all these years with an air of mystery. It is true, as one important commercial magnate said to me in Berlin, that one does not want to see a televised picture of a picture; but the Hungarian inventor whispers darkly that he has discovered 'new things'; but since we have heard this story before, I recommend readers to accept the rumours *cum grano salis*.

Moseley's opinion contrasted markedly with Aisenstein's thoughts. The patent, Aisenstein noted, had been taken out not in the name of Mihaly 'but by an artificial classification had been put in the group on cinematography instruments and was therefore little known between experts'. Aisenstein was shown the patent but the patent number and name had been covered with black paper. However the German patent referred to was 466,712 dated 30th May 1924.

Clearly White was not impressed by Aisenstein's urgent recommendations for a business relationship between his company and Mihaly and sought the opinion of Sir Ambrose Fleming, a very respected scientist of considerable standing, who had acted as a consultant to the Marconi Wireless Telegraph Company for many years. He was one of the few scientists in the United Kingdom who could give an unbiased opinion on the development of television from a commercial – as distinct from a domestic broadcasting – point of view, based on first hand knowledge of the subject.

Baird counted Fleming among his friends and often entertained him and others at his house in Swiss Cottage.[5] Whether Fleming told Baird of his commission from Marconis is not known, but it is interesting to note that Fleming's report to the company contained an idea which Baird had provisionally patented on 3rd October 1928. This patent[6] (324,029) had been published in its complete form approximately one month (on 3rd January 1930) before Fleming submitted his report to the Wireless Telegraph Company. Later Baird produced equipment which incorporated the ideas mentioned in the above patent and this apparatus was described in the November 1930 issue of the *Television* magazine.[7]

In patent 324,029 Baird claimed: 'means for transmitting a message or the like by television apparatus comprising a moving band, strip or other support upon which elements constituting the message are placed or formed, arranged in such relation to the exploring region of the transmitting apparatus and in such manner that the elements are slowly traversed past the said region'. The claims covered the use of characters and the like formed on the moving support (e.g.

by printing or typing), and the employment of similar characters retained in position on the tape by a magnetic force.

Fleming had observed in his report that it would be possible to transmit script such as Chinese, Arabic or Hebrew characters which could not be telegraphed. The inventor of the rectifier valve was thus greatly influenced by Baird's ideas, but possibly unfortunately for Baird and his company, the Marconi Wireless Telegraph Co. Ltd. took some considerable time to implement their consultant's suggestion. Yet, when the company eventually described its apparatus in 1932, it was so similar to equipment which Baird had produced in 1930 as to leave no doubt as to its provenance.

Initially, Baird utilised his invention to transmit what was popularly called Television Screen News. Until March 1930 only vision signals without sound could be broadcast by the BBC – pending the completion of the twin transmitters at Brookmans Park – and so the only way in which new items could be given on television was by the use of written words. Prior to April 1930 hand worked apparatus had been used for the purpose, but after that date an electrical device was introduced which 'functioned most satisfactorily and it is not too much to say that it made a most instructive and interesting addition to the daily programme.'[8]

By August 1929 a specialised form of the same equipment had been developed and constructed primarily for transmitting unorthodox characters and other printed matter by radio, when fading made transmission difficult and also when an ordinary creed undulator could not be used. One model was actually made for Prince Purachatra for use in Siam, but subsequent negotiations were never concluded. Baird seems to have been quite impressed by the possibilities which his ideas suggested, for he coined the rather unwieldy word 'telelogoscophy' from the Greek roots tele, logos and skopos, to refer to 'seeing writing at a distance'.[8]

Fig 9.3*a* illustrates Baird's apparatus. The message which was to be transmitted was typed, by a specially adapted typewriter, on to a white tape, and then scanned by an aperture drum scanner so that each hole of the drum passed vertically in front of the illuminated message. By using an aperture drum the curved scanning lines associated with the Nipkow scanner were eliminated, although initially this type of scanner was employed.[9] The beams of light emerging from the drum were focussed on to the uniformly moving tape by a suitable lens and the reflected light then allowed to fall on to one or more electric cells (the intermittent motion of the tape from the typewriter was overcome by having a loop in the tape between this machine and the framework which moved the tape uniformly before the scanning apparatus), see Fig. 9.4.

From a photograph of an early scanning disc which was published in the *Television* magazine, it would appear that a 60 lines per frame standard was used rather than the 30 lines per frame standard which Baird employed for his television broadcasts. A contemporary report of the equipment mentioned the 'excellent results which were obtained with an experimental receiver' and the simplicity of the arrangement 'as there is no graded light and shade to worry about but merely black and white contrasts.'

In an earlier patent,[10] 299,076 of 20th June 1927, Baird described a method

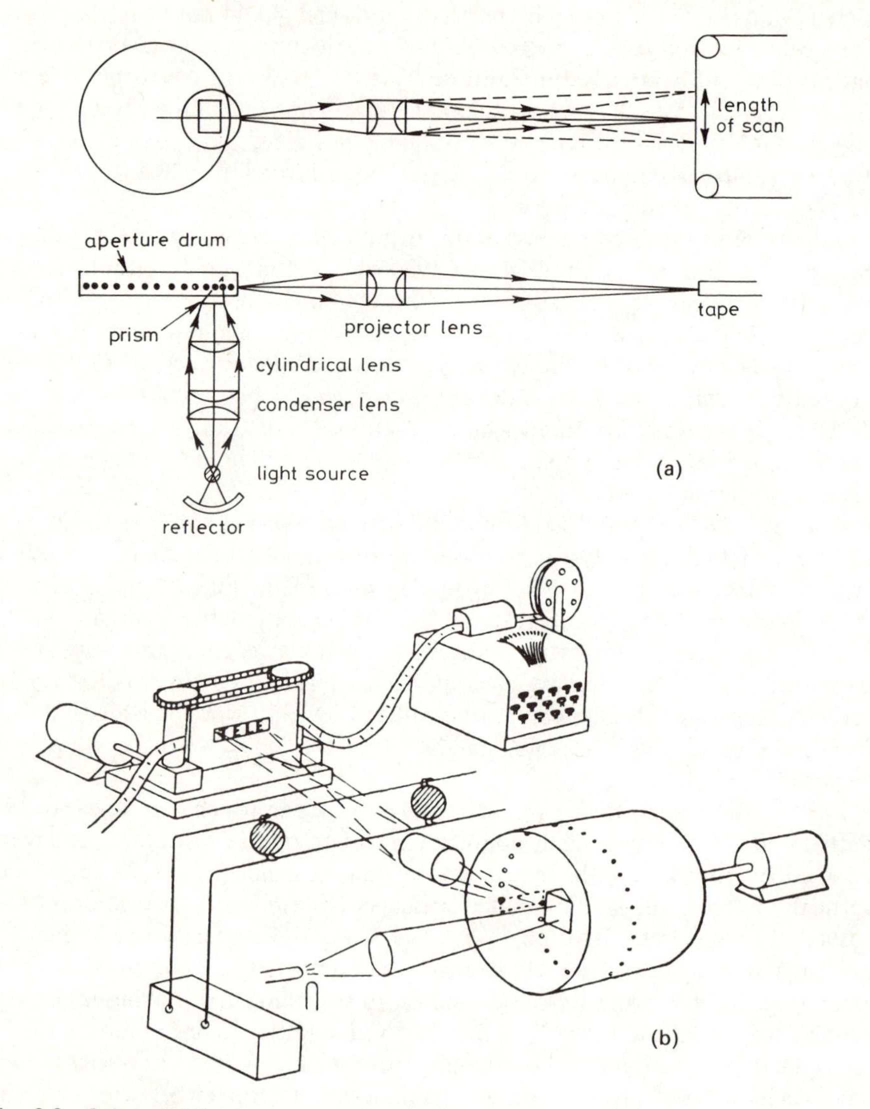

Fig. 9.3 *Schematic diagrams of telewriters*
a Marconi
b Baird

of transmission using strip scanning apparatus which allowed drawings, photographs, documents and the like to be reproduced at a distance. The drawing had to be cut into a series of strips which were arranged end-to-end to form a continuous band and the band then scanned. At the receiving end of the link the two-dimensional picture had to be reconstituted by cutting the receiver band into appropriate sections – the setions being placed side-by-side. Clearly this invention did not represent an advance on the several (Ranger, Siemens-

Fig. 9.4 *Baird telewriter*

Karolus, Belin, *et al.*), facsimile machines which were being introduced into the wireless telegraphy service at about this time.

Marconis' 'news by television' transmitter[11] is illustrated in Fig. 9.3b. A similar configuration of light source, condenser lens, reflecting prism, aperture drum and projection lens was used as in Baird's equipment, although the direction of scanning of the tape was different. Drums were provided for scanning the message in either 10, 15 or 30 lines per frame and the tape could be fed into the machine at either 6 feet per minute or 12 feet per minute. The transmitter was intended to be used in conjunction with a receiver giving a 'neon tube picture' of about 5 inch in length by about 1 inch in height and the dimensions of the typed letters were height of letter, 0.25 inch, breadth of letter, 0.12 inch. The number of letters to the inch was 5.

Apart from the lengthwise scanning of the tape, the Marconi scheme was very like the earlier system of Baird. The Marconi Company employed photocells situated behind the tape, unlike the later Baird model which used reflected light from the tape, but in a letter from Layzelle, the acting Secretary of Baird Television Ltd., to the editor of *Television*, the use of both opaque and transparent tape was noted. Indeed the use of transparent tape was not an original idea of the MWT Col., for Paule Ribbe, of Berlin, described in patent 29,428[12] (dated 31st December 1904) 'known telautographs and telephotographs, in which at the sending station a source of light emits a beam of rays of light through a lens and a small aperture in a screen, behind which a *transparent original*, such as a scripture, sketch, picture or the like, is reciprocated in a direction and fed after each reciprocation in another direction at right angles to the former direction, so that all the parts of the original are exposed one after the other to the beam of rays of light, which rays are altered or varied, before they further pass and act upon a selenium cell or other radiophone . . .'. The 'news by television' transmitter of the Marconi Wireless Telegraph Company thus did not represent any advance in the art of television and stemmed from Fleming's report on the possible applications of television to the company's interests, applications which were inspired by Baird – possibly based on Ribbe's work – and known to Fleming himself.

In his *Television* magazine account of the Marconi system, H. M. Dowsett made no reference to the historical background of his company's proposals, but stated simply:

> As it is the business of the Marconi Company to consider the various applications of wireless in the widest sense, its development of television has not been limited to the amusement field, and its research engineers have therefore been able to approach the problem from a different angle from that of broadcasting pictures.[11]

Following Layzelle's letter to the editor of *Television*, Dowsett[13] replied, somewhat unconvincingly, and mentioned that the Marconi apparatus was covered by British patent 371,288; but subsequently (in September 1932), Layzelle stated '. . . no patent of this number has been granted; all that exists is an application of this number, and this application has, in fact, been opposed by us'.[9]

The Marconi equipment for transmitting and receiving news messages was demonstrated for the first time during the meeting of the British Association at York in September 1932,[14] see Fig. 9.5.

Television images were transmitted by wireless on a wavelength of 750 m from the Marconi Works, Chelmsford, to St Peter's School, York (a distance of 180 miles), where the 15-line 'news' receiver gave a picture on a ground glass screen 25 inch × 3 inch. The receiver, which was of simple design (Fig. 9.6), consisted of one stage of high frequency amplification, an anode bend detector, one stage of low frequency amplification and a power amplifier. A sodium tube of the dumbbell type was mounted close to an aperture, and the modulated light was projected on to the screen by means of a mirror wheel, driven by a synchronous motor at 1200 rev/min to give 20 pictures per second. At the transmitter the lens drum and the tape transport mechanism were driven by a synchronous motor controlled by a tuning fork oscillator. Both picture and synchronising signals were radiated by the transmitter and the frequency band required by the apparatus was 13 kHz.

Fig. 9.5 *Marconi low definition television receiver shown at the British Association, York, in 1932*

Press reaction to the 'new television marvel.' seems to have been one of considerable surprise and delight. The *News Chronicle* reported:[15]

> Astonishing secrets of a new Marconi television marvel are out – with the result that:

> Messages typewritten in London can be read immediately on a screen thousands of miles away
> Cable services of the world will be revolutionised
> Business men may soon expect to see world prices flashed on a screen on their office walls
> News may be transmitted silently on home television sets
> And Scotland Yard wil be able to catch more criminals

The reporter of this news item clearly failed to realise that the apparatus was really only of use for the transmission of non-Roman scripts, for the telephone,

Fig. 9.6 *Internal view of the Marconi low definition television receiving apparatus. Horizontal movement of the spot is achieved by the apertured drum and vertical movement by the mirror lens*

facsimile, radio and teleprinter services could deal satisfactorily with the propagation of information employing Roman characters. Subsequent events showed a lack of enthusiasm for the invention, but at the time of the BA meeting considerable interest was engendered by the apparatus:

> When the schoolroom was opened, scientists crowded in to watch, and they broke into applause when in the darkened room the letters marked on a tape at Chelmsford began to flicker on to the television screen. The expected message came through with a thrilling clarity:

'Greetings from Chelmsford, England, to the BA at York, August 31 to September 7 1932.'[15]

Actually the televising of black letters on a white background posed fewer technical problems than the televising of objects and scenes possessing graded tones and essentially the method employed was comparable in difficulty of realisation to the transmission ot shadowgraphs, which Baird and Jenkins had achieved in 1925.

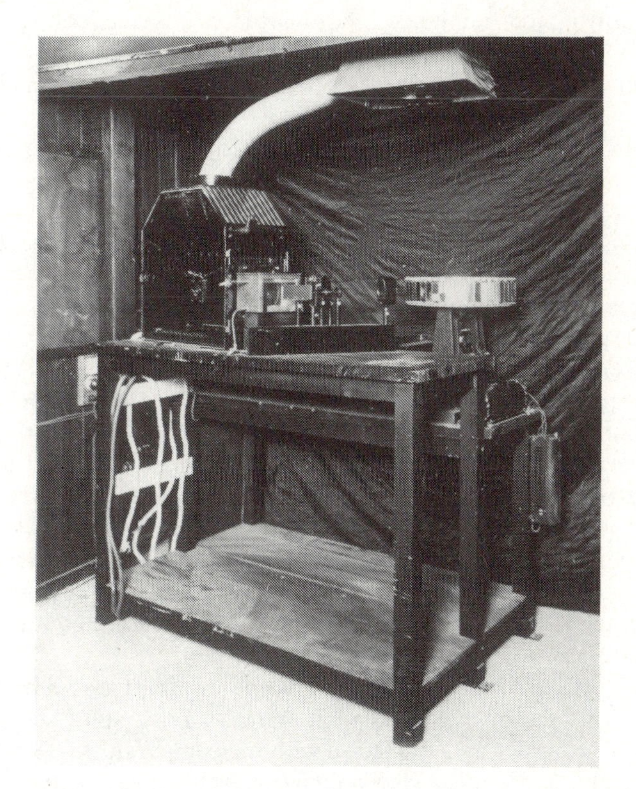

Fig. 9.7 *Marconi low definition television receiver used for large screen projection. A Kerr cell was utilised as a light modulator (1931/32)*

Possibly some of the enthusiasm for the demonstration was due to the emergence of the great Marconi Company as a rival to Bairds. For years Baird and Hutchinson and the Baird Companies had received a mixed reception from the press, but now, so it was thought, the respected Senatore Marconi, whose achievements and successes for the well-being of the human race were well known, was taking an interest in the new science of television. According to the *News Chronicle* report it was the inventor of wireless broadcasting who gave instructions to the engineers at the Marconi Research Station to the effect that they were to concentrate their activities on the commercial possibilities of television as distinct from the entertainment aspects. Dowsett, Plaistowe and Kemp were the engineers responsible for the development of the 'news transmitter and receiver' and the report stated that they had been experimenting for two years. This fact shows that Marconis must have initiated their research programme after the receipt ot Fleming's report.

Baird's reaction to the BA demonstration was certainly characteristic of him. On the 4th September 1932, the day after the *News Chronicle* report, the *Sunday Dispatch* mentioned that the Baird Television Company had announced that they were applying for an injunction for alleged infringement of patents against the Marconi Wireless Telegraph Co. Ltd.[16] The Scottish inventor had worked extremely hard for many years on the problems of television, had suffered much personal discomfort and abuse in his quest for recognition, had achieved many notable 'firsts', and consequently did not take kindly generally to anyone who attempted to steal his thunder.

It would appear that the Baird Company was not successful in its application, for at the Television Exhibition in the Imperial College of Science, London, in April 1933, the Marconi Company again demonstrated the 'news' transmission of letter-press over short land lines.[17] The receiver on this occasion was different from the one employed in the previous year and incorporated a multiple Kerr cell, with associated polarising and analysing optical devices, a high intensity arc lamp and a special double mirror wheel for projecting the two scanning beams through a curved diffusing screen of size 7×1.5, (Fig. 9.7).

During the year that heralded the Wireless Company's interest in television in its own right, negotiations took place for a merger between the Baird and Marconi Companies. According to some autobiographical notes left by Baird 'The Marconi Company got in touch with us in 1932 and were anxious to join forces. We had numerous meetings, I went up to Chelmsford and was shown round their television research department. Many meetings and luncheons followed and the whole stage was set for a merger.'[5]

Baird's too brief account on this matter is not only imprecise in its chronology but also gives no reasons why the great Wireless Telegraph Company should have wished to consider an association with the comparatively recently formed television company. Perhaps Marconis felt it lacked ideas in the new field of science; the company certainly had not shown anything which could be described as highly original at the British Association meeting in York – although the company's exhibits were excellently designed and engineered (see Figs. 9.8 – 9.11); perhaps the wireless firm found Baird's patents a stumbling block to its development of mechanical television systems. It may be that Marconis felt that a combination of their highly developed engineering skill and Baird's proved inventive ability would produce a powerful organisation for the furtherance of an industry which, in 1932, appeared to have an assured and prosperous future.

There seems little doubt that, had a merger taken place, the fortunes of the Baird Company would have been drastically changed. The Television Company would have had access not only to the patents and expertise which the Marconi

Wireless Telegraph Company had established in the fields of antenna, transmission line and transmitter design, but also to some highly important television patents, including the British patent 369,832, which covered a television trans-

Fig. 9.8 *Marconi 100-line lens drum scanner of the early 1930s*

mitting system using an iconoscope type of electron camera (Zworykin's invention). These patents would have been denied to Electric and Musical Industries Ltd. (which subsequently ousted Baird Television Ltd. from its position of supremacy), and together with their lack of expertise in designing transmitters might have retarded their television development programme.

In his book[18] on Baird, Moseley mentions that negotiations with the Marconi Company for a merger were begun before Sir Harry Greer joined the Board. An examination of some of Baird Television Ltd.'s letters (which list the Directors of the Company on its notepaper), shows that Greer became a director between 12th April and 9th June 1932. At this time Lord Ampthill was Chairman of the company but by 24th January 1933 his place had been taken by Greer.

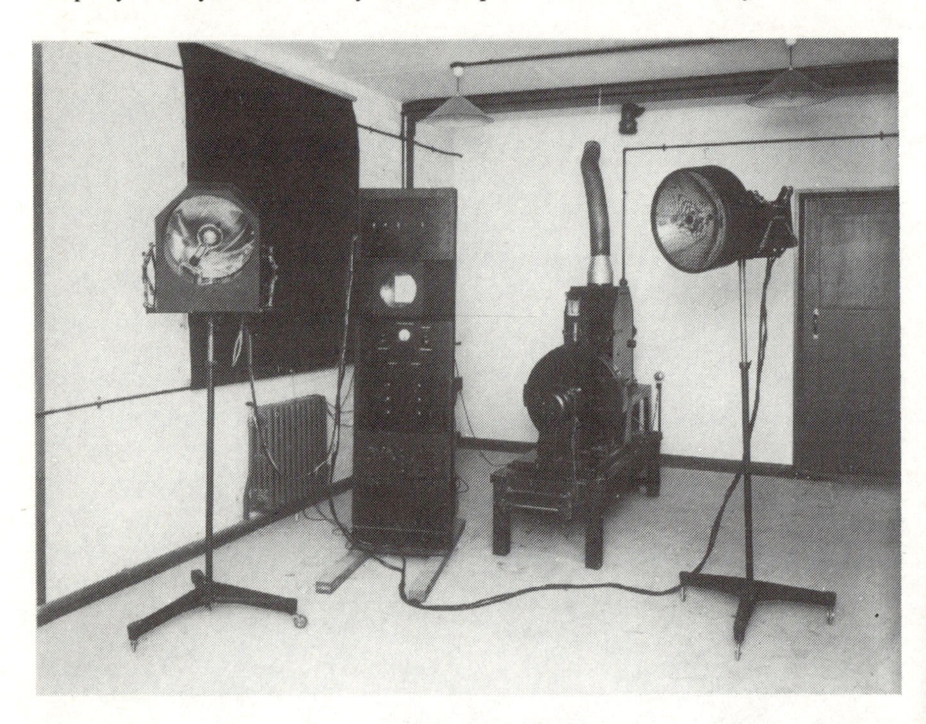

Fig. 9.9 *Marconi 100-line television equipment (1931/32)*

Sir Harry Greer, a director of several companies, and Mr Harry Clayton, an accountant who later became the Vice-Chairman of Baird Television Ltd., were the Gaumont-British Company's nominees on the Board of the Television Company – put there by Isidore Ostrer after he had taken control of the company in January 1932.

Moseley has written: 'It was obvious that Greer and Ampthill could not "mix". In fact, they were at variance from the start and, after two meetings, Lord Ampthill resigned from the Board.' Greer's succession to the Chairmanship (with Moseley still the Vice-Chairman), did nothing to further the well-being of the company. 'It was not long,' wrote Baird, 'before trouble began to blow up between Sir Harry and Sydney (Moseley), and this came to a head over

Fig. 9.10 *Marconi transmitter Type TT5. This used a lens disc for projecting the flying spot and an aperture disc for monitoring*

our negotiations with the Marconi Company which had reached an advanced stage.'

Moseley was 'heartily in favour of such a merger,' whereas it appears the new Board of Directors was more interested in selling receivers (which would yield

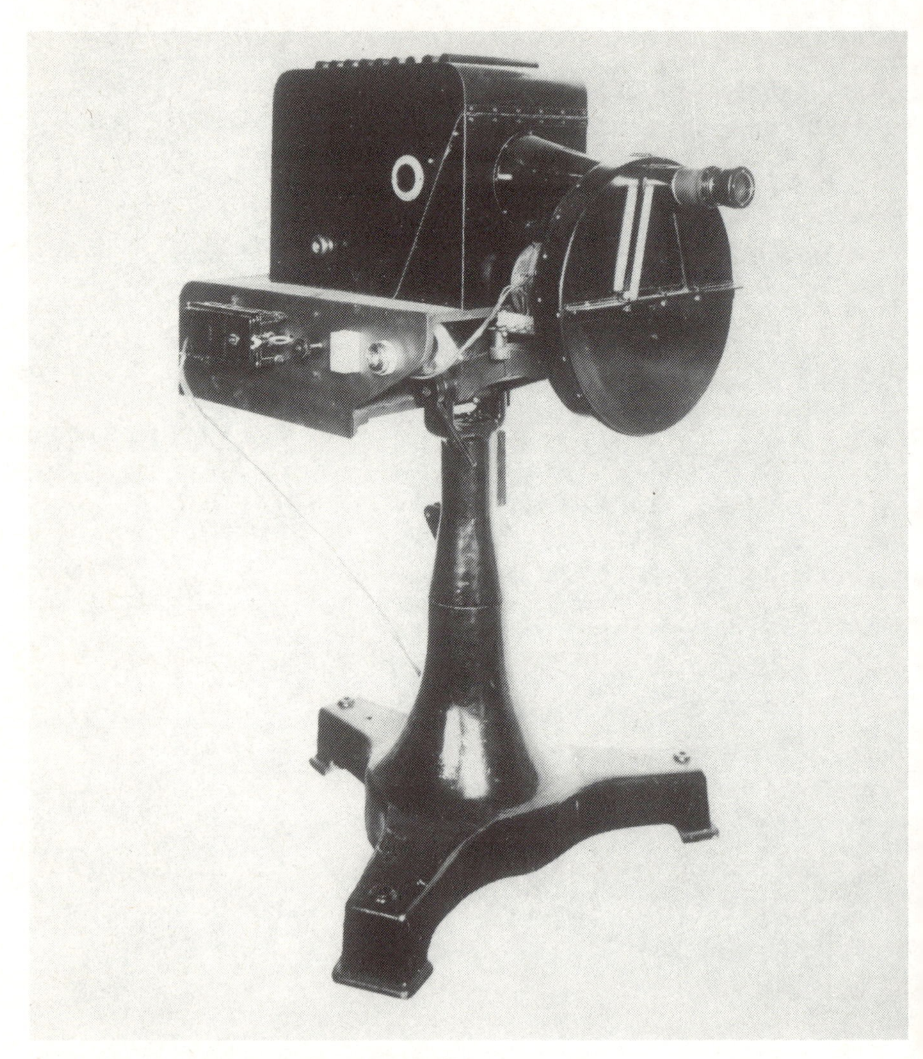

Fig. 9.11 *Marconi flying spot scanner (1932)*

profits from sales), than in reaching out into the unknown of visionary developments (which would consume large sums of money).

Moseley wrote:

> I only wish we could have reached (an agreement) for, if we had been able

> to join forces with the Marconi Company, Baird's future would have been assured, and probably investors in his company would not have lost their money. In my opinion, however, the Marconi people were never very serious about this proposition – at that stage anyhow. A number of visits had certainly been exchanged by the technicians – in our respective laboratories and I will not dispute my old friend's recollection that 'there had been a succession of meetings and dinings with Sydney in the foreground'. But, when he wrote that the Marconi Company was 'very keen' and that 'little remained to be done except to draw up the agreement', the wish must have been father to the thought.

Rather ironically, it was Moseley's success in obtaining the support of Ostrer in January 1932 which led to the eventual collapse of the proposed amalgamation, for, as noted previously, Greer and Clayton were Ostrer's nominations to the new Board of the Television Company.

References

1 FLEMING, Sir A.: 'A report on the Baird television system', February 1930, Marconi Historical Archives.
2 ZWORYKIN, V. K.: 'Improvements in or relating to television', British patent no. 369,832, application date (USA) 17th July 1929, (UK) July 1931.
3 AISENSTEIN, L. S. M.: letter to H. A. White, 28th June 1929, BBC file T16/67.
4 MOSELEY, S. A.: 'S. A. Moseley writes from Berlin', *Television*, June 1929, pp. 175–176.
5 BAIRD, M.: 'Television Baird' (HAUM, South Africa, 1974).
6 BAIRD, J. L.: 'Improvements in or relating to television and like apparatus', British patent no. 324,049, application date 10th October 1928.
7 Anon.: 'Studio topics', *Television*, November 1930.
8 Anon.: '"News" by television', *Television*, October 1932.
9 LAYZELLE, R. E.: letter to the editor, *Television*, September 1932, p.270.
10 BAIRD, J. L.: 'Improvements in or relating to facsimile telegraphy', British patent no. 299, 076, application date 20th June 1927.
11 Anon.: 'News by television. A Marconi development', *Television*, July 1932, pp. 164–166.
12 RIBBE, P.: 'Improvements in automatic photographing telegraphs', British patent no. 29, 428, application date 31st December 1904.
13 DOWSETT, H. M.: letter to the editor, *Television*, August 1932.
14 Press release (MWT Co.): 'Demonstration of Marconi television', 5th September 1932, Marconi Historical Archives.
15 Anon.: 'Marconi secret out', *News Chronicle*, 3rd September 1932.
16 Anon.: 'Television lawsuit', *Sunday Dispatch*, 4th September 1932.
17 Press release (MWT Co.): 'Demonstrations at Imperial College', 5th April 1933, Marconi Historical Archives.
18 MOSELEY, S. A.: 'John Baird' (Odhams, London, 1952).

Chapter 10

Financial difficulties, 1931–1932

In August 1931 Murray wrote to Moseley:[1]

> The experimental transmissions by the Baird process have been going on now for nearly four years. It seems to me that, at least at the beginning of next year, there should be some variation. Surely these transmissions have yielded all possible experimental data by now. If so, their indefinite continuance in the present form is a waste of money by the Baird Company, besides being the cause of some inconvenience and irritation to the Programme and Engineering Staffs of the BBC.

Murray was convinced that a better scheme could be evolved and introduced and thought a meeting would be desirable. He said that his colleagues and himself had been considering whether they might not discover new ways in which the BBC might help the Baird Company to expedite the progress of British television and went on to say that the BBC was naturally anxious that British television should retain and increase its margin of superiority.

Murray suggested the BBC should examine Baird's patents and technical apparatus in order to make a confidential assessment, and further that Bairds should give a demonstration under ideal conditions to show the actual progress made during the past year, as well as any new development that had emerged from research.

Moseley was naturally very pleased to learn of the BBC's desire to help Bairds:[2] 'I think it is vital in the interests of this country that the one great force which is able to help it should come to an arrangement with the Baird Company to safeguard the interests of British Television'. His enthusiasm for the BBC's new interest in the Baird Company's system was to be short lived. The following month the BBC adopted a policy of minimum concessions consistent with amicable relations[3] and Moseley was to write 'The venomous hostility of the former Chief Engineer has crystallised into a kind of cynical indifference'.[4]

Following the overtures contained in Murray's letter of the 12th August a meeting was arranged between the Chairman of the Board of Governors of the BBC and Baird on 17th August 1931.[5] 'He seemed extremely well disposed and anxious to help us', wrote Baird to Moseley,[6] who was in America at the time.

Following these discussions certain constructive proposals were put forward by Ashbridge and Murray,[5] it being understood, first, that all the Baird transmissions would continue to be experimental and described as such, and second that the company would defray all expenses incurred, and:

1 that the five half-hour transmissions per week in the morning be abandoned in favour of transmissions totalling not more than two hours at times not inconvenient to the BBC but more suitable to the Baird Company;
2 that one of the two midnight transmissions be abandoned;
3 that there be about once a week, when convenient, television transmissions of some routine programmes such as Jack Payne's band;
4 that, if possible, some television transmissions be given on the North Regional Transmitter;
5 that some transmissions be given on the Scottish Regional Transmitter later;
6 that while, of course, the BBC could not take over Bairds programmes. they could advise them how to effect considerable economies by eliminating a lot of expensive and unnecessary material which they were using at present.

Shortly after the above meeting Murray wrote a memorandum for the attention of the Chief Engineer of the BBC summing up the position *vis-à-vis* Bairds and the BBC at that time.[7]

> The agreed policy is to keep the pace as slow as is compatible with the maintenance of decent relations. It has been a considerable relief not to have to deal with a guerilla warfare on that front. At the same time we are careful not to concede too much for peace.

Murray's opening paragraph gave the statement of policy of the BBC towards Bairds and seemed in conflict with Reith's statement to Baird that he was asking those concerned to continue to explore various avenues of constructive co-operation.[8] In fact only two relatively minor concessions were accorded to the Television Company; first, the waiving of the extra charge which the BBC wished to make under the heading of engineer's expenses, and second the granting of facilities in connection with the company's portable transmitter.[7]

As for the 'absurd demands' which Lord Ampthill had made in his letter of 11th November 1930[9] – he had requested only a one half-hour transmission every evening (although the German Post Office had granted facilities for the broadcasting of television between the hours of 9.00 and 10.00 p. m. on five days each week and an hour at midnight every night except Saturday and Sunday – both the Chief Engineer and Murray definitely resisted the idea of intrusion into programme time.[7] The BBC was procrastinating, (Fig. 10.1).

Moseley, who was an intelligent, capable person, was aware of this and was to send an extremely strong letter to Murray complaining about the negative attitude of the BBC.[4] Murray wrote of Moseley being in one of his vicious periods[10] and requiring separate treatment and temporary segregation when the BBC was having discussions with Bairds.

The position regarding the portable transmitter which Bairds had developed was that the company was anxious to have it installed in a BBC studio for possible experiments in programme time. This proposal had been raised when Murray and Ashbridge visited the company towards the end of 1930. Subsequently the portable was installed in the large wharfside broadcasting studio in London, known as No. 10.[11]

Fig. 10.1 *Baird low definition transmitter and monitoring receiver (1929/30).*

Murray's account of this episode illustrates the rather dissimilar attitudes he adopted in relation to Bairds and the BBC. Referring to Bairds' request he wrote:[7]

> We fenced a bit, but indicated that we might do something about using the portable in one of our studios as an auxiliary to one of the ordinary television transmissions outside programme time. Subsequently I actually arranged for the portable to come over here during the first week of January. Then there was some technical hitch at their end. This I exploited to the maximum, with the result that several months elapsed before we actually accepted delivery. I liked your idea of the 'silent exhibit' in No. 10 partly because if their attitude was still purely political they would be satisfied. Not so however. I have discovered that at a recent Board meet-

> ing, the staff were taken to account for making the company appear ridiculous and amateurish. This being the case, I think we should take the initiative in moving the portable to Savoy Hill so that it may be used actually in connection with an ordinary Baird Transmission.

This policy of fencing and exploiting difficulties was in discord with Murray's views as expressed to Moseley a few months later:

> Since my conversion by you three years ago I have been intensely interested in improving the prospects of securing for Britain in television the same universal supremacy which the film has gained in America and I believe that this can best be attained by concentrated co-operation on behalf of the Baird process.[12]

John Baird later, on 1st September 1931,[13] met Murray and Ashbridge to consider the proposed changes in transmission times. Four points were noted but Control Board were to add a fifth:[3] 'That C. E. and A. C.(I), (Murray) handle . . . on the principle of minimum concessions consistent with amicable relations'.

John Baird received Murray's letter, which elaborated the points raised at his lunch with Ashbridge the previous week, on 9th September.[14] In writing it Murray felt it necessary to re-affirm the policy and attitude of the BBC towards the Baird Television Company. He listed four observations:[15]

1 the BBC does not admit the claims of the Baird Company to regular admission to programme time, nor does the BBC agree that television by the Baird or any other process, is yet within measurable distance of its service stage;
2 it is no part of the function of the BBC to concern itself directly with the development of commercial inventions, or to allow itself to be used by outside concerns as an instrument of research, unless the invention appears likely to become applicable to the service after a reasonable period of research;
3 subject to these general safeguards and provided that the regular broadcasting service to listeners is not prejudiced, the BBC is ready to do all it can to assist the development of television;
4 it is fundamental that the BBC would not be justified in entering into any financial arrangement for sale or partial control of any concern, which is in the position of the Baird Company.

Regarding the changes which Baird desired: first, the BBC could not agree to the present charges being reduced – it was felt that a substantial concession had been made as to the amount of expenses of the engineers, totalling about £300 a year; secondly, the BBC could not agree to more convenient times being allowed, say 10.30 p. m. on a weekday or 2.30 or 5.30 p. m. on a Sunday; thirdly, the BBC could not agree to Bairds having a rent free television studio at Broadcasting House; fourthly, the BBC could not agree to take over the whole of the Baird Programme Department under the BBC Director of Programmes. Following these rather hard decisions, which hardly accorded with the Director General's earlier special desire at a Control Board meeting that everything possible should be done to help Baird, Murray then went on to describe the concessions which the BBC was prepared to grant the Television Company (Fig. 10.2).

Fig. 10.2 *Baird copper lined studio, showing A. D. Calkin (1929).*

The concessions were explained to Moseley on his return from America.[16] He was furious with the BBC and sent an angry letter[4] to Murray pointing out that the development work of the company was being seriously hampered by the negative attitude of the BBC and that the results of Baird's meeting with Whitley were most unsatisfactory.

> Despite your personal assurances to the contrary, there is no doubt at all that the BBC regards television as a nuisance and would be glad to see it

'fade out', but we have no intention of obliging in this way. The venomous hostility of the former Chief Engineer has crystallised into a kind of cynical indifference. This being the case, effective constructive co-operation is endangered. The policy and action of the BBC reflect this attitude. Broadcasting House is actually being completed without any provision at all for television. Every resistance is advanced to proposals for better hours of transmission. Statements derogatory to the claims of Baird television are constantly made in calculated ignorance of the facts. Although the country is crying out for everything British, the BBC offers no active help either at home or abroad. What prospect is there that Baird Television will ever get a fair chance from the BBC? Therefore, after the proposed variations have been effected, we shall apply for separate broadcasting facilities, both aural and visual. We are assured of the definite support of influential MPs of all parties. Certain Cabinet Ministers are more alive to the need of developing this potential British industry than the BBC appears to be, and their interest in Baird Television has been continually active, even during the present emergency. They, of course, realise the vast potential value to British trade and prestige in the development and consolidation of the commanding position won for Britain by the demonstrated supremacy of Baird Television.

We hope the BBC will put no obstacle in our way. If we are such a nuisance to you, surely you will be disposed, if only on this account, to help us get separate transmitting facilities. If you oppose us you will be trying to work a grave disservice and, however, reluctantly, we shall have to re-embark upon another campaign to enlist popular support against what would be a tyrannous and indefensible attempt to extinguish us.

Murray's reply[17] was non-committal and referred to the concessions which had been made. Bairds dissatisfaction with the BBC was referred to at a Control Board meeting on the 6th October[18] together with his intention to apply to the Postmaster General for a licence and wavelengths to operate separately. Murray was instructed to again refer to the value of the concessions – particularly one and four – to Bairds and also the fact that the BBC had offered to co-operate with Bairds' engineers but this had not been taken up.[19] One point of information which was not to be disclosed to Bairds was that the BBC were experimenting with ultra short waves.[18]

The BBC was adopting an odd policy towards the Television Company: outwardly it had said it was ready to do all it could to assist the development of television; Reith had written to John Baird and told him there was the maximum goodwill towards him personally and Murray had told Moseley that the 'BBC was naturally anxious that British television should retain and increase its margin of superiority'; inwardly Murray had written how he had fenced and exploited difficulties. Control Board's policy to Bairds was based on the principle of minimum concessions and a reasonable request from Ampthill had been considered absurd. Now the BBC was to further implement its unhelpful policy by denying to Bairds any form of co-operation on the use of ultra short waves for television broadcasting although it had been apparent for some time that if high definition television was to be developed this development could only take place in the very high frequency bands. The BBC's defence was to be based on the fact that the offer of co-operation had not been taken up.

Certainly the Baird Company was well aware of this duplicity. 'I am sorry to say it', John Baird wrote to the Prime Minister, 'but it is not possible to rely on the promises of the BBC who have affected outward friendliness but have inwardly maintained a hostility which is difficult to understand, and I trust that it will be possible for you to take some action personally'.

Compared with the situation in America (see Table 10.1) the Scottish inventor felt his firm's progress was being hopelessly impeded through entirely inadequate broadcasting facilities.

In the United States of America, of the 22 experimental television stations which had been authorised by the Federal Radio Commission, (Table 5.1), nearly all were in operation.[20] Among the systems in use or projected was that of the Baird Company – as developed in Great Britain. Unlike the low definition transmissions in this country, most of the American stations operated on short waves and in most cases these were associated with the simultaneous transmission of sound and sound effects from a separate broadcasting station. The standard of the television transmissions sent out from the above stations was comparable with that broadcast by the BBC but considerable development work was taking place on utilising the cathode-ray tube for reproducing the received picture and ultra short waves for transmission. The National Broadcasting Corporation was at this time (September 1931) building a 5 kW, short-wave station in New York embodying an antenna 1250 ft above ground level (on top of the Empire State Building).

In Germany there was one main broadcasting company, the controlling interest in which (51 % of the shares), was held by the government. This company was a holding company and under it various subsidiary broadcasting companies were allowed to operate, the holding company maintaining some measure of financial, technical and programme control. Three or four of these subsidiary companies radiated television for about an hour in the middle of the day but the number of viewers was small and generally were technically interested parties.

Moseley's irate letter seemed to have some effect, for a few days after he wrote it, Ashbridge, the Chief Engineer, visited the Long Acres laboratories to see the company's recent developments – ostensibly, as he said, merely in accordance with an arrangement made some time ago that he should keep in touch with their development. He was favourably impressed.[21]

The company demonstrated their latest television reciever, which projected a picture on to a screen about 4 ft x 2 ft using a mirror drum and Kerr cell, and which was accompanied by ordinary conversation for head and shoulders images.

Table 10.1 *Columbia Broadcasting System, Experimental Television Programme – 3, Tuesday, December 29th 1931, W2XAB and W2XE New York*

2.00–6.00 p.m.	*Experimental sight programmes.* Card station announcements, and drawings of radio celebrities.
8.00 p.m.	*Hemstreet Quartet.* All-girl novelty group with Helen Andrews, soloist. Long shot group picture. White and silver backdrop curtains.
8.15 p.m.	*Grace Voss* – Pantomimes. Long shot and close-up. Silver backdrop curtain.
8.30 p.m.	*Senorita Soledad Espinal and her Pamperos* in a half-hour programme of Spanish and Latin American music and songs. Guitar Sextette and Mezzo Soprano. Group projection with various backdrop screens.
9.00 p.m.	*The Television Ghost.* Mystery character in weird costume enacts the murder mysteries in the character of one risen from the grave . . . the murdered!!
9.15 p.m.	*Hazel Dudley.* Song recital. Series of close-up pictures to be scanned.
9.30 p.m.	*Three-round exhibition boxing bout.* An experimental television demonstration of what the flying spot can do at a fight. Minature ring will be used. Blow by blow description by Bill Schudt. Dark backdrops. Long shot pick-up. Scanner will follow boxers around ring.
9.45 p.m.	*Gladys Shaw Erskine and Major Ivan Firth present.* Television novelties with visual illustrations. Alternate backdrops will be used.
10.00 p.m.	'*Tashamira*' introduces new German modernistic dances and technique. Extreme long shot focus. Close-up with varied backdrop curtains.
10.15 p.m.	*One man novelty band* with *Vincent Mondi.* Close-up against white backdrop.
10.30 p.m.	*Eliene Kazanova.* Violinist.
10.45 p.m.	*Grace Yeager*, song recital. Close-up shot of an artist singing semi-classic favourites.
11.00 p.m.	*Sign off*

Ashbridge noted:

> This picture was easily the best television which I have seen so far, and might be compared, I think, with a cinematograph 'close-up' say 15–20 years ago. It was quite easy to recognise the persons even after seeing them once and there was no difficulty in following facial expressions.

Fig. 10.3 *Dr. V. K. Zworykin with one of his camera tubes.*

The Chief Engineer was sufficiently impressed to observe that if it were possible for the ordinary public to buy such a receiver 'we should just have reached real programme value'. Such items as a well-known person giving a talk, or well-known actress performing and so on, would be of interest and entertainment value. Ashbridge's only stated reservation was that the apparatus was not in a fit state to be developed further commercially 'but there is, I think, reason to assume that this will follow in a comparatively few years time'. The equipment had the advantage that any number of people could view the picture (the only limitation being the size of the picture in relation to the distance), whereas in the case of some types of televisor it was necessary to stand in front of the screen.

> The next piece of apparatus I saw was an application of television to the 'singing arc'. The picture in this case, although large, was not very good, and I was not able to gather exactly what they have in view in attempting to develop on these lines. It may be purely a question of patents, or producing something cheap, but they were not very definite on this point.

One feature common to both these equipments was that they showed pictures which were grey in colour instead of the rather objectionable reddish colour which was obtained when neon lamps were being used.

Next Ashbridge was shown the latest model of the firm's ordinary commercial televisor. It was an instrument similar to the one installed at Savoy Hill 'but the picture that I saw was very much better, although fundamentally it had the same defects of ghost-like appearance and unnatural colour'.

The Chief Engineer's next demonstration was of a fairly simple type of mirror drum televisor – which, however, was not yet in a state to be marketed – which showed a picture about 9 inch x 6 inch directly and without magnification. This was fairly good, the colour was bluish-grey and it could be seen from any point in the room. Ashbridge felt that definite laboratory progress was being made, which should lead in a few years' time to television becoming of definite programme value.

Following these demonstrations Ashbridge had a talk with Moseley[21] about the general aspects of the situation concerning television. He reiterated his views that television was bound to take part in broadcasting eventually; that if the BBC adopted an obstructionist attitude then the Baird Company might lapse and television would be sold in this country on whatever terms the Americans liked to impose, that it was definitely up to the BBC to assist in the development of television in the interests of the trade and in its own interests, and finally hinted that the BBC might practically take over the Baird Company or alternatively waive all charges for transmissions and develop the technical side of their own laboratories. 'In fact', noted Ashbridge, 'he made practically every suggestion for combining the BBC and the Television Company that he could think of'.

This latter prospect had been known to the BBC for several months, for it had been suggested by both the Postmaster General and the Assistant Postmaster

General to Mr Whitley, the Chairman of the BBC's Governing Body.[22] 'They had suggested that the Baird Company was 'at its last gasp financially, and that the BBC might absorb it in liquidation, perhaps taking Mr Baird himself on the engineering staff of the Corporation'.

Ashbridge invited Moseley to suggest what he considered the BBC should do in connection with the transmissions of programmes. Again Moseley said that the BBC ought to televise items from the BBC's own programmes three or four times a week to begin with, without charge. He had in mind vaudeville, talks and plays.

> While not being impressed very much with what Moseley said, because a good deal of it was contradictory, I was much more impressed by the technical results which I saw, because I now think that a good deal of development has been made in the past nine months or so.

Ashbridge was coming to the conclusion that the BBC should not visualise the abandonment of television but that they ought to take steps to carry on with it, and was inclined to encourage the Baird Company to a reasonable extent in what they were doing. His view was based on the feeling 'that someone must develop television for broadcasting and if they do it adequately so much the better, if not sooner or later the BBC will be forced to do it at great cost to the listening public'. Additionally, if a Prime Minister, two Postmaster Generals and Members of Parliament were impressed with the Baird television of two years ago 'it was obvious that they would be very much more impressed by a demonstration of the latest equipment – which the Baird Company could give at any time'.

Both Murray and Ashbridge agreed that the situation had changed and that the tendency should be for the BBC to take the initiative, in some degree. If they failed to do this there was the possibility that the Post Office might grant the company facilities such as wavelengths outside the broadcast band, and this would not be to the BBC's advantage.

At last the BBC was wishing to take an interest in television. Whether the Chief Engineer's visit to the television laboratories was precipitated by the prospect of a further period of animosity and antagonism between the principals involved, as a result of Moseley's devastating letter, can be a matter for conjecture only, but following Ashbridge's report Murray submitted a memorandum to the Director General in which he argued for some form of co-operation to take place between the BBC and Bairds. 'One always had in mind that sooner or later there would arrive the right moment for the BBC to take the strategic initiative about television', he wrote.[23]

Murray continued:

> There is now a general re-awakening of interest to television. Fleet Street is alive to some of its possibilities: the *Daily Mail* is in treaty with the Baird Company for a publicity 'hook-up' beginning with a display at the election rally in the Albert Hall on the 27th inst. Politicians are notoriously

> amenable to the blandishments of Long Acre . . . One can count on the new interest being reflected in Parliament and for this among other reasons I think it would be a good thing for the BBC to take the initiative now, and to do it wholeheartedly.

Murray listed a number of suggestions in his memorandum. First, that three half-hours of television of the BBC's programmes should replace five half-hours in the morning; second, that a studio should be specially adapted for television work at Broadcasting House; third, that the charges made by Bairds should be discontinued, and fourth, that Ashbridge might find it worthwhile to keep a closer watch than before on what was happening at Long Acre.

The Assistant Controller (Information) thought the right direction for the future was for Bairds to concentrate on their commercial operations and for the BBC to look after the programme work. Once the initiative had been passed to the Corporation they would be in a position not only to counter any hostile use of television but also to determine the rate and manner of applying it to broadcasting, argued Murray.

Both Murray's and Asbbridge's memorandums were considered by Control Board on 27th October[24] and later by an Assessment Committee (2nd November 1931).[25] This committee concluded that the capital cost involved in a suitable programme of co-operation with the company was unlikely to exceed £1000 and might be much less, and that the revenue cost to the BBC for a one-hour programme on Saturday would be about £5500.

Following some prompting from Murray, who was anxious about any delay[26] in taking the initiative, Reith sent a letter to Lord Ampthill inviting him to pay a visit to Broadcasting House.[27] 'I have some suggestions to make . . . and feel that these should be done in the first instance to yourself alone'. On the same day that this letter was sent Control Board adopted the recommendations made by the Assessment Committee.[28]

Ampthill saw Reith on the 17th November and on the 19th[29] the Director General sent the Chairman of the Television Company a long letter detailing 15 proposals for a plan of collaboration between the Corporation and the company.

1 The company should provide and instal a complete television system in one of the studios at Broadcasting House – the equipment to be chosen by the BBC.
2 The BBC would pay a nominal hire rental for this apparatus.
3 The BBC would operate the transmitter in co-operation with the company's engineers.
4 The transmitter's installation would be subject to prior tests to ensure that no interference would be caused to other programmes by its operation.
5 The BBC would bear the cost for the necessary alterations to premises, installation of special lighting, extension of cables, etc.
6 The BBC would provide the necessary engineering staff at its cost.

7 The company would provide the maintenance and replacement of parts of the transmitter at its cost and would incorporate any improvements made by them.

8 The BBC would provide a weekly one-hour programme from 11.00 to 12.00 noon or 12.00 to 1.00 p.m. on Saturdays at the BBC's cost both with respect to programme staff and artists.

9 That:

a if Bairds desired they might continue the five half-hour transmissions as at present bearing the programme costs themselves – the BBC making a reduced charge for transmission costs;

b in this connection the BBC might later be prepared to arrange for two of the half-hour periods to be given at 12.30 p. m. on two nights a week, in which case they would undertake the programme costs of them;

c any programme which Bairds provided would originate in their studios.

10 The BBC's engineers would be permitted regular access to the company's laboratories with the object of improving any transmissions through the BBC's stations.

11 That if an experimental period, of whatever length it may be, should be regarded as having been satisfactorily concluded and additional television transmitters were required, Bairds would supply them at prime cost plus 10 %, no royalty being payable to the company in this connection.

12 That the transmissions be limited to Brookmans Park; vision signals being sent out on 260 m and sound signals on 356 m.

13 That the BBC should not bind themselves to continue the transmissions by the Baird system and should be able to give at any time an agreed number of months notice to terminate them either by abandoning the television transmissions or by carrying them on by some other method;

14 That the BBC should be free to give transmissions by other television methods whether the Baird transmissions were continued to not;

15 That if the Post Office agreed to institute an extra licence fee of 10 shillings for the use of television receiving apparatus, the proceeds less State charges would go to the BBC, the principle being that Bairds would make its money from the sale of television receiving apparatus and from licences to manufacturers.

At last, after much procrastination, the BBC was prepared to initiate a plan of co-operation for the development of television. Broadly they were to be responsible for the programmes and television transmission while Bairds would be responsible for technical development and the commercial exploitation of their receiving apparatus. Ampthill replied that such an arrangement would be agreeable to Bairds in principle.[30]

The company had incurred great expense in developing television (Fig. 10.4) and were obviously anxious to see the start of a definite service to the public so that they might recover some of their development costs from the sale of

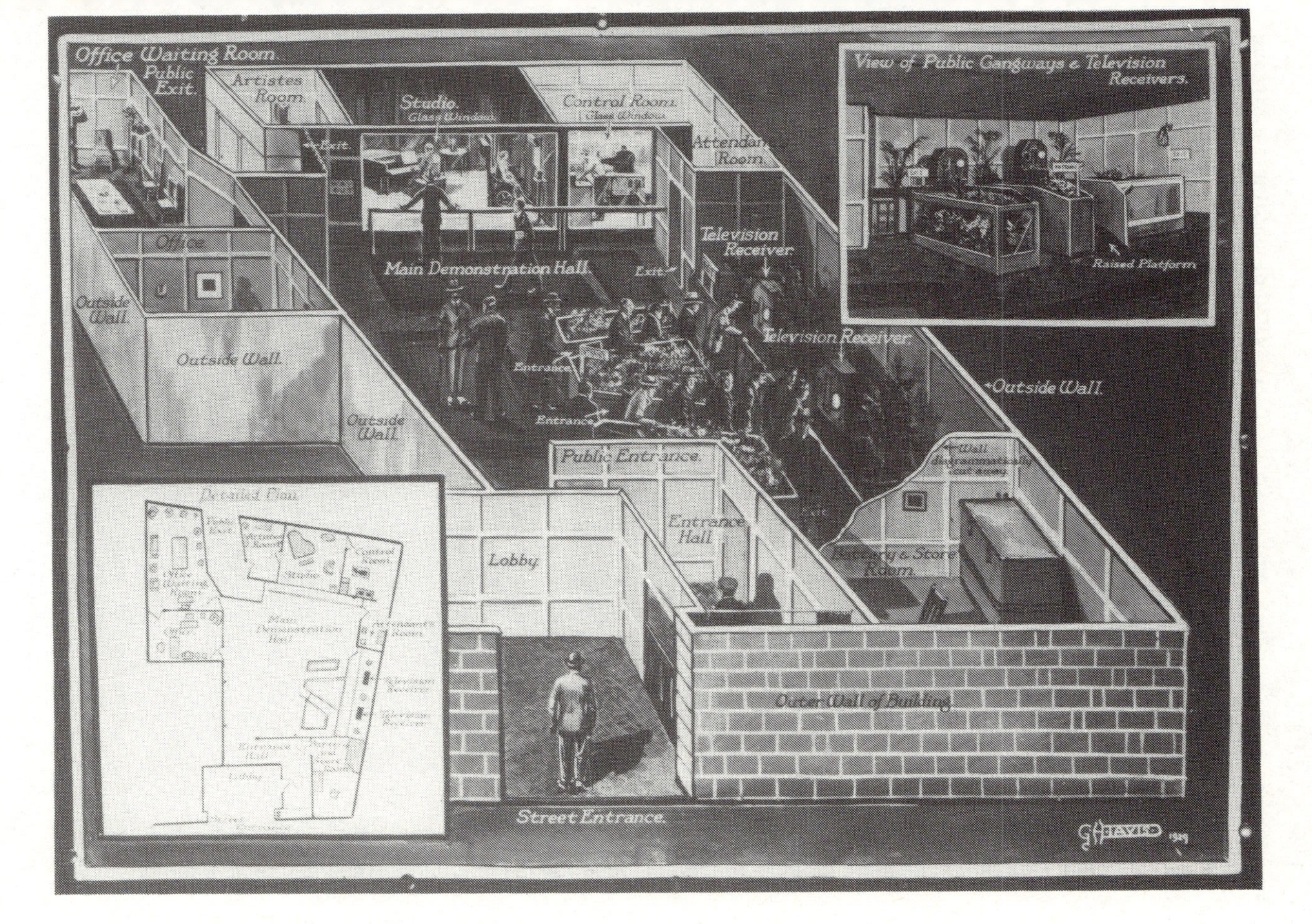

Fig. 10.4 *Exhibition of Baird television (1929).*

televisors. However, the public was unlikely to purchase television receivers unless it could expect a reasonable weekly television programme, and a one-hour weekly transmission was hardly likely to stimulate and maintain public interest or to encourage manufacturers. In addition it was essential for the trade to have some facilities for demonstrating televisors to prospective customers and for this purpose regular half-hour transmissions in the morning or afternoon would be desirable, as well as two or three evening programmes (not later than 11.00 p. m.

Ampthill wrote to Reith on these points but the Director General felt that the BBC's original proposal must stand.[31] The Chairman also raised the issue of some definite annual payment in consideration of what was in effect a service to the public. Possibly he had in mind that a certain portion of the licence revenue should be made available to the company for their services but Reith was adamant that in view of all the circumstances,which he did not wish to reiterate, they could not agree to this.

A further aspect regarding the sale of televisors was that the public would be unlikely to buy a set unless it could be assured that it had some security for what amounted to an investment. Similarly, manufacturers would be unwilling to establish production lines for televisors if the proposed service was to be limited to a few months only. Bairds thought an agreement for three or five years would be reasonable or failing this a period of one year for any notice of termination. The Corporation, however, did not feel able to guarantee a period beyond the minumum of six months.

Certain points raised by Ampthill in his letter were nevertheless agreed: first, that any improvement which arose out of Bairds own research or from co-operation with the engineering staff of the BBC would be made subject to a satisfactory agreement regarding patent rights; secondly, that the BBC's transmissions' allocation should not exclude the arrangement of special features such as the televising of the Derby, statesmen addressing the public at meetings and other events of public interest, subject to the BBC's discretion and with the cost borne by the Baird Company; and thirdly, that the transmissions should not exclude extension to the North Regional and possibly other stations as soon as conditions allowed – again subject to the BBC's discretion and in accordance with its assessment of development.

Having made these concessions Reith neatly terminated the correspondence between Ampthill and himself by stating that the matter appeared 'now to become one of consideration of detail and perhaps you will agree that the next step should be that discussions should be carried on between officials of the two concerns'.

Reith had given his views and did not want to be drawn into a long debate on the plan of collaboration; he did not feel that any purpose would be served at the present stage by a meeting of the Company's Directors and the Corporation's Governors as requested by Ampthill.[32]

Subsequently Baird, Moseley, Roberts and Boulding of the Television Company [33] met Ashbridge, Eckersley and Murray of the BBC[34] on 22nd January and 4th February 1932 to discuss certain points arising out of Reith's letter of the 22nd December.

Bairds were not wholly enthusiastic about the talks proceeding between officials of the two concerns and tried to impress on the Post Office the need to call a conference. Major A. Church, a director of the company, had written[35] to the Postmaster General, Kingsley Wood, and suggested that he should call a conference to discuss the television situation in this country. He pointed out that in Germany the Head of the Wireless Department, Herr Kruckow, had chaired a meeting to discuss the position in Germany. However, as Ashbridge mentioned,[36] the situation in that country was that there were several concerns interested in the broadcasting of television and therefore it was perfectly natural that a meeting between the concerns and the Post Office – which controlled the transmitting side of broadcasting – should take place. Here there was 'only one kind of meeting to hold, and that is between this one concern and ourselves' wrote Ashbridge. Church's ploy to use the Post Office to press the BBC for a high level conference did not succeed.[37]

During the meeting between officials Bairds agreed to accept sound transmissions on a wavelength of 399 m and vision on 261 m. This had the advantage that by using the new combination it was possible to transmit programmes at 11.00 p. m. on Monday, Tuesday, Wednesday and Friday, for 30 minutes each.[38] The programmes were to be an entirely BBC affair. It also emerged at the meeting that the Baird company did not desire to continue with the morning transmission, which could therefore be terminated. Another new and helpful point was that the company could undertake to hire as well as to sell sets on the understanding, in the latter case, that each set would be sold with a notice warning the purchaser that the transmissions by the Baird system might terminate at so much notice after a certain date. The acceptance of this point enabled the Corporation to give the Television Company rather more security than would be entailed by the six months notice agreement and the BBC suggested that the company should be assured that their transmissions would not be entirely discontinued until 31st March 1934 – discontinuance at that date or thereafter being by six months prior notice from the BBC. This did not imply the continuance of the four half-hours per week suggested or of an equivalent time but only that there would not be an entire discontinuance before that date. The representatives from the BBC also made it clear that the BBC was in no sense bound to refrain from televising by any other system at any time.

For Bairds the assurance about the continuation of the transmissions until March 1934 was an important concession, for television development was, by the beginning of 1932, accelerating rapidly. In this country the Gramophone Company had informed Ashbridge in December 1931[39] that they had effected an agreement with the Post Office for the erection of a transmitter for television experiments to be carried out on approximately 6 m wavelength and in addition there had been some press publicity that Marconi was going to undertake television developments in conjunction with some Italian broadcasters and that television experiments were to take place at Chelmsford. Later it appeared that Marconi had no new television development in view but Ashbridge had been informed in confidence that the company might, for publicity purposes, make some transmissions from Chelmsford to Australia.[40] Apart from the emergence of potential rivals in this country, Bairds faced stiff competition in the United States from the Radio Corporation of America, the American Telephone and Telegraph Company, General Electric and others. The British Broadcasting Corporation was aware of the various developments in the field of television and in addition realised the inevitability of this new form of communication. They had been much criticised for their lack of encouragement to the British contribution to its development and now felt that the time was opportune to acquire some knowledge and experience of transmission and programme techniques and at the same time appease the many critics of its former policy. However, although some agreement had been reached in January 1932, the details of the new plan of collaboration for the first public low definition television broadcast was not to take place until 22nd August 1932.[41]

Meantime further correspondence was exchanged between the company and the Corporation on their agreement. Bairds were not averse to anything in the nature of a warning notice being given when sets were sold but thought that instead of the notice stating that the television transmissions might terminate after the 31st March 1934, the notice should state that they would continue until 31st March 1934 at least.[42] The company was not being pessimistic about the possibility of high definition television being achieved using ultra short waves – indeed it had commenced development on the utilisation of these waves for this purpose.[43] It was keen, however, to sell receiving equipment, and this could only be done by giving purchasers some guarantee about the duration of the collaboration plan. Eventually, by 9th September 1932, a satisfactory wording for the notice relating to the sale of sets and kits of parts had been achieved.[44]

A further point associated with the warning notice was that the BBC had offered to give the Baird company a guarantee that television transmissions by their process would not be entirely discontinued until 31st March 1934. Bairds felt strongly that the word entirely should be deleted. Possibly they had in mind the eventuality that a rival organisation would produce a better television system before March 1934 and that the BBC would make available to the rival transmission time at the expense of the Baird transmissions; for Bairds television was a British invention which had been developed by British expertise in this country and the company did nor take kindly to any suggestion that television in Britain should be other than Baird television. However, there were ominous reports and rumours that this state might not prevail indefinitely. The company was keen therefore that the BBC should guarantee their broadcasts until March 1934 at least. The BBC was prepared to meet Bairds on this point to the extent of promising not to reduce the number of transmissions below two a week until 31st March 1934; discontinuance on that date or thereafter being by six months prior notice on either side.[45]

Bairds had, of course, hoped that the BBC would agree to refrain from transmitting television by any other system, but bearing in mind the purpose and spirit of the agreement they told the Corporation they relied upon them to refrain from transmitting by any other system unless that system was of a proved and unquestionably superior character. The BBC concurred with this view – providing that it was the judge on this point. This particular clause in the agreement was later to be used by Bairds against the BBC when the latter allowed EMI to install an ultra short wave transmitter in Broadcasting House.

Regarding the probable date for the commencement of the new agreement, the BBC thought Broadcasting House would be ready for the installation of Bairds' transmitter in May. In the meantime the BBC agreed to make an abatement of their charges for the experimental transmissions retrospective as from the 1st January 1932.

The meetings between Bairds and the BBC seem to have been conducted in a friendly and constructive way. The bitter outburst of Moseley against the BBC was smoothed over by the desire of the Corporation to contribute to the development of television, and by January 1932 Murray was writing privately[46] to Moseley and assuring him of his support to advance the hope that British television might achieve the same widespread standing which the film had acquired for the USA. Murray said that as long as he was with the BBC there would be no question of the Baird system of television being replaced by any other system unelss it was of a revolutionary advance. He promised he would continue unremittingly to make the BBC more television minded but realised this would be a task of unexampled difficulty which would require much patience, although it was one which he felt was not unsurmountable.

This attitude was certainly a change from his fencing and exploitation of difficulties and must have been received with considerable relief by the Baird Directorate. The prospects for a period of calm, cordial relations seemed very favourable, but they were almost jeopardised by Moseley's attitude during a visit by the Postmaster General, Kingsley Wood, to the Baird Laboratories at Long Acre in March 1932.[47] Ashbridge thought the demonstration was good but observed that Moseley delivered

> a long harangue about the development of a great British industry, suggesting that the whole question was more than a mere commercial proposition, and (was) concerned (with) the building up of a great public service. He suggested that there was a potential market of 4 million television sets, in fact most of what he said was hopelessly exaggerated . . . He hinted at some kind of Post Office or Government enquiry which, however, he left in a very nebulous state. He seemed almost to hint that there should be some enquiry similar to that which took place prior to the starting up of the British Braodcasting Corporation. The Postmaster General did not comment on this and I commented very sparingly.

In another private letter[48] from his home, Murray told Moseley that Ashbridge was furious with his attitude during the Postmaster General's visit and was disposed to break off relations. Ashbridge had obtained support for such an 'amiable' view and had received from a colleague an expression of great indignation at the nefarious policy of the Baird Company. Murray said he had endeavoured to ameliorate the situation by noting that Moseley had related nothing that was at variance with the honestly felt and well-known views of the Baird Company – the acerbity, if any, being a legacy from the impossible attitude of Eckersley.

While negotiations were proceeding between the two principals, a Television Committee was set up by the Director of Programmes to consider programmes, programme research, artists contracts and so on.[49] It consisted of Graves, in the chair, Wellington, Gielgud, Mose and Fielden. The first meeting was held on 22nd February and the committee agreed unanimously on six points: first, that a comprehensive demonstration of television be arranged so that its potentialities could be assessed; second, that an artists contract should be drawn up; third, that it was important, particularly from the artists point of view, to maintain the experimental and research nature of the work; fourth, that the television programmes should be kept entirely distinct from broadcasting; fifth, that the financial allocation for the four weekly half-hours should be maintained; and sixth, that while a particular individual might be made responsible for research on the programme side and for the preparation of detailed programmes, television reponsibility should rest with the particular member of the committee whose department would be primarily in charge of the work.

Regarding the cost of the four programmes, Gielgud thought that authority should be given for the programmes section to spend up to £100 per week, otherwise the standard of the television would not be in keeping with the general standard of the BBC's work.[50] Furthermore, the programmes had to be better than the experimental ones put out by the Baird Company, otherwise the BBC would be placed in a position where they could be attacked by Moseley. 'They have definitely got to be better than that and if we have to be restricted as much as we shall be on £3500 a year, I feel it is hardly fair to expect Mr Gielgud to maintain that standard which should be consistent throughout our work', wrote the Assistant Director Programmes.

While Baird was successful in his negotiations with the BBC, his attempt to secure some financial assistance from the Post Office was not. The company was without doubt in a difficult financial state, yet required money to enable it to compete successfully in the new fields of high definition reproduction, ultra short wave transmission and cathode-ray tube reception – fields in which their rivals were making progress. Unfortunately, whereas firms such as Electric and Musical Industries, Marconi Wireless Telegraph and the large American concerns had product sales to sustain their development of television, the whole of the revenue of Baird Television Ltd. so far as the broadcasting of television was concerned was derived from the very small profit on the receiving sets sold.[51] In addition, to compound their difficulties the firm was unable to extract any revenue from the numerous amateurs who built their own sets to see television.

However, the company had 'done all the pioneer work in this country at immense expense' and felt that it was entitled to ask in return for a small proportion of the licence fees to enable them to continue their research work on an adequate scale.

John Baird accordingly wrote to Kingsley Wood in April 1932[52] and suggested that a contrbution of 1 d out of the licence fee of 10 shillings should be made available to the company: '. . . this small sum though it is, would mean the difference between continuing our research work in a crippled and drastically restricted fashion, or going ahead vigorously'. Baird compared his company's inability to recover any revenue from amateurs with the position the BBC had found themselves in, in the 1920s, and pointed out that the position was saved for them by the government demanding a licence fee for the BBC.

Baird thought the prospect of the whole of the television industry falling to the Americans was a very real one, although the company had made strenuous efforts in Germany, France, the United States and elsewhere to advance British interests. 'I beg therefore', he wrote to the Postmaster General 'that you will discuss the situation with the Prime Minister who, I believe, is not unsympathetic. It is imperative that some action be taken immediately, otherwise I should not have addressed my request to you in these urgent terms'.

What Baird was really asking for in effect was an annual subsidy of about £20 000 per annum to aid his research and development programme. This sum could only come from the residue of the wireless licence revenue which accrued to the Exchequer and would be charged on the Post Office estimates and be subject to a parliamentary vote. Sir Evelyn Murray was not sanguine about this prospect:[53] 'I apprehend there would be many applicants with stronger claims than television'. Baird's analogy with the position of the BBC in its early days was a false one. The licence fee which was imposed on listeners-in was not to provide financial assistance for the manufacturers, whether for the purpose of research or otherwise, but to finance the maintenance of stations, the provision of programmes and so on, for which there was known to be a substantial popular demand. 'The group of manufacturers who put up the original capital in fact obtained nothing but a limited and modest rate of interest on their capital and its repayment when the company was liquidated', wrote Sir Evelyn Murray.

For Bairds the difficulty was that a popular demand for television did not really exist, although the company had gone to some considerable lengths to stimulate one. The retail cost of the lowest priced set was said to be 18 guineas if the buyer already possessed a high quality wireless set, or some £50 if he did not. Purchasers were unlikely to come forward in large numbers as long as television was limited to about two hours per week and prices were as high as these. Television was considered an amusement and an amusment de luxe and Sir Evelyn Murray could see no ground for financial assistance either from the Exchequer or the BBC: 'The contention that the looker-in should contribute by way of a licence fee as well as the listener-in may be perfectly sound, but there is no reason to expect a contribution, however small, from 4½ million listeners-in when not one in a thousand are lookers-in'.

Kingsley Wood replied[54] to John Baird's letter in appropriate terms but such was Baird's concern to gain some financial assistance that he wrote[55] again to the Postmaster General and asked to see him for a few minutes. Rather unexpectedly Baird found a sympathiser in Sir John Reith. For years the company had been something of a nuisance to the Corporation but now the Director General was writing[56] to Kingsley Wood and saying that the Corporation would support the idea of the company receiving some assistance from an arrangement similar to that by which the Opera Grant was given to the BBC.

Baird had acquainted Reith with the correspondence he had had with Kingsley Wood and was aware of the circumstances by which the Opera Grant was given and of the funds from which it came. Reith had told Baird that if some similar arrangement could be implemented it might necessitate some reconstitution of his company, for instance to the extent of a limited dividend (they had never paid any), but he said he foresaw no difficulty in complying with any alterations that might be required.

The opera subsidy was intended to stimulate the production of high class grand opera in England – 'an object which every person of culture would presumably accept as intrinsically desirable, though many might challenge the propriety of using public money for the purpose', but there was no analogy between this subsidy and a grant in aid of television. In the latter case the grant would be required not for production but for research, argued the Secretary,[57] and in any case television had no aesthetic and very little practical scientific value. An additional point was that grand opera had appreciable broadcasting value while television at that time had none. The position could be simply put: there were many more deserving objects of research if money were available, for example, medical research.

Following John Baird's discussion with Kingsley Wood on 31st May the inventor wrote another letter to the Postmaster General[58] pointing out the heavy expense the company had borne in developing television and that now the BBC was receiving the fruits of three years of unremunerated labour for which the Corporation was paying the company a nominal fee for the hire of the apparatus. This fee covered also free broadcast user rights in all the company's patents, inventions and developments, both in the past and anything the company might do in the future. Baird's latest suggestion was that the BBC should pay the Television Company £5000 per annum and that a further £10 000 per annum should be paid to them out of the proportion of the proceeds from the licence fees which went to the Post Office.

> In return for this the Post Office and the BBC would have the right each to appoint a Director to the Baird Company and the grant (would be) made upon the condition that the representatives (were) satisfied that the cash (was) spent on research likely to prove useful to the Post Office.

Baird softened his idea by suggesting that the arrangement could be subject to a yearly revision since the additional income likely to accrue to the BBC and Post Office might be considerable as television became more and more popular.

John Baird's personal wish was that the company should, in future, devote itself entirely to research work and that the broadcasting of television should be taken over by the BBC.

The Postmaster General's reply to Baird's request was as might have been anticipated:[59]

> I fear that it is quite impossible particularly so at the present juncture to ask the Government to add to the national expenditure on scientific research and even if funds were available you will appreciate that there would be many other claimants who from a practical point of view might seem to deserve a prior consideration. The question of a payment by the BBC is a matter for settlement between the Corporation and your Company and I am accordingly forwarding to Sir John Reith your letter of the 9th June.

However, Baird did not give up. He called to see Sir Kingsley Wood, the Postmaster General, on 21st June 1932 and asked him to put in a word with the BBC in support of his application for a more substantial payment from the Corporation. The prospect was bleak, however, and Kingsley Wood replied that he would probably have to call for further economies in the Department and elsewhere and hence could hardly ask the BBC to spend more money.[60–62]

Still Baird persevered. A further meeting was arranged for 12th July and this was attended by John Baird, Sydney Moseley, Noel Ashbridge and Gladstone Murray.[63] It was called to consider what action, if any, should be taken as a result of the decision by the Postmaster General not to give a government grant to the Baird Company. The latter reiterated their view that they had been treated unfairly from the beginning by the BBC and that it was now endeavouring to exploit their invention without adequate compensation. John Baird told the meeting that, following his conversation with the Director General, he had been left with an impression that if the government subsidy proposal failed, the BBC might support television to the extent of as much as £15 000 per year. However, Ashbridge and Murray made it clear there was no possibility of any such support from the BBC and that, in fact, the BBC would be justified in offering as little as 10 shillings per year for user rights, and further, the BBC would not tie itself to any one system of television. The discussion was rambling and somewhat acrimonious; however, there emerged a proposal that the BBC might consider 'paying the Baird Company £500 to £750 per year for user rights and rights in all Baird patents with breaks in March 1934 and March 1937 – the agreement to be terminated in 1942'. Nothing seems to have been concluded on this suggestion.

Baird had tried very hard to achieve some financial support and had failed. He had asked for a sum of money which was far bigger than the BBC could hope to pay; as Reith wrote at the time: 'I put that suggestion in its proper perspective by saying that that was what we paid for the whole of the wireless patents owned by Marconis, Standard Telephones, RCA etc'.[64]

During the activity to secure some recompense for his Company's efforts, Baird continued his experimental work: he was still anxious to seek publicity and secure world firsts for his inventions.

Ultra short waves were first used by Baird on 29th April 1932[65] when he gave a demonstration of television transmission using a wavelength of 6.1 m. This took place from the laboratories at Long Acre to an aerial mounted on the roof of Selfridges in Oxford Street – the scene of some of Baird's earlier experiments. Mr Selfridge was still a supporter of television and had a receiver working on the fourth floor of his shop. This first demonstration showed images of Dame Marie Tempest and Leslie Mitchell.[66] Mitchell later described the event:

> We were, I'm afraid, most inexpert, as at that time one had to stand rigidly in one spot, otherwise one's face bulged out of one side of the screen or the other, and one's forehead went over the top, or one's chin went below the screen. Dame Marie's comment indicated our reaction: 'Well Mr Mitchell', she cried, 'I must say that your voice was reasonably distinct'.

The test was of some importance because it not only demonstrated the use of ultra short waves but also because it used a new mirror drum type of home receiver which employed a Kerr cell.

Percy[65] mentions that Baird had decided to build six super home receivers giving a picture 12 inch x 6 inch and employing the mirror-drum/Kerr-cell combination in order to confound the critics who argued that pictures by the Baird system on television sets using neon tubes and discs were too small to have any entertainment value. The Kerr cells were actually made by Jacomb, and were built into test tubes, Jacomb and one assistant performing all the intricate assembly and glass work which was necessary. In addition to five receivers which were privately allotted in the London area, one was presented to the BBC and was installed in one of their press listening halls at Broadcasting House. Percy wrote that these sets were probably the ultimate in 30-line receivers, they were bright, presented a black and white picture and were exceedingly stable in operation.

The Selfridges demonstration marked a significant volte-face for Baird, who a few months earlier, while in America, had remarked to Errin Dunlop[67] the radio editor of the *New York Times*

> I observe that there is a movement towards the utilisation of ultra short waves in America. I am rather sceptical about their success in television because they cover a very limited area. I am of the opinion, based upon our tests in London, that the regular broadcasting channels are best adapted to carry the television pictures. We have not done much with the very short waves although I may later experiment with them using an aerial on top of the Crystal Palace.

As he did later.

Baird also rather rashly committed himself on the subject of the use of cathode-ray tubes for television during his American stay. 'He sees', said the *New York Times* report, 'no hopes for television by means of cathode-ray bulbs. He has developed what he calls a 'mirror scanning drum' which empowers him to cast images on the wall or screen . . . He asserts that the neon tube will remain as the lamp of the home receiver. For theatres he has developed a special arc light, which can be made to fluctuate rapidly in accordance with the incoming television signal'.

Fortunately John Baird quickly changed his mind about using ultra short waves and by April 1932 had obtained a licence for experimental transmissions from the Post Office,[68] much to the displeasure of the BBC, which had not been informed of the application.[69]

The use of a high frequency band for improved definition television had been suggested to the company in the letter which Sir Evelyn Murray sent to it in 1929.[70] If Baird had initiated experiments then on the utilisation of ultra short waves for the purpose of giving greater detail in television images he might have found himself in a better position than he did in 1932–33.

His view on the use of cathode-ray tubes had also changed by April 1932:

> We have also been conducting experiments with cathode-rays, and although at present mechanical methods definitely prove superior, cathode rays may in the future give us a television receiver of simple and silent operation. In the meantime I would, however, re-iterate that mechanical television is far superior.[71]

Shortly after Baird's demonstration at Selfridges the Television Company made plans to televise the Derby from Epsom on 1st June 1932.[66]

The broadcast was made through the BBC's transmitter and in addition there was a special transmission by landline to the Metropole Cinema, Victoria.

> I used the same van as the previous year for a much more ambitious experiment, and fitted up a large screen, nine feet by six, at the Metropole Cinema, Victoria. The transmitter was the same as that used the previous year and consisted of a large revolving drum. The picture sent out by the BBC was narrow and upright in shape, seven feet high and only three feet wide. To give a large picture at the Metropole I had three pairs of telephone lines from Epsom and sent out three pictures side by side. The three pictures thus formed one big picture on the screen at the metropole seven feet high and nine feet wide.
>
> The demonstration was one of the most nerve racking experiences in all my work with television, second only in anxiety to the Parliamentary Committee. The night before the show we were up all night putting finishing touches to the apparatus, and when the great moment drew near I remember literally sweating with anxiety. The perspiration was dripping off my nose.
>
> A vast audience had gathered in the cinema; even the passages were packed, and the entrance hall and the street outside were filled with a

> disappointed crowd, unable to get in. If the show had been a failure the audience would have brought the house down and I should have been a laughing stock.
>
> All went well. The horses were seen as they paraded past the grandstand. When the winner, April the Fifth, owned and trained by Tom Walls, flashed past the post. followed by Daster and Miracle, the demonstration ended with thunderous applause. I was hustled to the platform to say a few words but was too thrilled to say more than 'Thank you'.

The 1932 Derby transmissions were undoubtedly much better than those of the previous year. Two factors contributed to this improved performance; first, the use of zone television, and second, the much greater signal level used on the actual lines. The first Derby broadcast had been much impaired by interference and so Campbell, one of Baird's engineers, unofficially increased the permitted line voltage of 3 V to 30 V.

Meanwhile work was advancing on the installation of Baird's transmitter in Studio BB, three floors below Portland Place – a studio which had previously been reserved for danceband broadcasts and the like,[72] see Fig. 10.5. A BBC engineer, Douglas Birkinshaw, was appointed to the first BBC post in television, that of Research Engineer, and D. R. Campbell and T. H. Bridgewater were sent on loan from Long Acre to Broadcasting House. Between them they designed or constructed most of the apparatus which was used. Later both Campbell and Bridgewater were to be taken on by the BBC. It was soon noted that the initials of the surnames of these three engineers spelt BBC and this was felt to be a good omen. On the programme side, Eustace Robb, a former Guards Officer, was the first television producer and George Goldsmith was an advisor.

Although the first public transmision took place on 22nd August, Birkinshaw did not learn of the date until 3rd August.[73] 'To my amazement he was in ignorance of this and had not in anyway been informed of this important decision', wrote Robb in a strongly worded memorandum to Gielgud, 'nor consulted as to its practicability'. He continued:

> We are committed to four programmes a week to start with a fortnight's preparation work in rehearsals, auditions and experiments, a difficult task to start off with if we had a full working day on six days a week. We have been allotted 15 working hours a week, and are hedged around with serious hindrance from carrying out our work.

While appreciating the reasons for the hindrances imposed on them Robb had to consider the effect on his work and warned Gielgud that if the transmission started on August 22nd at the rate of four programmes a week he might be faced with the cancellation of an announced programme owing to lack of preparation and no time for rehearsals.

Still, the preparation went ahead and on 16th August[74] the BBC issued a press anouncement stating that a new series of experimental transmissions of television would take place weekly on Mondays, Tuesdays, Wednesdays and Fridays from 11.00–11.30 p.m. starting Monday 22nd August 1932; sound being broadcast from Midland Regional on a wavelength of 398·9 m and vision from London National on 261·3 m.

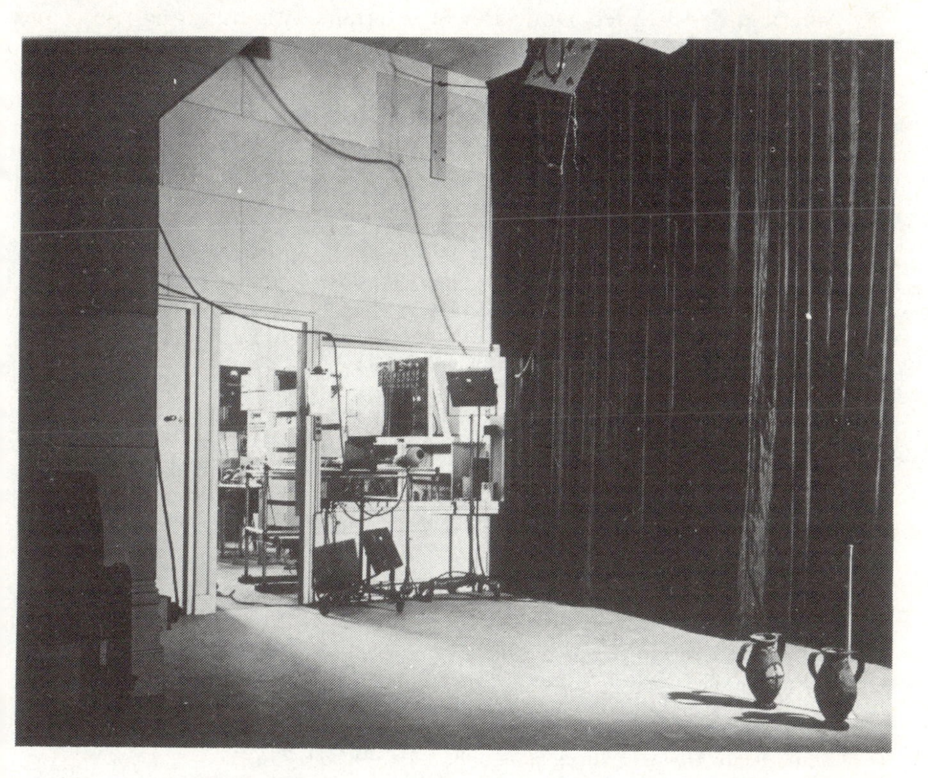

Fig. 10.5 *Television studio at 16 Portland Place looking towards the control room which can be seen through the projection window. The large black curtain screens the orchestra and their lights from the photocells. The cells are in position for close-up shots: one photocell is placed high in the ceiling for top lighting (1934).*

John Baird was most unhappy with this and sent a letter to Murray the following day.[75]

> The announcement omits the fact that the transmission is by the Baird process. I must confess to surprise and pain in seeing that the BBC's official announcement is not only very meagre but once more talks about experimental transmissions which I think is scarcely the case in as much as the BBC are beginning a regular period of transmissions as a result of the previous experimental transmissions being successful.

Baird's letter seems to have had an effect, for two days later the BBC issued another press announcement[41] stating that the inventor would be introduced to listeners as a preliminary to the first television transmission.

> The first programme of the new series of television broadcasts will be varied. Louie Freear, the comedienne who broadcasts for the first time, will sing 'Twiddly Bits', a number that she made famous in the 'Chinese Honeymoon', produced in the Strand Theatre in 1901. Betty Bolton will sing and dance; Fred Douglas will entertain with conjuring tricks and there will be light songs by Betty Astell.

The press release also announced that the Baird process would be used.

References

1 MURRAY, G.: letter to S. A. Moseley, 12th August 1931, BBC file T16/42.
2 MOSELEY, S. A.: letter to G. Murray, 12th August 1931, BBC file T16/42.
3 Control Board Minute, draft, September 1931, BBC file T16/42.
4 MOSELEY, S. A.: letter to G. Murray, 5th October 1931, BBC file T16/42.
5 Chairman's notes, 17th August 1931, BBC file T16/42.
6 BAIRD, J. L.: letter to S. A. Moseley, August 1931, personal collection.
7 MURRAY, G.: report to the Chief Engineer (BBC), BBC file T16/42.
8 Director General (BBC): letter to J. L. Baird, 15th January 1931.
9 AMPTHILL, Lord: letter to Lord Gainford, 17th November 1930, BBC file T16/42.
10 MURRAY, G.: memorandum to N. Ashbridge, c. August 1931, BBC file T16/42.
11 Anon.: 'BBC and experimental television. Portable transmitter in the studios', *Birmingham Post*, 1st May 1931.
12 MURRAY, G.: letter to S. A. Moseley, 10th January 1932.
13 Report on meeting attended by J. L. Baird, C. E., A. C.(I), 1st September 1931, BBC file T16/42.
14 BAIRD, J. L.: letter to G. Murray, 9th September 1931, BBC file T16/42.
15 MURRAY, G.: letter to J. L. Baird, 9th September 1931, BBC file T16/42.
16 MOSELEY, S. A.: 'Broadcasting in my time', (Rich and Cowan, 1935).
17 MURRAY, G.: letter to S. A. Moseley, 5th October 1931, BBC file T16/42.
18 Control Board Minutes: extract, 6th October 1931, BBC file T16/42.
19 MURRAY, G.: letter to S. A. Moseley, 8th October 1931, BBC file T16/42.
20 Anon.: 'Broadcasting in the U. S. A. and Germany', Minute 4004/33, file 18.
21 Chief Engineer (BBC): report to the Director General, and A. C.(I), 10th October 1931, BBC file T16/42.
22 Control Board Minutes: extract, 23rd June 1931, BBC file T16/42.
23 A. C.(I): memorandum to the Director General, 19th October 1931, BBC file T16/42.
24 Control Board Minutes, extract, 27th October 1931, BBC file T16/42.
25 Assessment Committee: report, 2nd November 1931, BBC file T16/42.
26 A. C.(I): memorandum to the Director General, 11th November 1931, BBC file T16/42.
27 Director General: letter to Lord Ampthill, 12th November 1931, BBC file T16/42.
28 Control Board Minutes: extract, 12th November 1931, BBC file T16/42.
29 Director General: letter to Lord Ampthill, 19th November 1931, BBC file T16/42.
30 AMPTHILL, Lord: letter to the Director General, 24th November 1931, BBC file T16/42.
31 REITH, Sir J. F. W.: letter to Lord Ampthill, 22nd December 1931, BBC file T16/42.
32 AMPTHILL, Lord: letter to Sir J. F. W. Reith, 11th January 1932, BBC file T16/42.
33 AMPTHILL, Lord: letter to Sir J. F. W. Reith, 21st January 1932, BBC file T16/42.
34 REITH, Sir J. F. W.: letter to Lord Ampthill, 13th January 1932, BBC file T16/42.
35 CHURCH, Major A. G.: letter to Sir Kingsley Wood, 22nd January 1932, BBC file T16/42.
36 Chief Engineer: memorandum to the Director of Information and Publicity, 28th January 1932, BBC file T16/42.

37 General Post Office: letter to Sir J. F. W. Reith, 25th January 1932, BBC file T16/42.
38 Statement on television, February 1932, BBC file T16/42.
39 Control Board Minutes: extract, 1st December 1931, BBC file T16/214.
40 Control Board Minutes: extract, 8th December 1931, BBC file T16/214.
41 BBC announcement, 19th August 1932, BBC file T16/214.
42 Baird Television Ltd.: letter to the BBC, 2nd February 1932, BBC file T16/42

Chapter 11

The emergence of a competitor, 1932–1933

The first, public, low definition television service started at 11.00 p.m. on 22nd August 1932 with the following introduction:[1]

> We are about to give the first of the new BBC series of television transmissions by the Baird process. Although television is still feeling its way, it has such potentiality that the BBC is lending as much of its time and resources as is consistent with the normal demands of the programmes.
>
> As it is the Baird process which is being used, we are glad to welcome its inventor in the person of Mr John L. Baird, who will now be televised.

Murray suggested that Baird should add the following acknowledgment:

> I wish to thank the BBC for inviting me here tonight and to express the hope that this new series of television transmission will lead to developments of broadcasting increasing its utility and adding to the enjoyment of the great listening public.

On the opening night both Ashbridge and Eckersley, who were present, appeared to be satisfied with the transmission (Fig. 11.1). Birkinshaw reported that everything went off without a hitch of any sort, either technically or in the programme. Considering the short time that Robb and Birkinshaw had had to arrange matters, this was very creditable. The apparatus was not installed in the television room until 8th August, but by Monday, 22nd August various items of equipment, including screens, had been obtained, certain modifications made to the studio and six test transmissions of vision sent out through London National. Baird considered the Tuesday evening transmission the best he had ever seen.[2]

Birkinshaw noted that there was a difference between the radio picture and the picture derived direct from the transmitter – the radio picture being inferior in quality – but remarked that, from the point of view of entertainment, the inferiority was more technical than apparent:

> The picture received by radio is not technically correct but is quite satisfying to the eye, as there is a good deal of contrast and the detail is quite

good, even in the extended position where we expected to get less detail. The movement of dancers, etc. can be easily followed and of course in a close up picture every expression of the face is exceedingly clear. The picture shows an absence of adequate l.f. amplification at the lowest frequencies, i.e. between 12.5 and 30 Hz. Also the cut-off at 10 kHz or thereabouts is of course manifested as a general loss of detail. Nevertheless, in my opinion, the result has considerable entertainment value and has certainly surprised everyone in Broadcasting House that has seen it.

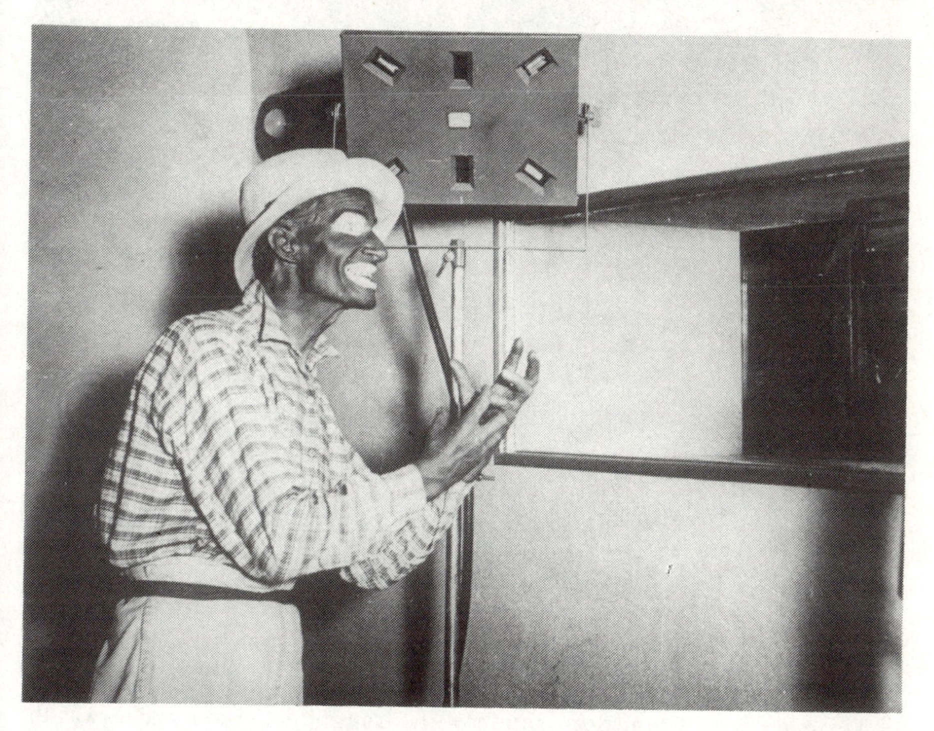

Fig. 11.1 *Fred Douglas being televised during the first programme of the BBC's 30-line television service (August 1932)*

The studio engineering operations were looked after by Birkinshaw, and Campbell and Bridgewater, the two Baird engineers.[3] They were taken on as maintenance engineers by the BBC from 1st September so that they could work with Birkinshaw permanently, as he found that there were three definite jobs to be carried out in the studio: first, the control of the projector; second, the manipulation of the cells in the studio, and third, the operation of the controls (Figs. 11.2 and 11.3).

The maintenance of the regular programmes and the necessary attendent rehearsals and auditions of new artists took up most of the three engineers' time, but they did find some opportunities to make a few improvements to the overall performance of the system (Figs. 11.4 and 11.5). A problem had occurred when a vertically rising curtain produced an additional spurious synchronising signal, but Birkinshaw corrected this by substituting a horizontal sliding screen. He later received appreciative letters from Rome and Paris indicating good reception of vision in those cities.[4]

Fig. 11.2 *D. C. Birkinshaw at the control panel in the BBC's television control room (1932)*

The main cause of the degradation of the transmitted vision signals was the line linking the studio to the Brookmans Park transmitter, but Birkinshaw prevailed upon one of the lines engineers to extend the equalisation of the frequency characteristics to 20 kHz.

The BBC was particularly fortunate in having two first-class enthusiasts for the posts of television research engineer and studio producer. Birkinshaw, the ex-Cambridge science graduate, and Robb, the ex-Guards officer, achieved some very considerable improvements in both programme and transmission quality, but unfortunately not without some difficulties and procrastination from more senior officials of the BBC. In a very interesting internal report[5] written nearly ten months after the start of the service, Robb listed 13 points which had given rise to problems: 'The majority of these needs were pointed out some ten months ago, but none had so far met with attention, that is to say though technical manipulation and studio technique have made leaps forward, equipment has not progressed at all.' Robb's list included studio, control room and dressing room accommodation, fading units and movable microphones, dual transmitting apparatus for picture fades, ventilation and other matters (Fig. 11.6). He seems to have had many good ideas – like Birkinshaw – but they were now encountering (as had Bairds for some time previously), a lack of enthusiasm and a rather slothful attitude from the BBC's oligarchy. Robb wrote:

> The wise man combines vision of the future with the facts of today. Can we not apply this precept to television. As I have so often pointed out the great danger is a policy of drift which will nullify the good work already accomplished by television, work which has kept us ahead of other countries. This advantage should enable this new art to play an important part in the reduction of unemployment and the increase of national wealth by the creation of a new industry.

Fig. 11.3 *Baird 30-line scanner installed in Studio BB, Broadcasting House. The operator is D. R. Campbell (August 1932)*

Both John Baird and Sydney Moseley had made similar comments on numerous occasions over a period of years. Now the BBC was hearing the same arguments from its own staff. Lance Sieveking had given the lead in 1930 but to no avail. For Robb, the future was clear:

> As certain as I write, if television is not helped to expand, it will suffer the fate of so many other new inventions which have become the valuable property of foreign countries who exploit them to their own very great advantage and to the very great detriment of our army of unemployed. Previously private people and organisations were to blame. In this case the blame would fall on the Corporation. I have often pointed this out previously. The danger is so great I must do so again.

Baird and Moseley would have been heartened to have read that.

Fig. 11.4 *Mirror drum scanner – the type used in the Portland Place control room*

Working under difficulties, Birkinshaw and Robb effected a number of improvements which Robb described in his report. As these were associated with the first British low definition, public television broadcasts, it is interesting to note the points made in its development:

> 1 I think I may claim that this invention of proper make-up to overcome the peculiar reactions to photoelectric cells and certain limitations of frequencies has succeeded in crispening the radio picture received. In any event, the attempts of any previous experiment were not availible to me so that present discoveries have been evaluated from zero.
> 2 The innovation of visual announcements and illustrated sub-titles.

3 The first use of suitable scenery and sets.
4 The introduction through the co-operation of the television engineers of continuous movements from extended to close-up position, thereby obtaining a perspective which was impossible at the outset.
5 The special study of costume and its modification for television use: a subject of extreme importance, as the wrong use of colour will nullify the picture, and an outstanding example was Madame Adeline Genés world farewell. Her brilliant white satin dress, which was striking on the Coliseum stage, destroyed the television picture. I had carefully to supervise its alteration to ensure successful transmission, otherwise her appearance must have been a complete failure. I mention the case of Madame Gené as an outstanding example, but such events are of daily occurrence.

Fig. 11.5 *Control room of the television studio at 16 Portland Place, showing the sound control engineer (T. H. Bridgewater) on the left and the vision control engineer (Mr Bliss) on his right (April 1935)*

6 Valuable work had been done bringing representatives of different schools of dancing to broadcasting and adopting their art to the exigencies of television, thus paving the way for the eventual expression of this art to the great radio public. I regard this move as one of great impor-

tance which can only have a cultural effect on a public too ignorant of the great beauty of dancing technique.

7 The transmission of art exhibitions by television is the beginning of an era when the public will be taught to appreciate great works of art, seeing them in their homes and at the same time that the finer points are demonstrated by an expert lecturer – in other words illustrated talks.

8 Animals, trick-cycling balancing acts, roller skating, all of which have been done, are useful for the light entertainment programme of the future as being the means by which the ear will be relieved of the intolerable strain of concentration by the eye. The appearance of Mr and Mrs Mollinson, Mr Maxton, MP, General Critchley and the experimental work now being carried out for the perfection of film transmission by television indicate the form that news bulletins will take in the not so distant future.

Fig. 11.6 *Studio scene: 16 Portland Place (March 1934)*

Robb was clearly improving the art of television – in spite of all the difficulties put in his way (Fig. 11.7).

Not only were some of these generated inside the Corporation, but Bairds too came in for some comments. Robb felt that an important factor in the development of television was the BBC's lack of knowledge of the Baird Company's plans with regard to putting the new Kerr cell mirror drum televisor on the market. He thought that the BBC might justifiably expect to be taken a little more into his confidence 'in view of the efforts we are making to put over attractive programmes for television enthusiasts'.

Fig. 11.7 *Television programme, 9th June 1933. The London Marionette Theatre*

Robb was particularly keen to have the public's reactions to his programmes, and, as the RMA had estimated that there were 5000–10 000 sets in operation, he was desirous of fulfilling their needs.[6]

Reith took up this point with John Baird and suggested that there should be no secrets whatever between them and in addition told him that his correspondence from lookers-in should go to the BBC.[7] The Director General also wanted to know what the manufacturing position was. Baird replied that he understood every help had been given to Robb in regard to programmes, but that matters of policy and so forth were surely not within his (Robb's) province.[8] However, he was quite ready to discuss the matter fully and frankly with Reith and a meeting was arranged for 2nd December 1932.[9]

At this meeting Baird accepted Reith's view about the need for complete frankness and Reith suggested that Bairds should send regular, (weekly, fortnightly or monthly), information covering:

1 enquiries for apparatus, whether from private persons or firms, and the part of the country;
2 correspondence about transmission showing what people think about them (indicating also the number of people who have receiving apparatus in use);
3 their plans for the future, e.g. on the matter of manufacture.[10]

Baird acknowledged Reith's letter[11] on these points and took the opportunity to raise another matter which had been troubling him and which was to cause him some anxiety for the next four years; namely, the progress which was being made by Electric and Musical Industries in the development of television.[12] Baird could not acknowledge – for some considerable time – that EMI's television system was being developed by British workers in a British factory using British resources. For him the Radio Trust of America, through its associated companies in London, was the mainspring of EMI's progress (a very considerable bitterness is evident in both Moseley's and Baird's letters during the early 1930s on the progress, and support from the BBC, of the Hayes establishment). Baird was always ready to point out that his company could match the progress towards high definition television which was being made by EMI and that therefore the pioneer television company, i.e. Bairds, should be supported.

In his letter to Reith, Baird mentioned:

> We have spent considerable time developing apparatus for use with ultra short waves, both with cathode ray and mechanical scanning, and have gone as far as 240-line scanning. This apparatus however, is entirely unsuitable for the wave band owing to the immensely high frequencies involved, but the pictures produced are of course immensely superior.

Baird raised this matter because he understood that EMI had been giving demonstration of apparatus under laboratory conditions and he did not want these demonstrations compared with those which the BBC was sending out on medium waves.

> Our own results in the laboratory are far superior to those which we are sending out through the BBC, but such results are only of academic interest until the ultra short wave channel is sufficiently developed to pass the very high frequencies involved.

Baird's concern to stress this point was evidently his belief 'that the American Radio Trust through its subsidiaries (was) endeavouring to create the impression that it (had) something which (was) superior to ours'. The inventor said his company could supply transmitters with from 90 to 240 lines as soon as the ultra short wave service had been developed by the BBC's engineers.

Essentially Baird was pleading for more co-operation with the BBC, and, in particular, thought that if the same co-operation could be developed between them as existed between the Reichpost and their affiliated company in Germany the time taken towards the achievement of high definition reproduction would be reduced. 'In the circumstances, may I express the hope that the BBC will concentrate on this problem and your engineers be stimulated to give all available support and encouragement to the pioneer television company, that is the Baird Company.'

Fortunately Bairds had a friend in the BBC at this time in the person of Gladstone Murray. Moseley and Murray had over the years established a considerable rapport between each other, and now, 10th December 1932, Murray felt it necessary to write a personal and confidential letter[13] to Moseley informing him that there had been some important meetings 'yesterday' at Broadcasting House on the subject of television. This meeting followed a demonstration which Ashbridge had seen at EMI.[14]

On the 29th November, Shoenberg, EMI's research director, had invited Ashbridge to a private demonstration both in the transmission and reception of television. 'In my humble opinion,' said Shoenberg, 'they would be of quite considerable interest to you'.

Ashbridge visited the Hayes factory on 30th November and was shown apparatus for the transmission of films using three times as many lines per picture as the Baird equipment and twice as many pictures per second. He was impressed and thought that the demonstrations represented by far the best wireless television he had ever seen and that they were probably as good as or better than anything that had been produced any where else in the world.[15] He went on:

> . . . there is not the slightest doubt that a great deal of development, thought and expenditure had been expended on these developments. Whatever defects there may be they represent a really remarkable achievement. In order to give some idea of the cost of such work, I might mention that the number of people employed is only slightly less than that in the whole of our research department.

This represented an ominous situation for the Baird Company which simply could not match the resources of Electric and Musical Industries. It was these tests to which Baird had alluded in his letter to Reith and which Murray was now following up in his letter to Moseley. 'Alfred Clarke on Tuesday had obviously followed up Shoenberg's tactics with Ashbridge on the previous week', reported Murray.[13] Baird's letter to Reith had been timely for the company's fortunes at that time as it upset calculations so far as these related to the early acceptance of the Hayes system.

Murray described to Moseley what he had ben able to secure, yesterday, which would help Bairds:

> 1 Ashbridge to avoid meeting Shoenberg again for the present and to ask Baird to demonstrate what he can do on ultra short waves, both direct and with film.
> 2 The BBC policy to be to bring together Baird and EMI in order that the whole apparatus required from the BBC end may be housed in one place. Ashbridge will sound Baird at their next meeting as to the attitude of his company; thereafter doing likewise with Shoenberg; but it should be noted that Ashbridge's present impression is that Shoenberg will have nothing to do with Baird.
> 3 The BBC to carry out its arrangement with the Baird Company in any event.

Murray's attitude now towards Bairds was very different from his fencing and exploiting difficulties stance of 1931. His letter of 10th December clearly indicated his desire to warn the company of the threat to their well being imposed by the new rival. Finally, Murray had some advice to offer Moseley on public relations:

> I think you should arrange without delay to put a set into Ashbridge's house: he probably has heard that I have one, and this would anger him. So, if there is not another available, I suggest moving mine to his place early next week. This is important.

Murray was certainly trying hard to be helpful to Bairds.

The practical aspects of EMI's choice of television standards were:[15]

1 transmissions could only be by ultra short waves;
2 the definition was very much greater than in the case of the Baird system on medium waves;
3 there was less flicker;
4 steady improvement in the system was possible.

The actual demonstration Ashbridge attended consisted of the transmission of a number of silent films over a distance of approximately 2 miles by means of an ultra short wave transmitter using a wavelength of 6 m and a power of about 250 W. On the quality of the images Ashbridge reported:

> The quality of reproduction was good, that is to say one could easily distinguish what was happening in the street scenes and get a very fair

impression of such incidents as the changing of the guard, the Prince of Wales laying a foundation stone and so on. A film showing excerpts from a play was in my opinion not so good. Also it was possible to follow what was going on all over the stage. On the other hand excerpts from a cartoon film were definitely good. I think they could have given a better demonstration had they been in possession of better films. The ones they showed had been in use for several years. The size of the screen is about 5 inch × 5 inch but they have a second machine which magnifies this by about four times in area. The quality of reproduction can be compared with the home cinematograph but the screen is smaller.

EMI had asked the BBC to take up this system – on an experimental basis – for about seven or eight months and then later for regular use on their programmes. On this issue the Chief Engineer had the following observations to make:

1 There is no doubt that the film is the only way in which we can develop the television of actualities. I cannot see any method developing in the immediate future so as to allow of the direct televising on a satisfactory basis of, say, the finish of the Derby or the Wimbledon tennis matches at any rate within the next few years.
2 Transmission by film would be entirely satisfactory for plays and sound could be added on another shortwave channel if it were not suitable for broadcasting.
3 The above remark (2) would mean that our technique would need to change considerably and that, if the system were established on a programme basis, it would mean practically the establishment of a further alternative programme and this would be extremely costly, particularly since it would be necessary to transfer everything onto a film at any rate at the present stage of development.
4 Provided that sufficient apparatus were available, a film could be made and re-transmitted within a matter of only a few hours.
5 It seems to me highly desirable to develop the system so that direct television could be carried out of a studio performance. There seems some doubt as to the immediate possibility of this.
6 No incident could be televised on this system which occurred at a distance from the transmitter on account of the wide frequency band involved, that is to say that ordinary music line cannot be used.
7 The above implies that the studio and transmitter should always be fairly close together, that is to say a matter of yards, not miles. The range of these ultra short waves cannot approach the ranges we are in the habit of reaching with our ordinary programmes, but it would be possible technically to erect a number of transmitters in a number of important towns and cities.
8 The cost of a receiver to take two shortwave channels, one for television and one for sound with ordinary broadcasting facilities in addition is

estimated by EMI at about £100 so that the service if established would have to be looked upon as a luxury service, possibly entailing a special licence.

9 If we took up the transmitting side we ought not to have difficulties in connection with development on the receiving side, having regard to the fact that EMI are prepared to make all the receiving apparatus and even manufacture their own cathode-ray tubes.

EMI were very keen that some form of television service should be started on ultra short waves, and following Ashbridge's visit to Hayes, Mr Alfred Clarke, the Chairman of the firm, paid a visit to the BBC to have discussions with Reith.[16] Clarke was anxious to see the Director General as he wanted to put to him a definite question of great importance, namely 'what wavelengths would be adopted for television, what number of pictures, and what number of lines?'. Clarke hoped Reith would say that the number of pictures per second and the number of lines would be 25–30 and 120–180, respectively. EMI would then have been in a position to have started an experimental service on ultra short waves early in 1933 and probably before Bairds were in a position to do so.

Reith raised the question of payment for the proposed service: the BBC's view was essentially the same as that which they had stated to Bairds on a number of occasions, namely 'why should (the BBC) bear all the expense when in fact for a considerable period (they would) have no benefit at all and eventually only a slight one compared with that accruing to the manufacturers of sets'. Clarke's idea was to licence almost any firm that applied to manufacture receivers in the same way that the Marconi Company did with the ordinary receiving apparatus. He was, of course, at a disadvantage in any talks of this type as Reith had obtained some considerable experience on such points in his dealings with the Baird Companies: Reith added confidently in his memorandum, '. . . I got him to appreciate my points of view'.

Clarke actually wanted the BBC to run the apparatus in Broadcasting House while his company continued to experiment and incorporate any developments in the equipment. However, the Chief Engineer, who had obviously been impressed by his visit to Hayes, did not have much objection to the BBC carrying the cost during the experimental period, at any rate. This would probably not have been a very heavy one to have borne and would have involved the cost of converting the transmitter, staff and power charges.

Ashbridge wrote[17] to Shoenberg on the same day that the Director General had given his report of his conversation with Clarke and invited him to put the outlines of a definite proposal in writing, that is, the financial details arising out of the proposed co-operation. He also mentioned that he would like Condliffe, EMI's research manager, to call at Broadcasting House to view the accommodation for EMI's apparatus.

However, although Ashbridge had been surprised by the developments of EMI and was entirely in favour of further research being carried out in television, he disliked the rush tactics of the company (Murray had called them shock tactics). 'They are too much like the Baird Company's tactics in the early days', he said, 'and the object I think may be the following: to try and rush us into establishing a service as quickly as possible so that they can get well ahead in the receiver market for television sets, although no doubt they would be prepared to licence other people when it paid them to do so.'[18]

Ashbridge thought the BBC should make it clear that it was not in favour of a regular service, mass production and all the rest of it being started next autumn (i.e. 1933), and felt that at least six months further experimental work was required. The position was essentially this:

> Let us assume that they persuade us to establish a regular programme service next September: it would cost us shall we say £20 000 a year to carry this on even in a limited way and they would probably try and make us pay for the television transmitter and no doubt we should pay for the two ultra short wave transmitters, further accommodation and so on: they on their side might sell a thousand receivers in the first year to a few people who made a fetish of buying anything new and have the money to spend. This might go on for several years, EMI selling a thousand or so receivers each year and ourselves paying £20 000 a year to keep the service going. They would be making a small profit and we should be making a large loss, even if an extra licence fee were paid for television reception. This is very unattractive from our point of view. On the other hand, if a very large number of receivers were sold, that is to say if the public took the scheme up with enthusiasm, then I think we should be more or less bound to go ahead, partly because we are a monopoly concerned and partly because in any case television may eventually help us in our main objective. What, however, I dislike is the nature of our paying out a very large sum of money in order to let EMI make a profit of a few thousands a year for a year or two with the possibility of making a much larger profit later on but with practically no risk whatever, except the danger of losing the sum they have spent on development. In any case, they would gain considerable publicity. However, you may say that we are entirely justified in making what may amount to a loss of £100 000 in order to establish television at the end of a period of three or five years.

Ashbridge's conclusion was clear: 'I feel very strongly against trying to dump a rather doubtful service on the public so soon as next autumn'.

Against this background of the possibility of higher definition television on ultra short waves, the BBC had to consider the position of the Baird 30-line programmes on ordinary broadcast wavelengths in the medium band. These programmes had been started, first, to determine whether a good producer could make anything of them from the programme point of view in relation to public interest, secondly, to gain experience of the technical and programme problems presented by such a service and, thirdly, to give Baird the opportunity of progressing further in his research.

It had been recognised for many years that 30-line television would never be a satisfactory long term solution to the broadcasting of television, and with the advent of medium definition (120–180 line) television reproduction, the BBC had to give careful consideration to the future of the Baird process and in particular the mass production of low definition receivers.

Ashbridge did not think the programmes which Robb had produced were such as would hold public interest on a permanent basis and he therefore thought the BBC's agreement with Bairds should be terminated in March 1934 – on the specified date – and that the company should be told. In addition, he thought they should concentrate on the production of 120-line pictures for transmission by ultra short waves and also on the production of studio scenes, as this had not been carried out by any other company in Britain.

On the same day that Ashbridge wrote his very important memorandum, Clarke sent a letter[19] to Reith pointing out that his company's experiments had been carried out under country conditions and that it was necessary for a further period of experimental tests to be undertaken in town in the interests both of transmission and reception. It was necessary that the transmissions should be standardised to allow of the development of the proper design of receivers and so Clarke said he would like to put a transmitter in Broadcasting House of the kind which Ashbridge had seen at Hayes.

With Ashbridge in favour of research being continued in the field of television, Clarke's letter was entirely satisfactory to him – so far as it went.[20] Reith, too, had no objection to EMI continuing their tests from Broadcasting House, but thought the company should be told 'what we feel about their autumn production idea'.[21]

Events now moved quite quickly. Shoenberg submitted a scheme,[22] with a blueprint of the television layout of the transmitting equipment, to Ashbridge on the 11th January and the Assistant Civil Engineer made preparations for the equipment to be installed on the 8th floor of Broadcasting House in the artists' waiting room (12th January).[23]

Baird followed up Murray's letter to Moseley about EMI's shock tactics with a communication[24] on 13th January to Ashbridge informing him that the company would be able to show him not a 120-line picture, as EMI had shown him, but one having 240 lines, in the second week of February. He also hoped to be in a postion to show his company's latest results using a cathode-ray tube, although he did not think it by any means superseded mechanical systems. Baird stated that they had been working extensively on cathode rays both in this country and in Germany and had the advantage of all the research work carried out in their associated company's laboratories in the latter country. He clearly wanted to show Ashbridge that the company was not going to rely on 30-line television indefinitely, but that like EMI it was carrying out research work in the two important fields of ultra short wave transmission and non-mechanical scanning methods. An additional point which Baird raised was his disquietude at the activity of the American Radio Trust. Ashbridge, who had now formulated his views on the policy to be adopted by the BBC with regard to both

EMI and Baird Television, replied that Baird and he should discuss the matter, as it seemed there was a good deal of misunderstanding on the whole question.[25]

The meeting was held on 27th January and was attended by Baird, Moseley, Ashbridge and Murray.[26] Its purpose was to discuss the proposed co-operation between the BBC and EMI in connection with the latter's method of transmitting film television by ultra short waves.

The main points in the Baird Company's argument that EMI should not be granted facilities to transmit television were:

1 according to the Baird Company–BBC agreement 'the BBC should not transmit by any other method than the Baird system unless the proposed new method showed an improvement of a revolutionary nature';
2 that if the BBC did so they would be 'dealing a heavy blow at British industry';
3 that Bairds 'could produce as good television of the high definition type as could EMI'.

On the first point the BBC contended that this did not apply to research transmissions, but only to regular advertised programmes. On the second point, it was not true to say that the EMI system was American, having regard to the fact 'that it had been entirely developed and wholly manufactured by an English staff in English workshops and that therefore it had already given employment to a very considerable number of workers in this country'. The BBC also stated that only 27% of EMI's capital was in American hands.

Regarding the last of Baird's observations, the BBC thought it was regrettable that, if Baird's claim was true, they had not given a demonstration of their ultra short wave higher definition system by the end of February.[26]

For Moseley the BBC–EMI television situation had an altogether sinister significance: he said EMI was contemplating a new share issue in the near future and the BBC was merely playing into the hands of American interests. So violently did Moseley disagree with the BBC's views that it was considered advisable to continue discussions on a technical level only between Baird and Ashbridge. Baird told Ashbridge he hoped the BBC would postpone the EMI transmission until his company had given their demonstration, but Ashbridge was doubtful on this and re-iterated his view that the tests were to determine the suitability of using ultra short waves as well as testing the method of producing television.

On the future of 30-line television, Baird stated to the Chief Engineer that the Television Company was going ahead with the manufacture of receiving sets. Considering Baird had said his company could demonstrate 240-line television this was a wrong tactic; the sets were bound to become obsolescent within a relatively short time, as the 30-line transmissions were only guaranteed until 31st March 1934. With EMI working at great pressure on the development of a higher television system, Bairds must have realised that the low definition transmissions could not last much longer, but for John Baird the decision had been made by his Board and he could not add anything further. The Chief Engineer[27] stressed his observation three days later when he wrote to Baird 'we feel strongly that it would be a mistake to take the somewhat irrevocable step of selling numbers of receivers to the public for television reception by a method which was already obsolescent', and of his meeting with the Baird Company, Ashbridge told Reith that the meeting achieved nothing but 'I suppose we can take it that there is bound to be a certain amount of further trouble'.[28]

On the same day that Ashbridge wrote this, Reith informed Clarke:[29] 'I think your proposed arrangement is in general quite satisfactory. You do not mention the question of publicity, but I am assuming that you will not wish to make any press statement in connection with these activities.' Subsequently Reith referred to the rumours of a new issue and told Clarke: 'I am sure you would never allow anyone to refer to the television experiments in connection with the issue without reference to us, and I cannot imagine your wishing to do so in any event'.

With memories of stock market jugglings still clear in his mind, Reith[30] was very anxious that the latest BBC venture in the field of television collaboration should be kept private and not used for publicity or any scheme which might cause the public to invest its money in a doubtful venture. He was to be reassured by Clarke, who told him:[31]

> Such a thing as an issue of any description has not been contemplated or even discussed by the Board. The spreading of the rumours which you have brought to my notice in your letter, together with statements which have appeared in the press as emanating from a quarter which can hardly be described as friendly to us, make it desirable that I offer you, as I now do, any information you may reasonable require on questions bearing on my Company or on its activities or intentions in the field of television. You will, I know, prefer to have such information directly from us.

Murray[32] naturally supported Reith on the issue of publicity, joint or otherwise, 'until such time as we were prepared to be fairly definite about the future. When the inevitable enquiries begin to be made by the press, the answer should be simply that the BBC conducts research and experiment continuously and that there was nothing new to say about television.'

Ashbridge was right when he forecast trouble. The day after Moseley's meeting with Murray and Ashbridge, Moseley sent a letter to the Postmaster General, Sir Kingsley Wood. Moseley was in good form: 'I wonder whether, in the welter of cynicism of modern politics, there is any sincerity in the plea for "British First?"'. He was strongly of the opinion that British pioneers should be encouraged, not by finance but by every other legitimate means.[33]

'I am not satisfied that the BBC realises its duty to the country in this respect,' he wrote and observed: 'The BBC which holds a monopoly by virtue of a Charter granted by HM Government, seems to me to be extraordinarily cynical where the rights of a British sister science are concerned.' On the question of the 'tentacles of the Radio Trust of America' extending throughout the world, Moseley feared that, if the PMG did not take steps immediately, one of them might 'force a means of "muscling-in" through the back doors of the BBC'. The issue was clear. If this was attempted, it would be a public scandal and Moseley would not hesitate to call a public meeting.

Phillips, who was always conscious of any political repercussions relating to matters affecting the Baird Company, immediately wrote to Carpendale asking for his observations.[34] However, Reith gave the BBC view that it had not undertaken not to carry out research on television by any method distinct from public transmission of advertised programmes.[35] Nevertheless, there was unnecessary alarm at the Post Office and they arranged for Gill to have a talk with Ashbridge on the 9th February.[36] Murray reported: 'Gill is going to represent that it would be better politically to conduct these experiments elsewhere than at Broadcasting House. Phillips said something about a suggestion from Downing Street to the PMG on this point.'

Following the Assistant Chief Engineer's meeting with Gill and Faulkner of the GPO on the 9th – at which the latter were told of the BBC's plans *vis-á-vis* the BBC and EMI – Bishop thought they went away satisfied.[37] Faulkner had asked why the work was to be carried out in Broadcasting House and was told it was necessary because Broadcasting House was a high building in the centre of the city. This was a perfectly reasonable explanation as no experiments on ultra short wave transmission with 120-line television modulating signals had been carried out in this country from a city centre site. The effect of other large buildings and objects on the characteristics of the propagation of the waves was a necessary investigation before a regular service could be started, and, of course, a high building was required in order to give an adequate range. It had been known for some time that the usable range of ultra short wave propagation was limited to the optical range and this was determined solely by the heights of the transmitting and receiving antennae.

Bishop later, 13th February 1933, accompanied Gill and Faulkner to Hayes to see a demonstration of television. 'They were very impressed indeed and considered that the work which HMV had done was an outstanding development in television which ought to be given every encouragement. I have no doubt that they will report to Mr Phillips in this sense.'[38] Gill was impressed and in his report stated:

> Reproduction was very good and very much in advance of anything which has hitherto been witnessed. There was a complete absence of any suggestion of scannning lines across the picture and the definition, while not as good as would have been obtained by direct projection on to a screen, was sufficiently good to enable details of the picture to be easily perceived.[39]

At the demonstration Shoenberg took the opportunity to put EMI's case to the Post Office and told Gill the Company was primarily interested in the sale of receiving equipment and that they had only ventured to develop transmission equipment because nobody else had produced a system which had a sufficiently good programme interest. This had been Eckersley's view from the outset of the Baird desire to establish a service and also accorded with Ashbridge's latest opinion of the 30-line service. From the BBC's position there really was no reason why Bairds should be concerned about its offer to EMI. 'If your claim to superiority over other systems can be demonstrated on a practical basis, I cannot see what you have to fear from the procedure explained to you at the meeting last Friday, 27th January,' Reith told Baird.[40]

However, Baird was not going to let the matter drop quite so quickly or so easily as the BBC would have wished. His colleagues and he had carried on protracted and often acrimonious negotiations and correspondence with the BBC in the past when rivals in this country did not exist. There was probably more at stake for the company in its present position than at any time in the past. Then, they could stimulate the public's imagination and interest with rather crude demonstration of television and good publicity – and no competitors: now there was a competitor who could demonstrably show better results than Baird had ever shown. Time was runing out for the company. EMI had established a lead which it would be hard for Baird to take over, and notwithstanding the 27 % American interest in EMI their system of television was British. The position of the television company was clearly a matter of some concern to Baird and his directors but Moseley and his colleagues had some valuable experience to draw on when it came to agitating for their company's welfare. EMI was a rival, and Bairds, apparently, were not averse to using any fact or rumour which might show the Hayes Company in a poor light. Moseley had raised the question of a new issue by the Hayes Company and now, 4th February 1933, John Baird was to continue this line of attack. He wrote to Reith mentioning that he had heard of 'an alarming report from the city which (stated) that an attempt may be made in May to raise money from the public on the strength of an arrangement with the BBC in regard to television. I do not know the company referred to, but it certainly is not ours' he added significantly.[41] There could only be one other possibility if the rumour was true.

Baird was well aware of the BBC's concern that the public should not invest their money in schemes which were essentially untried or insufficiently developed and raisd the spectre of Wireless Pictures to strengthen his point: 'May I remind you of the harm done to the BBC through a similar announcement made by Wireless Pictures, in which the public lost every penny it invested'. This line of attack really was unworthy of Baird: the case of Wireless Pictures was entirely different to that of EMI. In the former situation a company had proposed a system for transmitting a still picture which obviously would never have found a public demand when an alternative means of transmitting moving pictures was available – albeit in an undeveloped form – whereas, in the latter case, EMI had shown a system which could give a superior picture to any existing in Britain at the time.

Baird, in another letter of the same date,[42] suggested that a meeting would be desirable, as he did not understand what the BBC meant by the procedure explained to him at the meeting last Friday, 27th January. Meanwhile, Clarke's attention had been drawn to the rumour, and, in a telephone call to Reith, he thought that he had better visit Reith. Clarke said there was no truth whatever in the statement, but that there was obviously something funny going on and had Reith seen the *Daily Herald* of January 12th[43] (Moseley was associated with the *Daily Herald* as its radio critic).

Clarke was on safe ground. The relationships between the BBC and EMI were always cordial and correct, notwithstanding Shoenberg's shock tactics with Ashbridge the previous December. Unlike the Baird organisations, Electric and Musical Industries never engaged in open publicity or attempted to catch the public's eye. Indeed, so shrouded in secrecy were the company's achievements in the field of television that, when the 405-line system was revealed for the first time to the public in 1935, one newspaper described it as a Marconi Company system and completely ignored the contribution of EMI,[44] while another, anxious to put an individual before the public as an inventor, published a photograph of Senatore Marconi, alongside that of John Baird, as inventors of the two rival systems, although Marconi himself had never engaged in television work. Whereas the name of John Logie Baird was known to a large section of the public as the inventor of television, the press were never able to attach a similar appellation to an individual at Hayes. Even Shoenberg kept well away from newspaper reporters, although he could have gained some considerable publicity for his research team's efforts.

Accordingly Clarke told Reith the true situation about EMI[45] and received the Director General's observation: 'You can rest assured that irresponsible press comment on your company and its activities are appraised on their true value here.'[46]

On Baird's request for a meeting with Reith, Bishop, the Assistant Chief Engineer, advised Murray that it would be most undesirable at the present time since there seemed to be nothing more to discuss.[47] Bishop saw no objection to a memorandum being sent to Baird setting forth the BBC's version of what took place at the meeting held on 27th January 1933, and advised the DIP accordingly.

Sir John Reith replied to John Baird's letters of the 4th February and told him that no useful purpose would be served by having a further meeting.[48] Regarding the issue of the rumour Reith wrote: 'I do not quite understand why you say that you have just heard of an alarming report, since I noticed a reference to this effect at the meeting which was held here on 27th January. If it concerns the EMI Ltd., I am assured that there is not a vestige of truth in it.'

Baird had failed to corroborate his rumour and advice to Reith based on the need to safeguard the public interest. He was in a difficult position – not only because of EMI's possible superiority with ultra short wave television (Figs. 11.8 and 11.9), but also because the 30-line service had an uncertain future. Ashbridge had given his views on this and wanted to see an end to the service, but for Baird the service provided an outlet for his receivers – and a source of profit. 'With regard to the 30-line television receiver, it is not my opinion that this is already obsolescent. The ultra short wave transmission is still a long way from having reached the public service stage.'[49] Baird foresaw 'many difficulties still to be overcome before USW broadcasting (became) a service . . .'. He was right,

Fig. 11.8 *EMI 180-line mirror drum film scanning equipment (December 1933)*

Fig. 11.9 *Close-up of the 180-line mirror drum (December 1933)*

but as a pioneer he must have known of the great advantages to be obtained for development purposes by having an experimental service. Nevertheless he still pressed for a meeting with the BBC.

Fortunately for Bairds the company had some support for its cause from the unbiased, but always cautious, Phillips of the Post Office. Phillips, who advised the Postmaster General on matters relating to television, was well aware of the support the company could muster in the House of Commons and was naturally reticent about agreeing to suggest that the Post Office should give its blessing on any matter which might have political repercussions. In the past Baird publicity and lobbying had caused the Post Office to put pressure on the BBC: now Bairds were again trying to apply the same tactics.

Following a letter to Phillips from the Telvision Company, Phillips had a telephone conversation with Bishop[50] and told him the installation of EMI television equipment in Broadcasting House would be a severe and grossly unfair blow to the Baird Company. Phillips wanted to know:

1 whether it was true that EMI were in fact not particularly interested in the transmitting side;
2 whether the BBC could not hold up their arrangements with EMI if in fact it was the case that the Baird Company were able to produce a transmitter capable of handling a 120-line picture.

On the first point, Bishop said he could not express an opinion, and on the second point, told Phillips that it was a policy matter which was not for him to decide, but

> I pointed out that for some time we had been promised a demonstration of Baird wide sideband television, that this demonstration had already been postponed twice and it was now proposed for the end of February. I further pointed out that the Corporation had pressed Bairds for many months to give us a demonstration of wide side band television if they were in a position to do so.

Bishop's observation on the conversation was:

> Quite obviously Mr Phillips was afraid to face the issue and wanted to temporise by getting us to delay our negotiations with EMI.

Bairds sought assistance from other quarters besides the Post Office, and, on 21st February, John Baird sent a letter to His Royal Highness, The Prince of Wales, complaining that the BBC were crushing a pioneer British industry and were giving secret encouragement to alien interests.[51] Phillips was normally aware of letters of this type – whether written to the Prime Minister, Postmaster General or other influential persons. The Civil Service system was such that the private secretaries to such persons usually referred any communications, which they could not themselves answer, to the department responsible, for comments. This was the situation with letters which might be marked private and confidential. Thus on 1st March Phillips noted in a letter to Reith that the PMG had received representation from several quarters on the EMI–Baird situation.[52] 'The question arises as to whether it would be wise to defer a decision in regard to the installation of the EMI transmitter at Broadcasting House until the BBC and Post Office have witnessed Baird's proposed demonstration.' Phillips suggested a meeting and this was arranged for 7th March. The BBC representatives were Carpendale and Ashbridge.[53]

Subsequently, in a letter to Reith dated 13th March 1933,[54] Phillips reported that the PMG had 'received representations from Mr Baird who (had stated) positively that he (would) be ready to give a demonstration of his short wave apparatus by the first week in April.' Consequently

> . . . the PMG (considered) that it would be right to postpone a decision in regard to the institution of tests of the EMI apparatus at Broadcasting House until the demonstration of Baird's apparatus (had) taken place. The arrangement was that both EMI and Bairds would give demonstrations to be witnessed by the BBC and the Post Office and that a decision on the installation of EMI apparatus at Broadcasting House should be postponed until the results of these demonstrations have been considered.

Reith noted Phillips' view and in particular the inference that the Post Office did not propose to issue a licence for the EMI tests until after the demonstrations had taken place (Fig. 11.10). He added rather sourly: 'In these circumstances, we are unable to see why you should find it necessary to ask us to delay the tests with the EMI apparatus.[55]

The demonstrations were arranged for the 18th and 19th of April at Long Acre and Hayes, respectively,[56] and were seen by Admiral Sir Charles Carpendale, Mr Noel Ashbridge, Mr H.Bishop, all of the BBC, and Colonel A. G. Lee, Colonel A. S. Angwin, Mr A. J. Gill, of the engineering side of the Post Office, and Mr L. Simon, Mr F. W. Phillips, Mr W. E. Weston and Mr J. W. Wissenden of the administrative side.

The Baird apparatus was demonstrated over wires between neighbouring rooms in the Long Acre premises. It was described by Simon in the following terms:[57]

> The transmitting apparatus was of a makeshift type and, at the receiving end, pictures about three inches by three inches were produced in black and white on the broad end of a funnel-shaped cathode tube in two cases, and by Nipkow disc in a third. Films were fed into the transmitters: but the received pictures were in all cases indistinct, jerky and erratic. It was stated that arrangements were being made for the presentation of a picture nine inches by five inches. The best that could be said for the demonstration was that it was an interesting experiment in picture transmission with rather crude apparatus.

Baird still did not seem to appreciate the need to give professional demonstrations. In his favour it might fairly be urged that the apparatus had been assembled in less than two months; but on the other hand he had fixed the time limit for the demonstration, and in his letter of 17th February he had stated that his affiliated company in Germany had already supplied a similar transmitter to the German Post Office.[58]

Fig. 11.10 *EMI 180-line mirror drum film scanner television equipment*

Following the demonstration, on 19th April, Baird wrote to Ashbridge to say that all he was concerned in showing him was that they had a transmitter capable of sending a 120-line image with 25 pictures per second.[59] He seems to have been concerned that the demonstration might not have been impressive, for he added: 'Naturally the receivers do not yet show the perfect picture, as there is still work to be done on the reception side before this stage is reached.' This seemed to conflict with his earlier statements – particularly regarding 240-line television.

At Hayes the EMI apparatus was demonstrated by wireless transmission, the transmitting apparatus being at the works and the cathode-ray tube receiving set in a cottage 2 miles or so away. Simon observed that the complete receiving apparatus for sight and sound was complicated and involved the use of 25 valves; but it was claimed that the number of valves could be reduced and the apparatus otherwise simplified so as to reduce the cost of a television set to about £80 or £100[57] (Baird claimed that their receiving set could be produced in bulk for about £30 or £40). The demonstration consisted of the reproduction of films on a screen giving an image size of $6\frac{1}{2}$ inch each way. In one case the receiving set produced a black picture on a white background and in the other on a green background.

> The action on both pictures could be followed clearly throughout, without the guidance afforded by the accompanying speech; but the detail on the green background was superior to that on the white. A very high degree of stability was achieved. The company are experimenting with various substances on the cathode tubes with the object of securing a black and white picture without loss of detail. They are also experimenting in the further magnification of the received pictures and a demonstration screen picture nine inches square was shown. The Post Office engineers, who were present at the previous demonstration in February last, considered that marked improvement had been achieved.[57]

Clearly the EMI engineers were advancing rapidly in the art of television. They had excellent facilities at their disposal and a highly trained technical staff. 'In these respects they (had) a great advantage over the Baird Company.'

Following the demonstrations, a conference was held at the Post Office between the BBC and Post Office representatives on 21st April 1933. It was agreed that:[60]

1 the EMI results were vastly superior to those achieved by the Baird Company;
2 the results were incomplete because of the different transmission methods (line and wireless) used in the two cases (also the effect of electrical interference and absorption could not be tested);
3 further tests by wireless in a town area were essential to determine the range of reception and the effect of interference;
4 whatever system of synchronisation was adopted in the first instance for a public service might be liable to standardise the type of receiving equipment;
5 a test of one system could not be a reliable judgement on the results achieved by the other.

During the discussion the BBC said they were anxious to start trials of the EMI system, but considered the inability of the system at the moment to produce direct television a disadvantage, as the cost of a film – about £30 – might be prohibitive. When asked by the Post Office whether they would be prepared to give equal trials to both systems, the BBC replied that they would prefer to test the EMI system first and only give a trial to the Baird system 'when – if ever – Baird succeeded in producing results equal to, or better than, those of EMI.

No definite recommendations were made at the conference as to the future course of action to be taken. Television of high definition was now inevitable, although the BBC considered that progress towards a public service would necessarily be slow.

Unfortunately the Baird Company failed to realise that there was no future for low definition television and persisted in its efforts to advance the sale of 30-line receivers. Moseley sent Reith and others an invitation to attend a dinner to inaugurate the placing on the market of the first consignment of new 'Televisors'.[61] These had been produced by the Bush Radio Company, an associated company of the Gaumont British group. What Moseley hoped to achieve by such a move is uncertain: his policy certainly gave the BBC an opportunity to point out to the Post Office the ill-advised nature of Bairds policy.[62]

Naturally Reith could not accept such an invitation: 'I am afraid I cannot associate myself with the launching of a receiver by a particular company. Any public interest from officials here in the reception of broadcasting should be rather in connection with associations of manufacturers than with particular ones.' Moreover Reith felt it necessary to call the attention of Sir Kingsley Wood to the possibility of a difficult situation arising from the activities of the Bush Radio Company[63] as the BBC was anxious to stress that there was insufficient programme value to justify a permanent service of this type.

Moseley had further ideas about the publicity dinner – or so Bishop thought. He believed the Baird Company might seek to arrange a talk by some prominent person to form part of special transmission demonstration programme – to demonstrate the new television receivers to the press. Consequently Bishop wrote to Napier (the PMG's Private Secretary), and asked him for the PMG's views.[64] However, the PMG had met Baird the following afternoon, and, after declining Baird's invitation, asked him what the company was going to do. Baird told Sir Kingsley Wood they were putting 100 £70 television sets on the market.[65] He assured the PMG a warning would be given to the public about the possible discontinuance of the transmissions in March 1934, but did not say how it was proposed to warn the public.

On the question of shortwave television broadcasting, Sir Kingsley Wood said he would probably arrange for an experimental transmission by the two competing systems, but that Baird must face up to the possibility that 'the other system might prove superior, as there was some reason to believe . . .'.

Here was a clear warning to the company that their previously held position as the leader to British television developments could only be recovered by accepting a challenge: a challenge to produce a high definition ultra short wave television system preferably using cathode-ray tube receivers and cameras. Bairds accepted the position, and, by 20th November 1933, Colonel A. S. Angwin was able to report, following a visit to the Long Acre laboratories:

> A cathode-ray tube was used at the receiver, this being of a type specially developed by the General Electric Company, having a diameter of 12 inch and giving a picture of 10 inch × 8 inch in area. A very marked improvement in definition and stability was observed compared with that previously demonstrated by Bairds and it is now considered to be quite equal in quality to that shown at Hayes by Electric and Musical Industries Ltd.[66]

The demonstration had consisted of the transmission, over a short line, of three topical films at a definition of 180-line, 25 pictures per second.

This was quite an achievement by the company in such a relatively short time and showed that they possessed some expertise and endeavour, when faced with difficulties, to carry them forward. Effectively the company had reached this position after approximately nine months' work, although it had demonstrated ultra short wave low definition transmission the previous year – April 1932. If Baird had been less obstinate in his belief in the worth of the 30-line system and had concentrated his company's efforts on high definition reproduction, following Sir Evelyn Murray's suggestion in 1928, the company might have found themselves in an unassailable position in this country. Certainly companies such as EMI Ltd. would have faced some severe opposition in 1932–33 if the Television Company had established even an experimental 120-line service on ultra short waves. As it was, Bishop's conjecture that the Post Office wanted to temporise early in 1933 proved correct and Bairds were given an opportunity to make good their lack of foresight. Even at Baird's meeting with Sir Kingsley Wood in May, the inventor had stuck to his belief that 'Ultra short wave public broadcasting is still a thing of the future and, when it does arrive will have, in my opinion, a very restricted range and be only an auxiliary to the medium wave service – at least for some considerable time'.[65]

The demonstration given to Angwin had consisted of a transmission by line, whereas EMI's April 1933 demonstration used wireless transmission. However, the Television Company had made extensive tests on propagation characteristics using the Crystal Palace transmitter independently of the BBC. Mr West raised the possibility of the company extending its facilities at Crystal Palace and told Angwin that additional power and wavelengths were required.[66]

Following the April tests and the subsequent conference, at which no firm recommendations were made regarding the ultra short wave transmissions, Ashbridge tried to arrange for Baird and EMI to combine their efforts so far as the transmitting end was concerned and to arrive at some agreed arrangement about the use of one transmitter.[67] This was a sound suggestion, as it would have meant that each firm could then have developed the receiver side independently of each other – this was where profits could be made – and only one transmitter would be installed in Broadcasting House. Ashbridge reported that Baird afterwards rather came round to the idea, but Shoenberg 'would not listen to the proposal . . . (as) . . . Baird had nothing to offer'. The Chief Engineer then suggested to the Director General the principle of trying the two systems consecutively, starting with EMI, thus giving Baird ample time to get ready.

While these negotiations were going on, EMI was being delayed in starting its experimental service. This was matter of concern to the Governors of the BBC and so, on 25th May, Reith felt it necessary to report their view to the PMG:[68]

> . . . the Governors wish me to make a further representation to you. They feel it would be most unfair to the EMI still further to delay them. Nearly six months have elapsed since their first application and there is no difference of opinion between us as to the state of the Baird apparatus.

Reith urged Sir Kingsley Wood to allow the BBC to proceed immediately with EMI on the understanding that if

> at any time after two or three months Bairds were able to give a demonstration which, in the view of your engineers and ours, would be reasonably comparable with the EMI, experiments with his apparatus should be instituted.

The BBC Governors were not the only persons to be concerned about the delay. There was the company itself. Clarke saw Reith on 26th May about it and asked when it was going to end. Reith rather gathered from him that he would apply to the Post Office for permission to move his present transmitter into London 'with the idea of conducting experiments independent of us, which we, of course, do not want . . .'.[69]

Reith told Kingsley Wood that, if he was to decline to give Clarke permission to move, the situation could be rather embarrassing and suggested that the company might give publicity to what had already been effected at Hayes. 'They have abstained from any sort of publicity so far, and there are very few people who know that EMI have brought television to the stage they have. It would not be good for the Baird people if even this much were known.'

The Postmaster General, while appreciating the force of many of Reith's arguments and while always wishing to meet him, felt that to give EMI only access to Broadcasting House might give the *coup de grâce* to the Baird concern.

> I remain convinced that it will be in the best interests alike of the Post Office and of the BBC to treat the two competing systems on a footing of absolute equality. I am prepared to stipulate with Mr Baird that his apparatus shall be ready within a reasonable time, say two months.[70]

The PMG reiterated this view several days later, 27th May, following Reith's letters of 25th and 26th May, but with one important change. Baird had told Kingsley Wood on 24th May that his company could easily arrange a further test with about a fortnight's notice.[71] Nevertheless, following Bairds' assurances to the PMG on 6th June 1933 that considerable progress had been made,[72] Kingsley Wood wrote to Baird on 13th June and told him he was 'prepared to accept these assurances without a further demonstration in order to avoid any further delay in carrying out tests from Broadcasting House both with Baird apparatus and with that of the EMI. As it appears from your letter of the 24th May that a fortnight's notice would be sufficient for you, I propose to give the BBC permission to commence experimental transmission with both systems on the 1st July.'[73] Bairds were naturally pleased and eager to commence at once installing their transmitter at the BBC.[74] Likewise, Shoenberg was happy to receive the PMG's decision via Bishop of the BBC. 'He was not in the least concerned whether Bairds were in the building at the same time or not.'[75] The view which he held was that the best system, whichever it was, would win in the end, but, while a definite decision had at last been made, the matter was by no means at an end. Much further correspondence was to take place on the Bairds/EMI issue. The position was simply that Bairds could not accept the idea of their unique position in television broadcasting in Britain being usurped or challenged by rival, especially a rival with Amercian interests.

During the negotiations for the installation of the two transmitters at Broadcasting House, some important changes had been made in the Baird Company. One of the most important was the resignation of Moseley from the Board of Baird Television Ltd. There is no doubt that Moseley contributed a great deal indeed in furthering the aims and aspirations of the various Baird Companies. A powerful journalist, well known in Fleet Street – and with some considerable influence there – a born fighter for just causes, he was clearly the type of person the company needed in 1928 when it was experiencing difficulties with the BBC. In a letter dated 21st June to Moseley the BBC referred to his crucial influence:

> Although there has not always been agreement either in policy or in method, it should be recognised that your consistently active advocacy has been an important, perhaps the decisive, factor in the progress that Baird Television had made to date .[76]

Moseley wrote in his book on John Baird:[3]

> Many people have wondered why I resigned. They need wonder no longer. I felt certain that, if I left at this juncture, it would clear the way for a fresh beginning between the Baird company and the BBC. The Board seemed to sense this too. They had the impression that, if they were able to go to the BBC and say that the firebrand Moseley had taken his departure, they could begin a fresh, innocent friendship where they could receive fair play from the BBC. As a matter of fact, the very opposite happened. On the day a deputation from the Baird Company informed the BBC that I had

resigned from the Board, the comment attributed to my friend Noel Ashbridge was: 'Now that Sydney is out, Baird will be out very soon.'

The Board of the company was strengthened – as was the technical side – by the appointment of Captain A. G. D. West, who had previously worked for the BBC and the Gramophone Company.

On the 13th July 1933, West had an interview with Ashbridge and told him he was in effect Technical Director, having been put in by the Ostrer Brothers, who now virtually owned the Baird Company.[77]

> He added that Moseley was out of the concern altogether, and that Baird was in the process of going out, but would probably be retained in a kind of secondary capacity for detail research. He said that the past was to be wiped out entirely and he fully realised that the conduct of the company and its relations with the BBC in the past had been appalling.

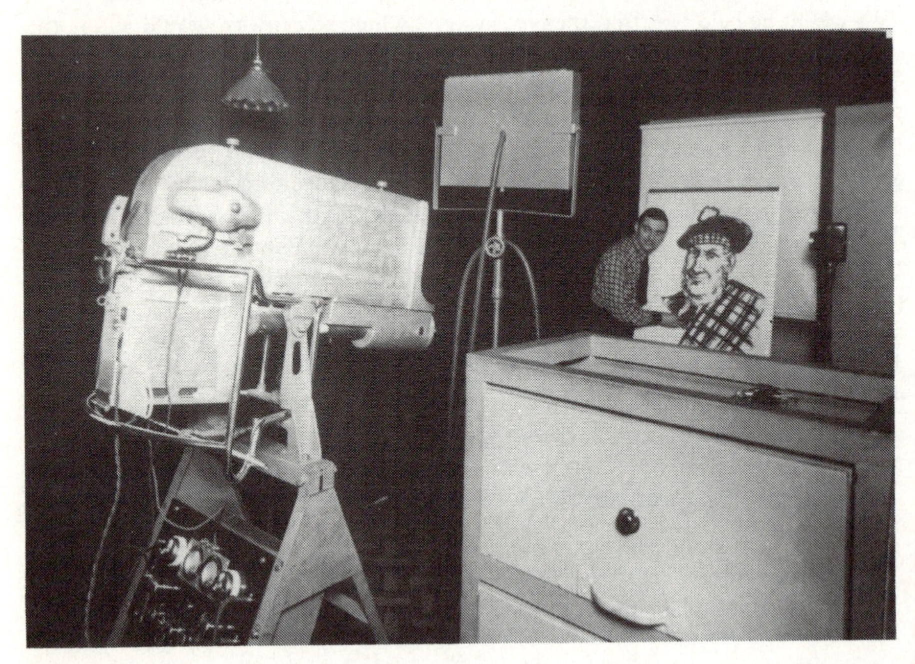

Fig. 11.11 *Baird Television Ltd. mirror drum camera used at the Ideal Home Exhibition (1933)*

West was able to throw some light on the proposed sale of 30-line television receivers. He mentioned to Ashbridge that the Board did not know anything about the letter which he wrote warning them of the danger of placing such receiving sets on the market, having regard to the obsolescence of the 30-line system, and had they known of this they would not have gone ahead through the Bush Company for the production of expensive receivers. 'They had made

£5000 worth of such receivers, which would probably be wasted.' West agreed there was no future for 30-line television, but told the Chief Engineer that somehow the company would like to try to sell the 100 receivers which they had made, with some arrangement by which they could be changed for ultra short wave multiline receivers when that system replaced the low definition one on medium wavelengths. (Fig. 11.11).

West further said the company was ready to install apparatus, for 120-line television, either for film or for direct reproduction from a studio, in Broadcasting House within a month. Ashbridge mentioned that this was subject to a demonstration and asked whether he would be prepared to let HMV carry out their tests first, and then follow on with his. 'Naturally he was a little doubtful about this, but did not turn the idea down, suggesting that possibly they might do studio work while HMV do film work or, alternatively, we might carry on with HMV here and with them (Bairds) at the Crystal Palace. This . . . was a little vague' noted Ashbridge.

In July of 1933 the Baird Company had taken a lease at the Crystal Palace, Norwood, and rented part of the ground floor and the south tower from the manager, Sir Harry Buckland.[3] The site was ideal for ultra short wave transmission experiments, as the Palace was sited on one of the highest points in London. When the company acquired the accommodation they transferred their work from The Studio, Kingsbury, and the Long Acre laboratories to it. The Postmaster General granted the company a research licence to operate on the following frequencies:

6040 kHz	at	500 W
1930 kHz	at	250 W
48 000–50 000 kHz	at	500 W

and this enabled them to send experimental transmissions to Film House in Wardour Street.[78]

References

1 MURRAY, G.: letter to J. .L. Baird, 19th August 1932, BBC file T16/214.
2 BIRKINSHAW, D.: memorandum to H. L. Kirke, 25th August 1932, BBC file T16/214.
3 MOSELEY, S. A.: 'John Baird' (Odhams, London, 1952).
4 BIRKINSHAW, D.: memorandum to H. L. Kirke, 11th October 1932, BBC file T16/214.
5 ROBB, E.: internal report to D. P., and D. I. P. and C. E., 10th May 1933, BBC file T16/214.
6 ROBB, E.: memorandum to the Director General, 11th November 1932, BBC file T16/42.
7 REITH, Sir J. F. W.: letter to J. L. Baird, 26th November 1932, BBC file T16/42.
8 BAIRD, J. L.: letter to Sir J. F. W. Reith, 29th November 1932, BBC file T16/42.
9 REITH, Sir J. F. W.: letter to J. L. Baird, 29th November 1932, BBC file T16/42.
10 Director General (BBC): memorandum to D. I. P., 2nd December 1932, file T16/42.
11 REITH, Sir J. F. W.: letter to J. L. Baird, 2nd December 1932, BBC file T16/42.
12 BAIRD, J. L.: letter to Sir J. F. W. Reith, 6th December 1932, BBC file T16/42.
13 MURRAY, G.: private letter to S. A. Moseley, 10th December 1932, private collection.

14 SHOENBERG, I.: letter to N. Ashbridge, 11th November 1932, BBC file T16/65.
15 ASHBRIDGE, N.: report on television demonstration at EMI, 6th December 1932, BBC file T16/65.
16 Director General (BBC): memorandum to the Chief Engineer, 1st January 1933, BBC file T16/65.
17 ASHBRIDGE, N.: letter to I. Shoenberg, 4th January 1933, BBC file T16/65.
18 Chief Engineer (BBC): report to the Director General, 1st January 1933, BBC file T16/65.
19 CLARKE, A.: letter to Sir J. F. W. Reith, 5th January 1933, BBC file T16/65.
20 Chief Engineer (BBC): memorandum to the Director General, 9th January 1933, BBC file T16/65.
21 Director General (BBC): memorandum to the Chief Engineer, 9th January 1933, BBC file T16/65.
22 SHOENBERG, I.: letter to N. Ashbridge, 11th January 1933, BBC file T16/65.
23 A. C. E.: memorandum to the Civil Engineer, 12th January 1933, BBC file T16/65.
24 BAIRD, J. L.: letter to N. Ashbridge, 13th January 1933, BBC file T16/42.
25 ASHBRIDGE, N.: letter to J. L. Baird, 19th January 1933, BBC file T16/42.
26 Minutes of a meeting held with representatives of Television Ltd., 27th January 1933, BBC file T16/42.
27 ASHBRIDGE, N.: letter to J. L. Baird, 30th January 1933, BBC file T16/42.
28 Chief Engineer (BBC): memorandum to the Director General, 30th January 1933, BBC file T16/42.
29 REITH, Sir J. F. W.: letter to A. Clarke, 30th January 1933, BBC file T16/65.
30 REITH, Sir J. F. W.: letter to A. Clarke, 3rd February 1933, BBC file T16/65.
31 CLARKE, A.: letter to Sir J. F. W. Reith, 8th February 1933, BBC file T16/65.
32 D. I. P.: memorandum to the Director General, 8th February 1933, BBC file T16/65.
33 MOSELEY, S. A.: letter to Sir Kingsley Wood, 28th January 1933, Minute 4004/33.
34 PHILLIPS, F. W.: letter to Admiral Sir C. Carpendale, 2nd February 1933, BBC file T16/42.
35 REITH, Sir J. F. W.: letter to F. W. Phillips, 3rd February 1933, BBC file T16/42.
36 D. I. P.: memorandum to the Director General, 8th February 1933, BBC file T16/42.
37 A. C. E.: memorandum to the D. I. P., 9th February 1933, BBC file T16/42.
38 A. C. E.: memorandum to the D. I. P., 14th February 1933, BBC file T16/65.
39 GILL, A. J.: memorandum to the Secretary (GPO), 15th February 1933, Minute 4004/33.
40 REITH, Sir J. F. W.: letter to J. L. Baird, 2nd February 1933, BBC file T16/65.
41 BAIRD, J. L.: letter to Sir J. F. W. Reith, 4th February 1933, BBC file T16/42.
42 BAIRD, J. L.: letter to Sir J. F. W. Reith, 4th February 1933, BBC file T16/42 (not the same as Reference 41).
43 Director General: memorandum to the Chief Engineer, 6th February 1933, BBC file T16/42.
44 Anon.: 'Television', *New Statesman and Nation*, 9th February 1933.
45 CLARKE, A.: letter to Sir J. F. W. Reith, 7th February 1933, BBC file T16/65.
46 REITH, Sir J. F. W.: letter to A. Clarke, 9th February 1933, BBC file T16/65.
47 A.C.E.: memorandum to the D. I. P., 8th February 1933, BBC file T16/42.
48 REITH, Sir J. F. W.: letter to J. L. Baird, 9th February 1933, BBC file T16/42.
49 BAIRD, J. L.: letter to N. Ashbridge, February 1933, BBC file T16/42.
50 BISHOP, H.: record of telephone conversation with F. W. Phillips, 18th February 1933, BBC file T16/42.
51 THOMAS, Sir G.: letter to R. H. Eckersley, 2nd February 1933, BBC file T16/42.
52 PHILLIPS, F. W.: letter to Sir J. F. W. Reith, 1st March 1933, BBC file T16/42.
53 Assistant Chief Engineer (BBC): letter to F. W. Phillips, 3rd March 1933, BBC file T16/42.
54 PHILLIPS, F. W.: letter to Sir J. F. W. Reith, 13th March 1933, BBC file T16/42.
55 REITH, Sir J. F. W.: letter to F. W. Phillips, 21st March 1933, BBC file T16/42.
56 PHILLIPS, F. W.: letter to Sir J. F. W. Reith, 10th April 1933, BBC file T16/42.
57 SIMON, L.: memorandum on Baird and EMI demonstrations, 27th April 1933, Minute 4004/33.

58 BAIRD, J. L.: letter to Sir J. F. W. Reith, 18th February 1933, BBC file T16/42.
59 BAIRD, J. L.: letter to N. Ashbridge, 19th April 1933, BBC file T16/42.
60 Notes of a meeting held at the GPO, 21st April 1933, BBC file T16/42.
61 MOSELEY, S. A.: letter to Sir J. F. W. Reith, 21st April 1933, BBC file T16/42.
62 REITH, Sir J. F. W.: letter to the Postmaster General, 27th April 1933, BBC file T16/42.
63 Chief Engineer (BBC): memorandum to the Director General, 27th April 1933, BBC file T16/42.
64 BISHOP, H.: letter to H. Napier, 22nd May 1933, BBC file T16/42.
65 Notes of a telephone conversation between the BBC and the PMG, 23rd May 1933, BBC file T16/42.
66 ANGWIN, A. S.: memorandum on visit to the Baird laboratories, 20th November 1933, Minute 4004/33.
67 ASHBRIDGE, N.: memorandum to the Director General, 12th May 1933, BBC file T16/42.
68 REITH, Sir J. F. W.: letter to the PMG, 25th May 1933, BBC file T16/42.
69 REITH, Sir J. F. W.: letter to the PMG, 26th May 1933, BBC file T16/42.
70 KINGSLEY WOOD, Sir H.: letter to Sir J. F. W. Reith, 22nd May 1933, BBC file T16/42.
71 KINGSLEY WOOD, Sir H.: letter to Sir J. F. W. Reith, 27th May 1933, BBC file T16/42.
72 BAIRD, J. L.: letter to Sir H. Kingsley Wood, 6th June 1933, BBC file T16/42.
73 KINGSLEY WOOD, Sir H.: letter to J. L. Baird, 13th June 1933, BBC file T16/42.
74 BAIRD, J. L.: letter to Sir H. Kingsley Wood, 15th June 1933, BBC file T16/42.
75 BISHOP, H.: telephone conversation with I. Shoenberg, 24th June 1933, BBC file T16/65.
76 BBC: letter to S. A. Moseley, 21st June 1933, BBC file T16/42.
77 Chief Engineer (BBC): record of interview with A. G. D. West, 13th July 1933, BBC file T16/42.
78 GPO: letter to the BBC, 27th July 1933, BBC file T16/42.

Chapter 12

Rival claims, 1933–1934

In the United Kingdom both Electric and Musical Industries and Baird Television were firmly committed to the development of medium/high definition television during the same period that the Radio Corporation of America was undertaking its development programme. Unlike the American company, which pursued a policy of publishing its findings in the journal of a learned society, the British companies maintained a policy of secretiveness and no published papers exist for the period 1932–1935 to show how their systems progressed. Fortunately for the historian, representatives from the BBC and GPO witnessed demonstrations of the two systems from time to time and their accounts are available for study.

Following Ashbridge's general discussion with West in July 1933, a demonstration of Bairds 120-line, cathode-ray tube reception television was seen by Colonel Lee on the 2nd August 1933.[1] Ashbridge had told West that the previous demonstration had been a failure, but now Lee was able to report 'the results of the Baird system are of about the same order of merit as those of EMI when seen at the demonstration on the 19th April'.

Lee also gathered that, in view of the progress made, the company would probably approach the Post Office with a request that the Baird system alone should be given a trial at Broadcasting House. He was right. On the 4th August, Major Church, one of the directors and a staunch supporter of John Baird, had a discussion with Sir Kingsley Wood on this point, but was told that the Post Office could not possibly rescind the permission already given to the BBC. The same day Church wrote a long letter[2] to Kingsley Wood and put forward another proposal:

> I suggest that it might be possible to carry out your promise to HMV and to answer any criticism that our understanding with the BBC has been broken by delaying introduction of HMV to Broadcasting House until March 1934, when our arrangement with the BBC finishes. This would allow a closer inspection and comparison of the rival system to be made than has at present been possible, and also to allow more time to study the effect on British Industry of the introduction of HMV into the transmitting situation here.

Alternatively, we ask you in the interests of British industry generally to allow our apparatus to be installed at Broadcasting House and tested before the tests with HMV apparatus begin; not because we are afraid of comparison with HMV, but because we are convinced that the HMV will make the utmost use of the facilities granted to them for propaganda in favour of the use of its own apparatus.

Fig. 12.1 *Baird 240-line telecine machine*

The latter strategy of the Baird Company as put forward by Church was somewhat unusual and would have tended to operate against the company's interests. They were proposing that Bairds should transmit first – and in the near future; but such an arrangement would give EMI extra time in which to perfect their apparatus and consequently produce a better picture than Bairds. As EMI had approximately a four months advantage in time as compared with the Television Company, the converse of Church's strategy would have been the better course of action to have followed. It may be that Baird was still obsessed with the desire to achieve world firsts or at least firsts in the United Kingdom, now that he had rivals not only abroad but in this country (Figs. 12.1 & 12.2).

West telephoned Ashbridge on the 10th August and enquired about the installation of apparatus at Broadcasting House,[3] but the Chief Engineer had to tell him that Church had recently had an interview with the PMG and that this

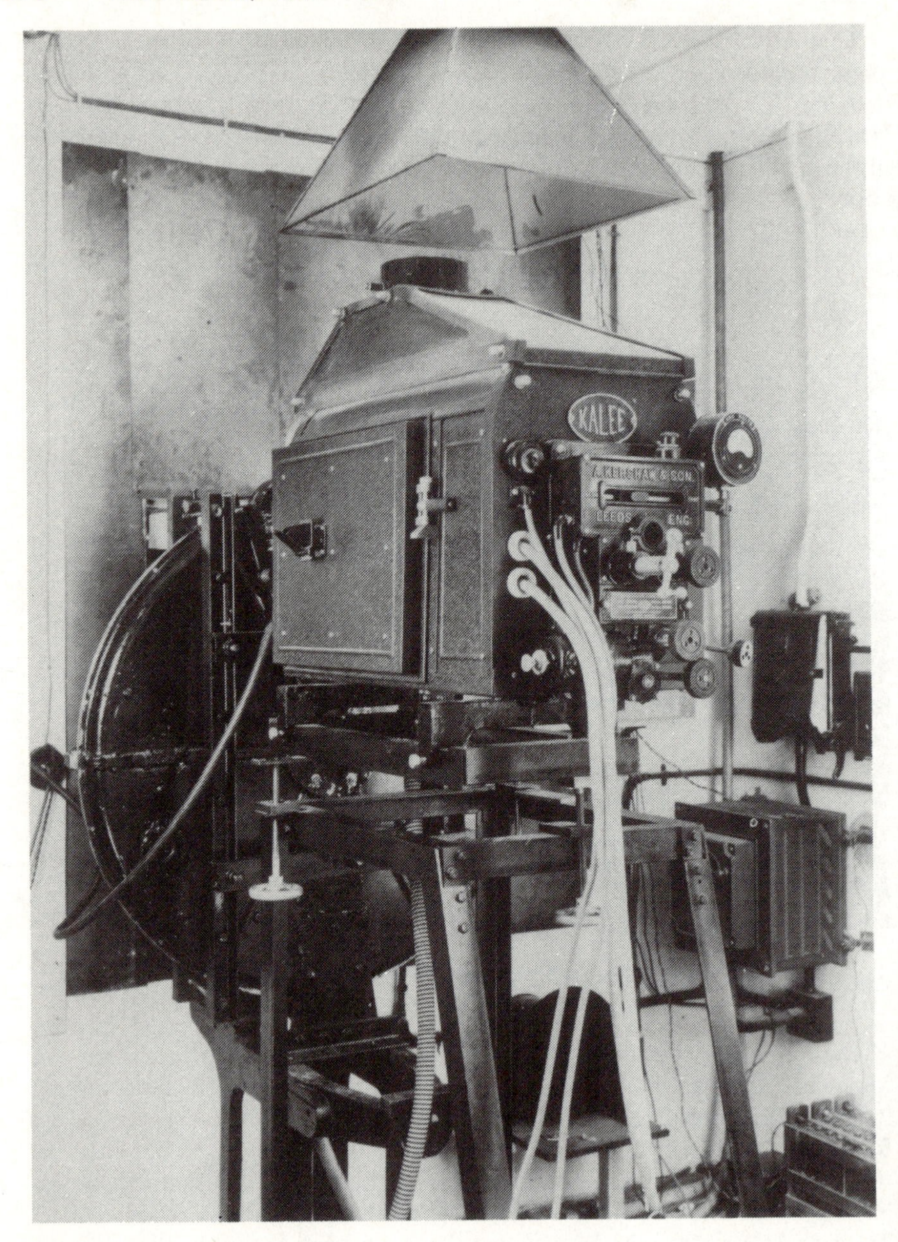

Fig. 12.2 *Baird spotlight scanner installed at Crystal Palace*

had complicated the situation. At a meeting[4] with Ashbridge the next day West was told it was very difficult to carry on successful co-operation with the Baird Company while accusations were being made that the BBC was not acting fairly in agreeing to carry out experiments with EMI. However, West learned that his company's demonstration had been sufficiently satisfactory to warrant transmission experiments from Broadcasting House. Notwithstanding West's previous comment that the past was to be wiped out entirely, relations between the BBC and the Television Company were still strained.

A few weeks earlier West had written to Ashbridge[5] about an announcement which had appeared in the *Daily Herald* (26th July) regarding the installation by Bairds and HMV of some new television transmitters at Broadcasting House. He told the Chief Engineer the announcement was inserted without any authority from them and went on to say that it had caused them great concern and was harmful to the interests of the company. Ashbridge acknowledged the letter and rather sarcastically observed: 'I don't know exactly how these leakages occur, but of course it is not the first instance either in connection with television or other subjects.[6]

The position was not improved by Sir Harry Greer's letter to Reith on the 16th August.[7] Bairds were still very unhappy about the EMI–BBC relationship and considered the proposed EMI transmission as in breach of the arrangements made between the BBC and themselves. The Corporation had a satisfactory answer which was well known to the company, but nevertheless Greer asked Reith to state in writing whether 'the system proposed to be used for the purposes of or in connection with the proposed transmission before mentioned is a system other than the Baird system and is also of a proved and unquestionably superior character'. The precise reason why Greer requested this information is not known, but John Baird had mentioned to Reith in a letter[8] dated 18th February 1933: 'What our legal redress would be, only legal action would decide'. Perhaps the company was now considering this possibility. If this was so their case presumably would have been based on the need for the BBC to show that the EMI system was, first, a proved system, and, secondly, an unquestionably superior system. On the first point, EMI could hardly have made such a claim, as they had asked the BBC for experimental test transmission facilities, and, on the second point, the Hayes Company was still using a mechanical method of scanning the film.

In reply[9] Reith gave Greer a mild rebuke for the tone of his letter and made the observation:

> An examination of the context of the relevant letters indicates that the word 'transmit' refers to regular programme transmissions for reception by the public. Our attitude has been perfectly consistent and we stand by the statement in our letter[10] of 3rd February last in which we said there can be doubt that the undertaking in our letter of 21st March 1932 did not apply to research or test transmissions.

The correspondence ended with a letter[11] from Greer, who was rather hurt by Reith's letter '. . . it seems to be clear that there have been misunderstandings'.

However, progress was slowly and rather painfully being made towards the implementation of the experimental service. At a meeting held on 23rd August 1933 between Shoenberg and Condliffe, from EMI, and Ashbridge, Bishop and Kirk, of the BBC, the possibilities of housing both the Baird and EMI systems in Broadcasting House was explored and also whether the same transmitter might be used by both companies on an alternate day, week or month basis.[12] The BBC had realised from the beginning that this was bound to prove difficult to arrange in practice, but a sharing arrangement had been imposed by the Post Office. EMI were opposed to this, as they intended making certain modifications to the transmitter, as a consequence of their research activities at Hayes, and they were not willing that the results of their work should be available to the Baird Company. Furthermore, they proposed to install monitoring arrangements in the transmitter room which would be special to their system and Shoenberg could not agree to this equipment being known to or used by the Baird Company. In addition, he stated it would be impracticable to remove his transmitter improvements in order that the transmitter itself might be used by competing interests on an alternate day basis. He claimed a prior right for the use of the transmitter, since he had given a satisfactory demonstration over six months ago, and genuinely wanted infromation concerning its performance when it was situated in the heart of a city, in order that the results could be compared with those he had already obtained at Hayes. There was no political significance.

Eventually, after discussion, both Ashbridge and Shoenberg agreed from a practical standpoint, it would be more satisfactory for one system to be tried out at a time. The BBC had always held this view, but had been overruled by the Post Office. Shoenberg confirmed that, if his company could have exclusive use until the end of the year, he would be satisfied and would be willing to remove his equipment at that time in order that Bairds or any other system might be given their turn. Ashbridge promised to take up the matter with the Post Office, on receipt of a letter from Shoenberg, in order that possibility of a modification of the existing restrictions might be reviewed. Meanwhile, Shoenberg stated he would hold up the installation of his apparatus in broadcasting House.

Events of the next fortnight were to hold up EMI's installation for more than four months. Bairds campaign to have the first use of the facilities at Broadcasting House was rewarded, and, following a discussion between Reith and Clarke, Carpendale wrote to Clarke[13] telling him 'we shall be ready to install your apparatus on 1st February (which I think was the date suggested by you to Sir John) for research experiments which would last for a similar period' (to Bairds). Clarke confirmed[14] they would be ready on the suggested date and mentioned: 'Our only object in agreeing to defer the work with you until you had completed your experiments with others was to ease a rather difficult situation.'

Actually EMI gained from the change of plan much more so than Bairds. Apart from having an extra four months to improve their equipment – and hence show up any deficiencies in the Baird system – their period of transmission would be during the critical period of February and March:

> The plan of course (was) to stage matters so that HMV ultra short wave would graduate to service transmissions once the Baird 30-line apparatus (was) scrapped at the end of March[15] . . . With HMV replacing Baird at the end of March it was proposed to allow Cossor to install their experimental plant replacing the HMV experimental work.

In an unsigned 'private and confidential' memorandum on[15] 'The BBC and television – the position as at September 22nd 1933' the writer commented:

> The anti-Baird attitude of the administration and technical branches of the BBC is as definite as ever but more dangerous, to the extent that previous disappointments have inculcated methods of greater subtlety and caution. Reith has recorded that the personal exchange of assurances between him and Greer and Clayton is well calculated to lull suspicions there. Secondly, Ashbridge openly boasts that he has West 'in his pocket', a declaration which the latter's damaging admission about 30-line has gone a long way to support.
>
> The Post Office, helpful last spring, when facilities defeated three successive attempts to displace Baird by HMV, is now obscure. Armed with West's admission about 30-line, Ashbridge has got the Post Office to agree to make no objection to the forthcoming giving of notice.
>
> There is an intention to reduce the transmissions by 30-line to the minimum under the agreement, that is from four half-hour periods to two or three fifteen-minute periods, this to be made effective in November. For obvious reasons this will not be mentioned in the notice letter, the plan being to drive the further wedge about the middle of October.
>
> It is important to recognise that the quality of the results of the Baird ultra short (wave) experiments during the next three months will not be the decisive factor in the battle with HMV. It has indeed been pointed out to HMV that, when they start in January, they will have not only the advantage of a further period of laboratory research, but also exact knowlege of what Baird has done and how he has done it. Baird is in effect faced with the combined resources and goodwill of HMV and the BBC engineering and administration. An outcome at all favourable to Baird will depend primarily on pressure in politics and publicity; in other words, on precisely the same instruments which have recorded what progress there has been in the past. There is the important difference that experience of past reverses has made the anti-Baird forces more careful and adroit. On the other hand, the adhesion of Gaumont-British to the Baird cause should be a counter-weight, but only to the extent to which it is used skilfully and intelligently.

This highly illuminating report of unknown authorship – but clearly written by someone in the BBC – ended with a short section headed 'Action':

> Prompt action probably will defeat the intention to reduce the transmissions to the minimum in November.

> Nothing can now prevent the communication of the notice about March 31st, but the right kind of action can turn it to such account that it may well have a boomerang effect, and will render the application of the notice most unlikely.
>
> The handling of the Baird experiments between now and the end of the year is of the utmost importance in mobilising the forces which alone can upset the Reith–Clarke agreement. Much depends on the skill and adroitness of Baird diplomacy during the next three months.

Bairds were told they could install their apparatus forthwith in Broadcasting House on 4th September and that they could continue their experiments at any reasonable times until Christmas 1933. The company was required to comply with eight conditions, follows:[16]

1 The experiments are to be regarded as research only and no action is to be taken which would lead the press, the trade or the public to believe that they are intended as regular programme transmissions.

2 No publicity regarding the tests is to be issued officially by any member of your staff except as agreed with the BBC. No announcement will be made as to the times of test except to persons agreed with your company and ourselves.

3 Your company shall indemnify the BBC against all costs in connection with alleged patent infringements which may arise as a result of this arrangement and against all claims, costs or expenses which may arise out of the use by the BBC of cinematograph film for these experiments. It is understood that the Baird Company will provide the necessary films for the experiments.

4 Authorised members of the BBC staff shall have access to your apparatus and shall be supplied on request with information as to its working. It is understood, however, that such information would be restricted to definite members of the staff and would be kept strictly confidential.

5 Any reasonable tests suggested by the BBC shall be carried out. The ultra short wave transmitter shall be operated by the BBC personnel and be entirely under the control of the BBC in this respect. Reasonable modifications required to carry out will be made by the BBC if found necessary, provided that such can be carried out with the facilities available.

7 The Baird Company shall keep their apparatus insured against all risks whatsoever.

8 The Baird Company shall provide the BBC with at least two receivers, in order that they may observe the results obtained from the transmission.

The Television Company agreed without any reservation to the conditions, but considered that in fairness two more should be added:[17]

> 1 Rival firms should make these tests for a like period as granted to us under conditions identical with those imposed on us.
> 2 On conclusion of these tests by rival firms, we should be given further opportunity to test our latest apparatus through the BBC, in order that a fair comparison may be made at the stage of the development reached at that time.

Bairds were obviously concerned that their equipment might not be superior to the EMI apparatus in the first instance and possibly they had in mind also the tactical advantage, *vis-à-vis* the present 30-line service and a new higher definition service, if they could terminate the series of tests with their equipment. However, the Corporation had other views, and while they agreed to the first point, Carpendale told the company the BBC could not agree to the inclusion of the second in their list of conditions.[18]

At last, after a great deal of procrastination and bitterness, installation of the Television Company's apparatus was commenced. Bishop, in a memorandum to C(A), 19th September, mentioned: 'The whole equipment will probably be working in about a week's time.'[19]

Three days later, 22nd September, the Director of Business Relations sent Bairds a letter[20] giving formal notice of the BBC's intention to terminate the arrangement between them on 31st March 1934, in accordance with the terms of their letter of 21st March 1932. However, this was not necessarily the end of the low definition era, for there was an additional point: 'Any other arrangements at which we may arrive with your company would be the subject of a fresh agreement.' Bairds acknowledged the letter[21] and requested a meeting to exchange views on this.[22]

The intention of the BBC to terminate the 30-line transmissions caused some consternation among British manufacturers of television sets in August, when the Corporation decided to conduct a census[23] of the people who received the television broadcasts. An announcer had asked lookers-in to send postcards to Broadcasting House marked 'Z'. The manufacturers felt that the BBC was engaged in the exercise to show that, if the response was not satisfactory, the Corporation was entitled to discontinue the transmissions.

A deputation of leading television manufacturers in the country thus decided, after holding a meeting at Olympia, to pass the following motion:[24]

> A number of British manufactuers of television receivers and components wish to record their deep resentment at the possibility of the BBC abandoning the regular television transmissions.
>
> Hundreds of thousands of pounds have been sunk in research in this branch of radio and it is felt that the interest of the public should not be allowed to die down, particularly at a time when it has been reawakened by the improvements shown at the Radio Exhibition at Olympia.
>
> It is clear that if lookers have any doubts as to the continuance of the service, the efforts of the pioneers will have been in vain. In any case, it is

> felt that the taking of a postcard census by the BBC at this season of the year is of very doubtful value, if not definitely misleading.

The survey showed that approximately 8000 televisors were in operation.[25]

By a curious twist in the history of publicity, it was now the turn of the Baird Company to accuse the BBC of leaking information to the press without first consulting Sir Harry Greer, the new Chairman of Baird Television Ltd. For on 28th September the *Daily Mail* published an article headed 'Television surprise – BBC ending agreement'.[26] Greer told Reith that this was prejudicial to the interests of his company and must have been from someone in the employ of the Corporation.[27] He referred to someone in the BBC Press Department, but Reith said that this person was not responsible.[28]

The policy of the BBC had always been to minimise the publicity of its experimental television services, and, during the months of discussion and negotiations, practically no information was given to the public. 'A discreet silence is maintained by the authorities at Broadcasting House' noted one newspaper. Rumours and leakages were the only sources available until 12th October when a press statement was released.[29] This followed some prompting by Greer, who wrote to Reith: '. . . the Press has been led to assume that the new experimental transmissions are by some other system than the Baird system, which gives the impression that another comany has been given preference over the Baird Company.'[30]

On the question of the continuation of the low definition television system, Greer wrote: 'We have taken no part in any organised campaign for or against the continuance of the 30-line transmission, but it is increasingly difficult to maintain an indifferent attitude . . .'.Bairds wanted a decision because they felt the uncertainty was seriously interfering with plans of the radio manufacturers. The position was that, while manufacturers had been very slow in putting receivers on the market, they were now ready to do so, but felt the market was prejudiced by the BBC's statement that the transmissions were going to stop. An additional point was that, while Bairds were required by agreement to put a notice on any sets they sold informing the public that transmissions might cease on a given date, firms such as Scophony and others were not so bound and there was nothing to stop them putting receivers on the market at once and selling them to an ignorant and unsuspecting public. The BBC of course was quite against lending its name to support such an adventitious demand. There was, however, one person in the Corporation who appreciated the manufacturers' point of view – the Director of Business Relations.[31] In a memorandum to Carpendale he stated:

> . . . if they have put a certain amount of money into the development of 30-line receivers and at the moment of marketing we stopped transmissions, they are badly placed. They will object to our policy and it will to some extent reflect itself in advertisements placed in our journals, i.e. if we have their goodwill we get more of their custom. Are we doing a service

> to television from a public point of view if we drop public transmissions for the year? There must be a percentage of the public who will pay for sets, even if only of value for a year or two.

The Director argued for the public interest in television to be maintained:

> I feel sure you will, at any rate, agree that if the policy of maintaining public interest in television is now right, no such administrative troubles, as lack of studio accommodation, should stand in the way. I know that this runs right across your view of dropping public programmes at the end of March next, but surely we are by now convinced that, much as one would like it from the pure engineering point of view, it is too late to thrust another pack into the laboratory state and that we must instead take the entrepreneur's point of view giving the public something which, admittedly imperfect, will develop either on existing lines or by new ones which will supersede them and take the public along with us, by telling them of the likelihood of such changes and set obsolescence.

Actually the Corporation had never taken the public along with them on this point. Indeed, Goldsmith had ordered in May 1933:[32] 'No reference is to be made in any of (our) periodicals to television without previous reference to Controller or Chief Engineer through me.' The order followed an article by Filson Young on 'Television: new problems of broadcasting reception' in the *Radio Times*[33] to which Reith had taken exception. Filson Young had prophesied that:

> Within a short time – a few years or even less – it would be possible for the millions to see in their own homes an image of events actually in progress elsewhere. The Grand National, the Boat Race, the Cup Final, now listened to in the form of descriptive ejaculations by an eye-winess, will be actually seen on the glass panel of some parlour cabinet a thousand miles away. Before you have learned to be perfect listeners, you will have to begin to learn a new technique – the art of looking.

Ashbridge supported a similar view:[34] 'It is probable that television will become a practical proposition within the next few years and will take its part in the daily programmes . . .', but this was too much for the cautious, conservative Controller, Admiral Sir Charles Carpendale, who in a note on Ashbridge's memorandum (Policy re television in our periodicals) wrote ' . . . this is too optimistic'. He added 'It is *possible* that television *may* become a practical proposition during the next few years', and then, as a cri-de-coeur 'but in view of all the graft, intrigue, etc. is it not best to drop the subject in our publications? I would prefer this course.' Reith agreed[35] and the editors of the *Radio Times*, *World Radio* and the *Listener* were told not to mention television at all without referring the proposed observations to the Controller or Chief Engineer.

Bishop replied to D. B. R.'s memorandum on the 12th October 1933.[36] He thought the situation was confused.

> We do not know how the high definition work will be applied. At present it is limited to the transmission of films. Studio scenes do not appear to be practicable, as our recent work at Bairds seems to show and this is not likely to continue. In fact we may find that, before the end of their three month period, Bairds will be in a position to demonstrate studio scenes. Eventually we have to consider what is the programme value of high definition television, assuming of course it is on present lines. We shall have to decide the point at which the system becomes practicable for reception by the public and, if and when it does become so, how we are going to provide permanent transmitting facilities on ultra short waves.

The Assistant Chief Engineer saw a number of problems associated with the implementation of a high definition television service. He did not see any objection to maintaining public interest in television, but submitted that, if the BBC was not careful, it would be maintaining it under false pretences. There were difficult problems to be solved concerning the distribution of a number of transmitters, as tests with the transmitter in Broadcasting House had shown the signals would be limited to a range of 5–10 miles. Then there were problems associated with the cost of providing and running them; these costs would be heavy. He did not think it was the Corporation's intention to stop all public transmissions for a year, but thought that, if the high definition experiment advanced at such a rate that it could be considered good enough at the end of March 1934, it might continue.

In replying to Bishop's memorandum, the Director of Business Relations reiterated his view for a greater openness in stating the position 'just as (the BBC) saw it and rather at length instead of in a cryptic press statement in which every word is weighed so as to have limited meanings.'[37] This latter comment probably referred to the press statement which was issued on 12th October 1933 and gave the position of Bairds and EMI *vis-à-vis* the BBC. Goldsmith's point certainly produced an effect, for Carpendale noted ' . . . alright, I suppose we can see them (Bairds) but let's meet and define our attitude'. Ashbridge concurred: 'I see no harm in ventilating the 30-line system and agree we should meet and clear the air between ourselves first.'

A meeting was arranged with Greer, Baird, West and Church for Monday, 30th October,[38] but prior to this date Ashbridge put forward some suggestions for anyone taking part in the meeting. He made three points:[39]

> 1 Some firms might deliberately place receivers on the market (without being bound to place a warning notice on the set concerning the continuance of television transmissions) with the intention of forcing the BBC to carry on with such transmissions – merely as a speculation on their part in connection with the sale of receivers. There (was) only one protection against such action and that (was) a press statement . . .
> 2 Apart from the question of protecting the public, it is a waste of time and money to go on transmitting by a system which in our opinion does not provide programme value.

3 Apart from financial loss on the part of the public, there would be resentment against us if the programmes which we put over did not give entertainment value, as the public would not understand that this (was) due to the limitation of the system and (would) think it (was) due to inefficiency on our part, both with regard to the technical and programme side.

The Chief Engineer concluded from these points that it would be better to have a gap between the cessation of 30-line transmissions and the beginning of public transmissions on ultra short waves by a 120–240 line system.

Phillips was acquainted with these views on 26th October, and the proposed meeting with Bairds,[40] and in effect said, 'good, there is no objection to you talking to Bairds on these lines, but I think you are being rather hard and I cannot imagine his agreeing it, and, of course, you won't let him think it has the Post Office's blessing'. Like the BBC's Controller, Carpendale, Phillips was cautious and conservative. However, whereas Carpendale was cautious in allowing Baird the use of BBC facilities, Phillips was cautious in any matter which tended to deny these facilities to the company. He had to advise the Postmaster General on matters relating to television and was always conscious of the possible political implications which might result from advice which tended to detract from the progress of television.

The Baird Company had many friends, not only among scientists and journalists, but also in the House of Commons, and Phillips felt that he had to proffer advice to the PMG which would prevent him from being 'shot at' by the proponents of television: and, of course, over the years that television had been developing, Phillips had been made well aware of the force which Bairds supporters could muster when required. In the present position, he considered the company would want something to keep the public interest alive in television, after 31st March 1934, and if the PMG closed down this interest he would be attacked. In addition, the wireless press was likely to attack the BBC if this were so. Phillips thought Bairds would probably say (at the conference) that they just on the point of getting out something new in 30-line television and needed to maintain the public's interest. He was right.

At the conference, the company claimed to be able to put on the market a cathode-ray 30-line set for use in the home, which could, at small cost, be transformed for high definition television up to 180 lines.[41] They said further they would be willing to inform the public of this and of the limiting conditions of the BBC's transmissions, so that there would be no misapprehension, and further offered that, if the sets in question, though inside the service area of 30 lines, were outside ultra short wave transmissions, they would take them back at a reasonable allowance. Finally, after an exchange of views on this and each side's general attitude to 30-line television, it was recommended:[42]

1 the BBC should make a public statement as soon as possible to the effect that the regular programme television transmissions of 30 lines through the Lon-

don National Transmitter would be discontinued as from 31st March 1934;

2 that if no decision had been made by 31st March to institute regular programme transmissions by some method giving high definition pictures (transmitted through an ultra short wave transmitter), then transmissions of television having a number of lines the same as the present 30-line system would be sent out through one of the ordinary broadcast transmitters: one period to be during the late evening and the other during the morning. The transmissions would be intended to assist experimentalists and would not be for a stated period, since their continuance would depend upon the rate of development of new high definition systems.

These points were contained in a press statement which was released on 5th December 1933, the day following another conference.[43]

As it was unlikely that a new high definition television service would be ready by 31st March 1934, Bairds had won an important concession. They would be allowed to continue with their low definition system and hence keep their name continually before the public. The unknown author of the memorandum on 'The BBC and television' dated 22nd September 1933 had given Bairds an important confidential summary of the prevailing policy regarding television at the time, but it was Goldsmith, the Director of Business Relations in the BBC, who, unknown to the company, had in large measure aided Bairds in their struggle to keep the public interest maintained in television. Together with some other employees of the Corporation, such as Sieveking, Robb and Birkinshaw, Goldsmith had been clearly concerned about the BBC's attitude to television and felt that the position should be clarified, and, additionally, that there should be a greater openness on the part of the BBC.

However, all was not well between Bairds and the Corporation. Although installation of Bairds television equipment had commenced in September and the Assistant Chief Engineer (BBC) had recorded on 19th September that the equipment would probably be working in about 'a week's time', Ashbridge noted more than a month later (on 24th October) that the first demonstration had still not taken place.[44,45]

Part of the delay was due to the fact that the Baird system transmitted a synchronising signal at 25 Hz, normally on the speech channel, but in the programme of tests no speech channel had been provided, as the purpose of the tests was to determine the characteristics of ultra short wave transmission of television. Bishop had recommended that the Boat Race transmitter, modulated by a 25 Hz oscillator, should be used together with an antenna slung from one of the BBC's masts for operation on about 100 m; however this transmitter was required in October for an outside broadcast and was wanted for a week before that for tests. Consequently it was not available for erection in the television room until the last week in October.[46,47]

On 6th December 1933 Greer wrote to Whitley,[48] the Chairman of the Governors of the Corporation:

> We have decided that it would be futile to install our high definition studio transmitter at No. 16 Portland Place on 15th December, in view of your refusal to allow the use of your radio transmitter after 31st December. Between 15th December and 31st December there will be practically no time to carry out experimental work required, on account of the delays necessary for putting up the apparatus and the intervention of the Christmas holidays. We have decided that we can better carry out purely line work without the use of a radio transmitter at our own laboratory, where we can work more quickly and in absolute secrecy. The position is serious because we believe that real television consists of the radio transmission of studio subjects, as distinct from the transmission of films. We asked for these facilities in September last, and subsequently, and naturally expected that they would have been available in ample time for demonstrating to you the latest stage in our development.

The mention of absolute secrecy in Greer's letter referred to an article which was published in the *Sunday Times* on 26th November, following private demonstrations given by the company, to a few of their important shareholders.[49] Greer complained to Whitley:[50] 'We have used our utmost endeavours to stop the publication . . .' etc. but Whitley had heard of press leakages on other occasions:[51] 'While, of course, your assurance of good faith is accepted, I feel bound to add that it is unfortunate that publicity of this kind continues and that it appears impossible to trace the author of such well-informed information on your recent developments.'

Ashbridge thought Bairds were reluctant to install their apparatus at 16 Portland Place because they were attempting to prepare a case that they had not been given sufficient facilities for the test, in order to cover up some lack of success with their apparatus. The position was that some weeks before 8th December, West had told Ashbridge he would shortly be ready with the studio high definition televisor and mentioned 15th December as a suitable date to begin installing. He also mentioned it would be necessary to use a certain arc lighting arrangement which took a very heavy current. The difficulty was West could not use the arc in the studio then available, BB, because the overheating due to waste heat from the consumption of electricity was already excessive, and it would not have been safe to have allowed the temperature to increase any higher. The BBC therefore arranged alternative accommodation and told West that 16 Portland Place could be used from 15th December 1933.[52]

Probably Bairds felt that they could not get the equipment adjusted and working before 31st December, that is shortly before Electric and Musical Industries were due to install their equipment. There may have been another reason: for although Bishop[53] had telephoned West and told him the BBC were willing to extend the date on which the Baird Company were required to remove their apparatus, from 31st December to 20th January, Greer, on 14th December, said his company could not avail themselves of the extension.[54] His reason was that they had been informed on 4th December that, by agreement, Electric and Musical Industries were to have the exclusive use of the BBC's ultra short wave transmitter from 1st January 1934, and, as a consequence, had had to make alternative arrangements 'which have involved us in commitments and decision which we cannot now revoke'. This may have been a convenient reason for the company to leave Broadcasting House. They had had two television transmitters (120-line and 180-line systems) installed in Broadcasting House[55] and Bishop had noted, that, whereas the former gave fairly satisfactory and consistent results, the results of the 180-line transmitter had not been at all consistent. 'The results have been most unstable as viewed on the checking receiver in the same room. The Assistant Chief Engineer conjectured: 'It is quite possible that they may seek to blame us for their slow progress . . .'.

On the other hand, Greer was correct when he said his company had other commitments. In July 1933 they had taken a lease on certain floor space in Crystal Palace[56] and, on 27th November, less than two weeks before Greer sent his letter rejecting the BBC's offer, Angwin of the General Post Office had informed Ashbridge[57] that he had received a letter[58] from West requesting consideration of a proposal to increase the power output of his company's transmitter and also permission to use an additional transmitter at the Crystal Palace. Bairds wanted an alteration in their licence to enable them to transmit 5000 W rather than 500 W and to radiate signals in a wavelength band of 6·0 to 6·25 m or 7·0 to 7·25 m or 6·75 to 7·0 m. The company put forward two reasons for their requests:

1 they wished to put synchronising signals on the sound transmissions 'to enable (them) to extend considerably the actual modulation of the television signal itself, with a corresponding increase of range';
2 they found 'a very marked increase of efficiency in going up from 6 m to 7 m due no doubt to the proportionately high capacity of the valves themselves in the h. f. circuits'.

Ashbridge[59] thought the reasons for the request were to determine:

1 the best way to couple a televisor to an ultra short wave transmitter;
2 the minimum field or the signal-to-noise ratio necessary for good reception of television of this type;
3 the behaviour of receivers.

He felt that, with a transmitted power of about 500 W, the firm could obtain the data required under all the above headings. 'Possibly the present Baird proposal is not unconnected with premature publicity, and it is for this reason that I feel it is undesirable for them to commence high power transmissions.'

A further point which had to be borne in mind was the narrow margin between experimental and regular programme transmissions. It may be the company hoped to use the strategy which they had employed in the early days of low definition television broadcasting by radiating experimental transmissions to a responsive public, and hence of course, because of the need to attract as many persons as possible to their cause, a large transmitted power was essential.

In addition to asking for an increase in power output, the Baird Company also arranged for the hire, at £22.10.0 d per month,[60] of three lines from the BBC control room to their laboratory in Crystal Palace. These were required for:[61]

1 sound plus a 25 Hz synchronising signal
2 control
3 a 3 kHz tone.

Whatever the reason for the company's rejection of the BBC's offer of an extension of time there seems little doubt all was not well with some of Bairds development work, particularly with the 180-line transmissions. The members of the Baird Company working on the apparatus and Birkinshaw were agreed that (in December 1933) the 120-line radio picture gave better results than the 180-line picture.[62] Birkinshaw noted that Bairds appeared to have some difficulty in getting a receiver to cope with an adequate frequency range.

> Towards the end of the week the receiver was brought round to Broadcasting House and a demonstration to yourself and the ACE was, as you know, given which you both regarded as unsatisfactory. Receiver tests are now proceeding at Clapham. Ample signal strength is being obtained, but it definitely needs much improvement.

Probably Greer was right when he said his company wished to pursue their researches in absolute secrecy – but not because of a likely leakage to the press, but rather because he did not want his company's results to be compared adversely with those which EMI would show in a few weeks' time. Captain West therefore gave instructions for the 180-line equipment to be dismantled shortly before 18th December 1933.

Birkinshaw's report was confirmed by another member of the BBC: 'The Baird Television people brought one of their new receivers to Nightingale Lane on Friday last, and carried out tests on Saturday. A picture of sorts was obtained.'

Bishop thought the company should make a statement of what work had actually been done.[63] He complained to Ashbridge: ' . . . we have received no invitation from Bairds to observe the results of their work while they have been in this building. I, and occasionally Kirk, have looked at some of their experimental transmissions by going into the 8th floor room without being asked.' He reminded the Chief Engineer it had been Bairds intention to provide the BBC with a number of television receivers, which could be installed at Clapham and at various private houses, and additionally that there had been a suggestion the BBC might put a receiver into a van and travel round London to observe results. However, he (Bishop) observed on 18th December that Bairds 'have only just produced one receiver and one only which they have sent to Clapham today'.[64]

Carpendale wrote to the Television Company about this point on 20th

December 1933:[65] 'We shall be glad to hear whether you have any comments to make on the experiments generally and whether you propose to give any demonstrations.' He drew attention to the fact that the company had been carrying out experiments in the BBC's building since October and yet they had not been invited to witness any tests transmitted by wireless, nor had they received any information from Bairds as to their progress.

This was too much for Sir Harry Greer, who immediately sent a long (four pages) letter to Carpendale giving the position from his company's viewpoint.[66]

On 3rd January 1934, Clarke[67] wrote to Reith, referred to the visit which he (Clarke) had made to the BBC on the 30th August 1933, and said he was now awaiting definite news advising him when the Corporation would be ready for the EMI tests. He gave some stimulus to his request by telling Reith: 'I know you will be interested to learn that we are successfully broadcasting television of high definition from Hayes and picking it up in London.' Coming after the recent strained relationship between the BBC and the Baird Company, this news must have been received with welcome relief by Carpendale and Ashbridge. Clarke invited Reith to see the equipment working, but he declined because of pressure of work.[68] However, the Director General mentioned that Ashbridge would be glad to attend a demonstration and consequently this was arranged for 12th January 1934.

Ashbridge was very impressed:[69]

> The important point about this demonstration is, however, that it was far and away a greater achievement than anything I have seen in connection with television. There is no getting away from the fact that EMI have made enormous strides . . .

The demonstration consisted of the transmission of films with 150-lines/per picture from the EMI factory at Hayes to the recording studios in Abbey Road, a distance of approximately 12 miles.

> This was by means of an ultra short wave transmitter with a power of 2 kW on a wavelength of approximately 6.5 m. The results were extremely good and there was no question in my own mind that programme value was considerable. The receivers used appeared to be in a practicable form and looked very much like large radio gramophones. On the other hand, it has to be said that the aerial arrangements were very elaborate, being directional in order to cut out interference.

Shoenberg told Ashbridge the policy of EMI was to develop television energetically and they believed there was a great commercial future for the firm which was first in the field with something practicable. EMI were spending £100 000 a year on research and, of this, a large part was allocated specifically to television. Ashbridge mentioned to Reith that there could be absolutely no comparison between the way EMI were handling the problem as compared with Bairds. 'Supposing therefore the BBC wished immediately to establish a television system, it would be almost unnecessary to consider the rival merits of the two firms from the point of view of who supplied the transmitter.' This of course excluded political considerations. EMI had clearly made good use of the extra time which had been made available to them by Bairds' insistence to have the first use of the facilities at Broadcasting House. Their continual desire to be first in any aspect of television development was now rebounding against them. In retrospect they might have been better advised if they had pursued television development work on ultra short wave transmission earlier and more vigorously, rather than to have devoted their energies to the continuation of the 30-line system.

Fig. 12.3 *Sir Isaac Shoenberg*

EMI's great success was undoubtedly due to the very powerful team which Shoenberg (Fig. 12.3), their Research Director, had assembled.

> Isaac Shoenberg was born of a Jewish family in 1850 in Pinsk, a small city in north west Russia. His great ambition in life was to be a mathematician, and it was this ambition that brought him later to England. But because of the difficulties facing a young Jew in Russia at that time, he had to settle

> for a degree in engineering; reading mathematics, mechanical engineering and electricity at the Polytechnical Institute of Kiev University. However, he was to retain his interest in mathematics throughout his life and, about 1911, he was awarded a Gold Medal by his old university for mathematical work.
>
> After leaving university, he worked for a year or two in a chemical engineering company, but very soon he joined the Russian Marconi Company and so – in the first decade of this century – became involved in wireless telegraphy. From 1905 to 1914 he was Chief Engineer of the Russian Wireless and Telegraph Company of St Petersburgh. He was responsible for the research, design and installation of the earliest radio stations in Russia.
>
> In 1914 he resigned his good job and emigrated to England where, in the autumn of the same year, he was admitted to the Royal College of Science, Imperial College, to work under either Whitehead or Forsythe for a higher degree in mathematics.

Unfortunately the outbreak of war brought this plan to a premature end and Shoenberg had to look for a job. 'He found one with Marconi Wireless and Telegraph Company at the princely salary of £2 a week.' His abilities and potential were soon noted and he became joint General Manager and head of the patents department.

Professor J. D. McGee, in his 1971 Memorial Lecture on 'The life and work of Sir Isaac Shoenberg, 1880–1963' has told how he joined the EMI organisation:[70]

> Shoenberg had no formal training as a musician, but he had an intense natural love of music; this made him a keen connoisseur of recorded music and the technique of recording. This in turn resulted in a friendship with Sir Louis Sterling, and an invitation to join the Columbia Graphophone Company and to put into practice his ideas on recording. He was clearly successful in this and soon the Columbia Company was competing effectively with the formerly almost unchallenged, prestigious firm The Gramophone Company or HMV. So began the association of Sterling with Shoenberg which was, to my mind, crucial in the field of television in the following decade. The shrewd business man and financier, Sterling, completely trusted Shoenberg, an engineer, scientist or applied physicist with a large dash of the visionary.

The company prospered and competed effectively with HMV during the record-selling boom of the mid- and late-twenties. However, when the depression occurred, both companies began to suffer and the result was the combining of the two firms in 1931 to form Electric and Musical Industries Ltd. with Alfred Clarke as Chairman, Louis Sterling as Managing Director, and Isaac Shoenberg as a director of research and head of the patents department. Shoenberg's Columbia team, which included A. D. Blumlein, P. W. Willans, E. C. Cork, H.

Holman and others joined the HMV group at Hayes. This group included the research manager, G. E. Condliffe, W. F. Tedman and C. O. Browne. As mentioned previously, Browne had earlier engineered television capable of producing a picture of greatly improved definition by effectively combining five separate channels, each using mechanical scanning. The equipment had been demonstrated at the Physical and Optical Societies Exhibition in January 1931, and had created quite a stir.

One of the first questions which Shoenberg and his team had to tackle was whether development should proceed on mechanical or electronic lines. Mechanical scanners had the advantage that they had been made successfully, whereas the electronic scanner did not exist as a practical device. On the other hand, electronic scanning had many potential advantages, as Campbell Swinton had indicated, particularly for high definition television. The HMV research team's exploratory work on television was based on the philosophy that they should aim to develop an effective cathode-ray tube picture receiver, but should leave any attempt to develop a television camera to others.

'The view had been taken that the company's business was in the field of receiving, not transmitting apparatus.' It was because of this philosophy that Browne, of HMV, had developed his signal generation equipment which could produce a 150-line picture. Tedham, also of HMV, had made a start on electronic television by developing a new type of photocell, discovered only in 1930 by Koller in the United States, which had a sensitivity about one hundred times greater than anything previously known. He had also begun the development of a hard (vacuum) cathode-ray tube to replace the gas-focusing CRT which was then the only available type. This tube was only satisfactory if the beam current did not vary greatly, as in the velocity-modulation technique developed by the Cossor Company.

Shoenberg has written how

> . . . remembering the vagaries and instability of the soft valves of the 1912–16 period, (he) decided against the soft cathode-ray tube and directed (their) research towards the development of a hard type with electron focussing. The efforts in this direction, particularly by Broadway and his group, turned what was thought a very speculative development in 1931 into the commonplace of today.[71]

The early research, 1931–1933, on hard cathode-ray tubes was undertaken by Tedham and McGee (from 1st January 1932). Later, in the autumn of 1933, when McGee was asked to lead the group charged with the development of a pick-up tube, this work was undertaken by a team headed by Dr. L. F. Broadway.

Professor McGee (Fig. 12.4) described in his Memorial Lecture the state of the art in 1931–32:

> 1 Photo-electricity: in spite of the fact that Einstein had made this an honest physical phenomenon some 25 years previously, we had still not

even begun to understand the physics of photosensitive materials. The physics of the solid state was still in its infancy. The technology of photo emissive and photo conductive materials was 'rule-of-thumb' – almost black magic. The first reasonably efficient photo-emissive surface, the silver–caesium (Ag–AgO–Cs or SI) photo-cathode, had only just been discovered by Koller in 1930; it was to be four years before Gorlich discovered the antimonide photo-cathodes;

Fig. 12.4 *I. Shoenberg with Dr J. D. McGee*

2 Vacuum technique was pretty primitive, but was just beginning to be adequate. Glass-to-metal seals were very unreliable, and Pyrex glass was just becoming available. For example, when I mentioned to a former Cavendish colleague – recognised as a leading vacuum expert – that Tedham and I were trying to make a hard CRT with a thermionic-oxide coated cathode, he exploded in rather rude laughter.
3 Electron optics had not been formulated as a subject. Although Busch had given the first clue in 1927, even he referred to the magnetic lens for a CRT as the 'concentrating coil'. Only with the publication in 1932 of the classic paper by Knoll and Rusks did we realise the complete analogy between light-optics and electron-optics. This was a fundamental requirement for both the transmitting and receiving cathode-ray techniques for television.
4 The physics of the solid state, on which the physics of phosphors depended, was still unformulated and hence the physics of phosphors was in much the same state as that of photo-electric materials, yet the

former were as important for the effective display of a t. v. picture on a CRT as the latter were for the generation of picture signals.

5 Secondary electron emission was known as a phenomenon, but very little was known about the important details of the effect. However, it was of fundamental importance in all cathode-ray tubes in which a stream of negative electrons bombard highly insulating targets or fluorescent screens. How does the target avoid being driven more and more negative until the electrons can no longer reach it? Only, of course, because of secondary electron emission.

6 Radio communication techniques were unbelievably primitive compared with today. Amplifiers of reasonable bandwidth, scanning circuits, pulse circuits, even smoothed h.t. power supplies, radio transmitters with bandwidths of more than a few KHz with a carrier frequency of 50 MHz and so on, all had to be invented as we went along.

In view of the state of the art in 1931–32, the task facing Shoenberg must have appeared daunting. Yet so convinced was he and his research team of the correctness of their approach to the television problem, that he was able to persuade Sterling to sink hundreds of thousands of pounds of his company's money into the speculative venture.

The development of the cathode-ray tube receiver could not have proceeded without a source of test signals, but fortunately Browne's 5-channel equipment was available and fulfilled this role in the early stages of development. A radio link was also required and so a 7 m shortwave transmitter was hired from the Marconi Company. Thus in approximately two years only the EMI scientists and engineers had produced a television system which so impressed Ashbridge that, after his visit to Hayes on 12th January 1933, he felt able to suggest to Reith, in a tacit way, that the BBC should consider the establishment of a high definition service.

Following the demonstration, a general discussion took place between Shoenberg and Ashbridge and the following facts emerged:

1 EMI no longer wished to install apparatus in Broadcasting House but to go directly to the next step beyond that and to experiment with transmitters of much higher power and not less than 4 kW as compared with the $\frac{1}{2}$ kW we have here.

2 They (were) still very anxious for (the BBC) to take the whole operation of the transmitting side and of course the provision of programmes when it (came) to regular transmissions.

3 They (were) prepared to loan to us not only the televising gear, but the wireless transmitter of 4 kW as well, if the BBC could find accommodation for it.

4 They (emphasised) that they (were) not interested in the transmitting side and (were) prepared to allow us complete access to every detail of information in connection with their transmitting apparatus.

EMI felt there would be sufficient business in the receiving side of television for them not to be particularly interested in the transmitting side. This was a point of view which Clarke had mentioned to Reith early in 1933. Their view seemed to change, however, two months later, for White, the Managing Director of Marconi Wireless Telegraph Co. Ltd., told Reith his company had concluded an agreement with EMI Ltd. to collaborate with them in the field of television transmissions.[72] In his letter of 23rd March 1934, White said:

> A separate company (Marconi–EMI Television Co. Ltd.) will be formed to supply apparatus and transmitting stations. This company will have the benefit of the extensive research work done by both organisations. The shares of the new company will be subscribed and held in equal parts by the Marconi and EMI.

Reith did not quite understand this position[73] but the most likely explanation was that EMI's high definition studio and receiving equipment was useless without a suitable ultra short wave transmitter capable of transmitting the wide sideband signals without distortion, and, as EMI themselves had practically no experience in this field, it was perhaps natural they should consider some form of co-operation with the leading manufacturer of transmitters in the United Kingdom. (Marconis' expertise had been demonstrated to Bishop on 3rd February 1933, when N. E. Davies of MWT Co. Ltd. visited Broadcasting House to modify the ultra short wave transmitter to deal with sidebands up to 500 kHz. The BBC's Assistant Chief Engineer seemed rather surprised at the ease with which Davis carried out his task: 'apparently it is quite a simple matter', he observed.)[74]

Ashbridge listed three points which for him were fairly clear:

1 Broadcasting House was not suitable for short wave television experiments on high powers, for example up to 20 kW;
2 the BBC would have to take other premises, suitable for future development;
3 the first television service would consist mainly of film transmissions and would be established first of all in London and afterwards in other places;
4 the problems of finance on both the programme and technical sides would require consideration.

On the technical side, the Chief Engineer thought the BBC should aim to install transmitting stations of about 20 kW power output, initially in London and later in Birmingham, Manchester, Newcastle, etc. Twin transmitters would be required to provide both a sound and vision channel. In addition he visualised the need for a system to rapidly distribute the films to each centre, since long distance lines could not be used. However, all of this would involve not only the BBC but also the public in very considerable expense, as the receivers could not cost much less than £80 to begin with. The service would have to begin as a luxury service and an extra licence fee of as much as £2.2.0 might not go very far in the early stages towards meeting expenses. Ashbridge suggested the financial difficulties might become so serious that the only way of getting enough revenue would be by having sponsored programmes, or possibly a separate service, not financed out of the 10 shilling licence fee but controlled by the BBC, which gave sponsored programmes with concessions to one or more firms like EMI on a territory basis.

On the programme side, Ashbridge considered the most attractive kind of programme he could visualise would be a sort of newsreel such as one sees in certain film theatres.

> If we could somehow obtain such films and give a performance each night for about an hour or an hour and a half, this might constitute the main programme feature; film plays would also of course be transmitted, but it would be difficult to obtain say three hundred different plays and probably the same play might have to be repeated several times accompanied by news and informative matter generally.

This type of programme would presumably be resisted by the regular film producing companies, both of the news collecting type and others, but Ashbridge imagined this resistance 'could be overcome by making (the films) ourselves or paying a large sum to a film company'.

Ashbridge stressed his report was intended to be merely an introduction to a discussion of the whole matter, but he had raised it because he thought that, if the BBC could not see their way to going on to make television an established service on a considerable scale, it would be better not to be involved in serious preliminary steps.

Undoubtedly, the Chief Engineer had been most impressed by what he had seen at Hayes and felt it necessary to consider the television situation in a new light, even although the development of a possible service might take some years.

All of this was in marked contrast to the prospects facing the Baird television system. With Ashbridge's glowing report of EMI's demonstration following so closely on Greer's letters, the anti-Baird faction at the BBC must have thought their problems were nearly over so far as Bairds were concerned. The latter's position was not helped by the demonstration which they gave on Monday, 12th March 1934[75] to the Prime Minister, BBC representatives, Colonel Lee, Colonel Angwin and others (but not Reith, who as usual could not attend).

Bairds showed their cathode-ray tube receiver and, as the tube was of GEC manufacture, GEC representatives were present. Tranmissions were given from the Crystal Palace using a wavelength of about 8.5 m for the vision signals and a longer wavelength for sound. The pictures transmitted consisted of an introductory talk given by a speaker, showing the head and shoulders, a violin solo by a lady violinist (also limited to head and shoulders), a talk on architecture illustrated by large scale photographs, and short extracts from two films.[76]

Colonel Angwin thought the standard of reception was about the same as that attained in the demonstration given on 20th November 1933, with some limitations due to the radio link. 'Some interference was obvious from electrical sources, but as far as the radio link was concerned, the conditions were fairly good. The receiving aerial was at the top of the four-storey building and fairly remote from motor car interference.'

The detail of the head and shoulder subjects televised and close-ups of the films gave fairly good detail. For larger scenes the detail was much improved. It was not a demonstration to inspire Ashbridge, who wrote of his disappointment. The Prime Minister, however, congratulated Mr Baird on the success he had attained and the very great improvement on his earlier attempts.

'The film transmission given by EMI is appreciably better than that shown by the Baird Company,' Ashbridge observed. 'On the other hand, however, no opportunity has been available so far to compare a demonstration under absolutely strictly comparable conditions. Moreover, the EMI Company have not so far attempted a demonstration with living objects.'[77]

Fig. 12.5 *J. L. Baird with Sir Harry Greer at the television demonstration held at Film House, Wardour Street on 20th March 1934*

In an attempt to settle the rival claims of the two companies, Reith, on 15th March 1934, wrote[78] to Kingsley Wood and suggested a conference 'between some of your people and some of ours to discuss the future arrangements for the handling of television'. Reith thought there were three aspects to consider – the political, 'using the term in a policy sense and for want of a better one', the financial and the technical. He nominated Carpendale and Ashbridge. Kingsley Wood agreed[79] and put forward the names of Phillips and Angwin. The decision to ask for a conference had not been precipitated by the Baird demonstration alone, for Phillips[80] had noted four days before this that the BBC would probably be seeking an interview shortly to discuss the whole question of the future arrangements in regard to television.

The extent of the Baird Company's knowledge of the EMI operations in the field of television can only be surmised, but it is likely they had some idea of the successes of the Hayes company. Kingsley Wood had told Baird early in 1933 he should face up to the possibility that there might be a superior system to the Baird system.[81]

Following the fifth annual meeting of Baird Television Ltd. on 20th March 1934 (Fig. 12.5), at which a number of shareholders criticised Greer's speech because it was devoid of any hint of hostility to the BBC, the Chairman told Reith[82] he had been asked by a number of representatives from various nesw-papers to attack the BBC for 'its apparent inattention to the efforts which we have made to bring television to the level of real entertainment value, and the apparent encouragement the BBC have given to a rival concern'. Perhaps Greer thought that a soft approach was the best tactic to adopt with the BBC after the previous year's rather strained relationship, particularly as Baird's campaign to prevent EMI from developing television had failed. In thanking Greer for his attitude, Reith informed him of the BBC/GPO conference.[83]

Whatever the reason first his action, Greer thought it prudent not to antagonise the BBC, although the BBC was being subjected to a sustained attack in the press for not pursuing the advancement of television with greater speed.[84]

References

1 PHILLIPS, F. W.: a memorandum, 2nd August 1933, Minute 4004/33.
2 CHURCH, Major A. G.: letter Sir H. Kingsley Wood, 4th August 1933, Minute 4004/33.
3 ASHBRIDGE, N.: record of telephone conversation with A. G. D. West, 10th August 1933, BBC file T16/42.
4 ASHBRIDGE, N.: record of interview with A. G. D. West, 11th August 1933, BBC file T16/42.
5 WEST, A. G. D.: letter to N. Ashbridge, 26th July 1933, BBC file T16/42.
6 ASHBRIDGE, N.: letter to A. G. D. West, 28th July 1933, BBC file T16/42.
7 GREER, Sir H.: letter to Sir J. F. W. Reith, 16th August 1933, BBC file T16/42.
8 BAIRD, J. L.: letter to Sir J. F. W. Reith, 18th February 1933, BBC file T16/42.
9 REITH, Sir J. F. W.: letter to Sir H. Greer, 21st August 1933, BBC file T16/42.
10 REITH, Sir J. F. W.: letter to J. L. Baird, 3rd February 1933, BBC file T16/42.
11 GREER, Sir H.: letter to Sir J. F. W. Reith, 25th August 1933, BBC file T16/42.
12 Notes of meeting in C. E.'s office, 23rd August 1933, BBC file T16/42.
13 CARPENDALE, Admiral Sir C.: letter to A. Clarke, 8th September 1933, BBC file T16/42.
14 CLARKE, A.: letter to Admiral Sir C. Carpendale, 9th September 1933, BBC file T16/42.
15 Anon.: 'The BBC and Television. The position as at 22nd September 1933', personal collection.

16 ASHBRIDGE, N.: letter to Baird Television Ltd., 4th September 1933, BBC file T16/42.
17 Baird Television Ltd.: letter to the BBC, 6th September 1933, BBC file T16/42.
18 CARPENDALE, Admiral Sir C.: letter to Baird Television Ltd., 8th September 1933, BBC file T16/42.
19 BISHOP, H.: memorandum to C(A), 19th September 1933, BBC file T16/42.
20 Director Business Relationship (BBC): letter to Baird Television Ltd., 22nd September 1933, BBC file T16/42.
21 Baird Television Ltd.: letter to D. B. R., 22nd September 1933, BBC file T16/42.
22 Baird Television Ltd.: letter to D. B. R., 27th September 1933, BBC file T16/42.
23 Anon.: 'Television census. Request for postcards', *Daily Telegraph*, 18th August 1933.
24 Anon.: 'BBC and television. Manufacturer's protest', *Daily Mail*, 18th August 1933.
25 Anon.: 'Television future. Looking-in on the look-out', *The Star*, 28th September 1933.
26 Anon.: 'Television surprise – BBC ending agreement', *Daily Mail*, 28th September 1933.
27 GREER, Sir H.: letter to Sir J. F. W. Reith, 28th September 1933, BBC file T16/42.
28 REITH, Sir J. F. W.: letter to Sir H. Greer, 30th September 1933, BBC file T16/42.
29 BBC: press announcement, 12th October 1933.
30 GREER, Sir H.: letter to Sir J. F. W. Reith, 27th September 1933, BBC file T16/42.
31 D. B. R.: memorandum to Admiral Sir C. Carpendale, 5th October 1933, BBC file T16/42.
32 D. I. P.: memorandum to the editors of the *Radio Times, World Radio, Listener,* 8th May 1933, BBC file T16/42.
33 YOUNG, F.: 'Television. New problems of broadcast reception', *Radio Times*, 28th April 1933, p. 204.
34 ASHBRIDGE, N.: memorandum to the Director General and Controller, 2nd May 1933, BBC file T16/42.
35 Director General: memorandum to D. I. P., 3rd May 1933, BBC file T16/42.
36 BISHOP, H.: memorandum to D. B. R., 12th October 1933, BBC file T16/42.
37 D. B. R.: memorandum to C(A), 20th October 1933, BBC file T16/42.
38 D. B. R.: letter to Sir H. Greer, 24th October 1933, BBC file T16/42.
39 ASHBRIDGE, N.: memorandum to C(A), 28th October 1933, BBC file T16/42.
40 Record of telephone conversation with F. W. Phillips, 27th October 1933, BBC file T16/42.
41 BBC: letter to Sir H. Kingsley Wood, 31st October 1933, BBC file T16/42.
42 BBC: press statement, 5th December 1933, BBC file T16/42.
43 Record of interview with Messrs. Greer, West, Church and Baird, and BBC Chairman, C(A), D. B. R., and C. E., 4th December 1933, BBC file T16/42.
44 BIRKINSHAW, D.: memorandum to S. R. E., September/October 1933, BBC file T16/42.
45 ASHBRIDGE, N.: letter to A. G. D. West, 24th October 1933, BBC file T16/42.
46 BISHOP, H.: memorandum to S. S. E., October 1933, BBC file T16/42.
47 S. S. E.: memorandum to Partridge, 9th October 1933, BBC file T16/42.
48 GREER, Sir H.: letter to J. R. Whitley, 6th December 1933, BBC file T16/42.
49 Press cutting from the *Sunday Times*, 26th November 1933, BBC file T16/42.
50 GREER, Sir H.: letter to J. H. Whitley, 24th November 1933, BBC file T16/42.
51 WHITLEY, J. H.: letter to Sir H. Greer, 28th November 1933, BBC file T16/42.
52 ASHBRIDGE, N.: memorandum to C(A), 8th December 1933, BBC file T16/42.
53 BISHOP, H.: memorandum to C(A), 9th December 1933, BBC file T16/42.
54 GREER, Sir H.: letter to J. H. Whitley, 14th December 1933, BBC file T16/42.
55 A. C. E.: memorandum to C. E., 18th December 1933, BBC file T16/42.
56 BAIRD, M.: 'Television Baird' (HAUM, South Africa, 1974).
57 ANGWIN, A. S.: letter to N. Ashbridge, 27th November 1933, BBC file T16/42.
58 WEST, A. G. D.: letter to A. S. Angwin, November 1933, BBC file T16/42.
59 ASHBRIDGE, N.: letter to A. S. Angwin, 29th November 1933, BBC file T16/42.
60 Baird Television Ltd.: letter to the BBC, 6th December 1933, BBC file T16/42.
61 BIRKINSHAW, D.: memorandum to the lines section (BBC), 1st December 1933, BBC file T16/42.

62 BIRKINSHAW, D.: memorandum to S. R. E., 18th December 1933, BBC file T16/42.
63 S. R. E.: memorandum to A. C. E., 19th December 1933, BBC file T16/42.
64 BISHOP, H.: memorandum to N. Ashbridge, 18th December 1933, BBC file T16/42.
65 CARPENDALE, Admiral Sir C.: letter to Baird Television Ltd., 20th December 1933, BBC file T16/42.
66 GREER, Sir H.: letter to Admiral Sir C. Carpendale, 3rd January 1934, BBC file T16/42.
67 CLARKE, A.: letter to Sir J. F. W. Reith, 3rd January 1934, BBC file T16/42.
68 ASHBRIDGE, N.: letter to A. Clarke, 5th January 1934, BBC file T16/65.
69 ASHBRIDGE, N.: report on television, 17th January 1934, BBC file T16/65.
70 McGEE, Professor J. D.: '1971 Shoenberg Memorial Lecture', *The Royal Television Society Journal*, 1971, **13**, No. 9, May/June.
71 SHOENBERG, I.: Discussion on 'The history of television', JIEE, 1952, **99**, Part IIIA, pp. 41–42.
72 WHITE, H. A.: letter to Sir J. F. W. Reith, 23rd March 1934, BBC file T16/65.
73 REITH, Sir J. F. W.: memorandum to N. Ashbridge, 26th March 1934, BBC file.
74 BISHOP, H.: memorandum to S. R. E., 3rd February 1933, BBC file T16/65.
75 GREER, Sir H.: letter to Sir J. F. W. Reith, 7th March 1934, BBC file T16/42.
76 ANGWIN, A. S.: memorandum, 12th March 1934, Minute 4004/33.
77 ASHBRIDGE, N.: report on demonstration to the D. G. and C(A), 12th March 1934, BBC file T16/42.
78 REITH, Sir J. F. W.: letter to Sir H. Kingsley Wood, 15th March 1933, BBC file T16/42.
79 KINGSLEY WOOD, Sir H.: letter to Sir J. F. W. Reith, 20th March 1933, BBC file T16/42.
80 PHILLIPS, F. W.: memorandum to H. Napier, 8th March 1934, Minute 4004/33.
81 Post Office memorandum, 23rd May 1933, Minute 4004/33.
82 GREER, Sir H.: letter to Sir J. F. W. Reith, 21st March 1934, BBC file T16/42.
83 REITH, Sir J. F. W.: letter to Sir H. Greer, 22nd March 1934, BBC file T16/42.
84 Anon.: 'BBC attacked by famous scientist. Sir A. Fleming and television hold-up', *Daily Mail*, 15th March 1934.

Chapter 13

The work of the Television Committee; 30-line television, high definition television, 1934

The Television committee, which was responsible for the historic document on television published in January 1935, which described the plan for the establishment of the world's first, high definition, regular, public, television broadcasting system, had its origin at an informal conference, held at the General Post Office on 5th April 1934, between representatives of the British Broadcasting Corporation and the General Post Office.[1]

Admiral Sir Charles Carpendale and Noel Ashbridge represented the interests of the BBC, while Colonel A. S. Angwin and F. W. Phillips and J. W. Wissenden provided the GPO's point of view. Phillips was chairman.

A number of general questions were discussed including:

1 the method of financing a public television service;
2 the use of such a service for new items and plays;
3 the relative merits of some of the systems available including those of the EMI, Baird, Cossor and Scophony companies;
4 the arrangements necessary to prevent one group of manufacturers obtaining a monopoly of the supply of receiving sets;
5 the possible use of film television to serve a chain of cinemas.

The use of film generally for television purposes was of some importance at the time of the conference, as only the Baird company had shown direct television. Both Electric and Musical Industries and Baird Television Ltd. had demonstrated the transmission of television signals using films with cathode-ray tube receivers, but EMI had refrained from showing direct television because they regarded its present developments as inadequate for commercial exploitation. They also considered that a film had a more lasting commercial value, whereas their view of direct television was that by its very nature it was essentially transient.

On the quality of the work being undertaken by the two rival British concerns,

Ashbridge felt Bairds apparatus was distinctly amateurish in construction and finish, whereas EMI had produced well constructed and workmanlike apparatus.

The BBC's Chief Engineer had been much impressed by the demonstrations he had seen at the Hayes factory, but his opinion of the Baird Company's product had possibly been soured by the years of friction between the Corporation and the Television Company. Thus he agreed with Colonel Angwin's observation that the cathode-ray tubes employed by the latter company (which were manufactured by the General Electric Company), were somewhat superior, especially in colour, to those utilised by EMI. Fortunately for the Television Company the conduct of the Post Office's officials in matters affecting the development of television had always been exemplary in regard to impartiality and fairness of view and policy. One Postmaster General and several important administrative and technical officers of this public service had made favourable comments from time to time on Baird's 30-line television development work and had often exerted their powerful influence and position to suport the company in its negotiations with the Corporation. The Post Office could certainly not be accused of duplicity, or dubiety in its relation with Baird or his companies and their was never any hint or suggestion of fencing or exploitation of difficulties by Phillips or any of his colleagues. Now, with the two rival companies campaigning for the establishment of television stations – EMI for a new BBC station and Baird Television for a station of its own – it was agreed by the conference that a committee should be appointed to advise the Postmaster General on questions concerning television. The BBC representatives were keen that this committee should be established at once as difficult questions were arising, and would continue to arise, and they felt it would be helpful for the BBC and GPO to have the weight of authority of a committee behind them in any decision they might take.

Subsequently, on 9th April 1934, Phillips[2] sent a memorandum to Simon and advised him of the conference's recommendation. Phillips proposed the following terms of reference for the new committee:

> To consider the development of television and to advise the PMG on the relative merits of the several systems and on the conditions – technical, financial and general – under which any public service of television should be provided.

Initially the membership for the committee was suggested as:

1 Chairman – the Assistant Postmaster General
2 Two Post Office representatives (one administrative and the other technical)
3 Two BBC representatives
4 One independent technical member to be nominated by the DSIR
5 One representative to be nominated by the RMA (who should not be connected with any manufacturers identified with one of the systems of television)
6 Secretary

Phillips recommended J. V. Roberts, a staff Officer in the Telegraph and Telephone Department of the Post Office, for the latter position and told Simon that he understood Sir Charles Carpendale and Noel Ashbridge would be the Corporation's nominees (Fig. 13.1). In addition, Phillips thought Colonel Angwin should be one of the two Post Office representatives.

Fig. 13.1 *Sir Noel Ashbridge (on the plinth), Lord Selsdon (with outstretched arm), Sir Charles Carpendale, F. W. Phillips and I. Shoenberg photographed at Hyde Park Corner, 1937.*

The composition[3] of the committee was:

The Right Hon. Lord Selsdon, KBE (Chairman)
Sir John Cadman, GCMG, DSc, FGS, MICE (Vice-Chairman)

Col. A. S. Angwin, DSO, MC, BSc, MIEE, AM Inst. CE, Assistant Engineer-in-Chief, General Post Office
Mr O. F. Brown, MA, BSc, Department of Scientific and Industrial Research
Vice-Admiral Sir Charles Carpendale, CB, Controller, British Broadcasting Corporation
Mr F. W. Phillips, Assistant Secretary, General Post Office
Secretary, Mr J. Varley Roberts, MC

The first meeting[4] of the Television Committee was held on 29th May 1934, and, apart from procedural matters, concerned itself with:

1 the characteristics of low and high definition television;
2 the need to assess the practical value, if any, of the 30-line system;
3 the desirability of the committee viewing demonstrations of the Baird and EMI systems before receiving evidence from the two companies;
4 certain questions – apart from those of a technical nature – which would have to be dealt with by the committee.

It was agreed that all the committee's meetings should be held in private and that reports of future meetings should not be issued to the press.

As part of the committee's brief was to advise the PMG on the relative merits of the several systems of television, it was necessary for the committee to consider not only British television developments, but also those being made in other countries, and so, on 18th May 1934, a letter[5] was sent to the Telegraph Administrations of France, Germany and Italy and also to Dr C. B. Jolliffe, Chief Engineer, Federal Radio Commission, USA, asking for information on a number of points concerning their country's television schemes.

The Television Committee worked with commendable speed and examined 38 witnesses (see Appendix 1), some of them on more than one occasion, and in addition sent delegations to the United States (headed by the Chairman, Lord Selsdon), and to Germany, (headed by O. F. Brown), to investigate and report upon progress in television research in those countries. By September 1934 the committee had so advanced its deliberations that it was able to commence the preparation of its report, and, but for the absence of Sir John Reith[6] (whom the committee wished to interview), towards the end of 1934, the committee's report would have been produced before January 1935 (the actual month of publication).

The first witnesses to be examined on 7th June were Major A. G. Church and Captain A. G. D. West of Baird Television Ltd. Previously the committee had seen the company's low and high definition television systems in operation on 31st May and 2nd June, respectively, and the EMI high definition system on 1st June.[7]

The questioning of Church and West centred mainly on certain aspects of television which had been put forward by the company in a précis[8] of a statement which had been submitted prior to the meeting and which covered the company's history and status; the 30-line television system; high definition television and the operation of a public television broadcasting service; non-domestic applications of television; and general and miscellaneous matters. Some of these points formed the basis of the committee's discussions with other witnesses and so it is convenient to consider their views in relation to the answers given by Church and West, rather than treat the opinions of the various witnesses in isolation.

13.1 30-line television

Perhaps surprisingly for a company which had fought long and hard for the establishment of a low definition television system, the Baird organisation now considered the 30-line transmissions had no commercial value. It recognised the limitations imposed by the 30-line standard on the type of subject matter which could be televised and, indeed, Church was rather unflattering in his opinion of the BBC's efforts to vary the images which they were transmitting. Head and shoulders shots gave rise to tolerable, but imperfect, images. When extended views were televised, the images 'became rather ridiculous' and the BBC's work on such scenes, Church felt, had really been a waste of effort. 'The BBC were trying to do something which was quite impossible', he observed.

In answer to a question from Ashbridge on whether the service had no use, Church thought the transmissions could be of interest to schools as a great deal of television technique could be learnt from the broadcasts.

Most of the Television Committee's witnesses were questioned by Lord Selsdon and his committee members on the desirability of continuing the low definition service. Possibly the remembrance of the ill-feeling which had been engendered by the BBC's procrastination and general unwillingness to commence the service caused the committee to enquire closely into the future viability of the 30-line standard. In addition, certain semi-popular technical journals, notably *Wireless World*[9] and *Television*,[10] had published articles and readers' letters advocating the retention of the low definition television broadcasts.

The *Television* magazine editor, in May 1934, decided to test the opinion of its magazine's readership by publishing a questionnaire on the subject. Somewhat naturally, as this readership was made up predominantly of home constructors, there was a strong feeling that low definition transmissions should be extended and encouraged until such time as high definition apparatus became available in a reasonably inexpensive form for use in any part of the country.[11]

The above point was taken up by the Television Society, both in its written evidence to the committee and in its oral evidence, which was presented on behalf of the society by Dr C. Tierney and R. R. Poole and W. G. W. Mitchell.[12]

The Television Society was formed on 7th September 1927 for the purpose of furthering the study and development of television and allied subjects among experimenters and private users of television.

Lord Haldane (the creator of the Territorials and an exponent of idealist philosophy), was the society's first president, but, when he died in 1928, Sir Ambrose Fleming, a staunch supporter of Baird, succeeded to the presidency. Notable vice-presidents were Sir John Smith Samuel, Sir James Percy and Professor Magnus McClean, and, for many years, J. J. Denton and J. Keay were the General Secretary and Treasurer, respectively.

The link between the Baird organisations and the Television Society appeared *prima facie* to be strengthened shortly after the first publication of the *Television* magazine in 1928. This was produced by Television Press Ltd., which had as its chairman Moseley. From the start of its activities *Television* devoted regular space to matters pertaining to the Television Society and so, to the unknowing public, the magazine seemed to be the official organ of the society.

Initially A. Dinsdale was the editor of *Television*, but in 1929 Moseley became the general editor and shortly afterwards the ownership of the paper passed to Benn Bros., who, in turn, sold it to Bernard Jones – a publisher who had made a considerable success of *Amateur Wireless*.

Television began with a circulation of approximately 10 000 readers, many of whom were home constructors and experimenters and therefore very much interested in 30-line television. Low definition television provided a practical, low cost outlet for the magazine's subscribers, which clearly could not be met by a high definition television system which required expensive cathode-ray tube receivers. The magazine therefore featured many articles on 30-line television (which, from 1929, was transmitted by Bairds), and so, with Moseley as its editor, *Television* was seen by the anti-Baird faction as being certainly pro-Baird.

At the time evidence was being taken by the Television Committee, the Television Society had a 'live membership' of 350 who paid an annual subscription of £1 per year. Dr Tierney described its membership as comprising independent users – persons 'who know what they are driving at and what they hope to achieve'.

The society had its early headquarters at the Engineers' Club, Coventry Street, but later moved to larger premises in Duke Street and then to the home of its Secretary in Lisburne Road, Hampstead. On the second Wednesday in each month the society held meetings at University College so that the papers could be read and technical discussions initiated. The lectures were rather more technical than popular and the society published some of these in its quarterly journal. Special features of the society's annual programme were its exhibition and the first lecture of each session, which dealt with the progress of television in Germany. The sole Honorary Fellow of the society until 1936 was John Logie Baird.

Thus both the publishers of the *Television* magazine and the Television Society had an interest in low definition television and in addition had (or had had), friendly associations with Baird and his companies. Clearly, the society and the publishers of the magazine were keen to state the views of their subscribers to the Television Committee and hence the two organisations arranged facilities for their members to make known their opinions on certain aspects of television, by means of questionnaires.

On 21st September 1934 the evidence of Tierney and his two colleagues was given. The society represented that the present length of the 30-line television programmes was insufficient to meet the needs of either the experimenter or the ordinary user and that, for the latter person, the two half-hour television programes per week were not considered to be a sufficient attraction to warrant much expenditure on a receiver by them.[13] Furthermore, the society's questionnaire showed that 93% of the respondents had expressed a desire for the extension of the present 30-line transmission, 56% had advocated the inclusion of 30-line television in the normal programme hours, 20% had proposed that television programmes should be transmitted outside ordinary sound broadcast hours in order that the latter should not suffer reduction, while 17% of the respondents had asked for an extension of the transmissions irrespective of the time of day. Of the members of the society who contributed their views, 45% expressed an interest in television from the point of view of entertainment, 80% were interested from that of experimental science, or, considered differently, 18% were interested in entertainment only, 55% in experiment only and 27% in both aspects. Consequently the society recommended the continuance by the BBC of transmissions of 30-line pictures, with sound, on the present channels, pending the complete development of a practical system providing a higher quality of vision. The present transmissions were felt to have a definite entertainment value and could be easily received on inexpensive apparatus. They served also to maintain public interest and afforded some facilities for amateur experimental work. All of this argued, of course, not only for the retention of the services but also its extension. The society suggested that, in the event of a considerable increase in the service, there would be little need in the initial stages to maintain a programme standard as high as that required for sound broadcasting, since the novelty would constitute a strong attraction, while the televising of artists and speakers taking part in ordinary sound broadcasts would involve little additional expense, apart from that due to the extra transmission channel. Then, if the number of users justified an increased expenditure, the programmes could be gradually elaborated.

On the issue of supplementary 30-line transmissions, the society strongly urged that these should be provided from the principal BBC stations, outside normal broadcast hours. These transmissions, it was argued, would serve primarily for experimental purposes and so the subject matter would be of secondary importance. Hence, recognising the cost implications of providing programmes and maintaining two transmissions, the society put forward the following alternatives as suggestions, in order of preference:

1 film with sound;
2 silent films;
3 living subjects without sound.

Another proposal which was advanced by the society was based on the responses to its questionnaire. This showed that 66% of the respondents approved the transmission of 30-line pictures of moving objects and geometrical designs, unaccompanied by sound. Such transmissions could be made late on Saturday nights and Sunday mornings, the society stated, when they would be most valuable to the amateur experimenter.

The enthusiasm shown by the *Television* magazine and the Television Society for the low definition television broadcasts was in marked contrast to the view advanced by Dr J. H. T. Roberts, on behalf of *Popular Wireless* and the *Wireless Constructor* (published by the Amalgamated Press Ltd.).[14] He held that the 30-line system had no real prospect of either entertainment or commercial value; indeed he thought the system was very crude and had no value other than a curiosity value. It is interesting to note that Lord Selsdon, during the interview with Roberts on 29th June 1934, confirmed that this was also his opinion. At this stage in its proceedings, the Television Committee had taken evidence from Baird Television Ltd., EMI and M–EMI only. Thus at quite an early stage in the committee's deliberations, and before many witnesses had been examined, the end of the 30-line system seemed assured.

Such an opinion was not unexpected, for *Popular Wireless* had Captain P. P. Eckersley, the former Chief Engineer of the BBC, as its principal radio consultant, and, for a number of years, he had been antagonistic to Baird's efforts in the television field.

Popular Wireless was established in 1922 and from its inception had paid great attention to the science of seeing by electricity. It had offered a prize of £500, in 1924, for television devleopment (which had resulted in Dr Fournier d'Albe producing the first practical suggestion for a non-scanning television method), and, in 1928, the publishers of the magazine had issued a challenge worth £1000 in connection with certain claims which they held were being made by Baird and his associates.

Prior to giving evidence to the committee on behalf of its readers, the editor of *Popular Wireless* had obtained the wishes of his subscribers by means of a questionnaire. This questionnaire was very carefully couched so that any supporters of the 30-line standard were not able to state whether such a service ought to be terminated or extended. Instead the readers were asked, *inter alia*, to specify the minimum acceptable size of a picture and say whether 'the picture needs necessarily to be as clear as a newspaper illustration. . .'. 80% demanded 90 squares inches or over, 10 inch by 10 inch being the approximate average. 75% answered the second question, above, with an emphatic affirmative. Hence, from this response and an elementary calculation, the publishers of *Popular Wireless* could assert (had they wanted to), that the majority of the magazine's readers were more interested in a standard of television better than that which could be provided by the 30-line standard.

In addition to the attitude of the technically-minded public on television matters, there was also the opinion of manufacturers of television sets to be considered by the Television Committee. Baird Television Ltd. had opined that the 30-line transmissions should cease, while Electric and Musical Industries Ltd., Scophony and A. C. Cossor Ltd. were much too involved with the development of high definition television systems to argue a case for the retention of the low standard system. However, for V. Z. de Ferranti and A. Hall (in their interview with Selsdon's Committee on behalf of Ferranti Ltd.) this system, from the point of view of performance, was considered to be really extraordinarily good. De Ferranti told Lord Selsdon: '. . . I was amazed that it was as good considering the limitations, and it brings home very forcibly to me how very much better these things are done by the BBC. . .'.[15] Towards the end of his interview, the Manchester engineer stated: 'I think it (the 30-line system) is absolutely useless at the present time of the day. If you could let us (the people in the north) have it – and I do not see how you can – at a time when it is of any use to us, I would like it very much.'

Ferranti probably realised that high definition television would not be available in the provinces until an experimental service had been well tried in the London area and therefore felt, for the interim period, the low definition system would stimulate and maintain enthusiasm for the new branch of science in the Manchester region. He informed Lord Selsdon: '. . . we come humbly to say: "Please do not forget us in the north". We have rather felt, even with these 30-line transmissions, that we have been right out of it.'

A similar plea was voiced by a W. Barrie Abbott, who represented many hundreds of Scottish constituents, when he gave evidence on 29th June 1934.[16] On the need to keep the transmissions, he observed that the system had 'this great advantage that it has long transmission'. Because of this there were 'people for instance in Newcastle who have got splendid results. . .'. Hence his objection to the suggestion to end the 30-line service was that it would deprive 'these long distance people from having anything until the high definition obtained a longer radius'.

Then there was the case of the small manufacturer of 30-line televisors to be heard. Such a manufacturer was Plew Television Ltd. The company had been in existence for approximately three months (when Dr G. C. Lemon gave evidence on its behalf), and was producing 30-line receivers at a price of 18 guineas.[17] Of course the company had a good practical reason for wanting the 30-line service to continue, but additionally their conclusion 'that the future of television (depended) upon the support of the large mass of the British public' and that this could 'only be obtained if that public (was) able to obtain their receiving apparatus at a price within reach of their pocket' was certainly pertinent.[18]

The various arguments, briefly mentioned above, advocating the retention of the low definition television system seem to have had some effect on the Television Committee, for whereas Lord Selsdon, in June 1934, had felt the service had no entertainment or commercial value, by the end of November he had modified his view and, apropos the discontinuance of the service, informed Sir John Reith:

> We were rather impressed not by the number of protests but by the fact that they did appear to exist. Personally I cannot conceive anyone attaching any value to the 30-line transmissions, but it does appear to us that there are a certain number of people who do, and we thought you would possibly expose yourself to more approbrium if you discontinued them altogether before even one high definition station was working than if some sort of transmission was still going on during the interregnum.[19]

Selsdon and some of his committee members had clearly been influenced by much of the evidence they had heard. He told Reith:

> . . . there is one journal at any rate in existence that devotes itself to television. I was surprised to see how well it was produced, and it would obviously be of a certain value to you later on when you were running the service. It would be a pity to put such a journal out of business altogether. . .

Phillips supported Selsdon:

> There is a service in Germany. There is no service except on experimental one in the United States, but there it is, we have had a service of a kind, and to abolish it before we have something else ready to take its place would in my opinion be a retrograde step.

In the face of the combined advice from Selsdon and Phillips, the BBC's Director General simply observed that the BBC would accept the committee's view.

When the report of the Television Committee was published it recommended, paragraph 34, that:

1. the existing low definition broadcasts be maintained if practicable, for the present;
2. the selection of the moment for their discontinuance be left for consideration by the Advisory Committee, with the observation that, if practicable so to maintain these broadcasts, they might reasonably be discontinued as soon as the first station of a high definition service is working.

13.2 High definition television

Much of the committee's time was, of course, taken up by the necessitity to formulate recommendations for a high definition television service. Effectively there were four competitors engaged in research on high definition systems:

1. Scophony Ltd
2. A. C. Cossor Ltd

3 Baird Television Ltd
4 Electric and Musical Industries Ltd. and Marconi–EMI Television Co. Ltd.

Not all of these organisations had reached the stage of finality with regard to their various designs and inventions, and, as demonstrations to the committee showed, only the Baird and M–EMI companies were in a position to offer complete television systems to the BBC.

Claims were made by Scophony for substantial improvements in the optical efficiency of their scanning devices both at the transmitter and at the receiver. These devices were electromechanical in nature, unlike those of the other three companies, which employed cathode-ray tubes in their receivers and in some cases also in their transmitters. The possibilities for obtaining improved efficiences (and flicker free reproduction which Scophony additionally claimed), were not conclusively demonstrated, although the committee recognised that some advance in this direction could not be completely discounted.[20] A further claim was made by the company for the possibility of adapting their 'stixograph' invention for television purposes, but this was not amplified in their statement and was only demonstrated in an elementary form as an application to the cinema.

There is no doubt that the Scophony engineers, Walton, Wikkenhauser, Sieger, Robinson and Lee, produced some of the most highly original devices ever seen in the field of television, but effectively the establishment of the Television Committee came about before some of these inventions were fully developed. The company was a small one and its financial position caused some concern. Sagall, the dynamic, forceful managing director of the firm, told Lord Selsdon his company had been labouring under great difficulties all the time.[21] They were in negotiations for more money and were apprehensive lest the committee recommended that 'television was not ripe at all for anything. We do not hesitate to say that such a situation would be disastrous from our point of view, it might mean that we should have to close the laboratory.'

Of the main ideas and claims which Scophony put forward, there was one which appeared, *prima facie*, rather incredulous. This concerned the possibility of transmitting high definition images (240 lines) via a very small bandwidth (15 kHz) transmission link. However, while some diminution of the bandwidth could not be disregarded, the committee felt 'the probability of such a drastic narrowing of the band. . . (did) not seem to be feasible'.

Another company which was not in a position to provide a complete television installation at the time of the hearings was A. C. Cossor Ltd. Their system, which was based on the velocity modulation principle,[22] utilised cathode-ray tubes, not only at the receiving end, but also at the transmitting end – the actual scanning device for the transmitter being the cathode-ray tube. It was thus fundamentally different from the schemes employing amplitude modulation, such as those of the Baird and M–EMI companies, and so television receivers based on the principle would not be interchangeable with those of other systems. Cossor did, however, demonstrate another scheme which combined velocity modulation with intensity modulation and, while being in the nature of a compromise, had the advantage of interchangeability with systems based only on intensity modulation.

As with Scophony, the Cossor Company was handicapped by a lack of time to develop their ideas. They were only able to demonstrate their scanners using films, although the firm claimed that these could be directly applied to the direct scanning of scenes and objects. Another difficulty was their inability to demonstrate their system using a high frequency transmitter, optical path and domestic receiver. Nevertheless, the committee seemed impressed by Cossor's work and recorded: 'The system has features of such promise as to warrant special consideration. One noteworthy facility is the automatic adjustment of the receiver to the number of scanning lines transmitted.'

In addition to the four British systems (mentioned above), Selsdon's committee also viewed demonstrations of the Loewe and the Mihaly–Traub systems which both used mechanical scanning at the transmitter.

The Loewe scheme corresponded to that which was being provisionally standardised in Germany (in 1934), and employed a type of tube and method of synchronisation comparable to those utilised by Bairds. No special features were incorporated in Loewe's apparatus which were not used by the British firms and the quality of reproduction was of the same order as that produced by other mechanical scanning systems in the United Kingdom.

A demonstration, with a laboratory model, was also given to the committee of the Mihaly–Traub system, but, as with a number of the other schemes previously mentioned, the development of the apparatus had not reached the stage of commercial production and the company had to state that further work was contemplated to remove some obvious defects.

The demonstration consisted of the reproduction of a 120-line image on a screen measuring 15 cm by 20 cm – the modulation of the scanning spot being effected by a Kerr cell. Mihaly–Traub claimed that 180-line scanning was possible but the Television Committee were not impressed by their method. 'The system is somewhat crude in its present form and the possibility of its application to high definition television is doubtful.'[23]

Thus, of the six systems of television examined, only those of the Baird and Marconi–EMI organisations were contenders for an immediate (1935) television broadcasting service in this country. Of these two systems, only Baird Television Ltd. could show the direct reproduction of scenes and objects as well as films.

It is interesting to note that, at the time the committee commenced its enquiry, both the Baird and EMI firms were employing mechanical methods of scanning a picture, but during the hearings the two companies engaged in intensive research in the further development of electron image scanning to television. EMI had adopted the Campbell Swinton method of 1911 and Baird Television Ltd. were making use of the image dissector principle of Farnsworth.

A point which emerged very clearly from Selsdon's questioning of the various witnesses was their complete lack of knowledge of the state of EMI's television work; the secrets of the Hayes factory were certainly well-guarded and only the members of the committee who had seen demonstrations of the equipment in June were aware of the tremendous progress which Schoenberg's team had achieved.

13.3 Bairds' proposals

Major A. G. Church and Captain A. G. D. West, both Directors, represented Baird Television Ltd. at their meeting with the Television Committee on 7th June 1934. Prior to the meeting, a long memorandum (50 pages) on the subject of television had been handed to the Chairman, Lord Selsdon, but unfortunately neither the Post Office Records Office nor the BBC's Archive Centre has a readily available copy. Certain other documents, dealing with the patent position of the company, also have been either lost or misplaced. However, the 60-page transcript of the shorthand notes taken during the oral examination of Church and West is available and this covers much of the ground elaborated in the detailed memorandum. In addition, the company submitted a further long memorandum on 'The position of the Baird Company with respect to the Postmaster General's Television Committee and the establishment of a television broadcasting service', to the committee on 14th August 1934, in amplification of certain points raised at the meeting on 7th June 1934, and this too can be studied.

Bairds considered that high definition television of 180 or 240 lines ought to be adopted at the commencement of such a service with a picture ratio of 4:3, scanned horizontally, with 25 pictures per second.[24] This standard was, perhaps surprisingly, similar to that put forward by EMI in a statement dated 7th June 1934, namely, 243 lines at 25 pictures per second. There was an important difference, however, between the scanning methods of the two rivals: EMI proposed interlaced scanning to give a flicker frequency of 50 frames per second, whereas Bairds specified progressive sequential scanning and hence a frame rate of 25 frames per second.

At the commencement of the committee's work, Bairds had shown 180-line pictures and were devoting much effort to producing good 240-line images. They had obtained images having the latter standard, but these were not better than the 180-line picture when West and Church met Lord Selsdon and his colleagues. West listed three difficulties associated with:

1 the video amplifiers;
2 the radio transmitter;
3 insufficient light spot intensity when employing the light spot scanner.

The third difficulty follows from an elementary consideration of the optical system employed in the light spot scanning. Assuming a light source of intensity

I and that kI lumens are collected by the condenser lens and projected on to the Nipkow disc, it follows that the aperture in the disc will pass $1/n$ of this, or kI/n lumens. Hence, as the numbers of lines per picture increases, the received photoelectric current must decrease for a given object and light source. In a paper on 'Optical efficiencies and detail in television systems', Levin and Walker of Marconi Wireless Telegraph Co. showed that a photoelectric current of 5×10^{-3} microamperes was possible for a 50-line image when using a light source of 15 000 candles (an ouput of $4\pi \times 15\,000$ lumens) and a photoelectric cell having a sensitivity of 10 microamperes per lumen.[25] Thus, for a 240-line picture, the current which could be expected would be of the order of 10^{-10} A. This value would be comparable to the noise currents associated with the first stage of the photocell's amplifier and hence would be unacceptably small for practical television systems. Bairds therefore proposed to adopt the intermediate film method.

The difficulties which the company were experiencing with its video amplifiers were connected with the need to obtain uniform amplitude/frequency and phase/ frequency characteristics. For television purposes video amplifiers must have a uniform response to periodic signals having frequencies from 25 Hz to approximately 1 MHz (although West mentioned 1.5–2.0 MHz). In 1934 the design of such wide bandwidth amplifiers was not the straightforward matter it is nowadays, and with the valves then available there were limitations to the gain-bandwidth products which were desired.

The second difficulty listed by West – the radio transmission problem – was described by him as being 'an exceedingly difficult one, particularly in respect of (the) phase change problem, and getting the radio transmitter to follow completely all (the) various components in phase'. Later, during the submission of oral evidence, West highlighted another problem associated with high power television transmission utilising short wavelengths, namely, grid circuit losses. West stated:

> The valves we are using now are Mullard valves and we have tried GEC valves for short wavelengths. We are getting 100% modulation for the picture, but run into many difficulties. It is all right if you keep (the) modulation low. We have not put out higher power experiments on the air at all. They do not go up to the full power we have mentioned that we require because we cannot get the power to drive them. We cannot get over the grid circuit losses. That is why we are trying the GEC valves, to try to get a wavelength for 240 lines.

Many years after the birth of high definition television in the United Kingdom, Sir Isaac Shoenberg stated: 'The man who deserves the real credit for television is Charles Samuel Franklin, who developed the (transmitting) valve. Without this my work would have been impossible.'[26] Franklin was also the engineer who designed the first high gain, wide band aerials used in shortwave transmission and these provided the basis of the design for the Alexandra Palace aerials.

EMI were particularly fortunate to have the expertise, experience and facilities of the Marconi Wireless Telegraph Co. Ltd. at their disposal, for by the early 1930s the latter company had already laid the foundations of shortwave and ultra short wave transmitter, modulator, feeder and aerial techniques. These formed an important contribution to the implementation of the 405-line system – a contribution which is sometimes overlooked by writers on the history of television.

During the period 1926–1929 the Marconi Company had developed the terminal equipment for a high speed facsimile transmission system for trans-atlantic experiments employing the shortwave beam service.[27] One of the Marconi engineers who was very active in the whole of this work (which involved the transmission of synthetic half-tone pictures by d.c. keying the radio transmitters), was N. E. Davis, who later became the company's chief scientific liaison engineer at EMI's Hayes factory during the period 1932–1938. It was Davis and E. Green who were essentially responsible for the design of the M–EMI Alexandra Palace 405-line television transmitter. Davis has written that this transmitter handled the 405-line television signals 'by substantially the same d.c. keying method as (had been) used for the high speed facsimile (system)'.[28]

The practical association between Electric and Musical Industries Ltd. and the Marconi Wireless Telegraph Company Ltd. commenced in 1931. In that year the Hayes firm had a small group working on a mirror drum film scanner, and later in the year Shoenberg invited the Marconi Wireless Telegraph Company to supply a low power ultra short wave transmitter, complete with a modulator, for use with the film scanner.[29] Davis was instructed to prepare a transmitter and modulator for this purpose and subsequently the units were sent to EMI in January 1932. The transmitter had an output of 400 W at a frequency of 44 MHz and employed grid modulation of the final stage (which utilised DA60 values). A frequency response extending to 500 kHz, a 3 dB point, had been requested by EMI and was achieved.

The licence which allowed EMI to establish a wireless sending and receiving station for experimental purposes at the Hayes premises of the company was sent to them on 8th December 1931. According to this licence EMI could operate within the following bands:

CW and telephony	62·01–61·99 MHz
television	44·25–43·75 MHz

and a power of up to 2 kW into the aerial was allowed.[30]

Davis was responsible for the installation of the above transmitter and aerial system – which consisted of a self-supporting half-wave dipole, Franklin feeder and terminating arrangement – in January 1932, and as the Hayes research group was completely without knowledge of USW transmitters, methods of modulation, etc. he remained at Hayes until April to advise and generally assist in overcoming initial problems. Thus commenced an association between the two companies which later was to lead to the establishment of the world's first, high definition television system.

Davis seems to have been well aware of the importance of the work which EMI were undertaking at that time and of the beneficial support which his company could provide, for he pressed for an assistant to remain at the Hayes factory during EMI's television developments. Initially a Mr Wassell was appointed temporarily, but in March 1932 a Mr W. S. L. Tringham succeeded him.

Marconi's former transmitter expert has stated that the first 120-line, 25 frames/second film transmissions were successfully transmitted during the first half of 1932, but that, in July 1932, Shoenberg complained to MWT Co. because he thought the transmitter and modulator were not reproducing faithfully the signals supplied to them by EMI's apparatus. Davis investigated this complaint and had the satisfaction of proving to Shoenberg that he was incorrect.

The success attained in these initial experiments led to EMI requesting a greater power output from the transmitter and an extension of the modulation bandwidth to 1 MHz, at the 3 dB point. Later in 1932 Davis therefore commenced an examination of the modulation characteristic of the transmitter used, an SWB4, to determine the limiting factors affecting its frequency response. At the same time he constructed a new experimental power amplifier, with CAT4 valves, and this was installed alongside the SWB4 transmitter in June 1933.

Davis's tests with the SWB4 disclosed a fundamental limitation to an extension of its video bandwidth (owing to an inherent unbalance of the neutralising arms of the push-pull bridge of the modulator stage), and consequently he decided to make provision for using this transmitter simply as a driver for the new CAT4 amplifier. He also designed a new modulator (consisting of three DA60 valves which worked into a non-inductive control impedance of 800 Ω and an input capacity of 322 pF), for operation from a signal of 200 V dap.

The new transmitter gave a peak power output of 4 kW, with a video bandwidth of 0–500 kHz – adequate for a 150 lines/frame, 25 frames/second television image.

Further development work was undertaken in the autumn of 1933 when the E402 valve (CAT15) was in experimental production. Its characteristics were favourable for a 2·5 MHz video bandwidth at a peak r.f. power output of 3 to 4 kW and hence Davis had them incorporated into a push–pull stage for use as an intermediate modulated power amplifier between the original SWB4 transmitter and the CAT4 final amplifier. The change over to the new vision transmitter started on 29th January 1934. It worked into a Franklin uniform aerial, which had been specially designed for operation on 44 MHz, and the arrangement of transmitter and aerial gave a peak output of 9 kW, with a video bandwidth of 1.4 MHz. This was eminently satisfactory for the 180 lines/frame, 25 frames/second television system then being investigated. A month later, on 23rd February 1934, an SWB4 sound transmitter, operating at a radiated signal frequency of 41·5 MHz, was delivered to the Hayes site.

Thus, several months before the establishment of the Television Committee, EMI had a complete experimental ultra short wave radio television system installed and operating for trials and demonstrations. They had benefited greatly from their contact with the Marconi Wireless Telegraph Co. Ltd., for their engineers had acquired a wealth of experience and expertise which could not be surpassed by any other UK company. EMI's engineers, while being of the highest technical competence, did not possess any comparable detailed knowledge and experience of radio engineering and were perforce indebted to the skills of engineers such as Davis. As G. B. Smith has written:

> Marconis had all of the v.h.f. transmitter, modulator, feeder, master oscillator, aerial and test equipment techniques, in fact we built the railway stations, bridges, signal box, etc., while Isaac Shoenberg and his team sometimes with our information and experience provided the camera chains and receiving means which is the locomotive of t.v.

Later the Marconi Company extended its range of products to include cameras and video equipment and by 1950 it was the only firm in this country which could offer complete television systems without calling on the assistance of sub-contractors.

The fortunes of Baird Television Ltd. might have been greatly improved had they had the support of MWT Co. Ltd., and correspondingly Electric and Musical Industries might not have achieved the supremacy they acquired during the critical period 1934–1936 without the Chelmsford company's advice.

Ostrer's lack of perspicacity was undoubtedly partly responsible for the downfall of the Baird company. 'If we had joined Marconi, we should have been with this combine, not against. Our policy of facing the world single-handed was sheer insanity: we should have made terms', wrote Baird.

There was also an attempt to merge with the General Electric Company Ltd., of which the Chairman, Lord Hirst, was a friend of Major Church – who was on the Board of Baird Television Ltd. On this Baird wrote: 'The immense importance of such a tie-up was very obvious to me. Although we got as far as having regular technical meetings and our two research departments were working in unison and, although the GEC combine were anxious for an agreement, at the last minute Isidore Ostrer turned the whole thing down.'[31]

In May 1934 the Marconi–EMI Television Company Ltd. was formed by the Marconi Wireless Telegraph Company Ltd. and EMI the two companies being the only shareholders, each holding an equal number of shares and entitled to an equal number of votes. The Board of Directors was composed of representatives of the two companies, and in 1934 were:[32]

The Right Hon. Lord Inverforth (Chairman)
Alfred Clark
The Marchese Marconi
I. Shoenberg
Louis Sterling
H. R. C. Van de Velde
H. A. White

The most surprising aspect of this merger was that Marconis initially did not wish to link up with EMI – it seems the Chelmsford firm preferred an association with Baird Television Ltd. – but Shoenberg was a shrewd, capable research director. He had been joint General Manager of MWT Co. Ltd. and knew well the wealth of talent, expertise and experience which existed in the company on transmitter, modulator, feeder and aerial engineering: he was aware too of his own group's undoubted abilities in research and development work; it seemed natural that a merger should take place. Consequently he put the proposition to MWT.

> At first they did not take very kindly to my suggestion, but at last we hired from them a transmitter, about 200 W, and they made us pay £1000 a year for it. After that we worked for about 18 months and, when we started getting results, we went again to the Marconi Company and suggested to them that, since we were each of us concerned with half the complete transmitter, obviously closer collaboration should prove beneficial to both parties. As a result of those conversations, the two companies decided to form between them a private company. . .

Thus, by a curious turn of the wheel of fortune, EMI's television activities were immeasurably strengthened and those of Baird Television considerably weakened. How ironic this episode in the history of television would have appeared to Baird had he known the full story. West's difficulties with his television transmitter would never have arisen had Ostrer shown greater far-sightness.

There was another advantage of the merger for EMI, for they were given a licence to use MWT's patents. With their other connections the Marconi–EMI Television Company consequently had exclusive rights to all patents relating to television which had originated with the General Electric Company of America, with the Radio Corporation of America, with Telefunken of Germany, with Marconi Wireless Telegraph Company and with Electric and Musical Industries.[33]

EMI agreed that the Secretary and Chief Accountant of MWT should act for the joint company and also that MWT should be engaged in the selling and contractual work.

On the relationship which existed between the two companies there was some suspicion at MWT that EMI charged-up their work more extravagantly than MWT did for their commitments and also that their liaison officer at Hayes never saw more than he was meant to see. Nevertheless the relationship was a very good one until after the Second World War.

Following the merger and the establishment of the Television Committee, Davis was asked to prepare certain documents and attend the committee as a witness to support and explain M–EMI's present and possible future developments. He subsequently gave evidence on 8th June 1934, together with Shoenberg, Condliffe, Blumlein, Agate and Browne. Of some significance were his comments on the output power which could be obtained from USW transmitters. Davis told Lord Selsdon the mean power from the transmitter currently being utilised at Hayes was $2\frac{1}{2}$ kW (telephony rating, 10 kW peak power); that MWT Co. were designing an output stage for three times that power, and in their development section the company saw 'no reason whatever, as valves (were) developed, why (they) should not be able to go considerably further than this'. This statement shows the importance of Shoenberg's remark in 1961 concerning Franklin's valve, namely 'without this my work would have been impossible'.

The new final power amplifier referred to by Davis was to include four CAT9 valves. This was to be added to the existing Hayes transmitter and the development programme also included the conversion of the SWB8 transmitter for use as a driving source in the ultra short wave band to replace the existing SWB4 transmitter. However, the four CAT9 unit was never employed.

On 4th February 1935 Davis and his assistants received an urgent request to visit the Hayes factory of EMI. Shoenberg had made his dramatic decision to depend absolutely on his company's emitron scanner (Fig. 13.2) and to offer a 405-line television system, to the Television Advisory Committee's Technical Sub-Committee, having a video frequency band extending from d.c. to between 1·4 and 2·0 MHz and a transmitter power output of 12 to 24 kW. For this purpose a change in the design of the new transmitter was necessary, and, instead of proceeding with the high power amplifier utilising four CAT9 valves, this was changed to two CAT9 valves to replace the existing CAT4 stage. Construction of the new amplifier was accelerated and delivery to Hayes took place on 15th March 1935. The demonstration was successful. Davis was confident that the transmitter at Hayes was the best that could be produced and would meet the BBC requirements for the London television service (Figs. 13.3).

Bairds at this time were employing Metropolitan Vickers' demountable type valves, but Davis thought these were 'a doubtful proposition' and could not compete at powers of 20 kW with the sealed-off type of Franklin valve.

It is significant to note that, although West felt compelled to inform the Television Committe on 7th June 1934 of the several difficulties which they were experiencing with their transmitter, EMI never at any time encountered problems with MWT's transmitting equipment. In a letter dated 5th April 1934, Marconi's liaison engineer, Tringham, wrote: 'We started giving demonstrations of television about sixteen months ago and in no case has any transmitter (fault) occurred during a demonstration.' Indeed, by April/May 1935, Davis considered radio transmitter developments had outpaced modulator developments. He ended his fascinating unpublished report with the statement: 'It (405-line television) was a triumph for Mr Shoenberg and a vindication of his shrewd judgement in bringing MWT to his aid in 1931. There was no other company in the world at this period which would have done the same.'

In one respect the task of MWT Co. during their period of co-operation with EMI was very much easier than for the latter company. Marconis had acquired their expertise and supremacy during many years of transmitter design, testing and application in the fields of wireless telegraphy, wireless telephony, radio broadcasting, facsimile transmission and low definition television, but EMI were effectively starting *ab initio* in many of their investigations. They had to explore and invent many techniques associated with the fields of vacuum physics, electron optics, wideband electronic circuits, pulse forming and shaping

Fig. 13.2 *Emitron camera, 1937.*

circuits, thin film deposition and so on. As a consequence, EMI's research and engineering effort during the 1930–1939 period far surpassed, ('probably of the order of 50:1') the corresponding labours of MWT. A useful yardstick to a comparison of the work of the two companies in high definition television can be obtained from an examination of the patents filed by MWT and EMI in the period 1930–1939:[29]

	EMI	MWT
television transmission	38	8
television other than transmission	105	28
electron camera tube	42	1
crt videoscopes	36	1
totals	221	38

Fig. 13.3 *Prototype for the Marconi–EMI vision transmitter final power amplifier, 15–19 kW, wavelength 6.8 m, as installed at Alexandra Palace for the BBC.*

The problem of good transmitter design was naturally linked to the requirement to obtain as large an area of reception as possible consistent with economic considerations. During the first half of 1934 EMI carried out a large number of tests on field strength using an antenna situated 180 feet above sea level and a mean power of 2.25 kW. However, as Shoenberg informed the Television Committee, if the radiated power was increased by a factor of three (and the MWT Co. were constructing a new transmitter for M–EMI with such an output at this time), and an antenna mast 300 feet high was erected on Highgate Hill, an approximate increase in field strength (for a given range) of ten times could be possible. Shoenberg observed that this increase could be utilised in one of two ways: either to enlarge the service area or to obtain 'a very safe signal within a limited area'.[32]

Fig. 13.4 *Final stage of glass blowing after assembly of an emitron, in the EMI research laboratories.*

The question of transmitter power output, antenna height above sea level and range for satisfactory television reception occupied the Television Committee and the two main competing companies in much discussion. It was felt that masts 1000 feet in height would incur objections from the Air Ministry and both Lord Selsdon and Shoenberg thought 400 feet would be an effective limit. Shoenberg assumed this value would lead to an approximate range, for reception, of 25 miles. When Selsdon sought confirmation on this point, Shoenberg's answer was significant: 'Yes, but for all I know it may turn out to be fifty miles. I could not give evidence on that because I have not the data.' This statement, based on the wealth of knowledge and experience of both Electric and Musical Industries and The Marconi Wireless Telegraph Company, gives some indication of the state of the art of propagating signals in the very high frequency band at that time. Of course this was before the emergence of radar, which later advanced the technology of very high frequency and ultra high frequency techniques by enormous strides.

Marconi–EMI's estimate of the power required to be radiated from an aerial situated 400 feet above the surrounding ground, so as to give a field strength of 5 mV/m at a distance of 25 miles was 10 kW. The figure of 5 mV/m was considered to be necessary for a picture under average conditions in an urban area while in very quiet circumstances in the country about 0.5 mV/m was sufficient.

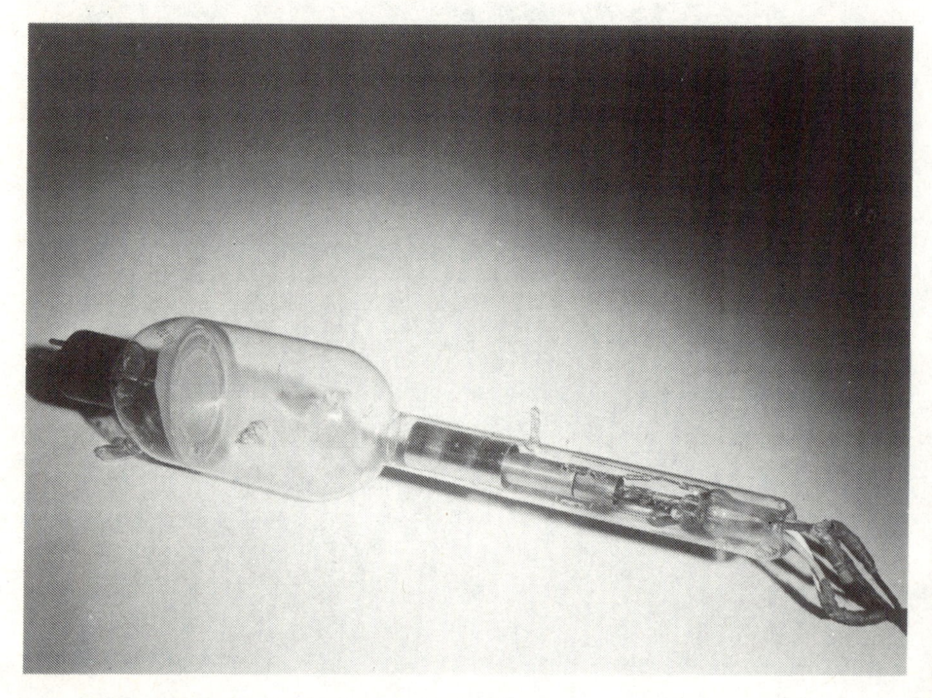

Fig. 13.5 *Early experimental emitron camera tube.*

Baird Television Ltd. also carried out many tests on the reception of television signals radiated by an aerial mounted on the top of the tower at Crystal Palace. They obtained pretty good results with a transmitted power of only 250 W within an area having a radius of approximately 30 miles. Their tests were thus carried out with one-tenth of the radiated power of the M–EMI installation, but, on the other hand, the Baird Company's antenna was almost 650 feet above sea level, whereas that of M–EMI was 180 feet above sea level. Bairds experiments clearly showed the desirability of having the transmitting aerial situated as high as practically possible above the surrounding ground. To cope with deep valleys and very difficult cases inside steel framed buildings, West thought a mean power output of 1 kW, going up to 3 or 4 kW on full modulation, would be sufficient for their purposes, but Church stated that the company would like to have 10 kW at their disposal. This was the value which the Reichpost had provided for television experiments in the Berlin area.

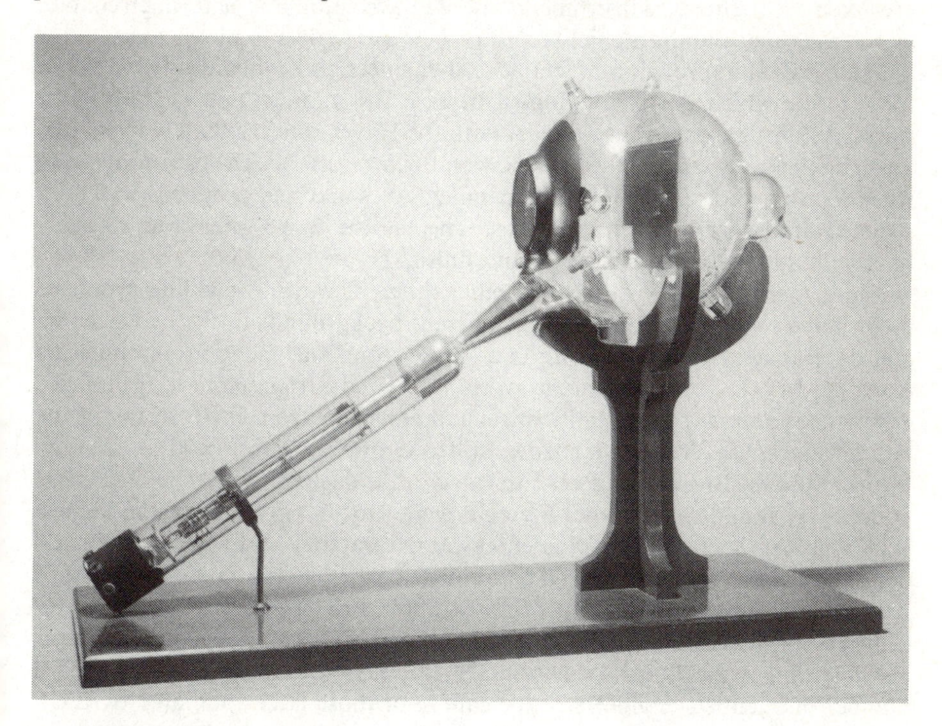

Fig. 13.6 *Completed emitron tube.*

Both M–EMI and Baird Television Ltd. conducted many field strength tests with their experimental stations: those of M–EMI appear to have been concentrated mostly in dense urban and sub-urban areas, whereas Bairds carried out a number of investigations in the regions between the Crystal Palace and Beachy Head to the south and Hitchin to the north. In one series of studies, an experimental transmitter was situated on Beachy Head – presumably to enable field strength measurements to be undertaken over the sea as well as over the ground. As Shoenberg stressed to Selsdon's committee, there was a shortage of data to check the theoretical calculations, and the determination of the effective service area corresponding to the proposed power output from a public service television transmitter had to be based on an extrapolation from low/medium power field trials.

The methods adopted by the two films in the reception tests were quite different and illustrated the scientific approach of the Hayes company and the more empirical, pragmatic style of Bairds. With the latter organisation, the testing procedure, according to Church, was as follows: 'Using an aerial, on a lorry, about 15 feet high, and the receiver inside the lorry, we tuned in and decided whether the picture was good or not and then moved on to the next spot.' Agate, for M–EMI, described how their measurements were made: 'They are taken with calibrated instruments and we have automatic gear which enables us to make the (photographic) record of field strength in a moving van. . .'

The M–EMI approach to scientific and engineering problems was typical of the attitude which the BBC adopted in its investigations, and possibly, as a result, Ashbridge felt a closer affinity with the Hayes concern than with Bairds. He had effectively summarised his view on the firms at the 5th April conference when he observed: '. . . EMI were manufacturers and had produced well constructed and workmanlike apparatus, whereas the Baird apparatus was distinctly amateurish in construction and finish,' (Figs. 13.4–13.6).

The disparity between these altogether different ways of tackling problems might have stemmed from the quite different backgrounds of the two research and development teams. The Marconi–EMI group was made up of engineers from the MWT Co. – a company with a vast industrial engineering practice experience to draw on – and scientists and engineers from EMI. Many of the latter, for example, McGee, Klatzow, Miller, Crowther, Lubszynski, Broadway, White, Stewart Brown, held research doctorate degrees, and others, such as Shoenberg, Blumlein, Browne, Blythen, were proven engineers of the highest standing. With such a wealth of talent it was natural that M–EMI's research and development methodology would be wholly rigorous. Bairds could not boast of having a personnel complement of this calibre: their *primus inter pares*, Baird, lacked a formal research training, as did many others of his organisation. It is possible that if the Baird Television Development Company had recruited a number of scientists/engineers of the calibre of those later appointed by EMI, the situation in which the company found itself might have been altogether different.

References

1 Notes on 'Conference at General Post Office, 5th April 1934', Minute Post 33/4682.

2 PHILLIPS, F. W.: memorandum to L. Simon, 9th April 1934, Minute 33/4682.

3 Report on the Television Committee, Cmd. 4793, HMSO, January 1935.

4 Minutes of the first meeting of the Television Comittee, 29th May 1934, Minute 33/4682.

5 Paper No. 3, The Television Committee, Minute 33/4682.

6 Minutes of the 10th meeting of the Television Committee, 14th September 1934, Minute 33/4682.

7 Minutes of the second meeting of the Television Committee, 7th June 1934, Minute 33/4682.

8 Notes of a meeting of the Television Committee held on 7th June 1934. Evidence of Major A. G. Church and Mr A. G. D. West on behalf of Baird Television Ltd., Appendix II, Minute 4003/1935.

9 Anon.: 'Broadcast television', *Wireless World*, 2nd March 1934, pp. 146–147, also 'Television: 30-line tests are wanted', *Wireless World*, 2nd March 1934, pp. 151–152.
10 See *Television*, 1934, pp. 119, 188–191, 228–230.
11 See *Television*, 1934, pp. 217, 236–237, 284.
12 Notes of a meeting of the Television Committee held on 21st September 1934. Evidence of Dr C. Tierney, Mr R. R. Poole and Mr W. G. W. Mitchell on behalf of the Television Society.
13 Paper No. 25, 'Evidence of the Society', The Television Society, The Television Committee, Minute 33/4682.
14 Notes of a meeting of the Television Committee held on 29th June 1934. Evidence of Dr J. H. T. Roberts on behalf of *Popular Wireless* and The Wireless Committee.
15 Notes of a meeting of the Television Committee held on 12th July 1934. Evidence of Mr V. Z. de Ferranti and Mr A. Hall on behalf of Messrs. Ferranti Ltd.
16 Notes of a meeting of the Television Committee held on 19th June 1934. Evidence of Mr W. Barrie Abbott.
17 Notes of a meeting of the Television Committee held on 21st September 1934. Evidence of Dr C. G. Lemon of the Plew Television Ltd.
18 Paper No. 2, copy of a letter dated 9th July 1934 from Plew Television Ltd., Television Committe, Minute 33/4682.
19 Notes of a meeting of the Television Committee held on 30th November 1934. Evidence of Sir J. F. W. Reith, Minute 33/4682.
20 The Television Committee: 'Description of television systems examined by the Committee in Great Britain', Report of the Television Committee, Appendix IV.
21 Notes of meeting of the Television committee held on 5th July 1934. Evidence of Scophony Ltd., represented by Mr S. Sagall and Mr G. W. Walton, Minute 33/4682.
22 BEDFORD, L. H., and PUCKLE, O. S.: 'A velocity modulation television system', *JIEE*, 1934, **75**, pp. 63–82.
23 See minutes of the 19th meeting of the Television Committee held on 3rd December 1934 and reference 20 above.
24 Baird Television Ltd.: memorandum dated 14th August 1934, Minute 33/4682.
25 LEVIN, N. and WALKER, L. E. Q.: 'Optical efficiencies and detail in television systems', *The Marconi Review*, No. 47.
26 PHILLIPS, P.: 'The shy genius behind your t.v.', *Daily Herald*, 20th July 1961.
27 See, for example: GRAY, A.: a memorandum to Senatore G. Marconi, 11th March 1927, Marconi Historical Archives.
28 DAVIS, N. E.: 'Marconi–EMI Television 1931 to 1937', appendix no. 2, 24th April 1950, Marconi Historical Archives.
29 SMITH, G. B.: 'Historical relationship between Marconi, RCA and EMI', 1st September 1950, Marconi Historical Archives.
30 WISSENDEN, J. W.: letter to EMI Ltd., 8th October 1931, BBC file T16/65.
31 BAIRD, M.: 'Television Baird' (HAUM, South Africa, 1974).
32 Notes of a meeting of the Television Committee held on 8th June 1934. Evidence of the Marconi–EMI Television Company represented by Messrs. Shoenberg, Condliffe, Blumlein, Agate, Browne, and Davis.
33 Notes of a meeting of the Television Committee held on 17th June 1934. Evidence of Messrs. Clark and Shoenberg on behalf of EMI and The Marconi–EMI Television Co. Ltd., Minute 33/4682.

Chapter 14

The work of the Television Committee; patents and finance 1934

During the examination of the two main UK television companies the Television Committee paid particular attention to the patent position. The committee was very much aware of the need to guard against any monopolistic control of the manufacture of receiving sets if a certain company's transmission system were chosen. Further, it was essential that nothing should be done to stifle progress or to prevent the adoption of future improvements from whatever source they came. The patent position was not an easy one on which generalisations could be made, as in regard to many of them there were conflicting views as to their importance and validity. One of these patents was held by Bairds, namely 269,658 of 1926.

Baird Television Ltd. presented to the committee a schedule of 124 British patents granted to the company during the period 1923–1934 and a schedule of 84 pending applications. Of these, the following were felt to be of primary importance:[1]

1 269658, dated 20th January 1926 on light spot scanning
2 269834, dated 21st October 1925 } on synchronising
3 275318, dated 3rd May 1926 }

Patent 269,658 described the principle of analysing a scene by means of a spot of light. West informed Selsdon: 'The light is transmitted either through something (like a film) or reflected by something such as a subject or a scene. That appears to be the basis of every practical form of television. I do not see how it can be done by any other method.' Because of this the patent was held to be of 'fundamental importance'.

Later in his evidence West stated:

> We have recently had a case of an infringement of the spotlight patent by a large company in this country who actually sold a transmitter for demonstration purposes. We have not proceeded against them, but have had a full admission from them and we have protected our patents.

At this time both Baird Television Ltd. and Electric and Musical Industries were contemplating using mechanical methods of scanning, and, in view of West's remarks, the committee may have attached some importance to them. Certainly it was a matter which was discussed with the Hayes company the day after Bairds gave evidence. But, Shoenberg was a patent expert, to use his own description of himself, and had obviously read the list of claims attaching to the light spot patent very carefully indeed.[2]

Claim 1 stated:

> In a television or like system using a high intensity of illumination of the object whereof an image is to be transmitted, the method of illuminating the object with an intensity so high that, if the illumination were continuous on any one part of the object it would be damaged (e.g. burned), which method consists in traversing over the object a spot of light of the desired high intensity.

EMI utilised this flying spot scanning principle in a demonstration of television before Selsdon and Cadman. However, Shoenberg's subsequent observation showed the weakness of Claim 1 above.

> When you were at Hayes I stopped this spot and made it rest on Lord Selsdon and Sir John Cadman, but I do not think that afterwards you found holes in your coats or felt any damage whatsoever. Therefore I do not think we are infringing that patent, and anyhow this is the only patent which can be dragged by the hair, so to speak, into a semblance of infringement by one part of our equipment, that is the disc for direct vision.

In any event the M–EMI Television Company was willing to give a complete indemnity to a future purchaser of its equipment in respect of actions for patent infringement. It had a pretty large patent department which carefully watched all patents which had any relevance to its business, and, further, the company had a rule that no piece of apparatus could be built in its works in any way whatsoever without the approval of the patent department.

As noted previously, both the Crystal Palace and Hayes organisations had superficially advanced somewhat similar proposals for high definition television to the Television Committee. The method of scanning was different in the two systems, but both firms thought that 240 lines/picture, or thereabouts, was suitable for a high definition service (indeed Shoenberg considered it was unlikely that more lines would be required for home receivers for many years);[3] both concerns were working in the 6 m wavelength band and both companies envisaged employing mechanical scanners for film reproduction and film cameras for outside and studio scenes. It was because of this dominance of non-electrical scanning methods at the transmitter that some importance attached to the 'light spot' patent.

As an aside, it is interesting to note Shoenberg's remarks (in June 1934) on the possibility of non-mechanical scanning transmitters (iconoscopes) being utilised: 'No one can say how long it may take – a year or five years – before another method – electric scanning – is perfected, which will probably allow direct television of outside events: but one does not know when that will come.'[4] Eight months later Shoenberg was to offer the Television Advisory Committee a 405-line system using the emitron camera.

In their evidence, Bairds mentioned two other patents which they felt were of vital importance to them, namely patents 269,834[5] and 275,318.[6] They also referred to patent 336,655[7] as a rigid patent. All three patents dealt with methods or apparatus for synchronising a receiver to a transmitter, but, as the latter two mentioned above could only be applied to mechanical receiver scanners, they were of no application in systems employing cathode-ray tube receivers – and both M–EMI and Bairds were now (June 1934) using these. Patent 269,834 described the use of 'a special image at the receiver which was separate from the picture, was of markedly different intensity therefrom and was transmitted from, or controlled by the transmitter' for synchronising purposes. Shoenberg made no reference to this patent in his evidence, and because cathode-ray tube receivers did not depend on special images for their operation and synchronisation, it seems strange that Bairds should have considered the above patent to have had some importance. Actually, as Baird had concentrated mainly on television schemes with mechanical scanners since 1923, his patents were of little importance to firms such as M–EMI which were working towards the goal of an all-electronic system. In reality, therefore, Baird's patent holding posed no threat to the Hayes company, and in retrospect, Bairds ought to have followed the example of the Marconi Wireless Telegraph Company Ltd. which acquired an extensive collection of patents, including some important patents on the iconoscope. It obtained these even though it was not actually engaged on developing the iconoscope camera at the time of their acquisition.

Marconi, whose contribution to the commercialisation of wireless made him more important as an innovator than as an inventor, surrounded himself at an early stage with many able and well qualified professional consultants and engineers. They contributed greatly to the success of Marconi's ventures. His companies followed a determined policy in seeking to obtain possession of many of the most important patents in the field of wireless communication. Thus the fortunes of the Marconi companies depended on the ideas and patents of many persons and the companies prospered.

When the Radio Corporation of America was formed in 1919 the immediate task which it faced was the establishment of a wireless point-to-point communication service. There were difficulties however. Many of the most important patents in this field were held by General Electric, Westinghouse Electric and Manufacturing Company and the American Telephone and Telegraph Company. Consequently, RCA had to take steps to acquire the ownership or the licence rights to many different patents to make its commercial operations a success. The acquisition of these was not easy, for no one person or firm controlled even a substantial percentage of them, and the corporation had perforce to establish cross-licensing agreements with the above named companies. Nevertheless, by June 1921, RCA had rights to more than 2000 patents in the radio field.

Neither Baird nor the Baird companies emulated to an appreciable extent the strategies of Marconi, Shoenberg and RCA.

On the other hand, M–EMI was in an immensely strong position with its access to the patents of five of the most important electrical equipment manufacturers in the USA, Germany and the UK. The following information gives some indication of the number of patents which Shoenberg had available to him for incorporation into the M–EMI system:

		EMI	American
1	d.c. working	6 (applications)	0
		(2 were being used and 4 covered alternatives)	0
2	synchronising	5 (applications)	1
		(2 used, 4 alternatives)	
3	interlaced scanning	2 (applications)	1
4	the film scanner	3 (published)	1
		6 (applications)	
		(8 used, 2 alternatives)	
5	the picture synchronising and mixing amplifiers	5 (published)	5 (published)
		1 (application)	1 (application)
6	the modulator and shortwave transmitter	5 (published)	3 (published)
		3 (applications)	
		(10 used)	
7	the transmitting aerial and feeder	6 (?)	0
8	cathode-ray scanner		
	a preparation of mosaic	1	3
	b correction of keystone effect	2	6
	c double-ended tubes	2	2
	d methods and circuits	0	4
	e CRTs with additional electrodes	3	0
	Totals	50	27

The supremacy of the British patent position in the field of d.c. working and transmitting antennas and feeders is noteworthy. In the former field, one of the most important patents was one due to P. W. Willans, namely patent 422,906. Shoenberg has written: 'A great controvercy raged for some months among my staff as to whether we should use d.c. amplification or a.c. amplification with the restoration of the d.c. component. In the early days P. W. Willans (of EMI) was the protagonist of the restoration method . . .'[8] Willans had been with the Marconi Wireless Telegraph Company and was probably familiar with their work on the transmission of synthetic half-tone pictures by d.c. keying the radio transmitters.[9] The patents of the second group mentioned above were all Marconi–Franklin patents. Hence in two of the fields listed above it can be seen that the Chelmsford company played an outstanding part. With regard to the transmitting antenna and feeder patents, Shoenberg observed: 'As an expert on patents, I permit myself the statement that some of these are master patents and I do not think it would be possible to construct an efficient short wave aerial without using them.'[2]

On the shortwave transmitting patents he stated:

> It is a group of patents which are of very great practical importance. In my opinion it would be very difficult to construct an efficient and reliable transmitter for the range of wavelengths we are dealing with, without making use of some of these patents: I think it would be next door to impossible to do so.

Zworykin applied on the 17th July 1930 in the USA for a patent covering an iconoscope type of electron camera. The corresponding British patent (369,832) was granted to the Marconi Wireless Telegraph Company, as were two patents (407,521 dated 24th February 1932, and 421,201 dated 30th July 1932) which gave protection to methods of producing mosaics for the electron cameras. EMI's first patent on an electron camera was applied for on 25th August 1932 (406,353). Thus, the Chelmsford company had at their command, by 1932, some important patents on iconoscope-type tubes which predated the first EMI patent in this field. All these patents placed the company in a vital position in any scheme for the development of high definition television.

After giving his evidence Shoenberg told the Television Committee that the reasons which guided his team in the choice of the fundamental features of their system appeared as convincing as the proof of a theorem in Euclid.[2]

This meant, of course, that the Television Committee had to ponder very carefully on the licensing system which it would recommend in its report. They discussed the issue not only with the main manufacturers of television apparatus but also with the Radio Manufacturers Association, and, at a meeting devoted entirely to the problem, with Sir William Jarratt, Secretary of the Trade Marks, Patents and Design Federation.

Essentially, as the system of transmission determined to a varying degree the type of receiver required for reception, the committee had to guard against any monopolistic control of the manufacture of receiving sets. It was this aspect of the new industry which would create profits rather than the sale of transmitting apparatus. As Shoenberg noted: 'I do not know that we contemplate licensing under our patents for manufacturing transmitters, because I do not think that is called for . . .'[3] He was prepared though to grant licenses for receivers:

> I would give him a licence and ask him to get on with it, in the same way as I gave licences when I was in the Marconi Company several years ago. That has been my policy from the beginning of broadcasting, and I maintain that the same policy will be the right one with regard to television.

When broadcasting started in the United Kingdom in 1922, the main patents needed to construct a valve receiver were held by the British Thomson-Houston Company and Marconi Wireless Telegraph Company. Initially Marconis handled the BTH patents under licence, but later they bought the entertainment rights in them. Then in 1928 the Chelmsford firm sold their patent interest to the Gramphone Company and a patent pool was formed.[9]

Before the formation of the British Broadcasting Company, manufacturers who wished to make use of the Marconi Company's patents negotiated licences individually with the company. The company had amassed a wealth of skill in the field of radio communications, and, in the course of its development, had accumulated a large number of patented inventions. In addition Marconis had fully appreciated the importance of a strong holding of master patents, and so, when it became clear that a broadcasting service would be started, they took all possible steps to obtain control of the significant patents. Negotiations were entered into, and agreements were made (or had been made earlier), with Telefunken in Germany and with the Radio Corporation of America, which gave the British concern control of all the essential patents in the above field within the British market. Their policy was to buy any radio patents offered to them.

When the formation of the BBCo. was being discussed, the manner by which patents were to be licensed was of some importance. Marconis proposed charging a royalty of 10% on the wholesale selling price of all receivers which were made by manufacturers licensed to employ the Marconi Company's patents. This was not acceptable to the other members of the 'big six' (MWT, BTH, GEC, Metropolitan-Vickers, RCC and Western Electric), but eventually towards the end of 1922 agreement was reached and general licences were available – the royalty under the licence being assessed at 12s 6d per valve holder.

The trade disliked the terms of the licence (which became the A2 licence in 1923), for it was felt that the royalty charged was excessive. When broadcasting commenced, the manufactured set carried a tariff to the British Broadcasting Company as well as the royalty payable to the Marconi Company. Subsequently the scale of tariffs was reduced, and then abolished, but in the meantime the existence of the two payments created a strong inducement to manufacturers to attempt to evade royalties by not taking a licence. Further difficulties arose with home constructors, and in 1924 Marconis stated that they would not enforce their rights against them because of the complication of doing so.

Naturally the patent licences were highly profitable to Marconis, although considerable evasion of royalties took place. As the price of sets came down, the relative burden of the royalties increased, and, as a consequence of the discontent of the trade with the pricing system, the RMA, in 1927, set up a Royalty Committee. It sent a deputation to MWT Company to persuade the company to reduce its royalties, but without success. Later in the year the RMA again complained to the firm and alleged that manufacturers in Holland, Portugal, Spain and Italy paid no royalties, while in Germany the maximum royalty was 2s 6d per valve holder. When no satisfaction was received, the RMA decided to question the Marconi licence, via the Comptroller-General and the courts, by means of a test case (involving the Brownie firm), and for this they relied on Section 27(e) of the Patents and Designs Act of 1919.

The Comptroller-General considered, in this and another case brought by Loewe, that Marconis were using their monopoly position to prevent the further development of radio in the one case, and to hinder the establishment of a new art in the other (August 1928). Marconis appealed against these decisions of the Comptroller-General, and, in November 1929, the appeal judge found that the company had not abused their monopoly position and were at liberty to carry on with its original royalty terms.

Following the success of the Marconi Company's appeal, the trade was in a state of considerable frustration and the Royalty Committee of the RMA again approached the company hoping to achieve by persuasion and discussion what it had not obtained by litigation.

The manufacturers were offered a new licence (known as A3), which covered all the relevant patents of the MWT, BTH and Gramophone Companies. This licence was issued by a patent pool, although Marconis undertook its administration and it continued to be called the Marconi licence. Under the terms of the licence, the royalty rate was reduced from 12s 6d to 5s per valve holder, but as a quid pro quo the licensee had to sign an agreement, running for five years from 28th August 1929, that he would pay this royalty on all sets produced, whether they used patented devices or not. This was because certain important patents were due to expire in October 1929, and in 1931, but with the above clause these patents would effectively continue to earn royalties until 1934.

The first challenge to the supremacy of the pool came from STC which held a number of patents on loudspeakers, push–pull amplifiers and super heterodyne receivers. Licences to use these patents were granted by STC to manufacturers on a royalty basis, but later in 1929 the company offered its patents to the pool in return for a share of the revenue of the pool. At first agreement could not be reached, thereby putting set makers in a dilemma regarding the licence by which they should manufacture, but by 1st April 1930 an enlarged patent pool was in operation arising from the merger of the Marconi Group with STC.

Further trials of strength between the patent pool and organisations outside its control came in 1933, when the Hazeltine Corporation, Philco and the Majestic Electric Company formed a new company in Britain called Hazelpat to licence manufacturers under their patents, and the Phillips and the Mullard Companies announced that they intended to grant licences under all the patents which they owned or controlled.

Hazelpat joined the patent pool in September 1933, but Phillips–Mullard did not become one of the grantors until 1938 – after an extended legal battle, involving an infringement suit, which reached the Court of Appeal in 1937, and finally, in 1939, the House of Lords.

These competitive ventures by powerful companies against the monopoly of the patent pool highlight one of the weaknesses of the pool system. Patents are not assets which hold for all time, but are of value for a definite period only: they represent wasting assets. Thus, when an important patent expires, the position of the patent pool is considerably weakened, and, unless patents can be continually introduced to reinforce the pool, it must eventually reach a state in which licencees can disregard it – otherwise alternative pools are formed. A patent pool can only maintain its position of power by expanding and embracing new patent holders, but with this expansion in the number of grantors in the pool, there must be a diminution of the share of the proceeds which go to the original grantors. As a consequence, the original grantors may tend to resist any widening of the pool, accepting it only when the need is essential. Apart from the constant requirement to maintain the strength of the patent pool by the acquisition of new patents, to annul the effects of expired patents, there was also a considerable pressure on the pool, from manufacturers, to further this aim so that they – the set makers – could deal with one body only.[10]

From the patent owners position there were two reasons which induced them to offer their inventions to the pool: first, the desire to obtain revenue from the exploitation of the patent, and secondly, the need to gain recognition as successful inventors. Most patent owners do not wish to risk an infringement of their patented ideas by, say, a large and powerful industrial organisation because such action would demand retaliatory measures via legal proceedings, and such proceedings involving court actions could be 'treacherous', to quote the RMA.[11] The RMA's view was that there was a tendency for all persons with patents of any validity and real application to broadcasting to negotiate and endeavour to enter the pool that way. Indeed one of the pool's difficulties was that so many of those people were not worth having in.

Certainly the large manufacturing organisations welcomed the stablishment of the patent pool: GEC considered it was the best thing that happened at the beginning of broadcasting and advocated its use when television broadcasting was started. Ferrantis, too, favoured a television patent pool:

> I think it would be very desirable, but I should hope they would learn something from running (the present) pool and run the next one very much better. For instance, they have not taken any steps to control the industry in any way. They licensed anyone to begin with. We want a well

> spread licence, but not an indiscriminate one, and they have taken no steps whatever to see that people sell at economic prices. I am all for price arrangements in general. In fact, in any business that lasts for any length of time you come to that.[12]

The Television Society and the Radio Manufacturers Association likewise advocated the creation of a television patent pool, but Bairds, as represented by Church and West, reserved their position on this issue. At one time the company had contemplated the pooling of television resources;[1] later in his evidence West mentioned '. . . we have not been fortunate in one or two approaches that have been made to us, and one or two conferences we have had with other interests'. This change of interest in the pool idea possibly stemmed from the outcome of Bairds negotiations with first Marconis, then GEC on a form of collaboration on television matters. Both these approaches were negated by Ostrer and his supporters on the Board of Baird Television and the general view of the company on licensing and cooperation was 'to get all the trade interested and to licence out everything so as to give the whole of the radio trade every opportunity of building sets, if they are using our system, to our licence . . .'. No monopoly position was contemplated by Bairds and their recommendation to the committee was that it should not be of great concern to the committee whether the commercial arrangements between the holders of television patents and manufacturers should be by direct negotiation between the individual parties concerned or whether holders of patents on receiving sets should form a pool – always providing that the royalties charged for patent rights were not exorbitant.

Subsequently the Selsdon Committee was to advise in its report the adoption of a similar viewpoint regarding the commencement of the television service.

One of the strong reasons for having a television patent pool, as evidenced by the working of the sound broadcasting patent pool, was that it would enable the trade to produce the best possible apparatus for the public, because manufacturers would not have to decide between choosing one monopoly or another – the patents of all participating pool grantors would be available to licensees and therefore sets could be designed to suit the most fastidious customer in addition to those with limited means.[13]

In 1933, the year before the Selsdon Committee met for the first time, the Radio Manufacturers Association was involved in negotiations with the then existing pool for the preparation of a new comprehensive licence (A4) to replace the A4 and RG2 licences which were in force.[9]

The RMA comprised within its membership 119 manufacturers engaged in all branches of the radio industry, including television, and in output and turnover comprised, at least, 90% of the whole production of the industry. The objectives of the RMA were threefold:

1 to ensure that receiving apparatus offered to the public was manufactured by British labour from British materials;

2 to promote a high standard of technical efficiency;
3 to act on behalf of the industry in negotiations with licensing authorities, broadcasting authorities, government departments, technical institutions, with a view to safeguarding the interests of the users and to securing the well-being of the industry in all matters affecting broadcasting reception.[14]

Probably one of the most significant points about the issue of the A4 licence was the degree of involvement and consultation of the set manufacturers, as represented by the RMA, with the pool grantors before the licence terms were determined and the provision of a Joint Consultative Committee with the RMA to decide on points arising from the operation of the licence. This was a far cry from the conditions under which the A2 licence was granted and reflected the growing importance of the set makers and the weakening of the pool. The former were now properly organised and included all the well known firms such as EMI, Cossor, GEC and so on. It was therefore natural that the RMA felt it had a definite part to play in the development of television.[13]

An essential point about the basis of the RMA was that it was a British basis[11] – it controlled the trade in the interests of the trade as a whole. Regulation 8 of the 'Constitution, Regulations and Bye-Laws' of the association prescribed an embargo on the dealing, and usage, by its members of radio apparatus and parts of foreign manufacture if a like type or class of goods was nationally available. To further this aim, the RMA co-operated with the Customs, the suppliers of materials and others to obtain returns, they published a monthly list of exports and imports of radio material and produced a bulletin. Because of the above basis of the RMA's being, it was very rare for, say, a German receiving set to be seen in Britain: there was an understanding between the different members of the patent pool that German set manufacturers would be licensed under its patents provided their sets were not sold in various countries, including the United Kingdom.[13]

Lord Selsdon's committee gave very careful thought to the issue of having a patent pool as a pre-condition to the commencement of a television service.[11] Their original idea was that this should be so, but later the committee had some doubts about the wisdom of such a policy.

The ideal solution, according to the committee, was that all television patents should be placed in a patent pool, the operating authority being free to select from this pool whatever patents it desired to use for transmission, and manufacturers being free to use any of the patents required for receiving sets on payment of a reasonable royalty to the pool. However, from evidence which the committee received they were 'convinced that, under present conditions, when the relative value of the numerous television patents was so largely a matter of conjecture, the early formation of such a pool would present extreme difficulty'.[15] Two of the experts who gave this evidence were Mr R. M. Ellis of the Radio Manufacturers Association and Sir William Jarratt, Secretary of the Trade Marks, Patents and Designs Federation. Ellis's opinion was quite definite: 'Neither the Government nor any body (could) give a list of patents which (were) necessary and sufficient for the establishment of a competent patent pool in television . . .' Again, the government had no power to compel an owner of television patents to put them into a pool against his will, and, furthermore, patent holders might find it exceedingly difficult to agree among themselves on a fair basis for charging royalties and sharing the revenue so obtained. There was thus a danger that a hastily arranged pool would in all probability end in failure.

Nevertheless, while the committee had to abandon its original idea, it was strongly of the opinion that it was in the public interest, and the interest of the trade itself, that a pool should be formed.

Following the Television Committee's Report in January 1935, manufacturers either negotiated an individual licence with a patent owner or produced without a licence. Later in 1938 the established sound broadcasting pool included television patents in its A5 licence.

14.1 Television operating authority

Three possibilities existed regarding the nature of the television operating or broadcasting authority. It could be:

1 a single private company;
2 a consortium of companies;
3 the British Broadcasting Corporation.

With the first two possibilities there was also the 'possible alternative of letting private enterprise nurture the infant service until it (was) seen whether it (would grow) sufficiently lusty to deserve adoption by a public authority'.

Bairds were in no doubt at all on this issue: television broadcasting should be done by private commercial enterprise.[16] They based their view partly on recourse to history and partly on their own experience with the General Post Office and the British Broadcasting Corporation and quoted two historical episodes to justify their opinion. The Telegraphy Act of 1868 and 1869 had prevented the development of telephony, the company opined, to such an extent that until recently this country was one of the most backward countries in the world in this respect. Again, the situation with reference to incandescent electric lighting caused by the Act of 1882 stifled it for six years and allowed other countries, particularly America, to get far in advance, until the amended Act of 1886 was passed.

Bairds referred to these examples to illustrate their contention that government officials were not receptive to new ideas and methods, and any participation by government in new commercial enterprises caused difficulties.

In addition to these objections there was the experience of Baird Television Ltd. in its endeavours to co-operate with the BBC and the GPO. The company's history, it felt, demonstrated the 'serious lack of understanding and apathy towards the commercial necessities of (the) new industry' shown by these authorities, and as an illustration Bairds highlighted some points which had caused them unease during the development of the low definition service. Again their attempts to secure a co-operation with the Corporation on the initiation of ultra short wave experimental transmissions had been rendered impossible, chiefly on account of the red tape methods and the ensuing delays which had resulted.

All of this argued for control by private commercial enterprise and naturally Bairds considered their expertise and facilities were second to none in the field of television. 'We consider that we are the only existing organisation which has the means, experience and power to inaugurate this service without delay' their memorandum stated. To support such a view the company mentioned the following points:

1 . . . no one (was) able to transmit actual objects and persons as they could, and if their competitors used the light spot principle to enable them to do this the company would obviously be compelled to apply for an injunction to stop it.
2 None of (their) competitors at present (could) transmit full size living subjects and scenes such as (could be) taken with an ordinary cine-camera. (Presumably Bairds were referring to the intermediate film method here).
3 None of (their) competitors (had) been able to demonstrate . . . the 240-line system.
4 No organisation (had) yet demonstrated high definition television on a screen as large as (they) had been able to demonstrate by means of Mr Baird's apparatus . . .'

Apart from these reasons the pioneer British Television Company no doubt considered their Crystal Palace installation, complete with studios, transmitting equipment and high antenna tower, a ready-made television broadcasting station: no other British company had a comparable set-up. Then again, no other British organisation, except Bairds, had had any previous experience of broadcasting and programme production: 'none (had) done the pioneer work in the unknown field of ultra short wave transmission, and its applications to television that the Baird Company (had) carried out'.

A further important reason (to Bairds) for entrusting the service to them stemmed from their association with one of the most important organisations in the entertainment industry – the Gaumont-British group. This group had given its assurance to Baird Television that they would have every facility for showing films in its possession or within its control, with the exception of the transmission of complete films within a reasonable period of their first release, at merely nominal cost. As it was thought by the committee that films would form a considerable proportion of the weekly broadcast programmes, this point was of some importance.[17] Bairds would certainly have been in a strong position if the Television Committee had recommended the establishment of a television service by a private concern for the company had acquired some valuable assets.

The Crystal Palace concern was the only one which put forward a strong case for the private control of television broadcasting. Marconi–EMI had sufficient capital to make it possible for them to think about it if the Television Committee was minded to farm it out to a private company, although Shoenberg opined:

> I personally would not think for one moment that anyone would compete with an existing organisation which has all its overheads paid in connection with an activity similar to that of television, and could introduce it merely as part of its programme. I do not think any private firm could possibly compete.[2]

Such a move as advocated by Bairds, and also (depending on the method of financing the service) by the General Electric Company, would have involved the granting of licences for the transmission of sound and vision to those firms interested in supplying a service. As a consequence, the principle of having only a single authority broadcasting a public sound service would have been negated, and any subsequent take-over by public interests would have been rendered costly owing to the growth of vested interests. The committee in addition foresaw 'serious practical difficulties as regards the grant of licences to the existing pioneers as well as possibly to a constant succession of fresh applicants'.[15] Bearing in mind these points, the excellent sound service being provided by the BBC, and, probably, the situation prevailing in the USA, the committee arrived at an obvious conclusion: 'The conduct of a broadcast television service should from the outset be entrusted to a single organisation, and we are satisfied that it would be in the public interest that the responsibility should be laid on the British Broadcasting Corporation.' With this conclusion Lord Selsdon could not see that any rational being could object and thought there could not be any objection in any quarter – except perhaps on grounds of private interests – to the BBC being entrusted with the task.[17]

Sir John Reith, the BBC's Director General, readily assented to this view. Indeed he hoped 'we should get as far ahead in this service as we are in our present service'. Phillips, who in the past had had to draft many replies to Bairds' missives regarding the need to further British (i.e. Bairds) television interests now rather unexpectedly adopted the Television Company's attitude in this regard.

> It is exceedingly desirable that the British system shall get well away as soon as, or if possible in advance of, its principal competitors which will certainly be the German system and the American system. Whichever system does get well away will bring a considerable advantage to the manufacturers in the country concerned. If the British system can get started successfully and be a success in the first six months, it would give

> British manufacturers a much better chance to get their system adopted in other countries which are not so forward as the above three countries.[17]

Baird would have been pleased to have heard that: he had made the same point on many occasions.

Discussion took place in the Television Committee on whether the initiation and early development of the new service should be planned and guided by an Advisory Committee. Most of the organisations which were asked to comment on this aspect of the committee's work advocated the establishment of a controlling body. The Newspaper Proprietors' Association, Scophony Ltd., GEC, Abbot, *Popular Wireless* and the Television Society all felt some form of control on television to be desirable.

Of the proposals put forward by the various witnesses, none was more carefully prepared than those of GEC. These entailed a two-tier structure comprising a Central Television Board and a Television Board Advisory Committee – somewhat analogous to the then organisation of the electrical industry with its Electricity Commissioners and Central Electricity Board.[10] The constitution of the CTB was to comprise representatives of the PMG and six to eight independent members selected from public life, industry and commerce and unconnected with any financial interests in television, while that of the TBAC was to include an individual from the GPO, the BBC, the DSIR, the approved television transmitting companies and the RMA.[18]

GEC's terms of reference for the proposed CTB were all embracing and covered every facet of television broadcasting, namely: the promotion, approval and control of the various experimental television transmissions; the arrangement for issuing licences (to companies), the recommendation of methods for collecting and disposing of licence monies; the supervision of television programmes and so on. On the other hand the TBAC was to keep the CTB informed on all matters relating to the development of television and to make suggestions to it for the conduct of the experimental television service. With the above structure, the CTB would have been in the position of judges and the Advisory Committee in the position of counsel and solicitors, to use a further analogy.

The recommendation of the Television Committee on the subject of control and the need for an Advisory Committee really represented an amalgam of some of the ideas put forward by GEC with the exception that Lord Selsdon's committee advised that normally the Advisory Committee would not concern itself with detailed financial allocations, or with business negotiations between suppliers of apparatus and the BBC, or with the compilation of programmes, the detailed construction of stations or their day-to-day operation, unless specifically invited to do so. In detail, the terms of reference of the Advisory Committee were to advise on the following:[15]

> 1 The performance specification for the two sets of apparatus . . . including acceptance tests, and the selection of the location of the first transmitting station.

2 The number of stations to be built subsequently, and the choice of districts in which they should be located.
3 The minimum number of programme hours to be transmitted from each station.
4 The establishment of the essential technical data governing all television transmissions, such as the number of lines per picture, the number of pictures transmitted per second, and the nature of the synchronising signals.
5 The potentialities of the new systems.
6 Proposals by the BBC with regard to the exact site of each station, and the general lines on which the stations should be designed.
7 All patent difficulties of a serious nature arising from the operation of the service in relation to both transmission and reception.
8 Any problem in connection with the television service which may from time to time be referred to it by HMG or the BBC.

Part of the Television Committee's object in suggesting a body of this nature was that, as Lord Selsdon put it, there might be moments when, frankly, it would be very convenient to the BBC as a buffer.[17] During the five-year period, 1928–1933, the Corporation's television policy had been under much attack, not only from the pioneer British company but also from the press, Members of Parliament and the general public, but with the likely (in November 1934) formation of an advisory body the BBC would no longer be subjected to the sort of onslaught it had suffered in the past. From industry's point of view the proposed constitution of the new committee (with representatives from the GPO, BBC and DSIR), would ensure that no single individual or individuals from a given organisation could prejudice the outcome of a new development in the art.

On the question of the life of the proposed Advisory Committee, there seemed to be an implicit assumption in its terms of reference (point 8), that the committee would simply expire if everything were going smoothly and there was nothing to refer to it. However, Reith preferred a definite term for its life: 'Some committees are not always ready to die when other people think their usefulness has come to an end' he observed. A period of five years (say) was subsequently prescribed for the committee's term of office.

14.2 Finance

The Television Committee had two suggestions to consider for the provision of the necessary funds for television development. These could be classified under two broad heads:

1 selling time for advertisements
2 licence revenue

The former could take two forms:

1 direct advertisements, for which time would be bought by the advertiser
2 sponsored programmes, for which the costs would be borne by an advertiser in return for the mention of his name in connection with the programme.

Direct advertising had previously been examined, in relation to sound broadcasting, and rejected by the Sykes Committee on Broadcasting in 1923.[19] The Newspaper Proprietors Association was vehemently opposed to any suggestion of raising revenue by this method, as was to be expected, and of the other witnesses and organisations which were asked to give evidence, only Baird Television Ltd. and *Popular Wireless* opined a view that direct advertising could have a beneficial effect. For *Popular Wireless*, if there was a choice between having an expensive service coupled with obtaining advertising revenue and having a very restricted service clear of advertising revenue, the former alternative was to be preferred, although it was not the way which the magazine hoped would be enforced.[20] Bairds advanced the method employed by the 'B' stations in Australia where advertising was allowed for 5 or not more than 10% at the most of the programmed time. 'We do not mean a sponsored type of programme like America, but something more like the higher class of publicity films at the cinemas where it is not blatant but restrained.'[1]

The Newspaper Proprietors Association's objection to direct advertising was based largely on the point of view of self-interest.[21] They were employers of appreciable labour forces, but if some advertising revenue were taken away from them and given to television, there was a very real danger that a number of newspapers would become bankrupt. For this reason the NPA did not wish to see the new service entrusted to private enterprise. 'Once you start farming this out to the EMI or Mr Isidore Ostrer, doing it with the sole right, it might put us all out of business.' The NPA stressed that newspapers did not have any prescriptive right to advertising ('advertising is on posters, by direct mail, on the films and thousands of other ways') but felt it was improper for any other organisation, which involved any form of monopoly even in what was essentially a public service, to hire out time or space for private advertising. Mr E. J. Robertson, of Express Newspapers, told Lord Selsdon they objected to advertisements in the *Radio Times*: this was because the publication canvassed for advertising when it had the opportunity, which was denied to any paper on a commercial basis, of advertising their own productions over the microphone. 'The profits of the *Radio Times* are phenomenal,' Robertson observed, 'that is a fee the newspapers are paying entirely, that has been taken from our particular field.'

Clearly the NPA was bitterly opposed to direct advertising and Lord Selsdon's committee wisely did not recommend this course for revenue purposes.

As regards sponsored programmes, the Sykes Committee saw no objection to their admission and they were (in 1934) specifically allowed under the BBC's licence, although the Corporation had only admitted them on rare occasions. Little discussion took place on sponsored programmes (as distinct from direct advertising) in the Television Committee, but both Baird Television Ltd. and Sir John Reith had no objection in principle to the sponsor system, in contradistinction to the opinion of the Newspaper Proprietors Association. The committee view was that there was no reason why the provision of these programmes in the existing licence should not be applied also to the television service and they also thought it would be legitimate, especially during the experimental period of the service, if the Corporation took advantage of this permission to accept such programmes.[15]

On the question of obtaining funds from licence revenue, there appeared to be four main possible sources:

1 the raising of the fee for the general broadcast listener's licence;
2 the issue of a special television viewer's licence;
3 the imposition of a licence upon retailers;
4 the retention of the existing listener's licence at 10 shillings, but with a contribution from the licence revenue for television during the experimental period.

In addition the following alternatives were put forward:

1 the use of the profits of the *Radio Times* (NPA);[21]
2 the provision of a government subsidy (Bairds);[16]
3 the obtaining of a subscription from trade interests (*Popular Wireless*);[22]
4 the taxing of the sales of all television receivers (*Popular Wireless*).[20]

Bairds suggestion was inspired by the intimation of the German Government, to the German television industry, of its willingness to subsidise the initial broadcasting of television by an amount far in advance of the amount which it was thought would be required to meet the first year's broadcasting deficit in Britain. The company further proposed that the grant should be recovered from the revenue obtained from broadcasting in subsequent years and they were confident this would be sufficient to meet the comparatively small amount required for the new industry. Bairds idea was thus a variation on the use of part of the existing licence revenue for television (point 4 above).

Of the alternative sources of revenue, the notion of a trade subscription was based on the principle of asking the trade to pay something towards the cost of a service because they would reap some benefit from the sale of sets. There had been a precedent for the industry providing a public service in the past and Dr Roberts, who represented *Popular Wireless*, told the committee he had in mind the manner in which the British Broadcasting Company was originally financed.[20]

The BBCo. was incorporated on 15th December 1922, and granted a licence (to run from 15th November 1922 to 1st January 1925), on 18th January 1923. Its capital consisted of 100 000 £1 ordinary shares on which the dividend was limited to $7\frac{1}{2}$%, and the company was guaranteed by the 'big six' firms, namely, the Marconi Company, BTH, GEC, Metropolitan-Vickers, RCC and Western Electric. These companies held themselves directly responsible for taking up 60 000 shares, and, when the company was wound up at the end of 1926, 71 536 shares had been issued, of which 60 000 were held by the above companies.

Outside the formal agreement, there was a definite understanding that the six companies were responsible for whatever money would be required to carry on broadcasting efficiently until the end of 1924, even beyond the £100 000 capital. The company agreed to establish eight stations and to pay to the Postmaster General a royalty of £50 per station. A tariff was payable to the company in respect of certain apparatus sold by a member and in addition the Post Office agreed to issue receiving licences at a cost of 10 shillings per annum – one-half of the proceeds being paid to the company.

Two Broadcasting Committees sat during the existence of the British Broadcasting Company, one in 1923 under Major General Sir Frederick Sykes, and the other, in 1925, under the chairmanship of the Earl of Crawford and Balcarres. The Sykes Committee came to the conclusion that the company was doing its job well and that the various allegations that the 'big six' used their position to browbeat smaller firms were not proven.[19] The second committee pointed out that in 1922 manufacturers of radio apparatus

> . . . were willing to subscribe the capital required, which might not have been forthcoming from other sources, they were prepared to conduct the service without cost to the taxpayer, and they had at their disposal technicians of the highest order . . . The scheme had the advantage that it tided over the initial period when the finance of broadcasting was highly speculative and established a system under the best technical auspices with a guarantee of adequate financial backing by responsible firms and a strict limitation of the operating company's profits. It was not intended to be more than a temporary arrangement, and the currency of the licence might be open to review when sufficient practical experience had been acquired.[23]

On 1st January 1927, all the shareholders of the company were paid out at par, and broadcasting in Great Britain became vested in a public corporation.

It was perhaps natural that someone would advance a method of financing a television broadcasting service on the lines which had been adopted in the early days of sound broadcasting, by which manufacturers would put up the capital needed, but such an arrangement would have involved the abandonment of the idea of a public corporation for the new service in favour of private concerns and this was something the Television Committee could not recommend.

In addition to the idea of a trade subscription, *Popular Wireless* also advanced the notion of a tax on all television receivers as a means of raising revenue. The imposition of a licence upon retailers of receiving sets, based on the number of sets sold, had another attraction (in the absence of a special TV licence), for it provided a means of keeping a tally upon the number of users, thus measuring the extent to which the service would be in demand. Lord Selsdon was keen that this should be carried out, and later, after the start of the London service, the RMA was asked to maintain a record of the number of receivers sold. However, apart from the administrative difficulties and the further difficulties which would inevitably arise later on when amateur constructors became sufficiently expert to construct homemade sets, this proposal had been previously considered, and rejected, for sound broadcasting, by the Sykes Committee. These points, and the argument which had been put to the Television Committee, convinced Selsdon and his colleagues '. . . the adoption of such a course would be vexatious to traders and detrimental to the development of the service'.

The above means of raising licence revenue had the advantage that only those persons who participated in viewing television, and hence derived entertainment value from it, contributed to its capital and running costs. This applied too to the issue of a special television looker's licence. However, the latter method had the disadvantage that initially the number of licence holders probably would not be large enough to avoid a costly licence. The licence would place a further burden on the viewer and might detract from the desirable objective of giving the service the best possible chance in its early days.

Ferrantis, a large Manchester electrical engineering company, saw no drawback to a separate licence and advanced it as a realistic way of collecting revenue. 'Like all these things, you cannot start on an economic basis straightaway. We never start to make anything without losing a lot of money for a long number of years until it gets on to a sound basis . . .' The licence could be compared to a subscription to a lending library for books, the firm noted '. . . and one feels that if television is good enough a 30 shilling or even £2 licence would not be excessive.'[12]

The Television Society and *Popular Wireless* also agreed that a special licence was feasible and satisfactory and Sir John Reith would not say, when he was giving evidence, that it should be abandoned: '. . . there are arguments in favour and against' he observed.[17]

In one respect there was an important difference between the cost of a licence at the start of the sound broadcasting service and the proposed television viewers licence. When broadcasting commenced in 1922, the price of a crystal set and headphones was approximately £2 to £3 and consequently the 10 shilling licence fee represented an appreciable proportion of the receiving equipment cost, but with the expected cost of a television set of £50 to £60, even a £2 licence was hardly a material consideration compared with this initial outlay.

Nevertheless the committee viewed the proposal, 'however logically justified', as having 'the fatal practical defect that, if the licence fee (was) placed high enough even to begin to cover the cost, it (would) strangle the growth of the infant service – while, if it was placed low enough to encourage growth, the revenue must for some time be purely derisory as a contribution towards the cost.' Their conclusion was obvious – they could not recommend any extra licence for the start of the service, but thought the question should be reviewed when it was seen to what extent the service had developed and when the costs of further extensions of it could be more accurately estimated.

Of the various possibilities for obtaining revenue from licences, the raising of the fee for the general broadcast listener's licence had the merit of certainty and simplicity. It was arguable whether an additional charge would seriously diminish the number of existing listeners, or even materially abate the normal rate of growth. On the one hand the public might expect to be provided with something at once for their increased licence fee, while on the other hand, if it was explained to them that the extra fee was really for research and development work in the new medium of television, and the point was appreciated, the public would probably support it. However, against the latter view had to be set the time interval between a service in, say, London, and a service in, say, Cornwall or Cumberland – regions of low population density. There was the possibility of the complaint from country listeners: why should we pay an increased charge for a service which only London or some other centres can receive. The Television Committee could see no adequate answer to this, nor to the further complaint, within areas served by television: 'Why should we people with restricted means pay this increased charge for a service which we cannot receive, because the necessary apparatus is at present so dear it is only within reach of the well-off?' The Television Committee could not, therefore, recommend the adoption of such a course.

As a result of all these deliberations, the committee was finally left with the conclusion that, during the first experimental period at least, the cost of the new service would have to be borne by the existing 10 shilling licence fee.

In 1928 the British Broadcasting Corporation's revenue was approximately £1 million, but by 1934 it had increased to nearly £2 million. This came from their share of the 10 shilling fee which was obtained from about 6.5 million listeners.[10] Three parties shared the licence revenue, the Post Office which took 1 shilling for the cost of collection and the Treasury and Corporation which together divided the remainder between them.[17] Effectively the tax payer was receiving a sum of £1½ million per year in relief of taxation from the licence revenue.

GEC's view of the problem of financing television was clear and simple: the Treasury had taken large sums out of the radio entertainment industry, not much appeared to have been put back into it for its future technical and commercial expansion, but now the time had arrived when some large sums were needed to cover the expenses of one or two years unremunerative working of this new venture.[18] Essentially the problem was no different to the development of a new manufacturing product by industry: it had to put back into the industry in the shape of research and development large sums out of its profits. Paterson, representing GEC, graphically described the commercial way of looking at the problem: '. . . the taxpayer is having this £1.5 million or whatever it is by way really of dividends; he is drawing too much out of the industry; he is bleeding the industry white and this money ought to be put back into the industry for its future development.'[10]

Actually it could be argued that, while the present licence fee of 10 shillings was perfectly fair when the number of listeners was 1 or $1\frac{1}{2}$ million, the fee had become excessive now that the number was nearly $6\frac{1}{2}$ million, and hence the existing licence revenue could carry the cost of the sound broadcasting service, a contribution to the Exchequer and still provide something for the introduction of television.

The Television Committee agreed with this view, and, in one of the drafts of the report, an opinion was given 'that the Treasury should shoulder a great part'. This clause was later modified (probably on Phillips' insistence), to read: '. . . the best course would be for a reasonable share of the amount to be borne by each of the two parties – the Corporation and the Treasury . . .'[15]

Reith seems to have been surprised by Selsdon's disclosure that a portion of the BBC's budget might have to be contributed for the start of the new service and observed: 'Sometimes *a* share may mean the *whole* share'. He saw no objection to the BBC providing a share – an enhancement share – when it was convinced that its present service would be materially enhanced, or begin to be enhanced, by a television service.

The committee anticipated three different types of objection to its proposed course:

1 the new service would, at first, enure to the benefit of a limited number of people in a limited area and hence it could be argued that it would be unfair to the general body of licence holders if part of their payments were diverted from the improvement of the ordinary programmes;
2 as regards a contingent Treasury contribution it might be held that the time was not yet ripe to cast a fresh burden upon the tax payer;
3 it might be said that there was no need to speed the start of the service and that this could wait until the BBC's charter was renewed when the financial question could be fully and finally settled.

Against these possible objections the committee put forward the following comments:

1 the existing programmes represented amazingly good value for one-third of a penny per day and because of this the general body of listeners might not unreasonably be asked to help, at no extra cost to themselves, in a national experiment which, if successful, will ultimately enhance programme values for a large part of their members;
2 the development of British television in addition to being of evident importance from the point of view of science and entertainment, and of potential importance from the angles of national defence, commerce and communictions (would) also directly assist British industries;
3 their enquiries convinced them that, apart altogether from any question of scientific prestige, any delay would be most regrettable.

The likely contributions which both the Treasury and the Corporation would have to make from their licence revenue provisions for the new service were, of course, intimately connected with the estimate of the cost of constructing and working a national television network, but here there was a difficulty: even with all the resources at its command, the Selsdon Committee were quite unable to do this. They therefore confined themselves to giving what they hoped would be a fairly close estimate of the cost of providing and running one station – the London station – during the period between the publication of its report and 31st December 1936 (the date on which the BBC's charter was due to expire).

Baird Television Ltd. was the only company which had had some experience of working a television station – albeit one giving a low definition picture and consequently one which was not capable of showing elaborate studio productions – and in a memorandum dated 14th August 1934, the firm gave their estimate of the costs needed for two hours programmes per day, from the Crystal Palace, for one year. Their figures showed that £100 per day would be required to cover the cost of artistes, films and electric power for lighting and radio transmission, or £36 500 per annum, and an extra £11 300 per annum would be needed for programme staff, properties and sets and £5000 per annum for rent and general overheads, thereby making a total of £52 800. If, in addition, an allowance, say £25 000, was made for contingencies, depreciation of plant, replacements and installations of improved apparatus, the overall total was £77 800.

Bairds estimate of £77 800 for two hours of programmes per day may be compared with the committee's estimate of £120 000 (engineering plus programme costs plus amortisation), for an hour's transmission in the morning or afternoon, and, say, two hours in the evening.

References

1 Notes of a meeting of the Television Committee held on 7th June 1934. Evidence of Major A. G. D. West on behalf of Baird Television Ltd., Appendix II, Minute 4003/1935

2 Notes of a meeting of the Television Committee held on 17th June 1934. Evidence of Messrs. Clark and Shoenberg on behalf of EMI and the Marconi–EMI Television Co. Ltd.

3 Notes of a meeting of the Television Committee held on 8th June 1934. Evidence of the Marconi–EMI Television Company represented by Messrs. Shoenberg, Condliffe, Blumlein, Agate, Browne and Davis.

4 See Reference 3, p. 54 and Reference 2, p. 33

5 BAIRD, J. L.: 'Improvements in or relating to television or like systems and apparatus', British patent 269,834, application date 21st October 1925

6 BAIRD, J. L.: 'Improvements in or relating to apparatus for maintaining rotating bodies in synchronism for television and like purposes', British patent 275,318, application date 3rd May 1926

7 BAIRD, J. L.: 'Improvements in or relating to synchronising means for television apparatus or the like', British patent 336,655, application dates 17th July, 14th September, 20th November 1929

8 GARRATT, G. R. M., and MUMFORD, A. H.: 'The history of television', *Proc. IEE,* 1952, **99**, Part IIIA, pp. 25–42, see discussion

9 STURMEY, S. G.: 'The economic development of radio' (Duckworth & Co., London, 1958)
10 Notes on a meeting of the Television Committee held on 12th July 1934. Evidence of Messrs. Paterson and Heather on behalf of the General Electric Company Ltd.
11 Notes of a meeting of the Television Committee held on 5th October 1934. Evidence of Mr R. Milward Ellis on behalf of the Radio Manufacturers' Association
12 Notes of a meeting of the Television Committee held on 12th July 1934. Evidence of Mr V. Z. de Ferranti and Mr A. Hall on behalf of Messrs. Ferranti Ltd.
13 Notes of a meeting of the Television Committee held on 11th July 1934. Evidence of Messrs. Burnham and Milward Ellis on behalf of the Radio Manufacturers' Association
14 Paper no. 16 of the Television Committee: Synopsis of evidence to be submitted by the Radio Manufacturers' Association to the Government Committee on Television
15 Report of the Television Committee, Cmd. 4793, HMSO, January 1935
16 Memorandum: 'General policy of the Baird Company', 14th August 1934, Minute Post 33/4682
17 Notes of a meeting of the Television Committee held on 30th November 1934. Evidence of Sir J. F. W. Reith
18 Paper no. 12 of the Television Committee. Written evidence of the General Electric Co. Ltd.
19 Cmd. 1951, HMSO, 1923, paras. 40–41
20 Notes of a meeting of the Television Committee held on 29th June 1934. Evidence of Dr J. F. T. Roberts on behalf of *Popular Wireless* and the Wireless Committee
21 Notes of a meeting of the Television Committee held on 11th July 1934. Evidence of Messrs. Lawson, McAra, Polley, Jarvis, and Robertson on behalf of the Newspaper Proprietors' Association
22 Memorandum: written evidence submitted to the Television Committee by *Popular Wireless*
23 Cmd. 2599, HMSO, 1925, pp. 18–19

Chapter 15

The report of the Television Committee, 1935

The Television Committee submitted its report[1] to the Right Honorable Sir Kingsley Wood on 14th January 1935. It had cost approximately £965 to prepare and had taken seven months to produce. In addition to its examination of the submissions of the 38 witnesses, the committee had had the benefit of consultation with members of various government departments and had received numerous written statements regarding television from various sources.

The report itself was quite short in length, a mere 26 pages, but this did not include notes of the formal evidence, which were presented in Appendix 2 (Volumes 1 to 4), because much of it contained secrets of commercial value which had been given in confidence and under promises of secrecy. For similar reasons, Appendix 3, which included reports on developments in the United States of America and Germany, and Appendix 4, which gave descriptions of each television system which the committee had examined, were not published. Appendix 5 dealt with certain financial details and was also considered to be of a confidential nature.

Of the television systems which were being developed in this country the committee thought that those of the Baird, Cossor, Marconi–EMI and Scophony companies were the most distinctive. It had seen a number of demonstrations of these systems and in addition various members of the committee had investigated and reported upon progress in television research in the USA and Germany.

The report was well received by the national press. 'The British Post Office which has shown so much enterprise, in all directions, deserves to be congratulated on the adoption of the first scheme in any part of the world to bring television within the reach of the public,' noted the *North Western Daily Mail*,[2] adding finally 'we think, in the circumstances, that the initiative shown by the Government will provoke more congratulations than grumbles.' For the *New Statesman and Nation* the report was 'well set out, concise, surprisingly brief and matter of fact. Indeed reading the short sensible recommendations it (was) hard to realise that there (was) another scientific miracle about to be let loose on the world.[3] The Television Committee has carried out an extremely difficult task with most commendable thoroughness.' 'The report (was) only an unbiased estimate of the present stage of development of television, but it has the merit, in addition, of being frank in its recognition of some of the difficulties,' observed the *Wireless World*.[4]

On the Stock Market the report was followed by a remarkable rise in the prices of Baird Television shares. In a few days the market capitalisation of the company was raised by £1·3 millions to £2·25 millions, as shown:[3]

	Amount issued	Price Jan 31st	Feb 5th	Market capitalisation
Baird Television 'A' ordinary 5 shillings	£50 000	31/-	80/-	£800 000
Baird Television 10% pref. ord. 5 shillings	£525 000	3/10½	7/-	£735 000
Baird Television def. ord. 5 shillings	£300 000	4/-	12/-	£720 000
	£875 000			£2 255 000

However, one observer noted: 'It (was) obvious that the "market" (had) either been too lazy to read the Television Committee's report, or had failed to understand it.' He thought it was premature to stage a boom in Baird Television shares: 'A market capitalisation of over £2 000 000 for any television company at this point (was) fantastic.'

In America the news of the publication of the Television Report

> fell like a bombshell on the ears of American radio engineers. Either the Britons are masters of the art of dissimulation or they are satisfied that they have something really good for Lord Selsdon and his staff did not indicate during their American tour that they thought England was ahead of this country in television development. On the contrary, those who met them socially and in the laboratories aver that they were effusive in their praise of comparative American advances.[5]

German reaction took the form of emphasising that Germany had not lost the lead. The Secretary of State responsible for the German Post Office television tests published an article pointing out that the right of being the premier country in television belonged to Germany: 'For the past year and a half,' he wrote, 'high definition television on ultra short waves (has) been broadcast in Berlin.'[5] He omitted to tell his readers, however, that reception of these experimental transmission was limited to the laboratories of interested firms and that suitable receivers were not available on the German market.

An interesting side effect of the report was the spate of suggestions which members of the public made to describe people who watched television. The report used the words 'looker' and 'looker-in' but many people found these words unlovely.[6] One correspondent, using the analogy of Bakerloo to describe the Waterloo–Baker Street tube line, suggested that a person who was at the same time a looker and listener should be called a 'lookener' and the act of looking and listening as 'lookening'.[7] Others suggested 'observer', 'perceptor', 'televist', 'watcher', and so on. In fact a large number of readers wrote to the *Daily Telegraph* and offered what they thought were suitable names:[8]

audobserver	telegazer	viewer
audoseer	teleseer	visioner
audivist	televist	visualiser
invider	teleobserver	visuels
radioseer	telspector	
beholder	opticauris	

For the apparatus itself and other purposes the following words were also put forward for consideration:[8]

airdvision	lustreer	teleite
bairder	mirascope	telescriber
endview	photelepathy	the apparatus
farscope	radion	the looker
farsight	radioscope	viseur
ingazer	telebaird	aeroscope
optiphone	radiovista	telver

Certainly the committee's choice of words was not one which appealed to readers of the *Daily Telegraph*: 'Only publicity and opposition can prevent the BBC from foistering "looker" on us.[9] If the word "looker" is officially adopted it will remain as a permanent reflection on the inventive ability of the BBC.[10] Other commentators mentioned:

> It is sad indeed to think that the combined efforts of the Television Committee, Mr Bernard Shaw and Professors Abercrombie and Lloyd-Jones can produce no better word than 'looker' and 'looker-in' to describe users of television sets.[11] The staple word 'observer' would be more euphonious and pleasing. I venture to prophesy that 'tele-critics' will describe 'lookers' as 'telefans' and the shows as the 'Teles'.[12]

Several writers mentioned 'viewer' and the BBC in at least two of their news bulletins used a new verb 'to televise'.[13] Sir Kingsley Wood also used this word in a broadcast programme on the report, but for a Mr Kenerdine, if 'television' meant anything then 'televise' must mean to receive a broadcast rather than to convey the meaning of 'to broadcast'. This view was probably based on the 'barbarous Graeco-Latin jumble' of the word television itself, namely, to see at a distance.[14] Sir John Risley felt strongly about the matter:

> Word makers in happier and less slipshod times gave us 'telegraph' and 'telegram' – not telescribe and telescript, still less telewrite and teleletter. They also gave us 'telescope' and telephone and similar correct compound words. Why, then should we now have 'television' thrust upon us . . .

The French, too, joined in the attempt to coin a new word to replace 'looker-in'. According to the Paris correspondent of the *Sunday Times* suggested names included 'televiste, televidiste, televoyonte, visiophile, televiseur, telespecteur and longemireur'.[15]

Following the decision by the government to commence an experimental television broadcast service from London later in the year (1935), the Postmaster General, Sir Kingsley Wood, found it necessary to broadcast an assurance that television would not involve the possibility of looking into other people's houses.[16] The belief that television enabled people to do this was, he said, a 'popular idea, which (had) gained a good deal of currency'. Sir Kingsley Wood observed that the Selsdon Committee had received a letter protesting against 'the invasion of the privacy of the writer's home by television'.

Another misconception was the fear that receiving sets would be rendered useless by the development of television, and Lt.Col. Moore-Brabazon, MP, President of the Radio Manufacturers' Association, felt the need to give listeners an assurance on this matter. 'Both the BBC and the radio industry,' he said, 'are anxious that the public should not be panicked by the wild statements calculated to give the impression that domestic television entertainment, on the scale that radio broadcasting has attained, is imminent or that it will supplant radio broadcasting as we now have it. However television may develop, it will always be supplementary to radio broadcasting.'[17]

Surprisingly the RMA, which might have been expected to welcome the birth of a new industry, viewed the recommendations of the report with some reservations: it declared that 'the time (was) not yet ripe for television to take the place of radio broadcasting'.[18] Possibly this caution reflected the boom which the radio industry was experiencing at the time. Philco Corporation, for example, had during the period 5th August to 31st December 1934 increased its company's sales by 78% compared to the similar period in 1933. 'We are living and doing business in the most prosperous country in the world'[19] the general sales manager of the Philco Radio and Television Corporation Ltd noted at a dealers' convention in Glasgow on 30th January 1935. 'A buying wave is sweeping over the whole of Britain. More people are buying radio than ever before, and still half the homes in Britain have not yet heard radio music' he continued. As a consequence of this boom, radio manufacturers were expanding their factory and plant facilities for the production of mediumwave and longwave receivers and the BBC was building three new transmitting stations for sound broadcasting. Clearly the industry did not want the bubble of prosperity to be pricked by the sudden invasion of a new entertainment medium, and the view that the buying public might cease to purchase wireless sets in the belief that they would become rapidly redundant was thus a very real one – and one that had to be discouraged. 'We shall not see television for at least two or three years (and) when it does come, it will be local. It will necessitate the building of booster stations all over the country,' Philco's general sales manager noted. 'It is only fair to the general public that this should be pointed out.'

However, whereas the trade was apprehensive about the possible outcome of the report on its current well-being, many newspaper columnists romanced on the prospects television offered. 'How splendid it would have been if last night's relay of Mascagni's new opera "Nero" from La Scala, Milan, could have been received in everyman's drawing room simultaneously with a television motion picture of the stage action.'[20] The public 'want the running commentary on, say, the Derby, or a centre court match at Wimbledon, or a magnificent state ceremony, or ski-ing in the Tyrol or Christmas Day on an Australian sheep farm illustrated on their television sets with a picture of the event.'

This view of the delights to come was encouraged by the opinion of the Television Committee: 'The time may come when a sound broadcasting service entirely unaccompanied by television will be almost as rare as the silent film is today,' it stated in its report – adding, probably as a sop to the industry, 'but in general sound will always be the more important factor in broadcasting'.

Not only was the radio trade apprehensive about the possible effects on its sales; the cinema and theatre industry too was much concerned. *Variety News* in an editorial dated 7th February 1935, asked:

> What is going to happen to the music hall when we can see as well as hear the performers in our homes by means of television allied with radio?[21] Experience suggests that such a transmission would keep millions of people at home, and consequently away from music halls and theatres, and from many other places of amusement. A televising of the English Cup Final might have a similar effect on the 'gates' of professional football matches. A televising of a Garbo film première might affect the takings of every cinema in the kingdom. So every form of public entertainment will be up against it when television comes.

The editor of *Variety News* quickly made up his mind about television:

> As likely as not it may prove to be an enemy. In that event, there is nothing like taking time by the forelock. No general goes into battle – or he should not – without ascertaining as quickly as possible the enemy's dispositions.

These views seemed to portend an ominous beginning for the new industry, but fortunately the writer, after discussing the wisdom of the cinema/theatre industry embarking on a programme of hostility, counselled 'co-operation with the menace' as the best plan of action. For, he pointed out, co-operation has made broadcasting helpful to living entertainment; it was co-operation that had brought music hall artists to the microphone and so enhanced their popularity that when they appeared in person in the music halls their value as a box office asset had been enormously increased 'and what the radio has done in that respect, television may do still more'.

Nevertheless, the Radio Manufacturers' Association – which stood to gain from a television boom – was very anxious about the wild talk and ill-founded sensationalism respecting television and felt the need to guide the public and protect a great, if youthful, industry from damage caused by such reporting and deliberate manipulation of public opinion for interested purposes.[22] Consequently their publicist sent a long letter to the press in which he attempted to play down the immediate impact of the future entertainment service. He listed seven points for consideration:

1 One of the functions of the Television Advisory Committee was to advise on television developments. 'That, in itself, suggests that television may remain in the experimental stage for five years.'
2 The Postmaster General had stated: 'If all goes well there should be a single station in London opened towards the end of the year.' Hence there were grounds for believing the experimental transmissions would not commence until well into 1936 and then serve only a percentage of the capital's population.
3 Provincial stations would not be constructed until the London station had operated long enough to justify the correctness of the adopted transmission system. Consequently 'as ten stations (would) not provide a service for even half the entire population, it will require all the Committee's five years to establish a complete national domestic service'.
4 Television would be limited to, 'at most', two hours per day; and a viewer would have to 'sit still in a darkened room, and concentrate attention on a glass oblong the size of a large post card'.
5 The new medium would use the shortwave band and so radio receivers would still be required to receive BBC and foreign programmes which would be the 'only broadcast entertainment available during the rest of the day'.
6 Television would be a rich man's hobby. 'It can be taken as certain that television receivers . . . will not be put on the market under £75 each.'
7 Programmes, initially, for about two years, would have a low entertainment value 'if only on account of the extremely small sum of money available . . . On the other hand the BBC (was) planning to develop their sound broadcasting service . . . and allocating more money for sound entertainment.'

Alligan's gloomy prognostications were later proved substantially correct and subsequently the Radio Manufacturers' Association were to offer suggestions to the Television Advisory Committee for the encouragement and improvement of the television service – and sales.

The principal conclusions and recommendations of the report are summarised below:

Type of service

1 No low definition system of television should be adopted for a regular public service.
2 High definition television has reached such a standard of development as to justify the first steps being taken towards the early establishment of a public television service of this type.

Provision of service

Operating authority

3 In view of the close relationship between sound and television broadcasting the authority which is responsible for the former – at present the British Broadcasting Corporation – should also be entrusted with the latter.

Advisory Committee

2 The Postmaster General should forthwith appoint an Advisory Committee to plan and guide the initiation and early development of the television service.

Ultra Short Wave Transmitting Stations

5 Technically, it is desirable that the ultra short wave transmitting stations should be situated at elevated points and that the mast should be as high as practicable.
6 It is probable that at least 50 per cent of the population could be served by 10 ultra short wave transmitting stations in suitable locations.

Patent protection

7 It is desirable in the general interest that a comprehensive Television Patent Pool should eventually be formed.

Initial station

8 A start should be made by the establishment of a service in London with two television systems operating alternately from one transmitting station.
9 Baird Television Limited and Marconi–EMI Television Company Limited, should be given an opportunity to supply, subject to conditions, the necessary apparatus for the operation of their respective systems at the London station.

Subsequent stations

10 In the light of the experience obtained with the first station, the Advisory Committee should proceed with the planning of additional stations – incorporating any improvements which come to light in the meantime until a network of stations is gradually built up.

11 The aim should be to take advantage, as far as possible, of all improvements in the art of television, and at the same time to work towards the ultimate attainment of a national standardised system of transmission.

Finance of service

12 The cost of providing and maintaining the London station up to the end of 1936 will, it is estimated, be £180 000. (See Appendix 2)

13 Revenue should not be raised by the sale of transmitter time for direct advertisements, but the permission given in the British Broadcasting Corporation's existing licence to accept certain types of 'sponsored programmes' should be applied also to the television service.

14 Revenue should not be raised by an increase in the 10 shilling fee for the general broadcast listeners licence.

15 There should not be any separate licence for television reception at the start of the service, but the question should be reviewed later in the light of experience.

16 No retailer's licence should be imposed on the sale of each television set, but arrangements should be made with the trade for the furnishing of periodical returns of the total number of such sets sold in each town or district.

17 The cost of the television service – during the first experimental period at least – should be borne by the revenue from the existing 10 shillings licence fee.

A minor recommendation referred to the low definition television broadcasts. These were to be maintained, if practicable, for the present and the selection of the moment for their discontinuance was to be left for consideration by the Advisory Committee.

Following the presentation of the report to Parliament, the Postmaster General rapidly established the Television Advisory Committee.

References

1 Report of The Television Committee, Cmd. 4793, HMSO, January 1935
2 Anon.: 'Television', *North Western Daily Mail*, 5th February 1935
3 Anon.: a report, *New Statesman and Nation*, 9th February 1935
4 Editorial, *Wireless World*, **36**, No. 6, p. 129
5 Current Topics, *Wireless World*, 22nd February 1935, p. 192
6 FLYNN, F. E.: letter to the editor, *Daily Telegraph*, 6th February 1935
7 DOWNING, R. C.: letter to the editor, *Daily Telegraph*, 8th February 1935

8 Letters to the editor, *Daily Telegraph*, 4th–9th February 1935
9 D. E.: letter to the editor, *Daily Telegraph*, 4th February 1935
10 SARE, R. J.: letter to the editor, *Daily Telegraph*, 4th February 1935
11 RAE, W.: letter to the editor, *Daily Telegraph*, 4th February 1935
12 A. E. P.: letter to the editor, *Daily Telegraph*, 4th February 1935
13 KENDERDINE, F. C.: letter to the editor, *The Times*, 9th February 1935
14 RISLEY, Sir J.: letter to the editor, *The Times*, 5th February 1935
15 Paris correspondent of the *Sunday Times*: a report, Current Topics, *Wireless World*, 22nd March 1935, p. 287
16 Anon.: 'Queer television myth killed by PMG. No possibility of looking into peoples' homes', *Daily Herald*, 1st February 1935
17 Anon.: 'Television and the BBC. Radio sets not obsolete', *Daily Telegraph*, 25th January 1935
18 Anon.: 'The coming of television, Radio manufacturers warning', *The Times*, 25th January 1935
19 Anon.: 'Television two years away. Radio chief says it cannot come before', *Daily Express*, 31st January 1935
20 Anon.: 'Television prospects', *Northern Whig*, 1st February 1935
21 Anon.: 'Is television a menace to the "halls"?', *Variety News*, 7th February 1935
22 ALLIGAN, G.: 'Television ramp', *Glasgow Weekly Herald*, 23rd March 1935

Chapter 16

The London Station, site and operating characteristics, 1935

The first meeting of the Television Advisory Committee was held on 5th February 1935 with Lord Selsdon in the chair. Colonel A. S. Angwin, Mr Noel Ashbridge, Mr O. F. Brown and Mr F. W. Phillips, who were members of the original Television Committee, now served on the new committee in addition to the new member Sir Frank Smith. Mr J. Varley Roberts retained his position as Secretary.[1]

There were a number of immediate items for the committee to discuss: the specification of the television apparatus and location of the London Station; the standard form of manufacturer's licence referred to in paragraph 56(d) of the report; and the various queries from both private individuals and industrial concerns.

The committee decided that its proceedings would not, in general, be made available for publication and that any statement which it might be agreed to make from time to time should be issued to the press solely through the medium of the Post Office publicity department. A technical sub-committee was set up with Sir Frank Smith as Chairman, Roberts as Secretary and Angwin, Brown and Ashbridge as members.

16.1 Site of London Station

The BBC undertook an exhaustive search for possible sites, in elevated positions, both on the north and south sides of London, following the establishment of the TAC in February 1935. There were four main possibilities to be considered:[2]

1 Alexandra Palace
2 Crystal Palace
3 Tudor House
4 Heath House

although numerous other sites had been investigated in Hampstead, Highgate,

Fig. 16.1 *General view of Alexandra Palace*

Sydenham and Shooters Hill but had been found to be unworthy of serious consideration.[3]

The best site from the point of view of television was Heath House in the Hampstead Heath district, the highest point in London, with a ground height 440 ft above sea level. Sufficient site area existed for the erection of one 100 ft mast but it was anticipated that any proposal for the erection of this or the installation of machinery in the neighbourhood would meet with great opposition.

Similar remarks could be applied to Tudor House, at 425 ft, which also adjoined Hampstead Heath, but it was feared that the disadvantages attendant upon the inadequacy of the accommodation coupled with the large expenditure which would be entailed in purchase and rebuilding together with the expected difficulties over amenities would more than outweigh the natural advantages presented by the site. The ground level at the Crystal Palace site was 350 ft above sea level and additionally the Palace possessed two 'pepper pot' metal and glass towers each 260 ft in height, giving a total height above sea level of about 610 ft.

Fig. 16.2 *Antenna used by Marconi–EMI Television Company Ltd., Alexandra Palace. The vision antenna is at the top of the mast and the sound antenna below it*

However, the Crystal Palace was viewed unfavourable on three specific counts: first although it was served by the Southern Railway, it was 'relatively inaccessilbe by car from Broadcasting House' (involving the negotiation of heavy traffic); secondly, the fact that Bairds were already in possession, and thirdly, it was so situated that a transmitter installed there would produce a 'signal distribution favouring those parts of London in which the proportion of residents who might be expected to acquire television receivers would scarely be as high as might be desired to promote the most successful inception of the service'.

The ground level at the Alexandra Palace site (Fig. 16.1) was 306 ft above sea level and a mast of 300 ft would give an aerial at a height of 606 ft above sea level. Phillips had learnt that neither the Air Ministry nor the Alexandra Palace authorities would object to the erection of such a mast.

There were three possibilities for the aerial:

1 to clamp a wooden pole to the brickwork of the south-east tower without disturbing the cupola superstructure, thereby giving an aerial height of 460 ft with a 70 ft pole above the 84 ft brick pier;
2 to remove the cupola completely and build on to the brick a 225 ft lattice mast, giving an aerial height of 615 ft;
3 to erect a 300 ft lattice mast in the open courtyard or in the Palace grounds.

Of these proposals the second proposal was the one which was most favourably considered (Fig. 16.2).

Alexandra Palace additionally possessed excellent accommodation which comprised suitable rooms for studios, *inter alia*, on both the ground and first floors:

Ground floor
1 67 ft × 50 ft
2 70 ft × 50 ft
3 57 ft × 40 ft
4 35 ft × 30 ft
5 40 ft × 30 ft
6 30 ft × 30 ft

First floor
7 30 ft × 30 ft
8 24 ft × 30 ft
9 65 ft × 30 ft

The BBC considered the accommodation provided by the first floor rooms was sufficient for two transmitter rooms, two studios, and two spare rooms, while the ground floor rooms provided space for the main studios. Access to Alexandra Palace was good and a car could make the journey from Broadcasting House in 20 minutes (without the negotiation of serious traffic). In addition the Governors of the Palace were ready to assist the BBC in every way within reason and the site was so situated that the service area for a transmitter at this point covered primarily 'that part of London in which the residents (might) reasonably be expected to be amongst the first to acquire television receivers'.

Further advantages which led the Television Advisory Committee to recommend to the Postmaster General that authority should be given to the BBC for the adoption of the site for the projected London Television Station were that the necessary cable link with the BBC Studio at Maida Vale would be about $3\frac{1}{2}$ miles shorter and therefore cheaper to provide from the Alexandra Palace than from the Crystal Palace:[4] also (possibly), Captain West had disclosed that the power supply at Crystal Palace was unsatisfactory owing to the old condition of the street cables and that Bairds were at present limited on that account in the power which they could employ. (On the cost of cables, Colonel Angwin stated at the sixth meeting of the TAC that cable suitable for carrying television signals up to a frequency of about 1.5 MHz over short distances could be produced at a cost of, say, £300 a mile, or including the cost of the duct and laying, say, £900 a mile. With higher grade cable capable of carrying frequencies up to 2 MHz over longer distances, the cost would be of the order of £1000 a mile.)[5]

The site recommendation of the TAC was made only after a careful study of a model of the relief contours of the London area, which had been prepared in the Post Office Engineer-in-Chief's department. This was made in wet sand and pieces of thin twine were stretched across to determine what obstructions existed for reception in various directions.[6] No practical field tests using portable transmitters were carried out owing to the lack of time available and the lack of equipment needed to make the tests.

Although the terms of the report to the PMG incorporating the TAC's recommendation were provisionally agreed upon on 12th March 1935, the report was not submitted to him until after the TAC's meeting on the 1st April 1935,[7] i.e. after the radio transmitting arrangements to be employed in connection with the television system had been clarified. The press notice which gave the choice of the site was issued on 3rd June 1935.

16.2 Specification of television apparatus

This matter was referred by the Television Advisory Committee, at its first meeting, to the Technical Sub-committee for consideration. The latter committee discussed the subject at considerable length at its first meeting[8] which was held on 11th February 1935 but decided that no definite proposals could be formulated until more precise information was available regarding certain EMI systems. It was accordingly decided to invite representatives of the two companies to meet the sub-committee at its next meeting which was arranged for 15th February.

Marconi–EMI were represented by Shoenberg, Condliffe and Blumlein.[9] They had prepared a list of 20 questions,[10] covering wavelengths, sound, common working of receivers, and dates and arrangements of the new London Station, for discussion, and additionally put forward a specification for their own transmitter and some suggestions regarding reception. On the transmitting side the company advanced the following proposals:

1 Scanning: 405 lines interlaced to give 50 frames/sec, frames to be scanned from top to bottom.
2 Modulation direction: high carrier to represent white, low carrier to represent black.
3 Background brightness components: d.c. working at transmitter. 70% of transmitter characteristic to be used for vision signals, the peak output to represent white, and 30% of the peak output to represent black. The suggested degree of perfection of d.c. working at the transmitter was to be such that during any one transmission the absolute carrier output representing black was to vary by no more than ±3% of the peak transmitter output.
4 Synchronisation: synchronising signals to be pulses in the blacker than black direction extending downwards from 30% of peak carrier to zero carrier; all pulses to be rectangular, the duration and shape being as shown in Fig. The synchronising pulses to extend substantially to zero carrier, residual carrier to be not more than 2% of peak transmitter output.

The second meeting of the Technical Sub-committee was the first occasion when Shoenberg put forward his company's intention to provide a system of television using 405 lines. It was a momentous decision to make.

J. D. McGee, a former member of Shoenberg's research team, believes that the most difficult decisions which Shoenberg had to face arose in the system as a whole.

> There was, for example, a long debate as to whether d.c. or a.c. amplification should be used – and then, if a.c., how could the d.c. level (black level) be re-established? And as the picture definition was increased the bandwidth increased, so that the radio transmission frequency necessary also increased, limiting the range to approximately line of sight. Would this be acceptable? Then, should positive or negative modulation be used? And, as the brightness of pictures increased, flicker became a serious problem so that the alternative of sequential or interlaced scanning had to be decided. These problems landed squarely on Shoenberg's desk; to paraphase a well known saying: 'the buck stopped there'.
>
> As the number of picture lines crept up from 120 to 180 and then to 240, the required picture-signal bandwidth increased; this increased progressively the problems of amplifiers, of transmitters, of tubes and of achieving adequate service area. It was clear that the quality of the picture increased very noticeably as the number of lines increased. Since it was possible to scan an electron beam at these much higher speeds, it was natural, indeed inevitable, that the advantages of the electronic system should be exploited to the maximum. But what was the practical maximum. It would clearly be difficult, if not impossible, to push the number of lines much further than 240. No one knew what disastrous snags we might meet if we attempted to reach still higher definition. Higher definitions were terra incognito, and not just the rather obvious line of development that it may now appear, 35 years later.

> To us, then young men, it was a challenge and an adventure, but to Shoenberg it must have been a very worrying problem. On him fell responsibility to the company and to his staff to make the right decision . . .
>
> At this point Shoenberg made what was probably the biggest – and I consider the most courageous decision – in the whole of his career: to offer the authorities concerned a 50 frame/second, 405 line/picture television system. Remember that this meant a 65% increase in scanning rate and a corresponding decrease in scanning beam diameter in the c.r.t., a nearly three-fold increase in picture-signal bandwidth; and – worst of all – a five fold increase in the signal/noise ratio of the signal amplifiers and this lists only a few of the resulting problems.
>
> The cynic may say that this was a piece of gamesmanship planned to overwhelm our competitors. But no one who knew Shoenberg or who was aware of the real state of technical development at that time would give this idea a moment's credence. No – it was the decision of a man who, having taken the best advice he could find, and thinking not merely in terms of immediate success, but rather of lasting, long-term service, decides to take a calculated risk to provide a service that would last . . .
>
> To us this decision was a stimulating challenge. To Shoenberg it must have been a heavy and worrying burden. In later years, he often recalled how colleagues in the higher management of the company had seriously questioned his decision, and had warned him that should he fail to fulfil his contract it would be disastrous for the company.
>
> Yet I cannot remember that he ever showed his worries to us at all obviously. The nearest perhaps was one day when things were particularly sticky. That day he finished up a rather depressing review of our progress with the comment 'Well gentleman, we are afloat on an uncharted ocean and God alone knows if we will ever reach port'.[11]

Mr S. J. Preston, formerly patents manager of EMI, has observed[12] that Shoenberg made his choice to adopt 405 line definition 'knowing that receivers available at that time could not be expected to deal with the full bandwidth which 405 line scanning would require but his view was that it was better that the early pictures should be somewhat lacking in definition along the line so that later developments in receiver design which he was sure would take place could be usefully employed without any change standards'.

The questions of the number of scanning lines to be used in a television system and line standards had been important ones from the time that Baird had first demonstrated television to members of the Royal Institution in January 1926. In the United Kingdom and some other countries, notably the United States of America, Germany and France, development of television broadcasting had initially taken place using transmitters operating in the medium waveband and consequently the bandwidth available for television transmissions had been severely restricted. Baird used a 7:3 aspect ratio, 30 lines/picture, 12.5 pictures/

second system standard which theoretically corresponded to a bandwidth of 13 125 Hz (using a formula derived by White in 1933).[13] This standard, with a change in the aspect ratio to 3:4 (vertical size to horizontal size), was later adopted by the German Post Office in 1929 so that the various companies which were experimenting in the field of television in Germany could market sets able to receive the Witzleben transmissions.

Previously, in 1928, a committee appointed by the Radio Manufacturers' Association of the United States had recommended that all 'radiovision pictures' being broadcast should be standardised so that 'one radiovision receiver with one disc would be able to receive any of them'.[14] The standard adopted was one C. F. Jenkins had used in Washington, namely 48 lines/picture, 15 pictures/second, and 1:1 aspect ratio (corresponding to a bandwidth of 17.28 kHz).

In Britain no standard had been officially proposed by either the BBC or the GPO for low definition television, unlike the German or American situation, for Baird was the only inventor able to demonstrate the transmission and reception of moving pictures.

It was not long after Baird's demonstration in January 1926 that some engineers felt the need for improved standards. Dr Schroter and Dr Karolus, of the Telefunken Company, considered the German standards were unsatisfactory, and argued that, if television was to be a complete success, greater picture detail would be required. Their company therefore embarked on a programme of research and development, using the 30 m and 80 m wavebands, in 1929, to achieve this objective.

The first published account on the factors affecting the choice of standards for commercial television appears to be a paper presented in 1929 by Weinberger, Smith and Rodwin of the Radio Corporation of America.[15] These authors found 'acceptable detail would just be obtained with an approximately 60-line scanning system and, while greater detail would naturally be obtainable with a greater number of lines, such a system would meet the entertainment requirement'. This conclusion was based on the results of a series of experiments using conventional half-tone pictures of various degrees of fineness of detail. After a correlation between the number of half-tone lines per inch and corresponding television scanning lines had been obtained, a number of half-tone letters and photographs had been prepared and their appearance compared with the television image, on a 48-line system, of the same original until apparent equality of detail had been obtained.

The RCA workers noted in their paper the agreement of their conclusion with the unpublished view of Dr F. Conrad of the Westinghouse Electric and Manufacturing Company.

Two years after the above work had been published, Gannett, of the American Telephone and Telegraph Company, gave the results of his researches on the quality of television images.[16] In his paper he illustrated the appearance of such images when reproduced with 625, 1250, 6250 and 12 500 picture elements. For this purpose Gannett used facsimile equipment, in which the received picture element was 1/100 of an inch square, and photographs, which were reduced in size until they contained, in their reproduced form, the appropriate number of elements. Subsequent enlargement to a given size thereby enabled photographs to be obtained having the number of elements stated above.

Gannett illustrated his work with pictures of a head and shoulders portrait, a footballer (full length), and a boxing ring (containing three people). He arrived at no firm conclusion for the optimum number of lines required in a televised picture but noted: 'It appears that when several objects are being observed, some psychological factor causes one to expect and be satisfied with fewer details in each subject. Another matter of interest is that in television a somewhat more pleasing effect is obtained than can be represented by still pictures because the eye, in following the movements in the television scene, is less aware of the streaks caused by dividing the image into rows of elements.'

Gannett's work was considerably extended by V. H. Wenstrom, of the US Signal Corps, in 1933.[17] He chose four photographs to 'represent approximately equal gradations of scene comprehensiveness from the lowest to the highest', namely:

1 a single face;
2 a small group (four people in three-fourth's view);
3 a musical comedy stage;
4 a general view of a football game.

These photographs were transmitted on facsimile machines so as to represent the performance of 60-, 120- and 200-line television systems. Approximate representations of 400-line television were obtained by means of suitable half-tone screens.

Table 16.1

Type of picture	Number of lines			
	60	120	200	400
face	fair	good	excellent	perfect
group	poor	fair	good	excellent
stage	poor to worthless	poor	fair	good
game	poor to worthless	poor	fair	good

Wenstrom's paper (of September 1933) seems to have been the first to specifically mention 400-line images. The author probably used this figure because he had previously determined that the definition of home cine films corresponded to approximately a 400-line picture. This was a good point for it was likely high definition television would be compared with home movies. Wenstrom was well aware that the method he had used had a serious disadvantage which might lead to an underestimation of the possibilities of television; namely, the apparent clarity of a scene in motion compared to a still picture owing to the integrating effect of eye on mind. Nevertheless, notwithstanding this defect and the possible enhancement of a television picture by the use of sound accompaniment, Wenstrom came to the conclusions shown in Table 16.1.

From these results and from observation of existing television systems he also arrived at certain general conclusions

> . . . which, if verified, may have the force of basic laws governing the performance of television:
>
> 1 The degree of definition required in any given television application is conditioned chiefly by the comprehensiveness of the scene to be portrayed.
> 2 The higher the contrast in a given scene, the lower the order of definition required to protray it.
> 3 While a lower order of definition may suffice for the momentary presentation of certain scenes, a higher order of definition is required if the observer's continued attention is to be held.

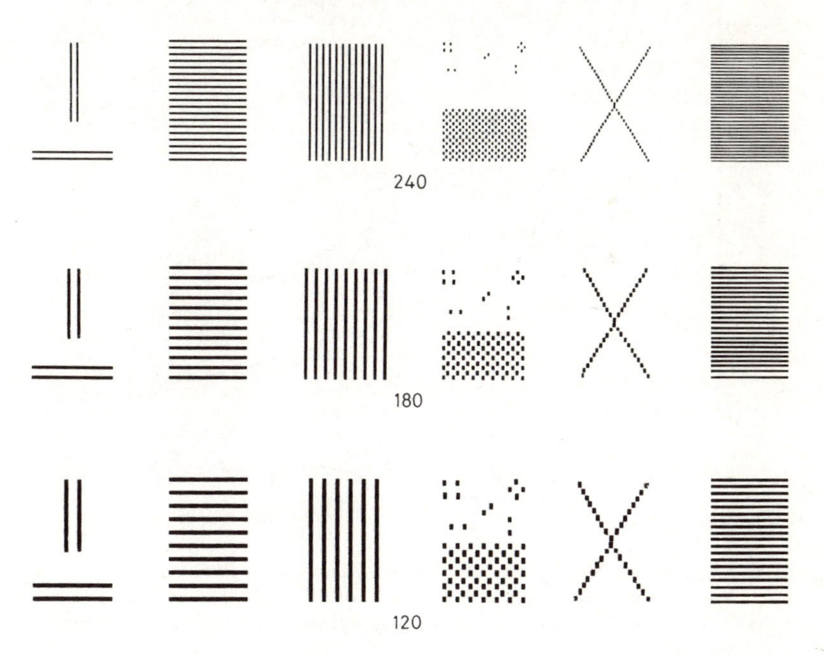

Fig. 16.3 *Form of test chart used by Engstrom in his study of television image characteristics*

Wenstrom's analysis thus showed that for continuous home television involving full theatre stage productions and outdoor spectacles a 400-line standard was required. However, his research had used still pictures and therefore it was clearly necessary that a definitive study of the problem using moving images should made.

E. W. Engstrom, of the RCA Victor company, undertook this investigation and made his results known in an important paper[18] published in December 1933. These findings were based on:

1 the resolving properties of the eye;
2 practical viewing tests of motion film specially prepared to give pictures having a
 (i) 60-line structure
 (ii) 120-line structure
 (iii) 180-line structure
 (iv) 240-line structure
 (v) normal projection print quality.

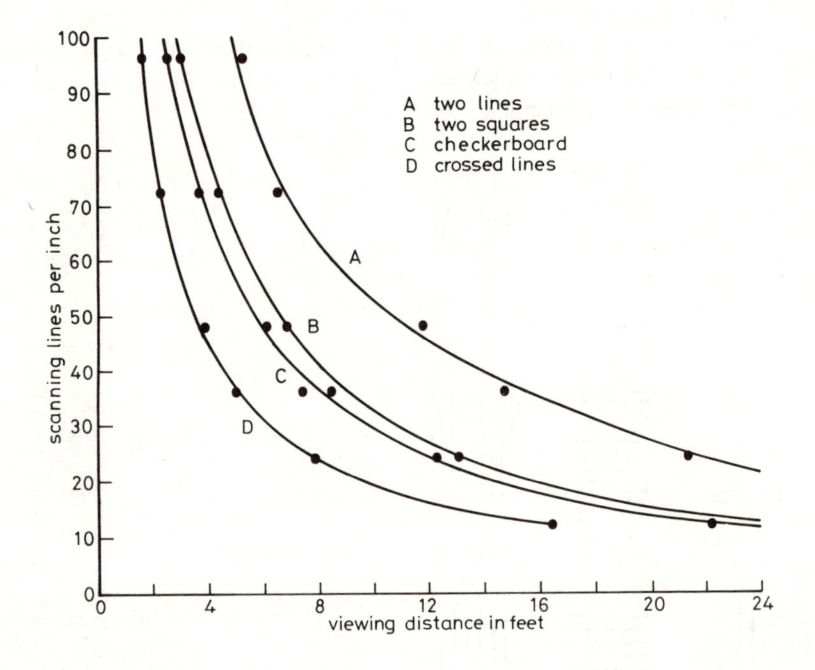

Fig. 16.4 *Experimental results of resolution tests (averages of three observers)*
For each type of 'scanning line detail' of Fig. 16.3, for example two squares, a viewing distance was chosen at which the details could just be resolved; graph B for the example given

For his elementary studies of some of the properties of vision, Engstrom used the chart shown in Fig. 16.3. This was placed at a viewing distance which just allowed, for example the two vertical lines of the pattern in the bottom left hand part of the chart (corresponding to a 120-line picture), to be resolved. By repeating this test for each of the various patterns Engstrom was able to relate the equivalent number of scanning lines per inch of picture height to the distance at which they could just be resolved for each of the different types of pattern. His results, which were based on the performance of three observers, are shown in Fig. 16.4. The theoretical curves corresponding to visual acuities (for that part of the field of view which falls on the fovea of the retina) of 0·5, 1·0 and 2·0 minutes of arc are shown superimposed on the observational data of Fig. 16.4 in Fig. 16.5. Thus, for example, an observer sitting at a distance of six feet from a television picture and having an acuity of vision of one minute of arc would be able to resolve a maximum of 48 lines per inch of screen. Consequently, a picture having a height of 9 inches, (a 15 inch screen), would require 432 lines to define it.

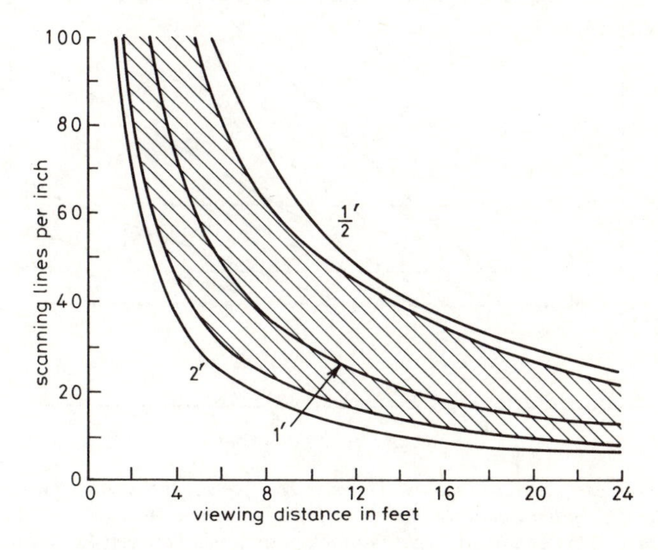

Fig. 16.5 *Theoretical curves for visual acuities of $\frac{1}{2}$, 1 and 2 minutes of arc superimposed on the observational data shown in Fig. 16.4*

In the second part of his investigation Engstrom used an ingenious technique to make cine films having a detail structure equivalent to television images. His films included:

1 head and shoulders of girls modelling hats;
2 close up, medium and distant shots of a baseball game;
3 medium and semi close-up shots of a scene in a zoo;
4 medium and distant shots of a football game;
5 animated cartoons,
6 titles.

Each of these films was made to correspond to line standards progressing from 60 to 240 lines per frame. Viewing tests were made with projected pictures of various heights, and, for pictures of a given height and line structure, observations were made for each type of subject matter on the film. In taking the observations, viewing distances were chosen at which the lines and detail structure became noticeable. At closer viewing distances the picture structure became increasingly objectionable while at the viewing distance chosen the picture detail was just satisfactory. It was noted by Engstrom that the type of picture subject did not influence the viewing distance chosen by more than 10%.

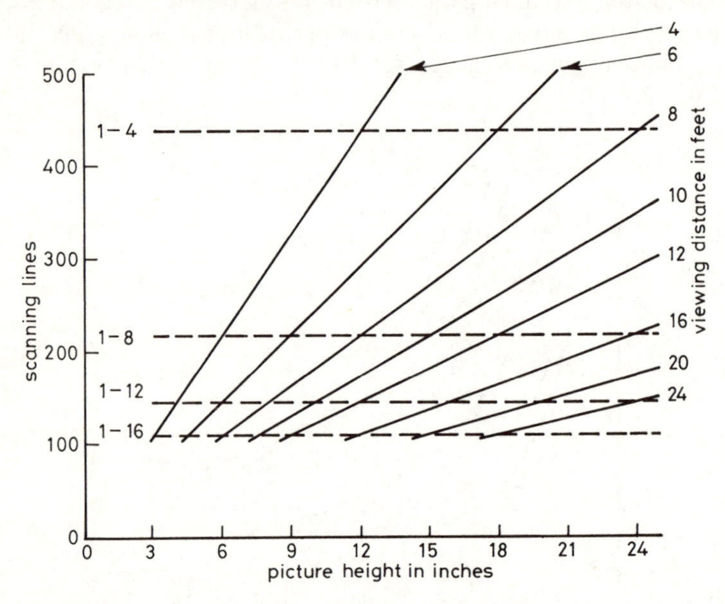

Fig. 16.6 *Relationship between scanning lines and picture size for several viewing distances. The broken lines indicate picture height to viewing distance ratios. Tests showed that viewing conditions were satisfactory for ratios between 1:4 and 1:8*

In this work Engstrom set as his standard the ability of the eye to see the elements of detail and picture structure. Another less exacting standard was the ability of images having various degrees of detail to tell the desired story. Taking as a standard the information and entertainment capabilities of 16 mm home movie film and equipment, Engstrom estimated the television images in comparison as:

60 scanning lines	entirely inadequate
120 scanning lines	hardly passable
180 scanning lines	minimum acceptable
240 scanning lines	satisfactory
360 scanning lines	excellent
480 scanning lines	equivalent for practical conditions

Engstrom's findings are summarised in Fig. 16.6 which includes all the necessary information to determine the number of scanning lines required if the viewing distance and picture height have been decided upon. His conclusions agreed effectively with those of Wenstrom, particularly with regard to the number of lines per picture needed to give excellent reproduction. Both investigators found that 360-400 lines/picture were required for this purpose.

It seems highly likely that Shoenberg's choice of 405 lines/picture for the Marconi–EMI system was influenced by the conclusions of the American workers. When the company was asked to give evidence to the Television Committee in 1934, it submitted a short memorandum, dated 7th June 1934, outlining the characteristics which would be suitable for a commercial broadcast television system. Of the eleven points listed, the only one which was later altered concerned the number of scanning lines per picture, namely 243. Blumlein told the Selsdon Committee:

> The reason for choosing 243 is that it is a multiple of 3 and would be convenient for electrical multiplication of frequency or mechanical multiplication of frequency. If an electrical signal is required at the transmitter, or electrical generation of frequency at any time, 243 is a number which allows a large number of things to be done, 241 would be a very awkward choice.[19]

Both EMI and Bairds thus had in mind, in 1934, almost identical line numbers. These were practically the same as the 240-line standard which had been investigated by Engstrom and very nearly similar to the 200 lines per picture which Wenstrom had used. However, by February 1935 Shoenberg had proposed the adoption of a 405 lines per picture characteristic.

No account seems to exist to show why this particular figure should have been adopted: however, the following points possibly had a bearing on the choice which was made:

1 Wenstrom had shown that the definition of home motion pictures corresponded to approximately 400 lines, and also that 400-line television was 'perfect for a single face, excellent for small group or detached objects, and good for full theatre stage or outdoor spectacles'.
2 Engstrom had concluded that 360 lines/picture gave an excellent image and 400 lines/picture was equivalent 'for practical conditions' to a home movie film.
3 343 lines/picture television had been demonstrated in 1934 by the Radio Corporation of America.
4 An odd number of lines per picture was required for interlacing.
5 The number of lines per picture had to be compounded from integers preferably less than ten for ease of signal generation.
6 Shoenberg was keen that a line standard should be adopted which would allow its full potentialities to be appreciated as research and development work made it possible for the system to be improved.

Taking the standard mentioned in 3 above as a base, the only odd integers from 343 to 405 (point 4) which can be factorised are as follows:

$343 = 7 \times 7 \times 7$ $\quad$ $375 = 5 \times 5 \times 5 \times 3$

$345 = 5 \times 3 \times 23$ $\quad$ $381 = 3 \times 127$

$351 = 3 \times 117$ $\quad$ $385 = 5 \times 7 \times 11$

$355 = 5 \times 71$ $\quad$ $387 = 3 \times 129$

$357 = 3 \times 7 \times 17$ $\quad$ $393 = 3 \times 131$

$363 = 3 \times 11 \times 11$ $\quad$ $395 = 5 \times 79$

$365 = 5 \times 73$ $\quad$ $399 = 3 \times 7 \times 19$

$369 = 3 \times 3 \times 41$ $\quad$ $405 = 5 \times 9 \times 9$

$371 = 7 \times 53$

Using point 5, the only numbers which can conveniently be used are 343, 375 and 405. Presumably Shoenberg did not wish to use 343 as he could have been accused of copying the RCA system. Consequently he was left with a choice between 375 and 405, and points 1, 2 and 6 favoured the larger number.

Unfortunately, Shoenberg, who died on 25th January 1963, left no published papers or books describing the achievements of the research team which he directed while at EMI. His only comments on the 405-line system were those which he made during a discussion on a paper presented to the Television Convention on 28th April 1952.

Rather surprisingly, perhaps, the minutes of the Technical Sub-committee do not contain any reference to the remarkably ambitious proposals which he put forward. Instead, the meeting seems to have been concerned with a discussion on whether certain parts of the M–EMI apparatus could be used by the Baird Company. Shoenberg said that, apart possibly from the feeder and aerial, no portion of his company's transmitting apparatus – whether radio or television – could be used in common with Baird's equipment. In the case of the aerial and feeder M–EMI were prepared, though with 'considerable reluctance', to consider common working with the competing company. Bairds, who were represented by Clayton, Church, West and Jarrard, were also prepared to consider using a common aerial with EMI 'if they were satisfied as to its efficiency for the Baird system'.

On the issue of the reception of the other company's signals, both M–EMI and Bairds said their receivers could receive both transmissions and both companies thought that a wavelength of the order of 7 m might be the most suitable.

It was decided by the sub-committee to send a questionnaire to each company 'with a view to eliciting information which would give a complete performance specification of their respective systems'. This was prepared by Ashbridge, agreed by the TSC on 19th February and despatched to the two firms in question on 20th February.

The replies to the questionnaire were discussed by the TSC, with M–EMI, at their fourth and fifth meetings on 26th February and 1st March, respectively, and with Bairds at the sixth meeting of the TSC on 8th March 1935.[20]

Ashbridge's questionnaire was searching and comprehensive and covered all the pertinent points relating to the proposed television systems.

Marconi–EMI put forward their 405-line system (with 202.5 lines per frame and 50 frames per second, giving 405 interlaced lines and with 25 complete pictures per second), for a picture having an aspect ratio of 5 (horizontal) to 4 (vertical) and scanned from left to right and from top to bottom. Baird Television Ltd. advocated a 240 lines per frame (progressively scanned), 25 frames per second picture with an aspect ratio of 4 to 3 and with the scanning taking place from right to left and from top to bottom.

Bairds recommended progressive scanning at the present stage of the art for the reason that it was not considered proved that interlaced scanning eliminated or reduced flicker, 'and even if interlaced scanning had advantages it (did) not follow that interlacing by scanning alternate lines (was) the most suitable mode of intercalation, of the many which (had) been and (were) being experimented with'. They stated that the process of interlaced scanning did not present serious technical difficulties to them but felt that the improvement in flicker, gained by this method, was not sufficient to warrant the extra complication involved at the receiver and suggested that the question should be deferred until they had been able to carry out further experimental work.

On the issue of the number of lines per frame to be used, Bairds considered the 240-line standard represented the economical and practical limit which could be recommended for standardisation for at least three years.[21] 'It (was) definitely true that the optimum definition for a picture transmitted and received by radio on a 240-line basis (had) by no means yet been demonstrated. When this (had) been done it (would) show a picture a long way ahead of what (had) been demonstrated by any system and (would) give ample definition, clearness and quality of all types of scenes as well as for close-ups' noted Bairds.

Bairds were very anxious that the number of lines should not be raised beyond 240, and when questioned as to their views on the adoption of 400 lines they stated a number of grave difficulties would be encountered both at the transmitter and receiver and indicated that all forms of mechanical scanning would have to be abandoned. The company felt that at the 'present state of the art, as known to them, a more pleasing result would be obtained by the faithful transmission of a 240-line picture rather than by transmitting a 400-line picture which would, of necessity, suffer during the process'.

They backed up this statement by giving a number of facts as a result of testing the reactions of a large number of people.

> With a bright 240-line cathode-ray tube picture 12 inch × 9 inch, with a picture frequency of 25 frames per second,

1 the lines are not seen at four feet or more;
2 at six feet or more the definition seen by the eye (was) equal to that of an actual photographic enlargement of the picture being televised – the dimensions, brightness and contrast being the same.

In addition as the number of lines was increased beyond 240:

1 the mechanical and optical problems associated with the design and operation of picture scanners for transmission became increasingly difficult;
2 the increased bandwidth required lowered the gain per stage, which meant more stages of amplification, 'with a consequential increase of interference noises';
3 the increased frequency range involved the use of a higher carrier frequency as it was 'generally accepted that the modulation frequency should not be higher than $2\frac{1}{2}$% of the carrier frequency';
4 the cost of receivers increased as more stages were required with a corresponding increase 'not only in the cost of components but in the cost of manufacture and inspection, particularly with regard to the "ganging" of the circuits'.

Also as regards the cathode-ray tube Baird felt it was already difficult to obtain in production a spot size small enough for the reproduction of a 240-line picture and furthermore, the illumination of the fluorescent screen of a cathode-ray tube decreased with increase in the scanning frequency.

A final point in the Baird memorandum concerned the lack of experience, of television problems, of most receiver manufacturers, even from an experimental point of view. In spite of this 'it was believed that standardisation on a 240-line basis would enable manufacturers generally to be in a position to produce receivers within the next nine months but there (was) no doubt that the receiving problems which (arose) with a greater number of lines than 240 would hold up the quantity production of television receivers for at least eighteen months'.

Bairds list of difficulties tended to highlight the tremendous technical advances which had been made by Marconi–EMI and particularly by Shoenberg's group. It was unlikely that the Baird company knew precisely the degree of success the Hayes company had acheived in high definition television as it was EMI's policy not to encourage publication of the results of their research work. However, Bairds knew that EMI were associated with RCA and Dr Zworykin of that company had recently published a paper on 'Television with cathode-ray tubes' in the *Journal of the Institution of Electrical Engineers*.[22] In this paper he had mentioned that some of the iconoscope tubes actually constructed had been satisfactory up to 500 lines 'with a wide margin for future improvement'.

This information, together with the ideas contained in EMI's patents and the Postmaster General's statement to Baird that he must face up to the possibility of there being a system superior to his own, should have led Bairds to form some conclusion, albeit a rather ominous one, concerning EMI's progress. This view is strengthened by the fact that Bairds never claimed that 405-line television was not possible but that it was not necessary and in any case there were serious manufacturing difficulties to be overcome with such a system which might delay the adoption of a high definition television system. Bairds said that they could design and produce such a system – even one going up to 700 lines – but in reality the film scanner and spot light scanner they proposed to use were not suited to 405-line working, and the Farnsworth electron image camera required excessive light levels. Effectively the company had to temporise and this could only be done by stressing to the Television Advisory Committee the need not to go beyond 240 lines for at least three years. However, the TAC and its Technical Sub-committee members had seen excellent television pictures at Hayes;[23] in the USA they had been shown 343-line pictures[24] and they had a very high regard for the work of Shoenberg's team. Still at this stage they were simply collecting evidence on which to base their final thoughts on the standards to be adopted; and there were other considerations in addition to the number of lines per picture.

EMI were led to consider interlacing when their cathode-ray tube receivers began giving brighter pictures and it was obvious that 25 Hz ficker would be unacceptable in practice.[25] Prior to the beginning of 1934 the company was working on a 180-line picture, with sequential scanning at 25 frames per second, using cathode-ray tubes which were not at all bright. With a poorly illuminated picture at the receiver, flicker was not very objectionable, but as the brightness of the screen improved the flicker could not be tolerated.

Persistence of vision is a *sine qua non* for any television system. Fortunately, by the time seeing by electricity became a reality, much work had been carried out on this aspect of vision and on flicker, and its characteristics were well understood.[26]

Many inventions based on the principle of persistence of vision were put forward during the Victorian era – the Daedelum or Wheel of Life, the Zoetrope, the Praxinoscope, the Viviscope, the Tachyscope, the Rudge projector and the Zoogyroscope amongst others – so that when inventors throughout the world turned their attention to the problems of combining photography with motion to achieve cinematography, in the 1885–1895 decade, the basic requirements for the solution of the problems were well known, and of course, similar applications of persistence of vision could be applied to the new science of distant vision.

Although the various contrivances mentioned above were capable of showing motion – of a sort – they also produced a sensory effect called flicker. When the retina of the eye (adapted to darkness) is suddenly exposed to a steady bright field of view, the sensation produced rises raidly to a maximum and then falls to a lower constant value. With the removal of the stimulus, the sensation does not disappear immediately but takes a finite time to decay below the limit of perception. The extent of the overshoot increases, and the time taken to settle down to the final value decreases with increasing intensity of the retinal illumination. Consequently, if the eye is exposed to a source of rapidly varying intensity, the effects of the definite rates of growth and decay of the sensory response (or the persistence of vision) may prevent flicker from being noticeable. Flicker is absent when the luminous variations are regular in nature and have a frequency above the critical flicker frequency.

William Henry Fox Talbot, one of the earliest of the English contributors to the science of photography, discovered the flicker effect and law which characterises it, namely, if the frequency of a varying light source is sufficiently high for flicker to be imperceptible the eye is able to integrate the brightness over the cycle of variations. The effect is as if the light for each cycle were uniformly distributed over the period of the cycle.

The highest frequency at which flicker can just be detected – the critical flicker frequency – is nearly a linear function of the logarithm of the brightness of the field (over the range appropriate for television), and the sensitivity of the eye to flicker is noticeably increased when the field of view is enlarged from a few degrees to an image of the size encountered in cinemas. The sensitivity to flicker is also greater for averted vision when viewing large fields of varying brightness. When the illumination is raised an increased flash rate is required to give the subjective impression of steady illumination.

Flicker is of importance not only in television reproduction but also in motion picture projection, and, as the implementation of the latter art considerably pre-dated the former one, some knowledge was available, on picture frame rates for flicker free viewing, when television advanced to the stage of a public broadcast service.

Edison seems to have fully appreciated the need to run cine films at a high speed through a projector for flickerless reproduction, for in his kinetoscope the images sped past the peephole at the rate of 48 frames per second.[27] This was a machine, developed in 1890, approximately four feet high and two feet wide in which the film ran with a continuous movement between a magnifying lens and a light source. A revolving shutter made it possible to view each individual frame momentarily and the capacity of the machine was limited to 50 ft of film – the length of the tables on which the film was manufactured. Consequently the combination of high picture rate and short film length limited the duration of the peepshow to a mere 13 seconds.

Some time later the picture rate was reduced to 16 frames per second, presumably to conserve film and reduce costs, and this rate remained roughly the norm throughout the era of silent films (with sound films the speed is 24 frames per second). However, the use of the above picture frequency and a single bladed shutter gave rise to an appreciable flicker effect and means had to be found to maintain the picture rate with a higher viewing frequency. The solution was to adopt a three-bladed shutter which gave three light–dark changes during the projection of each film frame, Fig. 16.7 and thus a viewing picture frequency of 48 flashes per second. A reasonably flicker-free image was obtained under normal circumstances, but, for many years after, the colloquial term 'the flicks' remained in popular usuage as a name for motion pictures. With the introduction of the 'talkies' and the adoption of 24 pictures per second, only a two-bladed shutter was necessary to give the same flicker-free frequency, Fig. 16.7.

In television the early workers were limited to the utilisation of a channel bandwidth of the order of 10 kHz, and, as the bandwidth for television reproduction is proportional to the square of the number of lines employed and directly proportional to the picture scanning rate, a compromise had to be decided upon between the need to use a high picture rate for flicker-free viewing and a large number of lines for good definition. Baird adopted a 30 lines per picture, 12.5 pictures per second standard, but naturally this gave rise to some image flicker. The Gramophone company also used the same picture rate in 1931, but when cathode-ray tubes began to be made and used for television reproduction in the early 1930's, capable of giving bright pictures, the low

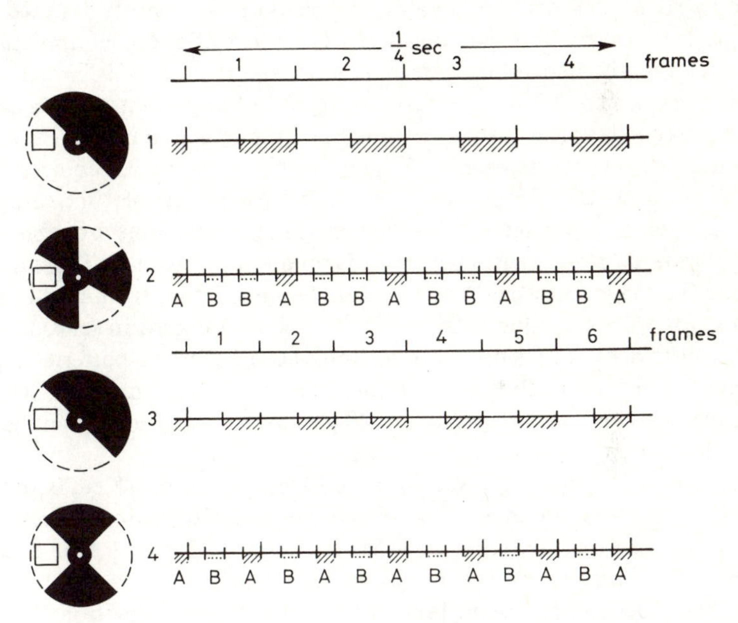

Fig. 16.7 *In early 16 frames per second cinematography the simple single-bladed shutter (1) used in the camera gave rise to a flickering effect on the projector screen. This effect was reduced by employing a 3-bladed shutter (2) and by moving the film in the projector during the dark intervals (marked A). With 24 frames per second cinematography a 2-bladed shutter (4) and an intermittent film motion (during the periods marked A) gave the same flicker frequency as in (2)*

picture rate was found to be unsatisfactory for prolonged viewing. Television engineers increased the rate to 25 pictures per second (or 30 pictures per second in the USA), and, in 1934, Electric and Musical Industries Limited proposed, to the television Committee, the adoption of interlaced scanning. This suggestion enabled the frame rate to be increased to 50 frames per second, while the picture rate was maintained at 25 pictures per second, and paralleled the practice which had been adopted in the motion film industry. However, there is an important difference between the projection of a cine film and the reproduction of a television image. In the former case the whole of the picture is shown at any instant of film projection, whereas in the latter the image is built up line by line. The employment of the technique used in the film industry to achieve a high frame rate clearly could not be used and so the principle of interlacing was adopted.

The wisdom of this choice was confirmed in April 1935 when Engstrom (of the RCA Victor Company) published his second paper on 'A study of television image characteristics' which dealt particularly with the determination of the picture frequency for television in terms of flicker characteristics.[28] Engstrom's investigations were comprehensive: he made measurements using a sector flicker disc and also a kinescope; he carried out tests on progressively scanned rasters and interlaced rasters and he considered the effect of room illumination, the persistence characteristic of the screen material and the spectral characteristic of the kinescope screen on his observations. These indicated that, when a cathode-ray tube screen was progressively scanned and the screen illumination was only one foot-candle, a picture rate of '38 pictures per second was required for just noticeable flicker, and 35 pictures per second for noticeable but satisfactory flicker: 28 pictures per second resulted in disagreeably objectionable flicker. Thus a standard of 24 pictures per second could not be justified. These data also indicated that 48 pictures per second would be satisfactory from the standpoint of flicker for values of illumination likely to be encounterd in television.'

For a system of television using an interlaced scanning pattern Engstrom concluded 'Satisfactory flicker conditions exist if each picture consists of two groups of alternate lines (equivalent to 48 pictures per second)' when the actual rate is 24 pictures per second.

In another RCA paper published in June 1936, the authors (Kell, Bedford and Trainer) found that an integer ratio between the alternating current mains frequency and the picture frequency was very desirable for progressive scanning and was almost imperative for interlaced scanning.[29]

Thus Shoenberg's decision in June 1934 to offer to the Television Committee a television system having interlaced scanning at a picture frequency of 25 pictures per second and a frame frequency of 50 frames per second was completely vindicated by later research. Since the commencment of the first, public, high definition service in the United Kingdom in November 1936, no change has yet been made in either the picture rate or type of scanning employed.

The scanning system which EMI adopted was that devised by R. C. Ballard of the Radio Corporation of America and patented by him in 1933 (patent 420, 391).

As an interesting aside it is pertinent to note the different attitudes of EMI and RCA to research publications. The Radio Corporation of America, like Bell Laboratories, was quite open with regard to its television interests and in an important series of papers published in the *Proceedings of the Institute of Radio*

Engineers from the early 1930s the company made its general television research activities and results available to the world. EMI's development plans, on the other hand, were kept strictly secret (except to members of the Selsdon Committee), and it was company policy that no papers or publications should be prepared or given by members of Shoenberg's team. Later the company agreed to the publication of some valuable papers, on its television work, in the *Journal of the Institution of Electrical Engineers.* These appeared in 1938, that is, after

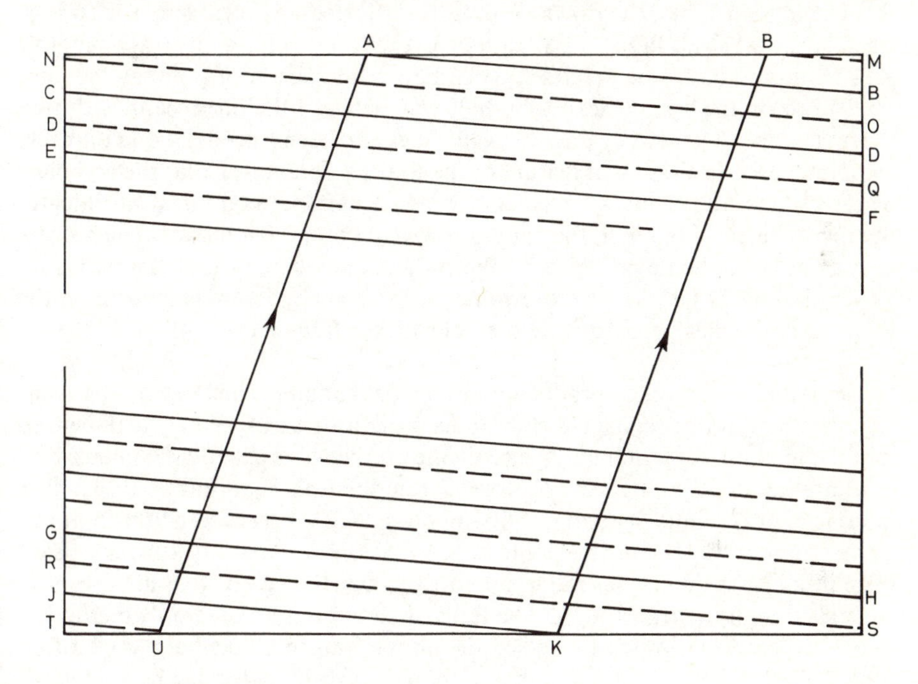

Fig. 16.8 *Interlaced frames of the 405-line picture*

the Television Advisory Committee had made a decision in EMI's favour. One outcome of this policy was a general lack of knowledge about EMI's contribution to television and in 1952 Shoenberg felt the need to comment on an IEE paper on 'The history of television' as he thought the authors had 'perhaps not done full justice to the pioneer work carried out in Great Britain'. Another possible effect of the policy was the erroneously held view that the M–EMI system was primarily based on that of the RCA Company. The similarity of the iconoscope and emitron tubes and the use by the Hayes company of the Ballard patent on interlacing did nothing to dispel this opinion.

In the EMI version of the Ballard system the picture was completely scanned in 1/25 second by means of two downward traversals or frames, each of 1/50 second duration, and the lines constituting one such frame were arranged to lie between the lines of the other frame so as to give good picture detail. In Fig. 16.8

the lines show the track of the scanning spot, which moved under the influence of a regular downward motion (frame scan), with a rapid return and a regular left to right motion (line scan), with a very rapid flyback.

Assuming that the commencement of one frame scan was at A the combination of the above motions caused the scanning spot to trace lines AB, CD, EF, . . . JK, when at point K the return stroke of the frame motion began and returned the spot to L at the top of the frame. $202\frac{1}{2}$ lines were scanned during this frame scan. Then the downward motion started again causing the spot to trace lines LM, NB, PQ . . . TV until at U the frame scan returned the spot to A. Because point L was arranged to be half a line ahead of point A the lines traced out on the second scan were half way between the lines scanned during the first frame. After two frame scans 405 lines had been traced out and the cycle recommenced. The essential feature of the Ballard patent was that the complete picture was scanned in two frames and as each frame contained an integral number of lines, plus a half, the two frames were bound to interlace. The process is analogous to the device used in film projection, whereby it is arranged that although only 24 pictures are shown per second, each picture is thrown on the screen 2 or 3 times so as to raise the light flicker frequency to 48 or 72 flashes per second.

The Ballard method of producing interlaced scanning[30] has the great advantage that both frame scans (which produce a picture) are absolutely identical 'so that the method of scanning at the receiving end remains exactly the same as for non-interlaced (scanning) and the receiver is blissfully ignorant as to whether interlaced or ordinary scanning is taking place.'[19] EMI considered this point to be very important, from the point of view of the universality of a television system, as it was most desirable not to tie down the receiver to any peculiar construction of the transmitter. The Ballard patent covered the only method of interlaced scanning which possessed the above feature. Shoenberg knew of no other: 'And if I knew of any other I should have patented it', he told Lord Selsdon.[31] 'I might also tell you that we have a pretty large patent department that watches all patents which have any relevance to our business and that no piece of apparatus can be built by the company in any way whatsoever without the approval of that patent department.'

The Ballard patent was certainly a master patent and EMI was fortunate in being able to use it. The utilisation of interlacing to diminish flicker was one of the three most important aspects of the M–EMI television system: the others were the transmission of the synchronising signals in the same channel as the picture signal and d.c. working.

Ballard's patent on interlacing was not the first on the subject. The terms interlacing and intercalation were introduced by Latour and Baird, respectively, simultaneously in 1926, and a form of this type of non-sequential scanning dates from 1914. Several inventors, including Hart, Stephenson, Walton, Latour, Baird, von Ardenne, and the Telefunken Company had proposed systems of non-sequential scanning for various purposes prior to the publication of Ballard's patent. The dates of Baird's patents show that by 1934 he had given some considerable thought to the problem of intercalation but when the Baird Company presented its evidence to the Television Committee it felt that the improvement in the quality of the picture gained by the use of interlacing was not sufficient to warrant tbe extra complication and suggested that the issue should be deferred until the company had been able to undertake further experimental work.

In any scene which is being televised by a television camera/scanner the camera/photocell tube generates signals having a value corresponding to the brightness of the various elemental areas which comprise the scene. The output from the camera/photocell tube is therefore unidirectional; albeit a fluctuating current above a certain zero datum line (corresponding to a completely black scene). This fluctuating, unidirectional current may be analysed into an alternating component which depends on the relative brightness values of the different parts of the picture and a direct component which conveys information on the brightness of the whole of the picture area. Thus the signals resulting from the televising of a completely black sheet of paper and a brightly illuminated white sheet of paper are the same except for the magnitude of the direct component in the two cases (assuming that the sheets of paper completely fill the fields of view).

The EMI team thought[31] it was essential to transmit the direct component so that the receiver would accurately reproduce the brightness values of the original picture although the practice in the United States of America was to use the alternating component only.[32] This practice possibly arose because of the difficulty in designing satisfactory stable high gain d.c. amplifiers. The consequence was that the resultant vision signal varied about a mean value rather than about a datum value which represented black. Now with an a.c. television signal the amplitude of the signal corresponding to black relative to the mean signal is a variable quantity depending on the nature of the picture. Thus the signal due to a white dot on a black background and that due to a black dot on a white background are as shown in Fig. 16.9 and from these it is clear that in the first case the signal corresponding to black is close to the mean line whereas in the latter case this signal is remote from the mean line. A change in the nature of the picture thus causes a wander of the amplitude of the black signal so that as the d.c. component has been removed the valves of an amplifier require to be biased in such a way that they can handle not only the apparent maximum amplitude (in the above example 10 V positive and negative) but absolute values in excess of the difference between black and white (17 V positive and 10 V negative using the same example). Indeed, a television waveform devoid of its direct component can drift about its datum line by approximately 60% of its amplitude. Such a signal requires larger and more powerful valves to handle it and the apparatus becomes unnecessarily costly, or, for a given available peak transmitter power, much smaller useful signals are obtained at the receiving point.

At the receiver it is necessary to separate the vision signals from the synchronising signals so that the latter may be used to control the scanning. Various circuits can be arranged to transmit signals beyond a certain amplitude in the black direction and hence achieve separation of the two signals, but if the black level is liable to wander at the separator it is necessary to employ large synchronising signals so that no possible amount of wander will cause either vision signals to be passed by the separator or synchronising signals to be missed by the separator. The size of the synchronising signal required at the receiver may be greatly reduced by ensuring that the signal at the separator contains a direct component, i.e. a signal having a definite amplitude for black irrespective of the nature of the picture. The direct signal may be obtained either by transmitting the direct signal or by converting the alternating signal to a direct signal by 'd.c. re-establishment' at the receiver.[25]

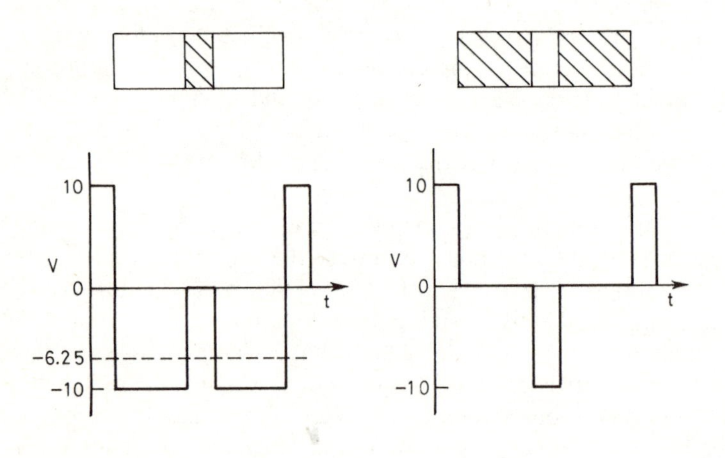

Fig. 16.9 *A black spot on a white background and a white spot on a black background lead to the production of line signals which have differing mean levels*

Shoenberg has written:[33] 'A great controversy raged for some months among my staff as to whether we should use d.c. amplification or a.c. amplification with the d.c. component. In the early days P. W. Willans was the protagonist of the restoration method and later on Blumlein advocated it.' Willans had been with the Marconi Wireless Telegraph Company and was probably familiar with their work on the transmission of synthetic half-tone pictures by d.c. keying the radio transmitter. It was Willans who proposed the solution,[34] in 1933 (patent 422, 906), to the problem of d.c. re-insertion following Shoenberg's decision to adopt d.c. restoration. In 1952 Shoenberg noted: 'It seems strange now that this matter should ever have been one for controversy, but at the time there were cogent arguments on both sides and in the end I made the decision in favour of restoration.' This decision meant that definite carrier values represented black, white and the synchronising signal. There was no mean carrier value as this depended upon the picture brightness and so the system of transmission was analogous to telegraphy rather than telephony.

In addition to stressing the advantage of transmitting the d.c. component in order to give a true picture, Shoenberg and Blumlein mentioned the gain in effective radiated power from the transmitter. Blumlein in particular gave the Television Committee an indication how this came about:[31]

> If one is not employing d.c. working one requires synchronising signals which are approximately twice the picture signals and there is a further requirement of extra characteristic room to allow for the wander of the values of the synchronising pulse and the values representing full white. That gives a total possible amplitude of the wave which is 3·9 times the useful picture amplitude. By using d.c. working we can work with synchronising pulses which are of the order of half or possibly one-third of the picture signal, and there is no wander whatsoever. A fair estimate would put the total amplitude required as 1·5 times the useful picture amplitude. That is a comparison between 3·9 and 1·5. The power required to handle this amplitude range is proportional to the square of the amplitude range. But there is a further advantage in d.c. working; we can afford to work over the curves of the transmitter characteristics, which are normally unusable, with synchronising signals because they are square top waves, and a distorted square top wave is still a square top wave. We can therefore use our transmitter very efficiently without working to keep exactly on the straight line portion. We use the last dregs of the transmitter right down to zero.

The result was, Blumlein stated, that the gain in effective radiated power was approximately eight by using d.c. working, although Shoenberg interposed 'ten or twelve times'. It was possibly because of this feature of the M–EMI transmitter that Shoenberg was so adamant about not allowing it to be shared with the Baird Television Company.[9]

It is interesting to note that Bedford and Puckle, of the Cossor Company, in the early stages of the development of their velocity modulation television system arranged their apparatus for d.c. transmission 'but found it of no value'.[35]

Both M–EMI and Bairds were in fairly close agreement on the wavelength to be adopted for the television transmitter, namely, 6·5 to 7·0 m and 6·0 m, respectively. Davis, the transmitter expert of MWT, considered 7·0 m was the longest wavelength which could be employed without the occurrence of multiple reflections during the worst sun-spot conditions.[36] Also the company advocated the use of the longest possible wavelength in the interests of easier operating conditions for the transmitting valves. Bairds confirmed this view when they mentioned that the efficiency of a transmitter decreased very rapidly with increasing frequency and they knew of no transmitting valves which worked efficiently below 5 m.[37] Consequently, as the carrier frequency was increased the effective range of reception of signals decreased for a given power supplied to the transmitter and also for a given power in the aerial, since the range was 'increasingly influenced by ground contours, earth curvature, steel framed and ordinary buildings, trees and so on'. They had standardised on 7·0 m as the best compromise for television transmission on 180 lines and between 6·0 and 6·5 m for 240-line transmission.[38] 3·0 m was required for the transmission of a 400-line picture in their view.

On the question of the wavelength to be adopted for the associated sound transmitter, M–EMI said the carrier frequency of this transmitter should be 4 MHz lower than the carrier frequency of the vision transmitter so that there would be no possibility of the sidebands of the two transmitters interfering.[36] An advantage in having the sound carrier reasonably close to the vision carrier was that a single receiving aerial could have a sufficiently flat frequency response to receive both transmissions simultaneously. Blumlein thought that the spacing could be reduced to 3 MHz if necessary.

Baird Television Ltd. recommended a wavelength of the sound carrier 'in the region of 7 or 8 m' (42·8–37·5 MHz), and said that the choice of a wavelength higher than than for the vision carrier signal was governed purely by the fact that it was less difficult to operate a transmitter at the lower frequency.

For the radiation of these wavelengths the two companies both preferred to use vertical aerials, thereby giving vertically polarised radiation. In the discussion on this issue Angwin enquired why M–EMI favoured the use of vertically polarised waves for transmission and asked whether the noise level due to automobile ignition was larger than that for horizontally polarised radiation.[36] Blumlein replied that although they had not obtained conclusive experimental data in connection with this point, they had found that the increase in noise level resulting from the use of a vertical aerial as against a horizontal aerial was not appreciable. On the other hand, he submitted that over flat country a better signal strength was obtained near the ground level when vertically polarised waves were used. This was apparently owing to earth reflection phenomena, which, of course, would be less noticeable in hilly districts. He further submitted that from the practical point of view it was probably much easier for the average listener to erect a vertical aerial, the more so on account of directional effects associated with some types of horizontal aerials. Shoenberg added that the EMI Company had carried out quite a lot of research work on simple aerial arrays suitable for use by the listener. Ashbridge thought that this was a 'point of extreme importance as most listeners appeared to show considerable reluctance in putting up outside aerials at all so that it was desirable that the erection of the aerial should be as far as possible a cut and dried job'.

On the question of the ability of people living on the lower floors of large blocks of flats being able to receive television transmissions, Shoenberg suggested that it would be extremely easy for the owner of the flats to erect a suitable array on the roof and then arrange for the signal to be amplified before distributing the output of the amplifier to the various flats by means of concentric cable. He explained that this could be done without much difficulty because only one wavelength would be employed for television and consequently the problems would be considerably less than those involved in ordinary sound broadcasting where many wavelengths were employed. In his discussion with the TSC Shoenberg made several references to the need to allocate a fixed wavelength to each district and hence have sets which could be permanently tuned at the factory.[2] The point raised in connection with aerials for large blocks of flats and the method of transmitting the signals in them strengthened Shoenberg's argument for fixed tuning.

An additional point which Shoenberg stressed was that 'it would be of little use for listeners to erect an aperiodic aerial' for the reception of television because the interference picked up on it would be of a very serious nature. In their experiments at Hayes, EMI found that it was necessary to employ 'a proper aerial array' in conjunction with a correctly terminated concentric feeder, and in some cases with a noise rejector in the feeder.

Bairds also recommended that vertical polarisation should be used and based their view on the considerable amount of experimental work they had carried out on both horizontal and vertical polarisation. They found, in common with EMI that the decreased interference obtained with horizontally polarised waves was not sufficient to justify the inconvenience of having a horizontal aerial with its proper directional feeders at the receiving end.

Aerial arrays were not favoured by Bairds when the transmitting site was reasonably central in the district to be covered, although they thought it might be desirable to design a particular array if the site chosen for the television station was outside the area to be covered. EMI preferred to use a Franklin vertically stacked array and agreed 'as a concession' that they would permit Bairds to employ their aerial array although the patents covering the use of this type of array were controlled by themselves. The maximum power to be radiated from this array which was considered to be economically feasible was between 12 to 24 kW peak for the vision signal, with a preferred value of 18 kW. Shoenberg felt that an increase in power above this figure would not appreciable extend the range of the transmitter, but as Ashbridge pointed out, while this might be so the signal strength within the effective area of the station would certainly be increased, a fact which might be of considerable importance in places where bad screening due to surrounding buildings and the like existed.[36] Bairds favoured a radiated power of 16 kW peak power from the aerial for the vision signal and 2 to 3 kW output power for the sound transmitter. The latter value agreed with EMI's recommendation.

At the 1952 Television Convention held at the Institution of Electrical Engineers, London, Shoenberg commented on a paper by Garratt and Mumford on the history of television, as follows:

> I think the authors, in their efforts to present an unbiased account of the

history of television, have perhaps not done full justice to the pioneer work carried out in Great Britain.

Although it may now seem to those with present day knowledge that the main features of the British television system developed by EMI are so self-evident that there could not have been much difficulty in choosing them, the picture becomes very different if it is examined in the light of the knowedge which existed when the system was being worked out.

In deciding the basic features of our system we frequently had to make a choice between a comparatively easy path leading to a mediocre result and a more difficult one which, if successful, held the promise of better things. Perhaps a few examples may be of interest and serve to illustrate this point.

When we started our work in 1931, the mechanically scanned receiver was the only type available and was under intensive development. Believing that this development could never lead to a standard of definition which would be accepted for a satisfactory public service, we decided to turn our backs on the mechanical receiver and to put our efforts into electronic scanning. The cathode-ray tube available used gas focusing and it was not possible to modulate the beam without losing focus. This fact did not deter one competent British organisation from attempting to make use of the tube, but, remembering the vagaries and instability of the soft valves of the 1912–16 period, I decided against the soft cathode-ray tube and directed our research towards the development of a hard type with electron focusing.

The efforts in this direction, particularly by Dr Broadway and his group, turned what was thought a very speculative development in 1931 into the common place to today.

Another problem which arose from our efforts to provide pictures of greater definition was the amplification and radiation of the greatly extended range of frequencies required. On the radiation side this involved abandoning the medium waves altogether and accepting the greatly reduced service area of the short wave transmitter.

Many thought that this limitation of the service area would be unacceptable but we saw no alternative. The transmission of a wide range of sidebands on short waves raised many new problems which had to be faced on the video side when one attempted to amplify wideband signals up to a level suitable for modulation of the transmitter. The amplification of the a.c. components alone would not have been so difficult and this was the practice in the United States at that time. We, however, thought it essential to transmit the d.c. component of the picture signals so that the receiver would accurately reproduce the brightness of the original picture.

Another difficult decision had to be faced in connection with the generation of the picture signals. In our early work we used mechanical scanners to generate our signals, but the limitation of these scanners – particularly

> in regard to line pick-up – were only too obvious. Dr McGee's group started work on the emitron pick-up tube in 1932 but the signals obtained from the early emitrons tended to become submerged in great waves of spurious signals associated with the secondary emission effects. There was a great temptation to continue with the mechanical scanner – as indeed another company elected to do. Instead, we decided that the potentialities of the electron scanning tube justified a great effort to overcome the problems it presented at that time. Improvements to the tube made these spurious signals more manageable, but although Blumlein and McGee invented at about this time the ultimate solution to the problem – namely cathode potential stabilisation – time did not allow it to be worked out in practice and we had to leave it to the circuit people to deal with the unwanted signal components. Great credit is due to Blumlein, Browne and White for the resourcefulness with which they devised tilt, bend and suppression circuits to combat this evil.
>
> Once the signals had become usable the way was open for a real high definition system. The choice as to the number of lines was no longer limited by mechanical considerations but by the bandwidths which could be dealt with at the time in the transmitter and receivers. Great differences of opinion existed in the laboratory, but finally, early in 1935, I took my courage in both hands and chose 405 lines. This may seem a low standard now, but at the time when I made the decision there were many who thought I had taken a great risk.
>
> At the beginning of 1935, I was thus able to submit to the Television Committee a fully detailed specification for a high definition system operating on 405 lines. This specification was subsequently adopted, with out any important modification in the Alexandra Palace transmitter.

M–EMI's ambitious proposals, while upsetting IMK Syndicate, Scophony and Bairds, were quickly taken up by the Chairman of the Technical Sub-committee, Sir Frank Smith. He thought it would be undesirable, if it could be avoided, that a standard should be fixed for the London Station which was inferior to the best that M–EMI were in a position to offer.

By the 14th meeting of the Technical Sub-committee held on 30th April 1935 some progress had been made in specifying some of the parameters of the system. The Secretary was instructed to confirm officially a statement which had been given orally to the Baird and M–EMI companies that the mean carrier frequency would be of the order of 43000 kHz and that the bandwidth should not exceed approximately 1·5 MHz, i.e. for each sideband.[39] This was a start but there still remained the problem of bringing the two companies together so that they could exchange technical information about each others' systems. This was necessary in order that a receiver could be designed which would be acceptable to either standard. Shoenberg made it clear that he preferred not to meet Bairds' technical representatives until after written particulars of the transmission characteristics of the two systems had been exchanged.[40] For his part Shoenberg was furnishing the TSC with his information but Bairds were procrastinating.

At the 30th April meeting it was agreed that:[39]

1 the information supplied by each of the two companies should be sent to the other company by post at once;
2 two representatives from each company should be invited to attend a Technical Sub-committee meeting . . . to discuss and clear up any doubtful points;
3 each of the companies should be asked to supply a receiver adapted to receive both types of transmissions for demonstration at the NPL on the afternoon of the 10th May;
4 the companies should be informed that the committee proposed to discuss at the meeting on the 7th May the possibility of reaching agreement on the adoption of a common number of picture lines for both systems of an order intermediate betweem the standards which the companies had proposed for their respective systems.

The last point had been raised by Angwin.[41] He expressed the view that as the sideband radiated could not in practice exceed 1·5 MHz, there was no advantage to be gained in fixing a transmission standard above double 150 lines, interlaced, 50 frames per second. Furthermore, there was the possibility such a standard might be attained by Cossors and even by Scophony, but the general feeling of the TAC was that if Bairds consented, without pressure, to adopt a standard of the order of double 200 lines, interlaced, 50 frames per second, that would be the best solution of the problem: otherwise there was much to be said for the imposition of Angwin's proposal. Captain West said that if it was thought to be in the general interest to adopt a common standard Bairds would be prepared to consider the suggestion favourably.[42] Shoenberg did not agree: he felt that each company should be allowed freedom to give the best service they could during the experimental period. This opinion was to prevail and the adoption of the two standard system was given in a press notice issued on the morning of the 7th June.[43]

On one point the two competing companies were in accord: they both preferred a side bandwidth of 2 MHz. Shoenberg in particular, for obvious reasons, was keen to have a large sideband. The sideband also determined the minimum separation between the sound and vision carriers. This had been provisionally fixed at 3 MHz but in a note[44] to the TAC Shoenberg recommended very strongly that the separation should be increased to 4 MHz as he felt that the smallness of the proposed separation would limit very severely the future development of television. With a 405-line system the sidebands of the vision signal could extend to 2·5 MHz on either side of the carrier frequency, but in addition there was the further difficulty that the phase requirements in the television system made it impracticable to operate with sharp bandpass circuits.

Consequently, while EMI did not contemplate that their transmitter would ever radiate much energy at 3 MHz from the carrier, they opined that for

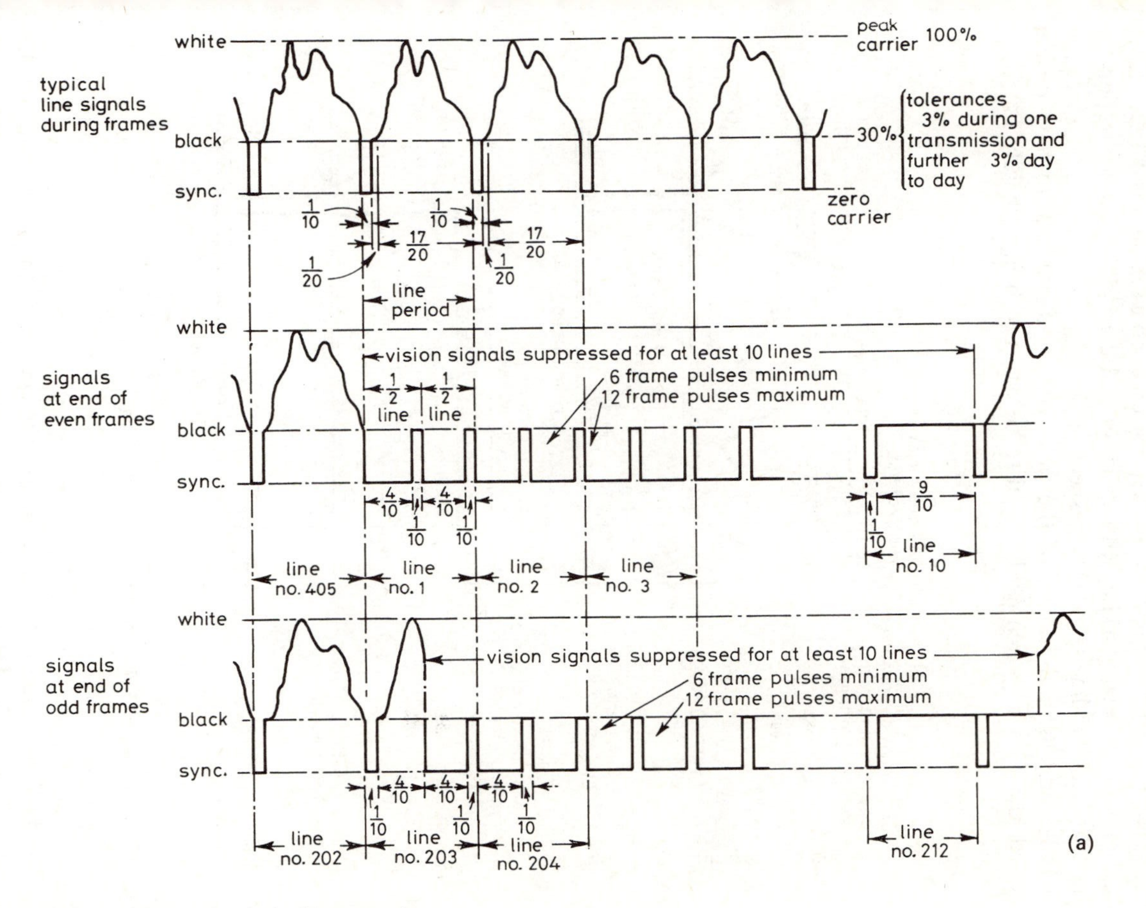

Fig. 16.10 *a) Marconi–EMI Television Ltd. video waveform*

reasonably clear sound reception the interfering vision sideband should be kept at a very low level indeed as compared with the sound carrier. The Technical Sub-committee agreed with Shoenberg's argument and increased the spacing to 3·5 MHz, notwithstanding Cossor's expression of concern on this matter[45] (they had previously been working with spacings of 1·2 and 2·0 MHz separation for the 240- and 405-line standards, respectively).

Following the publication of the press notice, invitations to tender were issued, the date by which tenders had to be returned being 4th July 1935.[45]

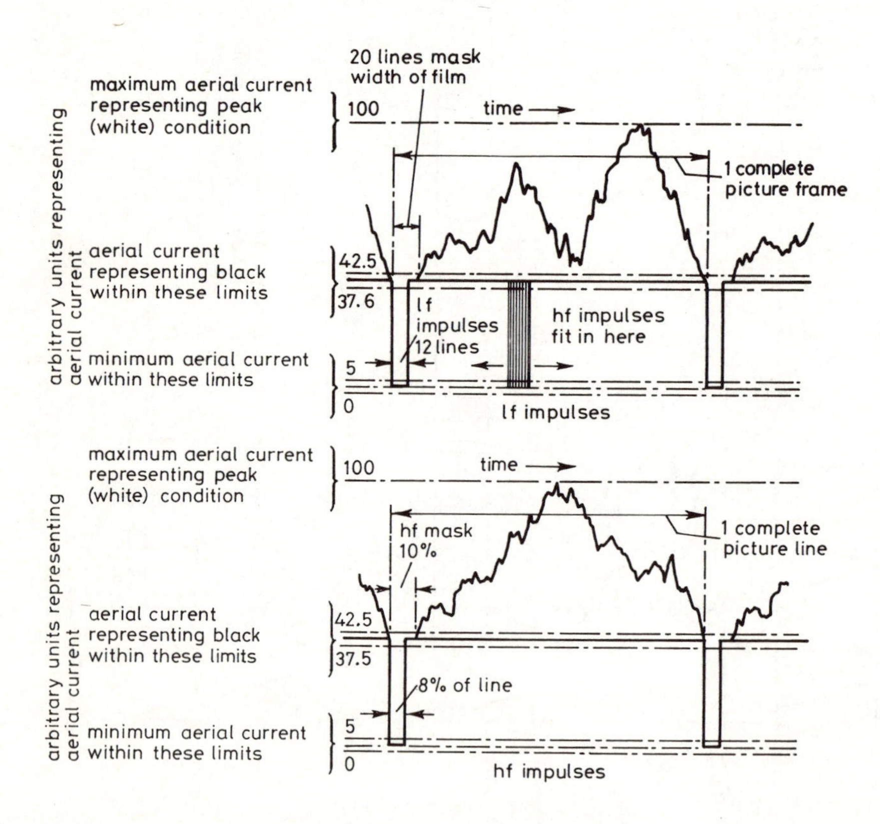

Fig. 16.10 *b) Baird Television Ltd. video waveform*

During their weighty deliberations on the standards to be adopted for the London Station, the Television Advisory Committee gave some thought to the discontinuance of the BBC's low definition television transmissions, and, in its Third Report to the Postmaster General, proposed that these should terminate on 10th August 1935. The last 30-line transmission was broadcast on 15th September 1935.

References

1 Television Advisory Committee: minutes of the 1st meeting, 5th February 1935, Post Office bundle 5536.
2 Television Advisory Committee: minutes of the 5th meeting, 5th March 1935, Post Office bundle 5536.
3 ASHBRIDGE, N.: 'Summary of information relative to suggested sites for London high definition television transmitter', Television Advisory Committee, 5th March 1935, Post Office bundle 5536.
4 Television Advisory Committee: 'Report to the Postmaster General', 12th March 1935, Post Office bundle 5536.
5 Television Advisory Committee: minutes of the 6th meeting, 12th March 1935, Post Office bundle 5536.
6 ASHBRIDGE, N.: 'Address on the occasion of the 8th annual meeting', *Journal of the Television Society*, 1936, **2**, part VI.
7 Television Advisory Committee: minutes of the 9th meeting, 1st April 1935, Post Office bundle 5536.
8 Technical Sub-committee: minutes of the 1st meeting, 11th February 1935, minute Post 33/5533.
9 Technical Sub-committee: minutes of the 2nd meeting, 15th February 1935, minute Post 33/5533.
10 SHOENBERG, I.: memorandum to the Technical Sub-committee, meeting held on 15th February 1935, minute Post 33/5533.
11 McGEE, J. D.: 'The life and work of Sir Isaac Shoenberg 1880-1963', *Royal Television Society Journal*, 1971, **13**, No.9 May/June.
12 PRESTON, S. J.: 'The birth of a high definition television system', *Television Society Journal*, 1953, **7**, July/September.
13 WHITE, E. L.: 'Modulation frequencies in visual transmission', *Proc IRE*, 1935, **21**, No.1, January, pp. 51–55.
14 Anon.: a report, *Nature*, 1st December 1928, p. 853.
15 WEINBERGER, J., SMITH, T. A., and RODWIN, G.: The selection of standards for commercial radio television', *Proc. IRE*, 1929, **17**, No. 9, September, pp. 1584–1594.
16 GANNETT, D. K.: 'Quality of television images', *Bell Laboratory Record*, 1931, **8**, April, pp. 358–362.
17 WENSTROM, W. H.: 'Notes on television definition', *Proc. IRE*, 1933, **21**, No. 9, September, pp. 1317–1327.
18 ENGSTROM, E. W.: 'A study of television image characteristics', *Proc. IRE*, 1933, **21**, No. 12, December, pp. 1631–1651.
19 Television Committee: evidence of Marconi–EMI Television Co. Ltd., 8th June 1934, minute Post 33/4682.
20 Technical Sub-committee: minutes of the 4th, 5th and 6th meetings held on 26th February 1935, 1st March 1935 and 8th March 1935, minute Post 33/5533.
21 Baird Television Ltd.: 'Additional information desired by the Advisory Committee', Technical Sub-committee, 8th March 1935, minute Post 33/5533.
22 ZWORYKIN, V. K.: 'Television with cathode-ray tubes', *J. IEE*, 1933, **73**, pp. 437–451.
23 Television Committee: 'Description of television systems examined by the Committee in Great Britain', Report of the Television Committee, appendix IV, minute Post 33/4682.
24 Television Committee: 'Systems of television demonstrated in the U.S.A.', Report of the Television Committee, appendix III A, minute Post 33/4682.
25 BLUMLEIN, A. D.: 'The transmitted waveform', *J. IEE*, 1938, **83**, pp. 758–766.
26 CORK, O.: 'Movement in two dimensions' (Hutchinson, 1963), Chapter 8, pp. 121–136.
27 Article on motion pictures, *Encyclopaedia Brittanica*, **15**, 1963, p. 851.

28 ENGSTROM, E. W.: 'Determination of frame frequency for television in terms of flicker characteristics', *Proc. IRE*, 1935, **24**, No. 4, April, pp. 295–310.

29 KELL, R. D., BEDFORD, A. V., and TRAINER, M. A.: 'Scanning sequence and repetition rate of television images', *Proc. IRE*, 1936, **24**, No. 4, pp. 559–575.

30 Marconi Wireless Telegraph Company Ltd.: 'Improvements in or relating to television systems', UK patent 420, 391, application date, 19th July 1933.

31 Television Committee: evidence of A. Clark and I. Shoenberg on behalf of EMI Ltd. and the Marconi–EMI Television Co. Ltd., 27th June 1934, minute Post 33/4682.

32 LEWIS, H. M., and LOUGHREN, A. V.: 'Television in Great Britain', *Electronics*, October 1937, p. 32.

33 GARRATT, G. R. M., and MUMFORD, A. H.: 'The history of television', *Proc. IEE*, 1952, **99**, Part IIIA, pp. 25–42, see discussion.

34 Electric and Musical Industries and WILLANS, P. W.: 'Improvements in or relating to signalling systems, such for example as television systems', British patent 422, 906, application date 13th April 1933.

35 BEDFORD, L. H., and PUCKLE, O. S.: 'A velocity modulation television system', discussion, *J. IEE*, **75**, p. 91.

36 Technical Sub-committee: minutes of the 4th meeting, 26th February 1935, minute Post 33/5533.

37 Technical Sub-committee: minutes of the 6th meeting, 8th March 1935, minute Post 33/5533.

38 Baird Television Ltd.: memorandum on 'Additional information desired by Advisory Committee', minute Post 33/5533.

39 Technical Sub-committee: minutes of the 14th meeting, 30th April 1935, minute Post 33/5533.

40 Technical Sub-committee: minutes of the 13th meeting, 15th April 1935, minute Post 33/5533.

41 Television Advisory Committee: minutes of the 12th meeting, 17th April 1935, Post Office bundle 5536.

42 Technical Sub-committee: minutes of the 15th meeting, 7th May 1935, minute Post 33/5533.

43 Technical Sub-committee: minutes of the 16th meeting, 5th June 1935, minute Post 33/5533.

44 SHOENBERG, I.: memorandum (paper no. 7) to the Television Advisory Committee, 27th May 1935, Post Office bundle 5536.

45 Television Advisory Committee: minutes of the 17th meeting, 14th June 1935, Post Office bundle 5536.

Chapter 17

The London Station, equipment provision, 1936

One of the most important of the issues to be considered by the TAC was one concerning the system of vision waveform generation to be used for film transmission, studio scenes, and outdoor scenes, and the lighting intensity required for each of these different modes of television. Whereas there had been a certain measure of agreement on some of the points associated with the radiation of the television signals, the responses made by M–EMI and Baird Television highlighted the great advances which M–EMI had made in producing electrical signals representative of different visual scenes.

The company proposed utilising a cathode-ray scanning tube camera for all three activities mentioned above and considered that adequate illumination for studio working would be produced by 50 W per square foot of floor using incandescent lamps from either an a.c. or d.c. supply.[1]

Marconi–EMI's preference for the emitron camera for vision waveform generation showed that they had advanced the development of this equipment since their meeting with the Television Committee in June 1934, for then Shoenberg had told the committee that direct television as practiced at that time did not satisfy him scientifically. 'We think that at the moment we have a cathode-ray scanner which can be used for direct vision with the same limitations as the disc is now used, that is, to say, within a prescribed area determined by the arrangement of the photocells, I am rather anxious not to over state it.' Later, at the same meeting, when referring to the M–EMI system, he mentioned that he had recommended interlaced scanning 'because I think finally, *in five years time, say,* the iconoscope or something of that kind is coming along which will enable you to do away with the intermediate film'.[2]

Bairds' response[3] to the above issue was much more extensive than that of their competitor for the simple reason that they did not have anything as flexible as the iconoscope in their scanning equipment. The company said that film transmission would be carried out first of all by means of disc scanning and then almost immediately by electron image scanning using an electron image tube (this was a reference to the employment of the Farnsworth electron camera).

Fig. 17.1 *(a) Farnsworth image dissector camera (1937). The image tube is on the right inside the deflection coils, (b) the image dissector tube*

For studio scenes, the spotlight method of scanning disc was recommended for the transmission of close-ups while for studio scenes of all types, the intermediate film apparatus was recommended.

'The minimum lighting intensity (required) in the studio (was) 150 ft candles for general lighting all over the scene and 350 ft candles for effect lighting.' D.C. lighting was preferred and the company gave examples of the power required for different situations – using the intermediate film method:

Close-ups	12 kW
Scenes including up to 8 people at full length, for instance a boxing match including an area of 12 sq. ft	21 kW
Larger scenes 30 ft wide by 16 ft high	35 kW

These values were considerably higher than those required by the iconoscope. Bairds hoped that they would soon be able to replace their cumbrous system of spotlight scanner, intermediate film apparatus and film scanner by Farnsworth's electron image scanner (Figs. 17.1), as it was more portable and flexible for studio operation, but unfortunately for the company the camera at that time required a minimum general lighting intensity of 1000 ft candles. However, Bairds stated that it appeared that the latest form of this type of camera, employing the electron multiplier, would operate with the same lighting as required for the intermediate film process, or less.

For outdoor scenes the company also proposed to use the intermediate film method and/or the electron image camera. They recommended the utilisation of the former method chiefly on account of the permanence and record so obtained and gave the cost of the film, complete with processing chemicals, as approximately £12 per hour, employing ordinary 35 mm stock, split down the centre. The sound track was to be recorded between the perforation and the edge of the film.

Following the publication of the press notice on the adoption of the two standard system, invitations to tender were issued, the date by which tenders had to be returned being 4th July 1935.[4]

Marconi–EMI's and Bairds' tenders for the vision transmitter and studio vision apparatus are summarised below:[5]

Vision radio transmitter:	M–EMI	Baird
1 Price of complete radio transmitter	£8035	£11 700
2 Price of two complete sets of valves	£1416	£1960
3 Price of complete water system	£940	£650
4 Price of power converting plant	£3706	£6130
5 Price of erection of all apparatus	£1560	£1910
	£15 657	£22 350
6 Price of aerial and feeder	£436	
7 Completion time from date of placement of order	26 weeks	26 weeks

Studio vision apparatus:	M–EMI
1 Studio equipment, vision, film scanning, and associated power plant	£17 166
2 Sound equipment, studio, sound on film and associated power plant	£6607
3 Installation and wiring	£503
4 Studio lighting	£254
5 Engineering charge for one year (all valves included)	£2567
	£24 530

1 *Studio television equipment*:		Baird
a Spotlight scanner	£2475	
b Electron image camera	£4500	
c Intermediate film scanner	£4950	
d Control room vision equipment	£2475	
e Vision modulation equipment	£2475	£16 875
2 *Studio sound equipment*:		
a Spotlight scanner sound	£340	
b Electron camera sound	£1240	
c Intermediate film sound	£1240	
d Control room sound equipment	£2475	£5295
3 Film scanning equipment, 2 units		£4725
4 Sound-on-film equipment		£600
5 Studio lighting equipment		£1740
6 Conversion plant		£2390
7 Installation and wiring		£3450
8 One complete set of valves and tubes		£940
		£36 015
9 List of suggested spares		£1490

Both tenders specified in some detail the transmitting and studio equipment to be supplied. In particular Marconi–EMI mentioned (in the section dealing with the studio and film vision apparatus) 'the contractor's offer covers the supply of six iconoscope cameras, four for studio use and two for scanning film. The two latter would be permanently associated with the film scanning machines situated in the amplifier room.'

Bairds reply in the same section was much more extensive and highlighted the cumbrous nature of the equipment *vis-à-vis* the highly mobile emitron camera.

> The contractor's offer covers apparatus for studio television by the following methods:

a Spotlight scanner This apparatus is suitable for the transmission of close-ups of speakers and announcers. It is proposed that about half the area of the room adjacent to the studio should be subdivided to provide a small studio, spotlight scanner room, and monitor room. The spotlight scanner cannot be used actually in the studio because it involves the use of a disc running at 3000 rev/min, and is noisy. It therefore works through a glass window. The monitor room also has a glass window so that the control engineer can see the performer and also the latter can see his own image on a monitor tube, and so avoid moving out of the spotlight beam.

b Electron image camera This would be used in the main studio for the television of larger scenes, and must be situated within 20 ft of its scanning and amplifying apparatus. It is proposed to form a small room, partly in the corner of the studio and partly on the colonnade to house the associated apparatus. The camera is mounted on a movable truck, which can be moved within a radius of 20 ft of the apparatus.

c Intermediate film apparatus This is also suitable for the television of large scenes, but it is of a bulky nature and requires connection with the water mains and drainage. It is, moreover, noisy and must be enclosed in a sound proof booth. It is suggested that an intermediate film camera room be built on the colonnade, with a sort of 'bow window' looking into the studio, so that the camera may be 'panned' to take in any part of the studio. By this means three sets could be effectively covered.

Thus two of the three cameras/scanners were noisy and static and required a soundproof room. On the other hand the emitrons were small, noiseless, easily portable, and, very important, fairly sensitive (Fig. 17.2). This was shown by the studio lighting requirements in the two tenders:[5]

Marconi–EMI:		
1 Roof lighting	18 kW	
2 Directional lighting	6 kW	Total = 24 kW
Baird:		
1 Supply arc for 2 telecine transmitters (Fig. 17.3) and intermediate film transmitter		28.5 kW
2 Supply arc of spotlight transmitter		31.5 kW
3 Supply studio lighting for electron camera and intermediate film transmitter		94.4 kW

An additional disadvantage of the intermediate film process was the cost of the 35 mm film stock and processing chemicals which amounted to £48 per hour or £12 per hour using split 35 mm film stock. Against this the cost of servicing the iconoscopes with tubes was £2.10.0d per transmission tube-hour.

The above tenders were considered at the 18th meeting of the TAC.[5] Selsdon and his colleagues found there were a number of points in each tender with which they were not in agreement and the two companies were accordingly invited to meet the committee on 24th (Bairds) and 23rd (EMI) July 1936.

In the case of the Baird Television Company the more important of these points were:

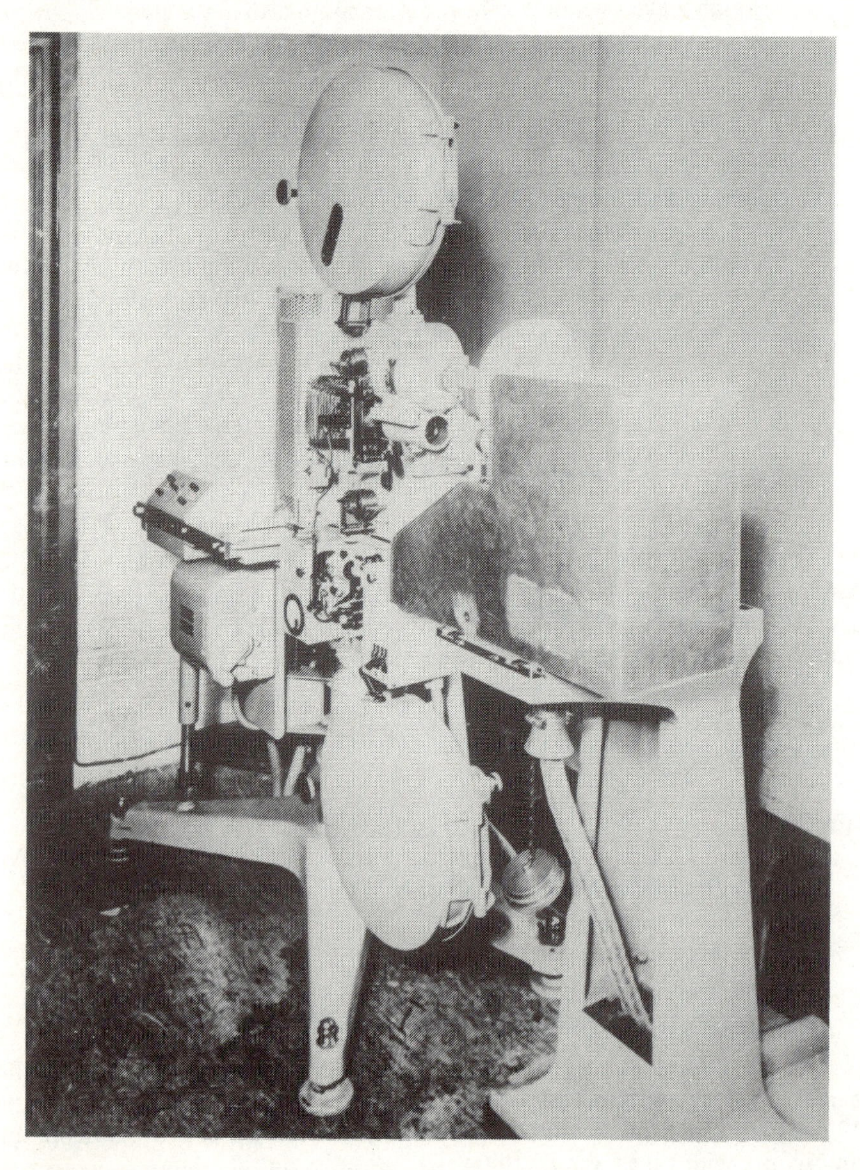

Fig. 17.2 *Emitron camera telecine at Alexandra Palace (1936)*

1 the price of £11 700 for the radio transmitter was considered excessive;
2 the price of £3450 for installation and wiring was also felt to be much too high;
3 the absence of a valve life guarantee;

Fig. 17.3 *Two complete telecine units installed in the Baird telecine area at Alexandra Palace (1936)*

4 the running cost of the intermediate film process;
5 the inclusion of an electron camera in the specification – the committee were not prepared to advise the ordering of this at present time.

The points to be discussed with the Marconi–EMI Television Company were:

1 the insistence of the committee that the Marconi–EMI Television Company procured from the Marconi Wireless Telegraph Company and EMI Ltd. letters which could be appended to the contract undertaking to give licences to responsible British manufacturers to use all present and future patents under their control, subject to the terms of a standard form of licence to be settled by negotiation with the RMA, or failing agreement by arbitration proceedings. The committee thought that the question of the persons to whom the licences were to be granted and the royalty to be paid and the perido for which the licences should be current would all fall within the scope of such negotiations or arbitration;
2 the absence of a valve life guarantee;
3 the company's proposals in respect of the supply of the camera tubes;
4 the company's conditions of supply.

Fig. 17.4 *Marconi–EMI sound transmitter at Alexandra Palace (1936)*

Ashbridge thought it would be quite impossible to settle all these points in time to allow the tenders to be finally accepted by the end of July but it was hoped that if the negotiations to be held on 23rd and 24th July were satisfactorily concluded it might be possible to place provisional orders with the companies by the end of the month in order to enable M–EMI and Bairds to proceed with the work, Figs. (17.4–17.7).

Captain West and Captain Jarrard represented Baird Television Company at the 20th meeting of the Television Advisory Committee on 24th July 1935. Lord Selsdon explained that as the committee had not seen a demonstration of the electron camera they were not in a position to recommend that it should be purchased at that time, although the question of purchasing a set for use at the London Station would be considered after a suitable test had been witnessed.

A lengthy and detailed discussion took place on Bairds' tender prices and eventually it was agreed that these should be amended as follows:[6]

1 Price of complete radio transmitter	£9000
2 Price of two complete sets of valves	£1800
3 Price of complete water system	£650
4 Price of power converting plant	£5000
5 Price of erection of all apparatus	£1550
	£18 000

Fig. 17.5 *Marconi–EMI television transmitter at Alexandra Palace*

This meant a reduction over £4000 on the original tender price. An additional saving was made by the non-inclusion of the electron camera and the consequential modification of the prices of the associated apparatus and of the cost of installation and wiring. The actual reductions agreed upon were:

1 Electron image camera	£4500
2 Control vision equipment and vision modulation equipment	£800
3 Electron camera sound	£1240
4 Electron camera	£870
5 Conversion plant	£300
6 Installation and wiring	£875
7 Valves	£512
	£9097

Bairds subsequently agreed to the total price for the studio vision apparatus being taken as £26 000, representing a saving of about £10 000, or more than

£14 000 on the complete tender. M–EMI were more fortunate than their competitors as none of its prices were altered.[7] However, even with the above cuts in the tender prices, the M–EMI quotations were still cheaper than Bairds – and the TAC had seen demonstrations of both companies television systems and had been very impressed with the Hayes firm's products and research methods.

Fig. 17.6 *Baird Television Ltd. 10 kW vision transmitter, Crystal Palace (1936)*

Both companies suffered some suggested cuts in the estimated running costs of their apparatus. Shoenberg agreed to consider reducing the charge for servicing the iconoscope tubes to £2 per transmission hour, with a minimum annual charge of £1000 per annum and to reconsider this charge at the end of six months. He estimated the life of such tubes to be 50 hours and thought it might be possible to evolve a lower grade of iconoscope for rehearsal purposes which would have a longer life.[7] However, this decrease of 10 shillings per hour in servicing costs was relatively minor compared with those which Ashbridge would have wished for in the operation of the intermediate film process. He viewed with grave concern the very high running cost of this system even if split 35 mm film stock was used and suggested that if the electron camera held out any early promise of success it might perhaps be expedient to consider its adoption for the public service in place of the intermediate film.[6] The difficulty was that this system was preferred to the camera from the point of view of both performance and reliability. Nevertheless, Bairds agreed to supply and install the camera free of charge in place of the film system when once they were

Fig. 17.7 *Baird Television Ltd. 240-line intermediate film apparatus for Alexandra Palace (1936)*

satisfied with its development. In the meantime Bairds said they would approach Ilfords and Kodaks to bargain for a reduction in the price of film: there seemed some hope of obtaining a reduction which would bring down the cost of intermediate film working to £9 per hour. It is interesting to note that Bairds should have quoted in the first instance for such an imperfect system as the Farnsworth camera.

The main point discussed between the TAC and M–EMI's representatives (Shoenberg, Condliffe, Blumlein and Dale-Harris), concerned the patents of Marconi–EMI's related companies and the terms of a manufacturer's licence for these patents. Shoenberg[7] assured the TAC he would give an undertaking with regard to the availability of these patents in a form which would be satisfactory to the BBC's solicitors but suggested that it should be laid down in the terms of reference for arbitration that the royalty payable for the use of EMI patents shold not in any case be less than 3% of the retail list price of sets. On the term of a manufacturer's licence for EMI patents, Shoenberg suggested a maximum of three years, provided M–EMI's apparatus was used at the London Station for so long, and a maximum of five years if its apparatus was chosen for a second television station. He was prepared to agree to a licence being granted in perpetuity in respect of any patent which might prove to be indissolubly associated with M–EMI transmissions, although he doubted whether any such patent existed.

The Television Advisory Committee's recommendations on various aspects of the two tenders were formally made known to M–EMI[8] and Bairds[9] in a letter signed by Carpendale dated 6th August 1935. Bairds concurred with the arrangements described in their letter[10] on 7th August and received confirmation that the British Broadcasting Corporation had accepted their tender for the supply, delivery and erection of certain equipment at Alexandra Palace in a letter from the BBC dated 20th August.[11] Shoenberg had some further discussions with the committee on 25th and 30th August but it was not until the beginning of October when the protracted negotiations were finished, and the modified conditions[12] offered to Bairds,[13] that the contract with the Marconi–EMI Company was signed and sealed. Bairds found the modifications objectionable[14] from their point of view but subsequently, 1st January 1936,[15] decided to accept these after a meeting with the TAC on 19th December 1935.[16]

The plan to establish a high definition television system was certainly a bold one and put Britain in the forefront of television developments anywhere in the world. It is perhaps pertinent at this stage to consider briefly the national progress which had been and was being made to advance 'seeing by electricity' in other countries.

In a world survey of 27 countries – comprising 20 European states, the USA, Canada, the USSR, Australia, New Zealand, Japan and South Africa – undertaken by Sir Noel Ashbridge, only five countries, the USA, the USSR, Germany, France and the UK, had transmitted by the beginning of 1936 some form of television broadcast. Brief details of these transmissions are as follows:

France

In Paris, two experimental television transmitters were in operation: one, a low definition transmitter working at a wavelength of 180 m and having a power output of 700 W, transmitted a 60-line picture at a rate of 25 pictures per second; the other, a higher definition system, operated at a wavelength of 7 m and had a power output of 1 kW. It broadcast a 180-line image (sequentially scanned) 25 times per second.

Nipkow discs were used for scanning and the pictures, which were not limited to films, were scanned horizontally, the aspect ratio being 1:1. Public viewing rooms were available on Sundays and the service was provided by the State Broadcasting Service. A typical receiver cost approximately three times that of a normal broadcast receiver.

Germany

Much television development work had been undertaken in Germany by January 1936 and demonstrations of television had been given at the annual Radio Exhibition since 1928. In March 1935 a public service was inaugurated, and, although no receivers were available to the public at the time, a number of public televiewing rooms were set up at various places in Berlin.

The transmissions, which aroused considerable public interest, were limited to three programmes a week, each programme consisting of film transmissions and direct television lasting one-and-a-half hours.

Two transmitters radiated 180-line images, sequentially scanned at 25 frames per second, at wavelengths of 6·7 m and 7·0 m. The transmitters were installed at the base of the Funkturm, Witzleben.

By January receivers were on sale to the public at a price of approximately 600 to 1800 RM.

United States of America

Although no stations were regularly transmitting television programmes in the USA, low definition transmissions had taken place irregularly from the late 1920s.

A great deal of research and development work had been carried out in various laboratories, particularly those of the Radio Corporation of America and the American Telephone and Telegraph Company, from circa 1925, and only economic considerations were responsible for there being no high definition service in operation.

For 1936 the National Broadcasting Company proposed to operate an RCA television system, using the iconoscope, for demonstration purposes, without the sale of equipment to the public. The system parameters were specified to be: 343 lines/picture, 30 pictures/second, two interlaced frames per picture, aspect ratio 3:4, vision frequency 49 MHz, sound frequency 52 MHz, peak power of the vision channel at maximum modulation 32 kW, power output of the sound channel 8 kW (Copenhagen rating).

The Columbia Broadcasting System, which had operated a low definition television service in the early 1930s, had no plans in January 1936 for a high definition service but had announced that it was keeping a close watch on progress and would await the outcome of the RCA experiment.

Union of Soviet Socialist Republics
The USSR were transmitting, using two longwave broadcasting transmitters, regular low definition – 30 lines per picture, 25 pictures per second – television programmes from Moscow at the beginning of 1936. Transmissions were made during the night and the programmes consisted of specially selected films, concerts, short scenes, etc. The format of the image was 3:4 and the direction of line scanning was horizontal.

Of the other countries mentioned, Italy was contemplating a television service based on a line standard of up to 300 lines per picture, and at the Philips Laboratories, Eindhoven, The Netherlands, experimental vision and sound transmitters had been installed although no public television had been formed. Prior to 1936 experiments had been made with 180-line and 360-line sequentially scanned images and it was anticipated that further work would be based on 375-line and 405-line standards.

During the considerations of the Television Advisory Committee in 1935 on the implementation of the recommendations contained in the Television Committee's Report, various suggestions were made by the committee members which showed their perspicacity for envisaging the future development of television. Lord Selsdon in April expressed an opinion that a claim should be put forward, when opportunity offered, for the allocation of a band of frequencies for television in the region of 2–3 metres.[17] The Chairman and some other members of the committee also felt that with a view to facilitating the televising of important personages some television studio accommodation in a fairly central position in London was very desirable[18] (Sir Noel Ashbridge later said that the studio would probably be located at 16 Portland Place).[12] Colonel Angwin raised the question of the provision of cable connections being provided between Alexandra Palace and the central studio. Outside broadcasts were also thought to be desirable and Ashbridge put forward an enquiry to both Bairds and M–EMI regarding the supply of portable apparatus for this purpose.[16] M–EMI replied that they were unable to undertake the necessary development work until such time as the London Station was nearing completion. At the 24th meeting of the TAC it was agreed that the proposed studio in Portland Place should be used for the purpose stated above and experimental purposes, and that, as Ashbridge thought that the space available would be insufficient to accommodate the bulky intermediate film equipment, the occasional broadcasts from it should use the iconoscope (during the M–EMI periods of transmission), and nobody outside would be any the wiser.

During the same meeting Ashbridge mentioned that the BBC were bearing in mind the possibility of establishing its own newsreel service, using special apparatus for the rapid developing of film, and that they proposed to arrange for film to be conveyed to the Alexandra Palace physically as and when it was considered desirable to do so. Another aspect of television which was discussed then was the question of using a lorry with a portable transmitting apparatus working on a very short wavelength. Thus, by the end of 1935, many ideas, which are now commonplace today, for a varied television service had been made. Some of these, of course, were not original: Baird had demonstrated the televising of outside events several years previously – albeit in a very crude fórm.

The completion date for the supply and installation of the television apparatus at Alexandra Palace was the end of February 1936[12] and the contract date for completing the erection of the mast was 15th March.[16] However, delays were occurring. Ashbridge gave the reasons for the delay in erecting the mast to the committee in December 1935, and in January 1936 told them that M–EMI anticipated completing their installation (but not the testing of it) by the end of March.[19] Captain West of Bairds thought his company's apparatus would be installed (but not tested) by the middle of March. At the 26th meeting[20] of the TAC in February, Ashbridge reported that the indications were that delivery in the case of both companies would be about two months behind the contract date and Lord Selsdon suggested that the Corporation should inform them that the committee 'took a serious view of the prospect of delay in the delivery of the apparatus and at the same time indicate to them that if one company's apparatus should be ready for use before the other, the Corporation would regard themselves as being free to begin a public service with one system only'.

The BBC's 20th March progress report contained a detailed account of the state of completion of each major item of apparatus: a summary is given below:

Baird Television Ltd.:
1 Studio sound apparatus; inspected and passed, 12th March.
2 Cinematograph film projectors; inspected and passed 16th March.
3 Spotlight scanner and intermediate film apparatus; to be ready for inspection, next week.
4 Radio transmitter; inspected and passed, 31st March.

Marconi–EMI Television Co. Ltd.:
1 Sound equipment; inspected and found satisfactory, 18th March.
2 Emitron scanning apparatus; to be ready for inspection, 25th March.
3 Vision modulator; to be ready for inspection, 'after Easter'.
4 Radio transmitter; inspected and found satisfactory, 28th February.

Marconi's Wireless Telegraph Co. Ltd.:
1 Sound transmitter; to be ready for inspection, 25th March.
2 Mast; completion anticipated by end of April.

Thus both companies were more or less in the same state regarding completion of their equipment at this time. On 2nd May Sir Noel reported:[21] 'In general whilst progress is not as satisfactory as had been hoped, the quality of the apparatus so far delivered is good.' Bairds had more apparatus on the site than M–EMI and a good many machines and switchboards had been installed, but the Chief Engineer of the BBC observed that the difference in progress might be more apparent than real as it was possible M–EMI were deferring installation until the last minute for fear of being copied. In addition to the above delays the mast was only completed to about half its height and was not likely to be completed until the end of May. The installation programme was running seriously behind schedule and Ashbridge did not think tests (i.e. initial tests of portions of the apparatus as distinct from trial programmes carried out with the installation as a whole), could begin seriously before July. The TAC's forecast was that trial programmes for the benefit of the trade would probably begin towards the end of July and would become more elaborate by degrees. It was thought the aim should be one hour's transmission in the morning and another in the afternoon, i.e. during normal working hours.

For the public television service Admiral Carpendale told the TAC the BBC proposed commencing the service with three hourly programmes per day, 3.00–4.00 p.m., 6.15–7.15 p.m., and 9.30–10.30 p.m., and to operate each of the two systems during alternate weeks. Phillips pointed out that the second hour suggested would overlap the beginning of the peak period of trunk telephone traffic at 7.00 p.m. and that it would be preferable from this point of view if the period be 6.00–7.00 p.m. Carpendale said 6.15 p.m. had been put forward so as to avoid clashing with the 6.00 p.m. news bulletin, which was very popular. This view was accepted by the committee.[16] However, this was looking ahead; there was still much work to be completed at Alexandra Palace, although both official and unofficial exhortations were being made to the contractors to expedite their work. Some of their excuses seemed reasonable though; in the case of the mast and aerial the reason for the non-delivery of steel was 'priority munition contracts'.[22]

Apart from the TAC's natural desire to see the results of their deliberations there was an added reason for a more rapid completion of the London Station – the forthcoming Radio Exhibition at Olympia which was scheduled to start on August 23rd with a press view.[23] The organisers of the exhibition, the RMA, had agreed that television exhibits should be allowed – although 'the general impression (seemed) to be that the RMA (were) half-hearted about the matter'[21] – and the BBC had proposed to provide television transmissions for reception at the exhibition if this was practicable. This meant, of course, that both M–EMI and Bairds' equipment had to be in working order not only for the period of the exhibition, 26th August–5th September, but also for the week before the opening in order that the programme staff could make the necessary trials.

The committee realised there was a considerable amount of work still to be carried out at Alexandra Palace but felt it was most important for the future of television that these demonstrations should be carried through successfully.[24] Regarding the nature of the programmes, since public interest was likely to be very great, it was felt the time available for each visitor to see a transmission must be a matter of a few minutes only and therefore nothing ambitious was planned. The proposal of Selsdon's committee was that two periods of transmission should be given each day, of about $1\frac{1}{2}$ hours each, and that each system should be used on alternate days. Ashbridge was asked to draw up two lists of transmitting periods A and B, which would be nearly equal in publicity value as possible, and then it was agreed the Chairman should draw lots to determine which firm should have the choice between the two series of times.[25] An added spur to each firm's efforts to complete their work by the third week in August was the TAC's decision that in the event of one system being insufficiently advanced to give a demonstration, or in the event of a breakdown of one system, the other should be used as and when it was deemed necessary.

Letters outlining these recommendations were sent out to M–EMI and Bairds on 10th July together with a timetable of the two schedules.

The Baird Company, which had always shown itself, in the past, very eager to demonstrate and publicise its activities, was now curiously unenthusiastic about the exhibition and the opportunity to show its results to the public. While Bairds did not refuse to participate, their response was certainly lukewarm: indeed their misgivings were such that a deputation consisting of Mr Clayton (the Deputy Chairman), Captain West and Captain Jarrard was received by the committee. Clayton told Selsdon that Bairds were frankly suspicious of the motives of the Radio Manufacturer's Association in the matter[23] and feared the public would not get a good impression of television from any demonstrations which could, in their opinion, be reasonably expected to be given from the Alexandra Palace on the 26th August. 'To create an erroneous impression would seriously injure a new industry at its inception.'

Their difficulty was that they were about to face their supreme test – one that could either make or break the company as a supplier of transmitting equipment. Whereas, from the initiation of Baird's experiments in 1923 until about 1932, the Television Company had not been subject to competition in the United Kingdom, now their television system was to be compared in a public trial with that originating from a powerful company which had achieved much in a short time – a company which to Bairds was an offshoot of the giant Radio Corporation of America.

Clayton urged that no demonstrations should be given unless the quality of reception was good and he also thought that it should be made abundantly clear that the transmissions were experimental. In their opinion Bairds considered the Television Advisory Committee should only agree to give transmissions on the condition that competent judges, appointed by the committee, would have the right to stop any transmission or reception if they were not perfectly satisfied that the general level of reception on the various sets shown was up to the standard known by that committee to be possible.

Selsdon agreed with this view and told Clayton the committee would be the judges of what was good enough to be shown. He assured the deputation that every effort would be made to prevent premature publicity, and Sir Stephen Tallents, the BBC's public relations officer, stressed that the experimental nature of the transmissions would be duly emphasised.

At this meeting of the TAC (the 30th held on 23rd July 1935), Lord Selsdon spun a coin in the presence of Bairds' representatives and invited Mr Clayton to call 'heads' or 'tails'. Clayton won the toss and undertook to let the Secretary know his company's choice the next day (Shoenberg did not send a representative for this ceremony).

Actually Ashbridge was not very sanguine about the demonstrations being given as the latest progress report from Alexandra Palace indicated that neither M–EMI's nor Bairds' apparatus would be ready. Owing to bad weather the aerial was still not complete and neither Bairds' intermediate film and flying spot apparatus nor Marconi–EMI's modulation unit and film scanner had been delivered. Nevertheless, it was agreed that Ashbridge, either Phillips or Angwin and either Smith or Brown should be appointed to assume responsibility on behalf of the TAC for judging whether or not reception on each of the receiving sets to be installed for demonstration purposes at Olympia – about twenty in number – was sufficiently good for public exhibition.

Tallents acquainted Strachan (or the RMA) of this decision the next day[26] but received what really amounted to an expression of concern from the Director of the RMA a few days later.[27] He felt the RMA should adjudicate on the merits or demerits of any particular apparatus in any exhibition it might organise. The BBC, however, controlled the transmissions from Alexandra Palace and the TAC's decision prevailed.[28]

Fortunately the two systems were ready in time for the exhibition. EMI had about two days in hand with their equipment but Bairds only installed their spotlight transmitter the day before the press view.[29] This was very well attended, about 70 pressmen turning up, and generally was a success. They saw demonstrations and the reaction in the papers was quite good.

Sir Noel Ashbridge's account[29] of the events on the opening day are interesting:

> The demonstrations at Olympia started at 12 o'clock on Wednesday, August 26th, with a programme on the Baird system using the spotlight apparatus and film projector. The first transmission was technically quite good but, of necessity, it had to consist of a lot of film, and I think the public would have liked more direct material. However it was a very smooth demonstration. In the afternoon there were frequent breakdowns lasting five or ten minutes at a time, but when the transmission was going it was good . . . The first EMI demonstration took place this morning (27th August), and was, on the whole very good. I think however that I have seen better EMI transmissions, but there was no definite defect. The programme was of course more interesting as it included outside scenes and a good deal of variety in the studio . . . We are of course gathering a great deal of information both from the crowd and our own observations and we have senior staff posted at every important point where information can be collected . . . On the whole the Press is favourable and helpful and is inclined to gloss over defects. There is of course a tremendous amount of notice being taken of television but I should not say at the moment that opinion is entirely unanimous that the transmissions are

> quite fit for regular programmes. What is more serious, however, is the high price of receivers. The lowest price seems to be eighty guineas, at the Exhibition at any rate. So far as the transmissions and programmes are concerned, I think that great strides will be made during the month of September because most of the difficulties are due to minor defects.

The Chief Engineer gave a short talk on 'Television' in the National (sound) programme on the evening before the official opening of the exhibition at Olympia. He told his listeners: 'We at the BBC are going to do everything in our power to ensure that television broadcasting in this country shall be second to none.' Apart from this pledge his talk was rather cautious because, as he said, he was 'most anxious not to raise any false hopes', but, he added, 'television progress was no further advanced anywhere in the world than it was in this country'.[30]

Programmes on the Baird system consisted, of necessity, mainly of films, with televised announcements by the flying spot method. These included: a new Paul Rother documentary film, 'Cover to cover' about books, which brought to the screen A. P. Herbert, W. Somerset Maughan, Julian Huxley, T. S. Eliot, Rebecca West and other well-known authors; 'Here's looking at you' a variety half-hour produced by Cecil Madden and G. More O'Ferrall – performed in the Alexandra Palace studios; a scene from 'As you like it' in which Elisabeth Bergner as Rosalind and Laurence Olivier as Orlando were seen in the Forest of Arden (this was the first British Shakespearean film and cost £150 000 to produce); an excerpt from the Alexandra Korda film 'Rembrandt' featuring Charles Laughton in the title role and Gertrude Lawrence, together with a number of other film extracts and the Gaumont British News.[31]

The variety show 'Here's looking at you' was television's first show of this type and was transmitted direct from Alexandra Palace using the Baird spotlight process. Unfortunately, although the Baird intermediate film apparatus was installed it was not used for the demonstration as it had not been properly tested by the start of the exhibition. This placed the Baird Company very much at a disadvantage, however, their film apparatus worked well although there were one or two breakdowns of a 'fairly serious nature' owing to mechanical difficulties and the breaking of film. The flying spot apparatus, which had been installed only a day or two before the start of the demonstrations, also worked fairly well, within its own limitations, but was suitable for head and shoulders only.[32]

Marconi–EMI used their emitron camera both for studio scenes, outside scenes and film scanning. The whole of the system's apparatus was employed, more or less according to plan, and reliability was fairly good, although Ashbridge thought it was not up to service standard. There was one bad breakdown during the Olympia press demonstration lasting about an hour, (Fig. 17.8).

The Marconi–EMI programme was much more interesting than Bairds, owing to its wider scope and included outside scenes of the Alexandra Palace grounds, variety acts in the studio, and, of course, a considerable amount of film. All this was a foretaste of what television would be like after the inauguration of the public service in November (Figs. 17.9 and 17.10). Throughout the demonstrations the BBC took a great deal of trouble to collect public opinion, both at the exhibition and in a demonstration room which they had established in Broadcasting House. At the latter a considerable number of the BBC's staff saw the transmissions as well as many people of importance from outside. It soon became obvious that the public generally were not very interested in film

Fig. 17.8 *D. C. Birkinshaw with the emitron camera in the grounds of Alexandra Palace*

excerpts, the comment being 'Oh; that's not television, that's pictures'. Ashbridge felt that 'on the whole one can say that the general reaction was that the demonstrations were a very remarkable technical achievement, but that it was not certain that the pictures had permanent programme value'. He put forward three reasons for this view:

1 The size of the picture.
2 Lack of definition.
3 The effort of concentrating on a picture which flickered to some extent, not merely due to frame frequency but also due to various other imperfections, and to the fact that the standard set by the cinema, was, of course, extremely high.

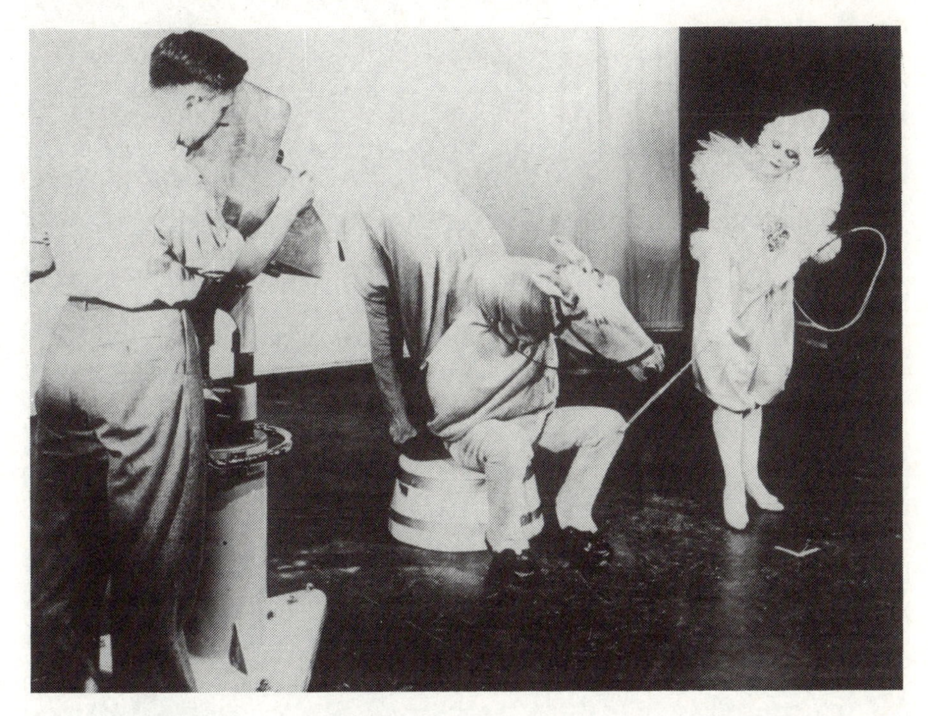

Fig. 17.9 *Television at Alexandra Palace. First transmission for the benefit of Radio Olympia Exhibition 1936. Miss Lutie is with the Griffiths Brothers comedy horse*

Seven manufacturers provided sets at Olympia but the public did not know the make of receiver they were looking at as they had been quite effectively camouflaged. The BBC's Controller (Engineering) placed the merit of the receivers in the following order: Cossor, Marconi–EMI, GEC, Baird, Ediswan, Phillips and Ferranti, and considered that the standard of reception generally at Olympia was very good indeed. From time to time the BBC's senior engineer on duty had to request the removal of one or two receivers when the results fell below a reasonable standard, but this did not occur very often and did not cause any friction.

One of the most disturbing features of these receivers was their extremely high price. There seemed to be an agreed preliminary price which stood at a minimum of about 95 guineas but this ranged up to approximately 150 guineas

Fig. 17.10 *Alexandra Palace. General view of a studio, showing the television orchestra conducted by Hyam Greenbaum at rehearsal (August 1936)*

depending on the size of the tube and whether the set was provided with facilities for receiving ordinary sound broadcasts.

There is no doubt that the RMA Exhibition enabled the BBC to gain much useful information about the way a television programme should be produced and which programmes attracted interest from the public. Ashbridge wrote:

> Perhaps the most interesting thing which emerged from the demonstrations was the lack of interest in films, and this I consider due almost entirely to the fact that insufficient detail is visible. Films produced especially for television could avoid this defect to some extent. One could observe that interest increased immediately the picture became a close-up, whether it was film or direct. Some people considered that films were better reproduced by the Baird system than the Marconi–EMI and that there was more clarity, and this is remarkable having regard to the fact that the Baird system has much more flicker, and theoretically less definition. There is no doubt that general opinion is that a good variety turn is better than an interesting film from the point of view of television. People were not so much impressed by outside scenes simply because the detail was insufficient and the general effect artificial. That is to say, trees were recognisable as trees, but they did not look like the real thing, and the light values were very obviously false.
>
> Another impression which I gathered was that the task of announcers is going to be extremely difficult. I heard numerous criticisms of the announcers, some of which I considered rather unfair. I came to the conclusion that people are much more liable to be irritated when they can see an announcer than when they cannot, while the nervousness on the part of the lady announcer was very apparent when you could see her, whereas nervousness on the part of unseen speakers is very often unnoticed. Of course, it is hardly necessary to say that the newness of the technique was very apparent. I don't consider that the best had been done with make-up, and the dresses worn by various people were, to my mind, not suitable, but all this of course, is only to be expected. On the other hand our programme people took unlimited trouble and were working under considerable difficulties, due to the newness of everything.

Nothwithstanding Ashbridge's reservations on certain points and the need for further technical improvements the television exhibition was considered to be a success technically although public reaction was somewhat disappointing. Sir Noel had an idea moreover that the Radio Manufacturer's Association had a feeling that television, from their point of view, was a long way off, that commercially there was not yet much in it, and that it was even unlikely to affect the sound service for many years to come. 'Whether or not they are intentionally keeping the price high remains to be seen, but it is difficult to see quite how receivers could be sold at the moment for very much less than the prices quoted.'

In addition to their own reports, the Television Advisory Committee had the benefit of a useful report by A. C. Cossor Ltd.[33] As designers and manufacturers of radio receivers they were particularly concerned with any departures from the published waveform standards of the two companies, as possibly the general public might attribute picture defects to the set manufacturers rather than the contracting authorities.

The Marconi–EMI waveform was excellent according to Cossor, except for three effects:

1 variable tilt on the iconoscope;
2 an effect analogous to low frequency distortion which produces unpleasant effects;
3 an occasional appearance of too much high frequency compensation producing harsh edges.

The first defect, owing to tilt, was inherent in the iconoscope/emitron: however, EMI had developed special circuits to overcome its effect. In the case of the second fault listed by Cossor, the appearance on the screen of a television receiver was, 'for instance, when transmitting the words BBC demonstration to Radiolympia etc., those parts of the background which were scanned by lines which also passed through the script appeared to have a different degree of illumination from those in which the scanning lines passed through no letters'.

Notwithstanding these observations, Cossor found: 'With two exceptions, the operation of the Marconi–EMI equipment was all that could be desired.'

'Unfortunately,' said the receiver manufacturer, 'there (were) several rather important errors in the wave form transmitted by the Baird system'. One of these concerned the line synchronising pulses, which, instead of being rectangular pulses having leading and trailing edges of short response times, had trailing edges which returned to the black level in an exponential manner after an initial rise to the black level followed by another fall.

The effect of this was to make line synchronisation difficult in some receivers (but not the Cossor) and to cause a large black strip to appear down the left hand side of the picture, thereby cutting off some of the transmitted intelligence.

Another defect was that the frame pulse lasted for 23 lines instead of the specified 12 lines and the black barrier was not present after the frame pulse for most of the time.

Again the picture/synchronising signal ratio was not the 60/40 ratio specified but was for the most part between 40/60 and 50/50. The effect of this discrepancy was to cause insufficient contrast in the received picture.

A surprising defect of the Baird apparatus was the lack of equality of hole spacing in the scanning disc. This resulted in the 'image being displaced by small random amounts along the lengths of successive lines' so that it was impossible to obtain sharply defined vertical edges. Cossor thought that the quality of the reception was 'seriously marred' as a consequence and that the 'good intrinsic definition of the picture (was) wasted'.

Other faults which Cossor listed concerned a transmitted 100 c/s ripple, high frequency overcompensation, sound on vision, occasional bursts of oscillation during the spotlight transmissions and a power output which was some 8 or 10 dB down in relation to that emitted by the Marconi–EMI transmitter.

Unhappily, some of these defects were to persist after the establishment of the first high definition service in this country, and Cossor felt compelled to write to the TAC in November about the discrepancies between Bairds published waveform specification and the actual transmission.[34]

The comparison which Cossor made between the two contracting companies' studio equipment was also ominous from Bairds point of view.

> The direct transmissions were rather a revelation. We were extremely surprised to find so much entertainment value obtainable in this way. It is, however, our opinion that the spotlight studio is extremely inflexible and that since it involves almost perfect stillness on the part of the artist, we think that its value for television transmission is almost nil. On the other hand, the use of the iconoscope for indoor and outdoor sets presents great flexibility and is capable of providing first class results. Particularly were we struck with the extremely neat and pleasing results obtained by fading from one iconoscope to another. The depth of focus, range of vision and ability to pan in any direction which the iconoscope possesses makes it an extremely valuable piece of apparatus, and in our opinion makes the spotlight studio a complete anachronism.

All of this was known to the BBC, but as it came from a radio and television set manufacturer, the report lent added significance to Ashbridge's observations. It seemed, at the time, that the result of the forthcoming trial period between the two competing companies was a foregone conclusion.

One effect of the exhibition was to cause Bairds to have further thoughts on the desirability of increasing the frame rate and of introducing interlacing. For the first time since television was introduced the Baird system had been compared with a rival company's product and one of the conclusions was that a high frame frequency was necessary to reduce flicker. Captain West[35] had expressed it as his opinion that their present picture frequency standard of 25 per second was inadequate and that he would like to increase it to 50 per second, but was undecided whether to introduce interlacing or not. The BBC thought, however, it was too late to change and no record exists which would show that Bairds seriously proposed the change to the TAC.

With the exhibition behind them, the British Broadcasting Corporation and the Television Advisory Committee could now turn their attention to the forthcoming public service. This was to be preceded by trial programmes for test purposes during the month of October when half the transmissions would be on the Baird system and half on the Marconi–EMI system.[35] Selsdon proposed that the system to be operated during the first week of the regular public service should be selected by the tossing of a coin and that the two systems should, as far as practicable, be operated thereafter week and week about until the 1st

April 1937, at the end of which period the arrangements would be reconsidered. Both companies were informed of this arrangement on 24th September and subsequently, as on the previous occasion, Bairds won the toss.[36] Again Shoenberg declined to attend.

The opening of the public service was timetabled for the 2nd November, but there was an ominous cloud on the horizon. At the 32nd meeting of the TAC on 15th October, Sir Noel Ashbridge said the working of the Baird spotlight system at the Alexandra Palace did not justify him in recommending its employment for the opening ceremony and the intermediate film method, although capable of giving good results, had so far not proved reliable. In the circumstances the committee proposed[37] to witness a demonstration of the reception of television transmissions by the intermediate film process on a Baird receiver on 26th October and to decide on the results whether or not that method could be regarded as being satisfactory for use for the opening ceremony. It was clearly important that the public service should not be allowed to start with an inferior production of television and therefore the committee had to consider the possible use of the Marconi–EMI system if the Baird tests proved unsatisfactory.

The results of this demonstration, at which the electron camera as well as the intermediate film method were tried, confirmed Ashbridge's suspicions, for they were not entirely satisfactory.[38] Bairds were definitely opposed to the idea that the use of their apparatus for public service should be deferred until a later date, and so, as the committee considered it would be unsafe to rely wholly on the Baird system for televising the opening ceremony, it was agreed that this should be televised twice – first by the Baird system and afterwards by the Marconi–EMI system. Bairds appeared to be unhappy with this decision but eventually agreed to it. In making their judgement the committee no doubt took into consideration some representations which had been made to Selsdon when he had visited the Hayes works.[39] Tactfully the committee did not indicate, possibly, the real reason for their action to Marconi–EMI and equally tactfully the BBC in its press announcement[40] on the subject referred to the need to secure that the honours of the day were shared between the two systems. The stage was now set for Bairds greatest test.

References

1 Electric and Musical Industries: response to questionnaire of the Technical Sub-committee on 'Proposed vision transmitter', minute Post 33/5533

2 Notes of a meeting of the Television Committee held on 8th June 1934, evidence of the Marconi–EMI Television Co. Ltd; represented by Messrs. Shoenberg, Condliffe, Blumlein, Agate, Browne and Davis

3 Baird Television Ltd.: response to questionnaire of the Technical Sub-Committee on 'Proposed vision transmitter', minute Post 33/5533

4 Television Advisory Committee: minutes of the 17th meeting, 14th June 1935, Post Office bundle 5536

5 Television Advisory Committee: minutes of the 18th meeting held on the 18th July 1935, Post Office bundle 5536
6 Television Advisory Committee: minutes of the 20th meeting held on 24th July 1935, Post Office bundle 5536
7 Television Advisory Committee: minutes of the 19th meeting held 23rd July 1935, Post Office bundle 5536
8 CARPENDALE, Admiral Sir C.: letter to Marconi–EMI Television Co. Ltd., 6th August 1935, minute Post 33/5536
9 CARPENDALE, Admiral Sir C.: letter to Baird Television Ltd., 6th August 1935, minute Post 33/5536
10 Baird Television Ltd.: letter to the BBC, 7th August 1935, minute Post 33/5536
11 BBC: letter to Baird Television Ltd., 20th August 1935, minute Post 33/5536
12 Television Advisory Committee: minutes of the 23rd meeting held on 2nd October 1935, Post Office bundle 5536
13 Secretary of the TAC: letter to the Secretary of Baird Television Ltd., 4th October 1935, Post Office bundle 5536
14 GREER, Sir H.: letter to the Secretary, TAC, 7th December 1935. Post Office bundle 5536
15 Secretary, Baird Television Ltd.: letter to the Secretary of the TAC, 1st January 1936, Post Office bundle 5536
16 Television Advisory Committee: minutes of the 24th meeting held on 19th December 1935, Post Office bundle 5536
17 Television Advisory Committee: minutes of the 12th meeting held on 17th April 1935, Post Office bundle 5536
18 Television Advisory Committee: minutes of a meeting held on 30th April 1935, Post Office bundle 5536
19 Television Advisory Committee: minutes of the 25th meeting held on 20th January 1936, Post Office bundle 5536
20 Television Advisory Committee: minutes of the 26th meeting held on 20th February 1936, Post Office bundle 5536
21 Television Advisory Committee: minutes of the 27th meeting held on 2nd May 1936, Post Office bundle 5536
22 BBC: notes on 'London television station progress', 26th May 1936, Post Office bundle 5536
23 Television Advisory Committee: minutes of the 30th meeting held on 23rd July 1936, Post Office bundle 5536
24 Secretary of the TAC: letters to Marconi–EMI Television Co. Ltd., and Baird Television Ltd., 10th July 1936, Post Office bundle 5536
25 Television Advisory Committee: minutes of the 28th meeting held on 24th June 1936, Post Office bundle 5536
26 TALLENTS, Sir S.: letter to G. Strachan 24th July 1936, minute Post 33/5536
27 STRACHAN, G.: letter to Sir S. Tallents, 30th July 1936, minute Post 33/5536
28 Secretary of the TAC: letter to G. Strachan, 4th August 1936, minute Post 33/5536
29 ASHBRIDGE, N.: letter to J. Varley Roberts, 27th August 1936, minute Post 33/5536
30 BBC announcement: 'The Chief Engineer on Television', 25th August 1936
31 BBC announcement: 'Experimental television Programmes for Radiolympia', 25th August 1936
32 ASHBRIDGE, Sir N.: 'Report on demonstrations of television at the RMA Exhibition at Olympia, 26th August–5th September 1936', 7th September 1936, minute Post 33/5536
33 A. C. Cossor Ltd.: 'Notes on the BBC television transmissions to Olympia', c. September 1936, minute Post 33/5536
34 A. C. Cossor Ltd.: letter to J. Varley Roberts, 13th November 1936, minute Post 33/5536
35 Television Advisory Committee: minutes of the 31st meeting held on 14th September 1936, minute Post 33/5536

36 Television Advisory Committee: minutes of the 32nd meeting held on 15th October 1936, minute Post 33/5536

37 VARLEY ROBERTS, J.: letter to Baird Television Ltd., 16th October 1936, minute Post 33/5536

38 Television Advisory Committee: minutes of the 33rd meeting held on 26th October 1936, minute Post 33/5536

39 VARLEY ROBERTS, J.: letter to Electric and Musical Industries Ltd., 27th October 1936, minute Post 33/5536

40 BBC press announcement: 28th October 1936

Chapter 18

The trial, 1936

The opening ceremony for the world's first, high definition, regular, public television broadcasting system was a singularly modest affair and represented rather an anticlimax to the months of strenuous activity and preparation which had been undertaken by the two contracting organisations and the Television Advisory Committee. A low cost, low key approach was adopted for the proceedings which would be remembered in the future as an historic occasion, 'not less momentous and not less rich in promise than the day, almost exactly fourteen years ago, when the British Broadcasting Company, as it then was, transmitted its first programme from Marconi House'.[1]

The opening programme, arranged by the BBC and approved by the TAC, lasted hardly one-quarter of an hour and consisted of a few short speeches, lasting four minutes each, by Mr Norman, the Chairman of the BBC, Major Tryon, the Postmaster General, and Lord Selsdon[2] (Fig. 18.1).

Both the chairmen of the Marconi–EMI and Baird Companies were allowed half a minute each to say, literally, a few words, and after an interval of five minutes (sic) the programme continued with the showing of the British Movietone News and then Variety (Fig. 18.2). While Mr Norman's and Major Tryon's speeches tended to be, somewhat naturally, platitudinous in character, Lord Selsdon in his speech endeavoured to give some assurance to the future purchasers of television sets regarding the stability (for at least two years) of the transmission standards.[3] He was keen to point out that the service would cover the Greater London area, with a population of about ten million people, and said he was unwilling to lay heavy odds against a resident in Hindhead viewing the Coronation Procession. For all three speakers the future of the new medium was bright and held 'the promise of unique if still largely uncharted opportunities of benefit and delight to the community'. Norman remarked on the foresight which had secured for Great Britain a national system of broadcasting and which promised to secure for it also a 'flying start in the practice of television. At this moment,' he said, 'the British television service is undoubtedly ahead of the rest of the world. Long may that lead be held. You may be assured that the BBC will be resolute to maintain it.'

None of the three main speakers referred to Baird's pioneer activities: he was not even a member of the platform party. Margaret Baird, his wife, was to write many years later:[4] 'John, not invited to sit on the platform, remained in the body of the hall, furious at the slight by the BBC.'

Fig. 18.1 *The Postmaster General, Major G. C. Tryon M.P. with Jasmine Bligh, R. C. Norman and Lord Selsdon*

The experiences gained by the British Broadcasting Corporation of the operation of both television systems under service conditions from 2nd November to 9th December were described in an important report[5] written by Gerald Cock, the BBC's Director of Television. It proved highly damaging to the Baird interests; indeed it meant the end of the company as a supplier of television studio and transmitting equipment for the Corporation's stations and studios. Cock stated that the Marconi–EMI equipment had proved capable of transmitting both direct and film programmes with steadiness, and a high degree of fidelity.

> Its apparatus being standardised throughout, reproduces a picture of consistently similar quality and requires only one standard of lighting, make-up, and tone contrast in decor. Its studio control facilities are convenient and comparatively simple. It has proved reliable, and has already established a large measure of confidence in producers, artists and

> technicians. Outside broadcasts and of multi-camera work have added considerably to the attractions of programmes. With improved lighting, additional staff and studio accommodation, single system working by Marconi–EMI would make a service of general entertainment interest immediately possible.

Alas, the Director of Television found it difficult to say anything complimentary about the Baird system. The programmes were being transmitted under practically experimental conditions and the prospects of anything approaching finality in the studio stages of transmission seemed remote.

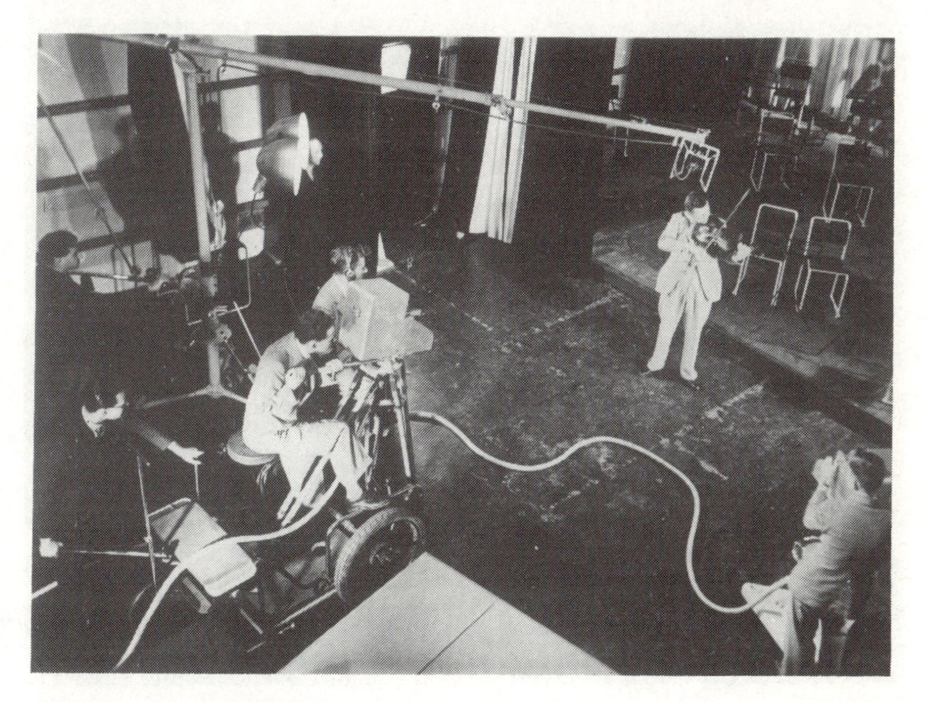

Fig. 18.2 *Studio A showing two emitron cameras in use*

> Alterations in apparatus were constantly taking place. Breakdowns, with little or no warning, and, even more serious, sudden, unexpected, and abnormal distortions are a frequent experience. In such cases, it is difficult and embarrasing to make a decision to close down, since there is always the possibility that faults may be corrected within a short time. This inevitably leads to criticism of television by those who may only have observed it in adverse conditions.

In studio operations (Figs. 18.3 and 18.4), the Baird system made use of the spotlight scanner, the intermediate film method, the electron camera and the telecine scanner for 35 mm commercial film. The first three systems each required a different technique in lighting and make-up to add to the difficulties of the producers and cameramen. Cock's views on these methods were as follows:

1 *Spotlight scanner*
This apparatus is limited to double portrait reproduction. Distortion of picture tone and shape still appears to be intrinsic and unavoidable. No reading is possible in a spotlight studio, nor could any artist depending upon looks and personality be expected to televise by this method. The result is a caricature of the image televised.

Fig. 18.3 *Baird system in operation at Alexandra Palace (December 1936) 'The art of self-defence' programme is being produced*

2 *Intermediate film method*
This is extremely intricate, and depends upon so many processes that it causes continual anxiety. It is inflexible and rigid in operation, being confined to 'panning' in two planes. Changes of view can only be effected at the cost of lens changes and 'black outs'; otherwise the picture is static (Fig. 18.5). Its quality is variable; the delay action is extremely inconvenient for timing and other production purposes. The maximum continuous running time is at present limited to approximately sixteen minutes, which adds to the difficulty of arranging programmes. The cost

for film alone is £12 per hour (rehearsal or performance). Sound, recorded on 17·5 mm film and subject to development at high speed, is invariably of bad quality, whole sentences having occasionally been inaudible. It is consequently an unsuitable method of presenting any programme item in which quality of music or speech is important. Mechanical scanning produces line bending and twisting in a variable degree. Black and white contrast is however generally good.

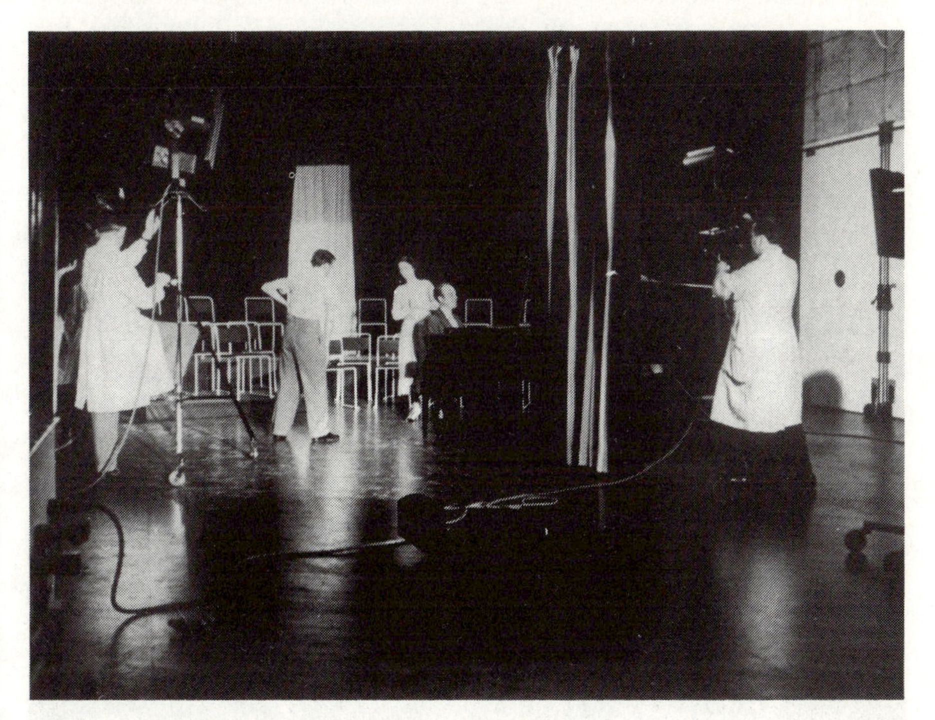

Fig. 18.4 *Baird studio, Alexandra Palace (August 1936)*

3 *Electron camera*

These cameras have quite recently been improved, but future progress is likely to be seriously handicapped by destruction of the Baird research plant and technical research records in the Crystal Palace fire. Bending and twisting of lines are pronounced. The cameras are in a somewhat primitive stage of development and are still without facilities for remote (outside) or 'dissolved' work. Breakdowns have been frequent. Electron cameras do not appear likely seriously to compete with emitrons at any rate for a considerable time. They have advantages over other Baird apparatus in being instantaneous in action; in permitting good sound transmission; in mobility; and in the elimination of mechanical scanning. At present their operation seems somewhat precarious.

4 *Telecine*

Originally, this apparatus gave the best picture obtainable by the Baird system and possibly (flicker apart) was better than the Marconi–EMI for reproduction from standard film. Line bending and twisting have since been noticeable due perhaps to mechanical scanning and the difficulty of maintenance in first class condition. Its contribution to programme is limited by the present restricted use of 35 mm film.

Fig. 18.5 *Baird intermediate film studio*

This very damaging report, from Bairds point of view, was discussed by the Television Advisory Committee at its 34th meeting on 16th December 1936.[6] They had to decide whether the time had now arrived for them to make a definite decision on the question of transmission standards. The contracts made with both companies contained a reference to the London Experimental Period which was defined as terminating on the date on which a decision is reached concerning the system to be employed at the London Station. In the light of Cock's adverse account of the working of the Baird system (with the possible exception of the telecine equipment) the committee had to consider whether further expenditure on the development of a system which appeared most unlikely to survive was prudent. Failing this it could suggest that the system should be temporarily suspended until the apparatus could be handed over in a reliable and efficient state – and within a reasonable time limit. Cock foresaw that artists and celebrities might refuse to appear in Baird programmes thus

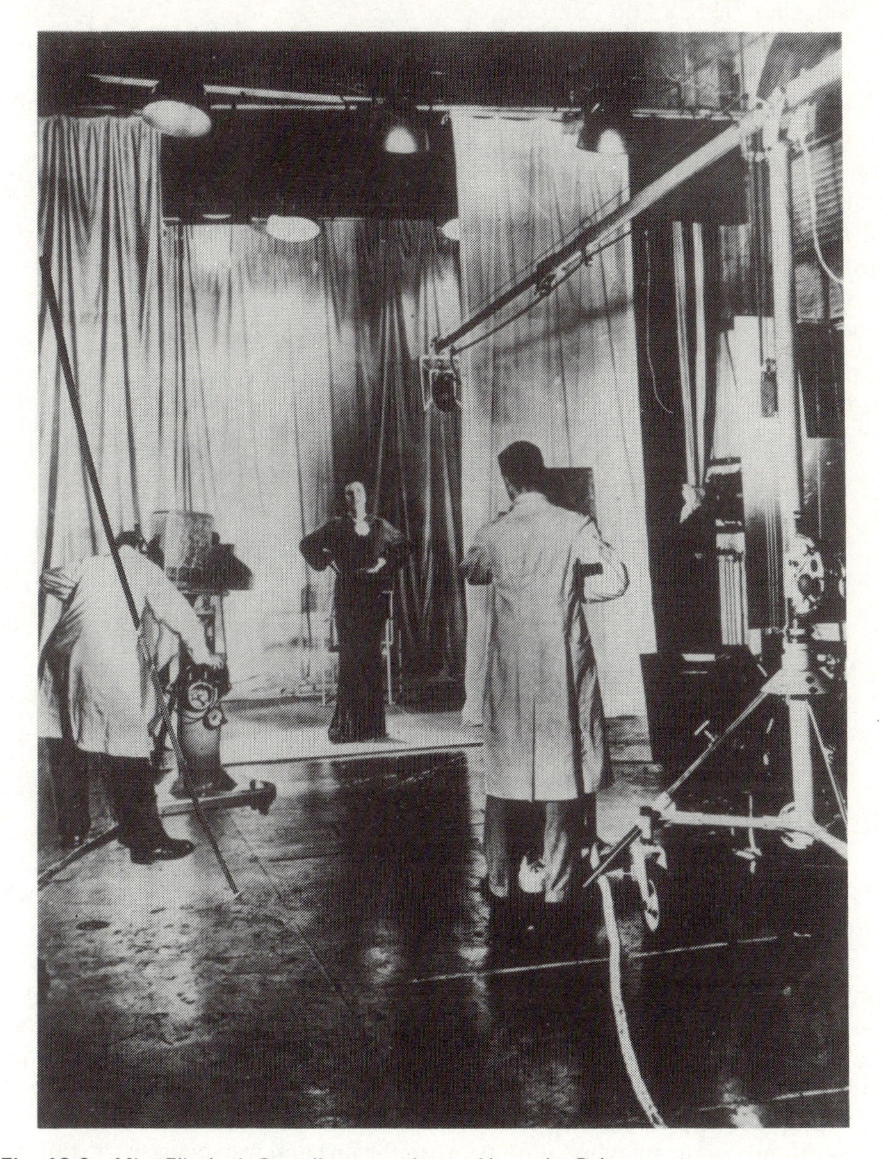

Fig. 18.6 *Miss Elizabeth Cowell announcing at Alexandra Palace*

making the problem of producing programmes still more difficult. There was no doubt the uncertainties and limitations of the equipment were having a deplorable effect on the production staff whereas with the Marconi–EMI system the apparatus was sufficiently advanced and reliable to enable interesting and entertaining programmes to be devised and transmitted with a 'high degree of reality and with complete confidence on the part of the producers and studio organisation' (Figs. 18.6–18.9).

Cock's report was not the only one that the TAC had received on the operation of the two systems. A. C. Cossor Ltd., which had developed a velocity modulation television system in 1933 after a remarkably short period of research activity, had produced, for the Television Advisory Committee's benefit, a number of constructive and hepful papers on various aspects of television. The company had prepared a most useful account of television at Olympia in September and in November[7] had written to the TAC on the Baird and Marconi–EMI waveforms. Cossor's introductory comments on the former wave form set the scene: 'We should be very glad to know by what date we may reasonably expect this system to be operating on its published waveform.'

Fig. 18.7 *Photograph taken from the screen of a television set showing Miss Jasmine Bligh*

According to the data which had been published, the frame synchronising impulse should have been of 12 lines duration followed by a black mask of 8 lines, but at the time of the Cossor letter, the system was operating with a frame pulse of approximately 27 lines duration while the black mask was usually missing altogether or came prior to the frame synchronising pulse where it served no useful purpose.

Additionally, the phase relation between the line and frame pulses, instead of being fixed, changed in a random manner with an excursion of about half a line. For the receiver manufacturers these difficulties were objectionable, as it meant they had to allow an artificially long black mask at the top of the received picture in order to permit the line synchronism to recover after the frame pause.

The effect of this was plainly stated by Cossor: '. . . we estimate that the useful number of scanning lines falls below 200, so that it might even be questioned whether the system at present conforms to the original stipulation of the Television Committee as to the number of lines.'

In the case of the Marconi–EMI waveform the published data stipulated only a minimum of 20 'lost' lines per complete picture. However, Cossor noted that M–EMI's system appeared to have settled down to a very satisfactory standard with a total of 38 'lost' lines per complete picture made up of 15, 8, 15 lines for the bottom mask, frame signal and top mask, respectively.

Fig. 18.8 *Leslie Mitchell, announcer, as he appeared on the screen of a television set*

With Cock's and Cossor's reports in mind, the TAC quickly came to a conclusion: they would recommend to the Postmaster General the adoption of Marconi–EMI's transmission standards as the standards for the London Station, at any rate for the next two years.[6] Bairds were to be given the opportunity of making representations, before the public announcement, to the committee on 23rd December.[8] The discontinuance was to date from 2nd January 1937.[9]

Bairds reply[10] to the TAC's letter of the 16th December was surprisingly moderate and low key in tone, far different from the letters which Moseley used to write on behalf of the Baird Companies when events were not in Bairds favour. Apart from a mild reference to the very short time which had elapsed since the start of the transmissions and an expression of the company's belief that the committee could not have made its recommendation on the basis of any defects or inferiority of the Baird transmitter, the letter contained no objection to the committee's decision. Indeed, the Baird Directors were in principle in accord with the view that one standard was preferable to two provided certain conditions were satisfied. So comparatively mild was their reply that it gives the impression that they were not dissatisfied with the recommendation. The Directors, of course, would have been well aware of the broadcast results of their system *vis-à-vis* those of Marconi–EMI, and while Cock's report was private and confidential to the BBC and the TAC, nevertheless Bairds' engineers must have given some indication to their directors that all was not well with their Alexandra Palace equipment. Unhappily those engineers had their difficulties compounded by the disastrous fire which had occurred on 30th November at the Crystal Palace. John Baird wrote:

> In the fire we lost a hundred thousand pounds' worth of apparatus. True, a good deal of the value of this was recovered from the insurance, but the immense disorganisation, loss of time, and loss of valuable records . . . occurred at the most awkward possible time and interfered seriously with our transmission from the BBC. Spare parts and apparatus which we were about to install were destroyed in the fire.[4]

This meant that Bairds could not offer immediately any equipment at the new standard.

Fig. 18.9 *405-line image showing a cookery demonstration*

Bairds' view that the period of transmission since 2nd November had been inadequate for forming a conclusive opinion about the two systems was one which had been mentioned by the Postmaster General but Selsdon observed that it had to be borne in mind that the two systems had been under close observation for some weeks before the opening of the public service in November.[11] An additional point on the speed at which it was proposed that the announcement should be made was that it would preclude any possibility of the matter being raised in Parliament. Members of the Baird Board had shown themselves singularly adept in the past in mustering support in Parliament and the press when required. Perhaps Selsdon, or the BBC representatives Ashbridge and Carpendale, had this in mind at the 31st meeting of the TAC.

Margaret Baird, in her biography of her husband has written: 'The Board was relieved by the turn which events had taken. Clayton, Ostrer's accountant, observed that the transmissions through the BBC had done nothing but lose money, the Company's only hope of making money being the sale of receivers.' This may indicate the reason for the mildness of the Baird Company's letter of 22nd December. It is interesting to note Baird's view after the Crystal Palace fire: 'However the fight was not yet over and had my friend Sydney (Moseley) been with us, I think we might have won, combine or no combine. As for myself, I at least had realised the position but had not the ear of Isidore Ostrer, the one man who controlled our destinies – and without that I was impotent.'

The Baird letter did contain a rather half-hearted request that the committee should recommend that facilities should be given to their company 'notwithstanding any patent difficulties to modify the plant and apparatus supplied by (the) company to make it operate in accordance with the proposed new standard . . .'. However, the minutes of the meeting between Sir Harry Greer, Mr. H. Clayton and Major A. G. Church and the TAC do not indicate that the Baird representatives pressed this point home with vigour. Possibly the above statement was made in their letter as a face-saving exercise. Moseley's book on J. L Baird[12] contains a sentence which tends to support this view. He wrote: 'The Directors argued that transmissions were not of much consequence: it was the sale of Baird receivers that mattered; *that* brought in the money.'

Certainly Sir Harry Greer and his colleagues were very concerned about the desirability of guarding against any monopolistic control of the manufacture of television receivers. In particular they were apprehensive regarding the position of their company relative to the Ballard patent 420,391 employed in the M–EMI system of 405 lines, in the event that the committee's decision to adopt the standard was approved by the Postmaster General. So concerned were Bairds on this matter that they took the precaution of seeking Counsel's opinion on the subject before their meeting with Selsdon and his colleagues.

Patent 420,391 was to assume great importance in the development of television – comparable possibly to the famous four seven's patent of Marconi. Bairds' Counsel (W. A. Jowitt and R. Burrell) came to the following conclusions:[13]

1 The adoption of the proposed standard would result in any transmitter which operated according to it coming within the ambit of one or other of the claims contained in the patent.
2 Once this standard was adopted at the London Station, its extension to other stations seemed inevitable.
3 The alteration of the Baird transmitter or the supply of new apparatus to conform to the standard, would be impracticable without the consent of the proprietors of the Ballard patent since no transmitting authority would

contemplate the use of such apparatus unless a licence had been granted under the patent.

4 The validity of the patent could only be tested if proceedings for infringement were to be taken by the proprietors of the patent, or if a petition for revocation were presented to the court.

5 Receivers adapted to receive the standard transmission of the M–EMI Television Company could plausibly be said necessarily to come within the scope of some of the claim of the Ballard patent.

6 'It follows that the adoption of the proposed standard (would), if the Ballard patent were valid, and if such plausible construction be held to be the true one, result in establishing a monopoly in the manufacture of receiving apparatus.' This was contrary to paragraph 51 of the Report of the Television Committee.

7 EMI's letter of 19th August 1936 on the patent was of very little value. (This stated that 'it (was) not in fact the intention of the writers to take any action under patent 420,391 in respect of television receivers such as you describe, to restrain their use for the purpose of receiving for entertainment purposes television transmitted from the transmitter supplied to the BBC by the M–EMI Television Co. Ltd.'). It was observed in the first place 'that it dealt merely without intention. In the second place, it covered only the use of the receiver for the reception of transmission, and did not extend to the manufacture of the receiving sets themselves. Thirdly, it relates only to the particular transmitter supplied to the London Station.'

8 The Ballard patent would 'if tested in legal proceedings, be held by the court to be invalid on the grounds of ambiguity, insufficiency, absence of patentable subject matter and probably also on anticipation'.

9 'In all the circumstances the decision to adopt the standard proposed (would) result in all manufacturers of receiving sets being placed in a position of serious uncertainty.' There was the risk that manufacturers might be forced into the position either of taking licences, involving possibly the payment of large sums by way of royalties, or of incurring the risk and expense of contesting the validity of the patent in legal proceedings.

The importance of the Ballard patent arose from the fact that it stated the only way of achieving interlaced scanning, unless a specialised receiver was used for reception from a specialised transmitter. As noted previously Shoenberg told Lord Selsdon: 'This is the only way I know of, and if I knew of any other I should have patented it'.[14]

Shoenberg was a patent expert and the Television Company had the exclusive right to all patents relative to television which originated with GEC of America, RCA, Telefunken, Marconi Wireless and EMI.[15] All these patents were available to Shoenberg. He stated before the Television Committee '. . . the technical value of the system does not after all depend on whether it is covered by patents or not. If the system is the most logical that could be chosen at the present stage it is necessarily the best, and we have tried to cover it by patents.' To this observation Lord Selsdon said: 'I think the answer is that they have picked the plums out of the pudding, in so far as they can find any plums.'[14]

With Jowitt's weighty conclusions before them the Television Advisory Committee had to give very careful consideration to the problems raised by patent 420,391. It was thought Shoenberg might be asked a number of questions pertinent to this patent and in addition Angwin undertook to obtain the opinion of a patent expert regarding the validity of the Ballard patent; first, as a whole, and second, with special reference to television receivers. The questions were as follows:[11]

1 and 2 Would the Marconi–EMI Company be prepared to give an undertaking that they would not take any action for infringement of the Ballard patent 420,391 or any other patent concerning interlacing owned or controlled by them, against any manufacturer or user of:

(i) a television receiver designed to receive television transmissions involving interlacing and emanating from any BBC station?

(ii) any item of transmitting apparatus, not supplied by the Marconi–EMI Company which might be incorporated in a television system involving interlacing operated at any BBC station?

3 If the BBC should at some future time desire to incorporate in a television system operating on a standard of 405 lines (interlaced) at one of their stations items of other than Marconi–EMI design and manufacture, would the Marconi–EMI Company be prepared to quote for the supply of such items of their design as the BBC might require in connection with such a system?

At the 36th meeting of the TAC Shoenberg thought the answers to these specific questions would be:[15]

1 The company would not at any time take action for patent infringement in the circumstances postulated; and they would be prepared to give a written undertaking to this effect.

2 The company would give an undertaking so far as the London Station was concerned, not to take any action for patent infringement in the circumstances postulated during the next two years, and they would further undertake not to do so at any time in respect of any such item of apparatus which might be installed in the London Station during the next two years.

3 The company would retain liberty of action to quote or not to quote for the supply of such apparatus.

Shoenberg appeared to make his company's position quite clear: on the one hand EMI had no wish whatever to stand in the way of progress and would not attempt to frustrate the introduction of any apparatus, no matter from what source it might come, which was definitely superior to their own, while on the other hand EMI, having spent large sums on television research, considered that they were entitled to reasonable royalties on their patents. Naturally Shoenberg hoped that his company's system would be favoured by the committee – and they desired above all that a decision should be given one way or the other without delay – but he told Selsdon that apart from the understandings suggested in answer to question two above, his company would regard the employment of their interlacing principle in connection with transmitting apparatus other than their own as a violation of their patent rights. Shoenberg felt that a decision should be given quickly as no orders for the supply of apparatus could be expected from foreign countries and additionally it was very desirable from a national point of view that an extensive and highly specialised television research organisation such as they had built up should be kept in being.

Marconi–EMI's position, as stated by Shoenberg, seems perfectly reasonable: the company had assembled a superb research team and was spending at least £100 000 a year on research and therefore were entitled to a fair return by the way of royalties. After Shoenberg's departure from the above meeting Sir Noel Ashbridge observed that 'it was clear that the introduction of interlacing was going to cost the BBC something extra' but thought a definite advantage would result on the receiving side. 'In these circumstances he thought that there might be some justification for admitting a degree of monopoly on the transmitting side.'[15]

Following the 36th meeting of the TAC the Secretary prepared a draft letter to the Marconi–EMI Company in order to obtain from them confirmation of Shoenberg's answers to questions 1 and 2 mentioned previously. The TAC attached great importance to this matter and particularly had in mind the terms of paragraph 51 of the Report of the Television Committee. Clearly they were loathe to give a decision on the adoption of the standard to be chosen for the London Station until the Hayes company had given some assurance to the committee on this subject. However, of course, the company did not wish to make all its hard won expertise and knowledge freely available to any manufacturer of television apparatus. Furthermore, the TAC wanted an undertaking to the effect that if at any future time it should come to the conclusion that the use of an item of equipment which was not of Marconi–EMI design and manufacture, and which would result in a material improvement in the television service, the item could be used without the company withholding their co-operation in the supply of other items of apparatus.

The draft letter to M–EMI was seen by Sir Raymond Woods, the Post Office's solicitor. He thought the undertaking

> . . . in order to be effective must be in a form which (would) be enforceable in the courts by the Postmaster General or the British Broadcasting Corporation, who (would) not be directly concerned in such a claim . . . An effective agreement could be framed upon the lines that in consideration of the adoption of the EMI apparatus as standard, the company

> covenants not to claim royalties, etc. under the Ballard and interlacing patents from manufacturers or users of receiving apparatus making use of these patents, or alternatively, to grant free licence to such manufacturers and users. Such an agreement should be in a form in which it could be enforced by injunction at the instance of the other party to the agreement.[17]

In view of the importance and complexity of the agreement Sir Raymond Woods advised the TAC that it would be well to have the draft agreement settled by Counsel who was familiar with the practice of the Courts of Equity by which injunctions were granted.

Bairds' representatives, Mr Clayton and Major Church, were given some indication of the committee's discussion with Mr Shoenberg on the 4th January 1937.[18] Clayton agreed that so far as receivers were concerned an undertaking on the lines suggested would be very helpful. He did not seem to be interested in the transmitting side and neither he nor Church made any objection to the proposed termination of the 240-line, 25 frame per second standard. This was to be effective tentatively from 16th January 1937.[18] Again both Clayton and Church told Selsdon and his colleagues that it would be a good thing to get on to a single standard of transmission quickly. The only note of a very slight dissension from the TAC's proposals concerned their inclusion of the 240-line, 25 frames per second standard in the suggested press announcement. Bairds were prepared to have this deleted – probably because of its association with their company – but Lord Selsdon thought it would be necessary to include it, in fairness to the owners of the sets, in order to make the announcement quite unambiguous.

Bairds future interest in receivers rather than transmitters was confirmed when the BBC was given approval to enter into negotiations with the company, after the decision on the London standards had been announced, for the return of the Baird apparatus at Alexandra Palace to the company, subject to an undertaking that no further claim for payment in respect of it would be made beyond the sum already paid to the company, namely, £19 000.[19] The company told Ashbridge they did not wish to take any of it back.[20]

At the above meeting Major Church asked whether permission could be given for the company to make experimental transmission over 'the ether from the Crystal Palace or elsewhere in the neighbourhood' in connection with their television research and Clayton referred to the large screen (visual public address) experiments which Bairds were conducting at the Dominion Theatre. Cinema television had been an interest of Baird for many years (he had given a demonstration of large screen television in 1930), and now that Baird Television had the backing of the powerful Gaumont British Company it was obvious that the Television Company should turn its attentions to this form of entertainment.

Baird wrote:[4]

> It seemed to me that now we should concentrate on television for the cinema and should work hand-in-glove with Gaumont-British, installing screens in their cinemas and working towards the establishment of a broadcasting company independent of the BBC for the study of television programmes to cinemas. I reported this view to the Board, but it was rejected. However, in my little laboratory at Sydenham within narrow limits I had a free hand and had built up a big screen and projector. This, with Ostrer's consent, had been installed in the Dominion Theatre, so that some sort of start has been made. The BBC decision was a blow to Ostrer and he was thoroughly dissatisfied and was even thinking of withdrawing his support from the company when, by a heaven sent opportunity, I was thrown in contact with him at the Television Exhibition of the Science Museum. I was filled to exploding point with enthusiasm for cinema television and let him have it in full force.

The exhibition was held from June to September 1937 and thereby gives an approximate date in Baird's account which is otherwise imprecise in chronology. It does however give significance to Clayton's and Church's references, to the TAC, to ultra short wave experimental transmissions and large screen experiments: the former were presumably required to enable television transmissions to be sent from the Crystal Palace to cinemas in London independently of the British Broadcasting Corporation. Selsdon thought every facility would be granted to the company for conducting experimental work 'consistent with the public interest and the requirements of the BBC'.[18]

Further consideration[21] of the draft letter to M–EMI and the Post Office solicitor's amendments was given by the TAC at its 38th meeting on the 11th January 1937. Sir Frank Smith did not agree with the solicitor's suggested addition and thought it improbable that the company would give such an undertaking.[22] He felt that the position was adequately covered by the provisions of the Patents and Designs Acts and circulated extracts from these acts, with his comments, to the committee members: he was strongly of the opinion that it was unwise to antagonise a company upon whose co-operation in television research 'we should inevitably depend to a great extent in the future'. This was a perfectly reasonable point of view but unless adequate safeguards were secured the company would have been put in a very strong monopolistic position – and in the light of past experience with the Marconi Company this was to be avoided if at all possible.

Sir Raymond Woods was inclined to favour reliance on the Arbitration Acts (as suggested by him in his amendments[21]), rather than on the Patent Acts, as under the latter compulsory licences could not be granted until three years had elapsed from the date of the sealing of the relative patent, unless the withholding of licences was prejudicial to trade as a whole or that it was in the public interest that they should be granted. Woods argued it might be difficult to establish this point.

After further discussion the committee decided to modify the solicitor's original wording – while at the same time including a reference to the Arbitration Acts. They also included an incentive to reasonableness on the company's part: 'If the company agrees with these proposals and will signify their willingness to embody them in formal undertakings, the committee will tender advice to the Postmaster General who will then no doubt consider in what way contractual effect can best be given to the foregoing arrangements.' Acceptance of the proposals was to be regarded as constituting a gentlemen's agreement and would allow a public notice regarding the standards to be adopted without further waiting for the conclusion of the legal formalities.

The above mentioned letter was considered by Marconi–EMI Ltd., the Marconi Wireless Telegraph Company Ltd., and Electric and Musical Industries Ltd. and while they could not commit themselves to undertakings in quite such broad terms as those indicated in this letter, they were prepared to give certain assurances which they thought should be satisfactory to the committee, namely:[23]

1 (See question 1) EMI Ltd. were willing to give an undertaking that they would not take any action against any person for infringement of patent 420,391 'by reason of the manufacture, sale, or use of a television receiver intended to receive television transmissions involving interlacing and emanating from any BBC station and used for entertainment purposes – for the whole life of the patent.
2 (See question 2) Marconi's Wireless Telegraph Company Ltd. were prepared to give an undertaking not to take any action for infringement of patent 420,391 against the manufacturer (or the BBC as user), 'of any item of transmitting apparatus not supplied by the Marconi–EMI Television Company which within two years from a date to be agreed is incorporated in the television transmitter supplied by the Marconi–EMI Television Company for the London Television Station but only while it remains so incorporated'.

With regard to the request that the company should give an undertaking to the effect that, if it was decided to install a system composed in part by the company's system and 'embodying parts of other systems', they should give their full co-operation in respect of their equipment at certain prices, Shoenberg and Van de Velde, the signatories to the letter said 'such an undertaking would go far beyond what was necessary to comply with the general assurance that (they had) no wish to stand in the way of progress . . .' They pointed out that such an undertaking would mean in practicable terms that they would be under an obligation to help competitors who were unable to produce as good a system as theirs, by supplying them with the important parts which they were unable to supply themselves. M–EMI could not, therefore, give this undertaking. However, they had no intention of being unreasonable and said that if such a case arose they would not wish to resist an application for a licence on satisfactory terms.

At the 39th meeting of the Television Advisory Committee held on the 22nd January, Lord Selsdon and his colleagues discussed Marconi–EMI's letter and agreed[19] that the suggested undertaking regarding receivers and the incorporation in the transmitting apparatus at the London Station, during a two year period, of any item not supplied by the company and the renewal of the item might be accepted. The proposals made in the remainder of the letter were not however acceptable to the committee for three reasons:

1 after the expiration of the two year period at the London Station and at any time elsewhere, the company 'reserved any right they might have' and gave no indication that they would be willing to grant a licence on payment of royalty 'of reasonable amount';
2 the company were not giving the suggested undertaking relative to the supply of apparatus, or alternatively the granting of licences, in the event of it being desired at some future time to use the company's apparatus in conjunction with apparatus not of their make;
3 the company (did) not accept unreservedly the committee's assumption 'that no question of royalty would arise in regard to any tests of items of apparatus which the BBC might conduct at any station by means of purely experimental transmissions effected outside the hours of the regular service and without notification to the public'.

Subsequently, in order to speed the completion of the draft agreement, Selsdon saw Shoenberg, on behalf of the TAC, and on 25th January the two spokesmen arrived at a mutually agreeable wording for the draft.[24] This was confirmed by Marconi–EMI[25] a few days later and on 5th February a press notice was released announcing the adoption of the company's standards for the London Station.

The BBC's last transmission using the Baird system was sent out on 30th January 1937.

> And so, after all these years, we were out of the BBC. The fact that it was the RCA system, imported from America, the scanning used being covered by the RCA–Ballard patent and the transmitter being the iconoscope of Zworykin and the research department, did not hinder the Marconi Company proclaiming the system all British. The iconoscope was now called the emitron. Ballard was ignored, and in an amazingly short time the Marconi publicity department had established it in the public mind that Marconi had invented television, (Baird).

The Television Advisory Committee's recommendation, approved by the Postmaster General, was a 'terrible blow' to Baird. 'It seemed that he had been forgotten by the world', his wife noted in her biography of him. 'He bore up with practically no mention of his troubles to anyone and presented himself to his assistants and the press with his usual calm exterior, but I knew that inwardly he was seething.'

Baird, however, had been a sufferer in adversity for many years and did not intend giving up his life-long interest. If the BBC would not allow him to transmit from Alexandra Palace, it seemed to him that now 'being out of the

BBC (he) should concentrate on television for the cinema and work hand-in-glove with Gaumont British installing screens in their cinemas and working towards the establishment of a broadcasting company independent of the BBC for the supply of television programmes to cinemas . . .'

References

1 BBC Chairman: speech at opening ceremony, 2nd November 1936
2 Television Advisory Committee: minutes of the 32nd meeting, 15th October 1936, Post Office bundle 5536
3 SELSDON, Lord: speech at television opening, 2nd November 1936
4 BAIRD, M.: 'Television Baird' (HAUM, South Africa, 1974)
5 COCK, G.: 'Report on Baird and Marconi–EMI systems at Alexandra Palace', TAC paper no. 33, 9th December 1936
6 Television Advisory Committee: minutes of the 34th meeting, 16th December 1936, Post Office bundle 5536
7 A. C. Cossor Ltd.: letter to the Secretary (TAC), 13th November 1936, Post Office bundle 5536
8 Secretary (TAC): letter to Baird Television Ltd., 18th December 1936, Post Office bundle 5536
9 Draft of public announcement, Post Office bundle 5536
10 Secretary (Baird Television Committee): letter to J. Varley Roberts, 22nd December 1936, TAC paper no. 34, Post Office bundle 5536
11 Television Advisory Committee: minutes of the 35th meeting, 23rd December 1936, Post Office bundle 5536
12 MOSELEY, S. A.: 'John Baird' (Odhams, London, 1952)
13 JOWITT, W. A., and BURRELL, R.: 'Joint opinion', 22nd December 1936, TAC paper no. 35, Post Office bundle 5536
14 Television Committee: evidence of Clarke, A. and Shoenberg, I., on behalf of Electric and Musical Industries Ltd. and The Marconi–EMI Television Co. Ltd., 27th June 1934, Post 33/4682
15 Television Advisory Committee: minutes of the 36th meeting, 29th December 1936, Post Office bundle 5536
16 Secretary (TAC): draft letter to Marconi–EMI Television Co. Ltd., paper no. 36, Post Office bundle 5536
17 WOODS, Sir R.: Post Office Solicitor's advice, 6th January 1937, Post Office bundle 5536
18 Television Advisory Committee: minutes of the 37th meeting, 4th January 1937, Post Office bundle 5536
19 Television Advisory Committee: minutes of the 39th meeting, 22nd January 1937, Post Office bundle 5536
20 Television Advisory Committee: minutes of the 40th meeting, 26th February 1937, Post Office bundle 5536
21 Television Advisory Committee: minutes of the 38th meeting, 11th January 1937, Post Office bundle 5536
22 SMITH, Sir F. E.: comments on the draft letter proposed to be sent to the Marconi–EMI Television Co. Ltd. by the Post Office, TAC paper no. 38, Post Office bundle 5536
23 Marconi–EMI Television Co. Ltd.: letter to the Secretary (TAC), 21st January 1937, TAC paper no. 39, Post Office bundle 5536
24 SELSDON, Lord.: letter to Phillips, F. W., January 1937, TAC paper no. 42, Post Office bundle 5536
25 SHOENBERG, I. and VAN DER VELDE, H. R.: letter to Lord Selsdon, 27th January 1937, minute Post 33/5536

Chapter 19

The service, 1936–1939

Following the inauguration of the television service in November 1936 a wide variety of programmes were produced, both in the studios on the premises of the Alexandra Palace and in the surrounding park. The studio programmes included extracts from West End productions, revues, variety, ballet and illustrated talks and demonstrations, as well as a weekly magazine programme of topical interest called 'Picture Page'. From outside the studio came demonstrations of golf, riding, boxing and other sports. Because of the limited facilities for rehearsal, it was seldom possible, in the early days, to attempt original productions in the studio, but valuable experience was gained in methods of presentation over a very wide field. Transmissions were limited to two hours per day, excluding Sundays.

Some of the programmes which were televised during November and December 1936 were:[1]

Studio productions:
- 'Citizen soldiers of London', a pageant reconstructed from the Lord Mayor's show
- 'Mr Pickwick', extracts from a new opera by Albert Coates
- 'The tiger', scenes from the play by Reginald Berkeley
- 'Murder in the cathedral', scenes from the play by T. S. Elliot
- 'Fascade', a ballet performed by the Vic-Wells Company
- Traditional plays and songs by the Children's Theatre Company

Variety:
- 'Cabaret cartoons', a vaudeville presentation with simultaneous drawings by Harry Rutherford
- 'Animals all', a programme by animal impersonators from the pantomimes
- 'Old time music hall', by veterans of variety

Talks:
- 'Inn signs through the ages', by Montague Weekly
- 'The modern house', a discussion between John Gloag and Serge Chermayeff
- 'Ships', a Royal Institution lecture, by Professor G. I. Taylor
- 'The pattern of 1936', a review of trade, finance etc., by Professor John Hilton

Outside broadcasts and topical programmes:

A tour of the North London exhibition at the Alexandra Palace

'Model theatres', from the British model theatre and puppet club exhibition

Armistice day programme

Anti-aircraft defence display by the 36th Middlesex anti-aircraft battalion, RE (TA), and the 61st (11th London) anti-aircraft brigade, RA (TA)

'Diary for 1936', a reconstruction of outstanding items in television programmes

Serial features:

'Picture page', a weekly magazine programme to which the contributors have included Jim Mollinson, Roger Quilter, David Low, Will Hay

'Starlight', including Bebe Daniels and Ben Lyon, Manuela del Rio, Lisa Minghetti, Lou Holtz, Sophie Tucker, Noni, Frances Day and George Robey

'London characters'

'Friends from the zoo', animals from the London zoo, introduced by David Seth-Smith

Several of the programmes were considered to be disappointing and the television correspondent of the *Daily Telegraph* wrote on 24th December:

> A visit to a local cinema brings into sharp relief the pathetic triviality of the BBC programmes. Television, with its champion cockerels and prize fishes, seems to offer no comparison as entertainment.[2]

Specific programmes which the correspondent commented upon included a 15-minute programme showing the stages in the construction of radio transmitter valves, which was given during an afternoon and then repeated in the evening; another programme devoted 15 minutes to a description of the mobile Post Office by an official who retailed a catalogue of 'dreary technicalities'.

Films, apart from newsreels, were all of the semi-educational documentary type made for a full-sized cinema screen, wrote the reporter, and some, for example, 'The Land of the Nile' and 'Fisherman's fortune' were each shown on four separate occasions.

Art programmes, too, did not escape his criticism. A number of televised examples of pottery, art and sculpture depended for their appeal on exquisite colouring and were 'meaningless in the black and white of television; others were of interest only to the connoisseur and the dillettante'.

Moreover vaudeville had not been strongly represented; but champion birds, beasts and fishes were being brought to the studio.

On the credit side 'Picture Page' and the television magazine were felt to be promising and successful features, and 'Starlight', which brought stars to the studio, was another, even though there seemed to be a shortage of front rank artists willing to broadcast for BBC fees.

The *Evening News* on the 10th December 1936 mentioned[3] that there was still a need for slicker timing of the television programmes to avoid frequent intervals, and the lighting and make-up problems had not been entirely solved though considerable progress had been made.

Mr J. H. Thomas, the general manager of A. C. Cossor Ltd., was reported on 21st December to have said that the television programmes were 'footling' and that unless speedy improvements were effected the programme material would 'not give television a chance'.[4]

Much of this criticism arose because of the limited facilities and money allocated to the service. At Alexandra Palace the staff had to produce shows without a rehearsal studio, and sometimes a performance had to be given with only an hour's rehearsal in the actual transmission studio. Producers, assistant producers and a larger technical staff were required: the BBC were prepared to appoint them but was faced with the problem of finding the money. Gerald Cock, the Director of television, was undoubtedly handicapped. Of the £180 000 granted by the BBC and government for television £110 000 had been spent on the station and another £20 000 was being allocated to the purchase of a mobile unit – only £50 000 remained for programmes and maintenance.[3] The cost of the service was being borne by the Corporation and until the television service was firmly established there seemed little likelihood of a separate television licence fee being approved. Potential viewers had to be stimulated and encouraged to purchase sets.

When the Television Advisory Committee agreed to the adoption of the Marconi–EMI system it was hoped that one of the consequences would be a simplification of receiver design and construction and hence a reduction in the cost of television sets. Lord Selsdon was very anxious to do everything within his power to develop television and had been rather concerned about the high cost of receivers. He did not have long to wait after the announcement of the adoption of the single standard for a reduction to take place, for on the 9th February 1937 *The Times* reported a decrease in the prices of the Marconiphone and HMV television sets – from 120 and 95 guineas to 80 and 60 guineas, respectively.[5] In addition it was stated that those persons who had bought sets at the original prices would be compensated by the makers.

This sudden drop in prices rather mystified Selsdon and Ashbridge[6] but it came about because EMI found the sales of sets, at the high prices then prevailing, so stagnant that they decided to make an immediate cut. Sterling, the managing director of EMI, hoped that Selsdon and his committee could somehow increase the number of hours television was being shown, without thereby incurring very much extra expenditure, so as to stimulate further sales.[7] He told Selsdon that if the sales showed good results it was EMI's definite intention to produce a much cheaper television set in the autumn. It was clear that the factors which led to increased sales were lower prices and more broadcasts. The Baird companies had found during the era of low definition television that the public did not rush to buy receivers when the total number of hours of television per week was low and/or the programmes were arranged at inconvenient times.

These facts were known to the BBC and the TAC but now there was an incentive, the prospect of a licence fee, for the BBC to increase its television transmissions, if possible. With these considerations no doubt in mind, Selsdon suggested to Ashbridge that an extra hour's reception for the trade should be provided between 4.00 p.m. and 5.00 p.m. 'The programmes,' he said, 'should consist largely of a continuous film which need not be varied frequently but which ought to be carefully chosen as giving good reproduction and in part by occasional appearances of the announcer with any small items, e.g. a pianist. The cost of such a demonstration programme would only be a fraction of the regular programme and I attach great importance to it from a sales point of view.'[7]

Marconiphone's and HMV's initiatives certainly produced immediate results. Osborne, the sales manager of the former company, was able to write to the press controller of the GPO on 19th February 1937 as follows:[8] 'May I say at once that while television sales at the moment show most excellent promise, there still seems to be some doubt in the minds of both trade and public concerning the stability of the present transmission system, and until this is finally removed I feel that it will have a detrimental effect upon future business.' Again Bairds had experienced a similar situation during the 30-line definition period. Public reaction to the new invention, while being one of considerable interest, manifested itself initially in a wait-and-see attitude. They were lothe to pay a high price for a television set which might become redundant in a short time and in effect they required an assurance that the 405-line standards would be in force for some appreciable time. The press announcement of 5th February 1937 had stated that the transmission standards would remain unaltered until the end of 1938 but Osborne thought this could be emphasised. 'The publication of this would, I am sure, react most favourably upon the buying public and would enable the industry as a whole to embark upon its production programme with the feeling of perfect security,' he wrote.

This point was discussed by the TAC on 26th February:[9] Phillips felt that a repetition of the press statement might wait until a press notice was required on some other aspect of the television service. Selsdon agreed but said that in the meantime it might be a good thing if arrangements could be made for a friendly question in the House to be put to the Postmaster General on the point. The secretary of the committee[10] consulted Shoenberg about this but he and Sterling were of the view that more harm than good would be done by ventilating the matter in this way. On the other hand they definitely regarded an extension of the television service on weekdays and the opening of a service on Sundays as being of great importance and expressed the hope that a favourable decision on this matter would be reached and publicly announced before Easter.

The BBC was of the opinion that the sales of television receivers should be monitored so as to give some indication of the growth of the new industry and in 1936 the Radio Manufacturers' Association was asked to compile such a record. After an initial comment that they were not in a position to do so in the absence of a special television licence, the RMA subsequently advised the TAC that they had requested their members to send to the auditors of the Association, Messrs. Derbyshire and Company, quarterly returns of the number of television sets sold. The first return was to be made up to 31st December 1936 and thereafter quarterly for one year only.

For the period to 30th June 1937 the returns showed the following numbers of sets had been sold:[11]

quarter ended 31st December 1936	427
quarter ended 31st March 1937	670
quarter ended 30th June 1937	347
	1444

Shoenberg reported that the number of sets sold by EMI to August 1937 was 1107, but he estimated that one half of these remained in the hands of the dealers. However, after the Radiolympia Exhibition in September, 64 EMI sets had been sold and this 'represented a considerable improvement on sales during June, July and August'.[11]

The comparatively poor response of the public to the purchase of receivers did not deter manufacturers from producing receivers and at the Radiolympia Exhibition 14 firms showed their products, namely:[12]

Baird	Pye
Cosser	Ecko
Ferranti	Halcyon
GEC	Invicta
Haynes	Kolster-Brandes
HMV	Marconi
RGD	Ultra

Prices ranged from 35 guineas to 170 guineas, with the cheapest being the GEC set which gave a picture 6 inch × $4\frac{1}{2}$ inch. The next cheapest sets were an Ultra at 38 guineas, with a picture measuring $7\frac{3}{4}$ inch × $6\frac{1}{4}$ inch, a Halcyon at 40 guineas (picture size 9 inch × 7 inch), and a Pye at 42 guineas (picture $7\frac{1}{2}$ inch × 6 inch).

The usual size of picture was 10 inch × 8 inch, and prices ranged between 50 and 60 guineas for a set giving television only; between 70 and 80 guineas for all-wave sets, and between 95 and 170 guineas for all-wave sets with autoradiogramophones. Bairds had a set with a picture $13\frac{1}{2}$ inch × $10\frac{3}{4}$ inch at 85 guineas.

The RMA's figures of sales show a marked decrease for the quarter ending 30th June 1937 compared with those for the previous two quarters. This fall may have been a seasonal one or it may have been owing to the TAC's decision to close the station at Alexandra Palace for a few weeks during the summer. The initial suggestion to do this had been made by Ashbridge on 26th February 1937.[9] He told the committee that unless the staff at the London Station was considerably augmented, which would have involved heavy additional costs, he did not see how the present staff who had been working under great pressure could be granted holidays while the station was in operation, and therefore suggested that the best thing might be to close the station during the month of August. Ashbridge anticipated that, in any case, very few people would wish to view television programmes in August. Phillips correctly forecast the probable effect: he was 'inclined to think that as television had been inaugurated as a regular public service it would give a bad impression to close it down for a period as proposed; the wireless trade would be antagonised, and owners of sets, (especially recent purchases), would have a grievance'.

Selsdon thought the Post Office would have to consider carefully how a step such as that proposed by Ashbridge would be viewed by the general public, but at the next TAC meeting[13] said that having regard to all the circumstances it would be justifiable to close the station in 1937 for 3 weeks at the end of the London season, say, from the 26th July to the 14th August. A further reason for closure was that it was necessary to carry out 'extensive reconstruction operations at the station and consequential rearrangements of studios and equipment which it would be inconvenient to do whilst the station was in operation'.

Phillips did not press his objections but Brown, on behalf of Sir Frank Smith and himself, dissented entirely from the proposal. However, because of the exceptional nature of the case and the small number of television sets in use it was decided to notify the RMA and Marconi–EMI so as to allow them an opportunity to draw to the committee's notice any objections to such a course of action, or any suggestions in regard to dates, b efore any public announcement was made. The committee also agreed that the announcement of an additional television period (in the middle of the day), should be made concurrently with the above notice. The Postmaster General, too, felt some unease about the issue, for he agreed in principle only,[14] subject to the committee reviewing the position in the event of strong opposition from either the RMA or M–EMI.

Phillips's misgivings were soon confirmed by Shoenberg: 'I need hardly say that our sales people will not be very delighted when the news of the suspension is finally announced, so I am sure they will be pleased with the extra lunch-time transmission.'

The BBC's press notice was released on 13th May 1937, and, although it referred to the need to carry out an overhaul and for certain internal adjustments to be made to the Alexandra Palace equipment, the real reason – 'to give the television staff a holiday' – was not lost on newspaper correspondents.[15] The *Wireless Trader* was not happy with the interlude '. . . three weeks without any television is a long time and the BBC must make every effort to restore the service in a shorter period . . . Thus, only a few weeks after the Coronation success of last week, both the trade and those members of the public who have bought receivers would be deprived for a substantial period of making sales and receiving entertainment respectively.'

With these and other reports critical of the suspension, it was clear the BBC had to make some arrangements to appease the television trade. The Corporation's solution[16] was to provide two transmissions each weekday, from 11.00 a.m. to 12.00 noon and from 2.00 p.m. to 3.00 p.m. during the period 26th July to the 14th August. The programmes consisted of the television demonstration film, short magazine films and newsreels, accompanied by appropriate sound, and exterior shots from the balcony with music from gramophone records. Each transmission commenced with a display of an announcement caption stating: 'BBC Television: Test Transmission for Radio Industry'. By using film material the BBC was able to reconcile the rather conflicting requirements of providing test transmissions for the television industry and markets and giving a vacation to the hard pressed staff at Alexandra Palace: as *Popular Wireless* noted '. . . the television holiday of the BBC (was) to be a half holiday only'.

This magazine, which in the past had been antagonistic towards the Baird system and had had Eckersley (the former Chief Engineer of the BBC), as its radio consultant, was now sharply critical of the BBC's recent suspension policy: '. . . only Gilbert or Lewis Carrol could do justice to the BBC 1937 . . . We must not have a close season for television – enough.'

The most outstanding outside broadcast television transmission during 1937 was the televising of the Coronation procession, at Apsley Gate, Hyde Park Corner, on its return journey from Westminster Abbey. Ashbridge had had outside broadcasts in mind for some time and in December 1935 reported that he had put forward enquiries to both Bairds and EMI regarding the supply of portable apparatus for this purpose.[17] Possibly he rememberd the success which Baird had achieved in 1931 and 1932 with the televising of the Derby, using low definition equipment. Now there was the prospect of high definition outside television broadcasts with the portable and versatile emitron camera.

Marconi–EMI were unable to undertake the necessary development work at that time because of their heavy involvement with the installation of the Alexandra Palace equipment,[17] but in June 1936 Ashbridge was able to inform the TAC that the company had quoted a figure 'of the order of £14 000' for the supply of an outside broadcast unit.[18] This would comprise three emitron cameras with associated scanning apparatus and a four ton lorry, but excluded sound and power equipment (Fig. 19.1). The quotation and specification mentioned that the cameras would be capable of operating at a maximum distance of 1000 ft from the lorry. On the basis of this information the Corporation was given approval to enter into negotiations with the Marconi–EMI company for the acquisition of an outside broadcast unit. Later in December the company tendered a quotation for the supply of equipment to provide an ultra short wave link between the OB unit and Alexandra Palace, but the price was regarded by Ashbridge as very high. A reduction seems to have been successfully arranged.

In the meantime Lord Selsdon had suggested that a small sub-committee consisting of Post Office and BBC representatives should be formed to discuss the transmitting arrangements for the outside broadcasts.[19] Two issues immediately arose: first, the nature of the events which were appropriate for television and their locations; and second, the means for sending the signals to Alexandra Palace for radiation from the main antenna. Westminster Abbey was suggested as a suitable site (particularly because of the forthcoming Coronation ceremony), together with Victoria Railway Station, Wimbledon and Wembley.

Fig. 19.1 *Marconi–EMI mobile television unit which comprised three emitron cameras and associated electronic and communication equipment*

Two means existed for propagating television signals from one place to another; either the use of a radio link or the use of a cable link. As previously noted the cost of a cable suitable for carrying video signals having frequencies up to 1·5 MHz was £300 per mile but in addition there was the cost of laying the cable in a duct and the total expenditure for such a cable link was thought to be about £900 to £1000 per mile for a cable capable of handling signals up to 2 MHz.

At this time (1935–1936), although Marconi–EMI had given some excellent demonstrations of their high definition system, there was a little doubt as to whether the 405-line system could really produce the best results of which it was capable. Because of this, Angwin of the GPO questioned the necessity to provide a coaxial cable designed to propagate 2 MHz signals, and felt that for all practical purposes a lower cost cable would suffice. He thought it might be found to be technically possible to utilise ordinary junction circuit cable for the transmission of signals up to 2 MHz.[20] In this case such cable was available for the televising of events in the Metropolitan area either independently or in conjunction with a ring of coaxial cable provided specially for television around London. Ashbridge seemed rather dubious about this prospect and emphasised the importance of obtaining good results from any outside television broadcast.

Meanwhile EMI had also been giving this matter some thought. Shoenberg and his brilliant research team at the Hayes factory had realised that outside broadcasts would take place, that radio link transmissions would be difficult, that a cable[21] was the best answer and that the Post Office had no experience of transmitting television signals. With characteristic foresight and courage Shoenberg had initiated a research programme to evolve a cable and associated terminal equipment which would transmit video signals over the 7 to 8 miles from the Coronation camera positions to Alexandra Palace.

The cable was manufactured by Siemens Brothers and taken to the EMI laboratories at Hayes where the TAC was invited to witness television pictures before and after the signals had been transmitted through the cable and terminal equipment. No appreciable loss of quality was observed and Angwin was able to report that the demonstrations had given excellent results and that so far as the cable link was concerned there was every reason for anticipating that the projected Coronation television broadcasts would be successfully sent out. Subsequently the cable was pulled into ducts by the Post Office and became their property. It was rented by the BBC.

The cable was of the twin or balanced pair type, and comprised two 0·08 inch diameter conductors each located centrally within a tube of paper insulation having an external diameter of 0·91 inch. The two tubes were twisted together and had copper screening laid around them: the whole structure was encased in a lead sheath. Over the frequency range between 10 Hz and 3 MHz the attenuation varied between 0·036 and 8·0 dB per mile, the velocity of propagation changed from 18 000 to 180 000 miles per second, and the characteristic impedance altered from 3000 to 190 ohms. Fig. 19.2.

The cable could be used in lengths of up to 8 miles without the need to employ repeaters, and, as the length of the cable route from Broadcasting House to Alexandra Palace was 7·25 miles, the headquarters of the BBC became aconvenient repeater station for OB transmissions into Alexandra Palace.

One of the problems which had to be solved before a satisfactory very high frequency radio link, for use with outside camera units, could be designed, related to the frequencies which could be utilised for this purpose. Fortunately this matter had been considered by the Wireless Telegraphy Board at an informal meeting held on 13th July 1934.[22] Angwin had represented the Television Committee's interests on the board as well as those of the Post Office and had proposed that the following sub-division of the waveband below 10 metres should form a working basis during the preliminary development of the VHF services:

Frequency range	Number of channels	Type of service
30–32	4	Television experiments and mobile
32–35	5	Mobile
35–38	5	Fixed and mobile
38–48	5	Television
48–50	20	Broadcasting
50–56	8	Fixed
56–60	–	Amateurs and experimental
60–85	20	Fixed
85–88	20	Broadcasting
88–100	5	Television
100–120	12	Mobile
120–130	–	Amateurs and experimental
130–150	–	Fixed
150–300	–	Unreserved

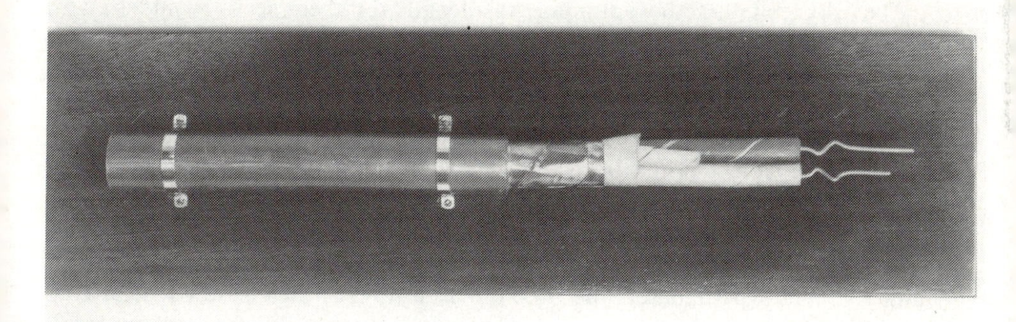

Fig. 19.2 *Twin-wire balanced television cable*

No recommendations for a sub-division below 1 m were made as there was 'very little data available regarding spacing of channels', although several manufacturers in this country had produced equipment operating on 17 cm, 26 cm, and 65 cm.

Provisionally it was agreed that television broadcasting should be confined to the 38–48 MHz and 88–100 MHz bands. The service could not have unlimited access to the 1–10 m band as certain important bands had to be reserved for aircraft services. The civil aviation industry was expanding rapidly and wireless facilities for navigation and communication were essential for its development.

In the London area, for example, about 20 aerodromes had to be provided with communication equipment and each one needed at least one separate frequency for the guidance of aircraft in bad weather. Other large towns having airports required similar frequencies. The safety of aircraft was dependent on the efficient operation of a number of very high frequency beacons which demanded a considerable number of independent channels; no other safety means were as suitable as these.

Confirmation of the allocation of the 38–48 MHz band for television was given by Angwin on 2nd October 1935. Later the band was changed to 40·5–52·5 MHz so as to clear the 30·83–40·5 MHz band which the ICAN had put forward, at a conference held in Paris, for use by blind landing systems.[23]

It is interesting to note that Angwin, in 1934, estimated that five television channels could be accommodated in the 38–48 MHz band. This was based on the assumption that 1 MHz would be needed for the bandwith of the signal and 1 MHz to cover frequency drift. However, the frequency separation of 2 MHz between channels was not, of course, suitable for the high definition 405-line system. A spacing of 7 MHz was necessary and consequently a frequency band was not available for an outside broadcast television link in the 40.5–52.5 MHz frequency range. An increase in the bandwidth was not possible as the Services were 'extremely reluctant to countenance the appropriation for television purposes of any ultra short waves outside the band already allocated' for this provision.

The BBC felt that the optimum wavelength for their needs would be approximately 5 m as experiments seemed to show that lower wavelengths of the order of 3 m were not at all satisfactory for point-to-point working where only low antennas were at hand at the transmitting end.[24] They had discussed this question with Marconi–EMI and had been told that to obtain an adequate performance from the link transmitter and the Alexandra Place transmitter it would be desirable to operate the two transmitters with a frequency separation of at least 7 MHz. A carrier frequency in the neighbourhood of 60 MHz was wanted for the link but this was in the 56–60 MHz band which was reserved for amateurs and experimenters. Because the utilisation of the OB transmitter was presumed to be very occasional and moreover was definitely of an experimental nature it was thought wise to raise the matter with Lt. Col. Miles, Chairman of the Wireless Telegraphy Board.
Miles wrote to Angwin on the 10th November 1936.[25]

> The Wireless Telegraphy Board are concerned to learn that the existing television band of 12 MHz is only sufficient for the operation of one television service in the London area. The actual wave chosen would interfere with the wave of 60 MHz allotted to and used by the police service who have a prior claim. The Board will agree to a wave in the (64–65) $\pm$ 2.5 MHz band for use by the BBC radio link transmitter on the understanding that the use is regarded as a temporary and experimental

one, that only a minimum power and mast height are employed and that the Board are advised if any alterations are proposed.

The radio link television transmitter which the BBC used on Coronation Day was intended for standby purposes only as the coaxial cables had been laid between the mobile control unit and Alexandra Palace.

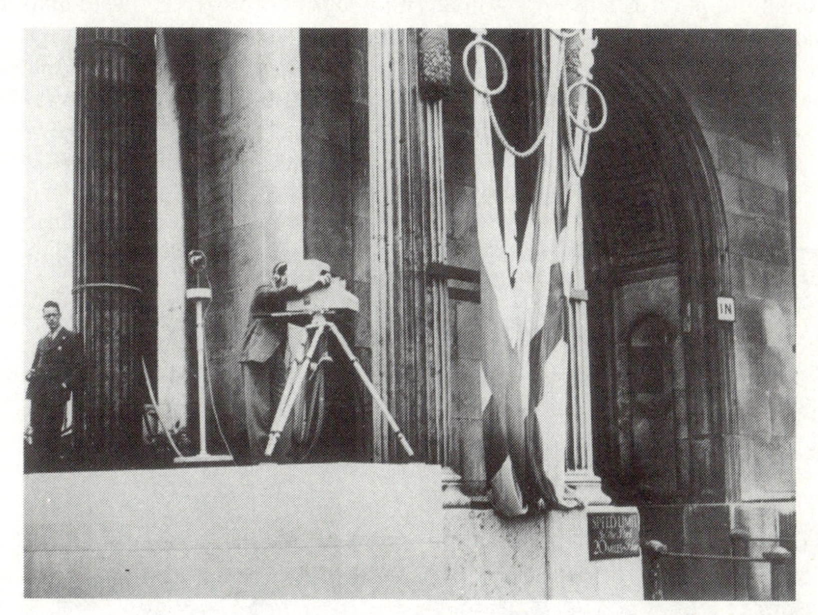

Fig. 19.3 *Emitron camera at Apsley Gate, Hyde Park for the Coronation. The commentator's microphone is visible on the left*

The earliest reference to a suggested route occurs in the minutes for the 26th meeting of the TAC. Selsdon commented that the cable might follow the route which included Regent Street, Trafalgar Square, Whitehall, Palace Yard and terminate at a point behind St Margaret's church.[26]

Three emitron cameras were utilised to televise the Coronation procession: two were mounted on a special platform at Apsley Gate and were fitted with telephoto lenses to obtain distant and mid-field shots of the procession and the crowds to the north and south of the Gate, and a third camera was installed on the pavement to the north of the Gate to give close-range views of the Royal coach and other important parts of the procession passing through the Gate (Fig. 19.3). The cameras were connected by about 50 yards of special cable to the mobile television unit behind the park keeper's lodge, whence the sound and vision signals were propagated by cable to Alexandra Palace.

The mobile television unit[27] comprised three large vans each about the size of a Green Line coach (Fig. 19.4): one contained the control apparatus and scanning equipment; one housed the power plant, and the third was fitted with the VHF radio link transmitter (1 kW). Marconi–EMI's system was entirely self-contained and most comprehensive in its specification. The control room unit held two rows of equipment mounted along the sides of the vehicle so as to leave a clear aisle for the engineers operating the controls. Each row comprised six racks each 7 ft 6 inch high and $19\frac{1}{2}$ inch wide: the total weight of the vehicle and apparatus was approximately $8\frac{1}{2}$ tons. The engineers were able to view the televised scenes on a monitor fitted into a compartment over the driver's head and if necessary make adjustments by means of controls mounted on the front panels of the equipment racks. In addition to the vision units the vehicle also included sufficient faders and amplifiers to complement the four microphones which were used to pick up local sounds associated with the scene being televised and the commentary given by F. H. Grisewood.

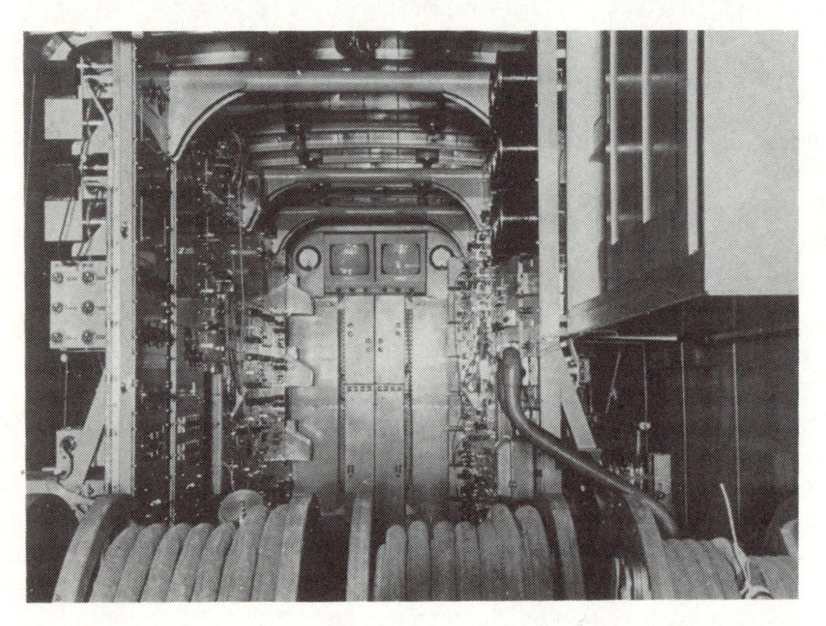

Fig. 19.4 *Interior of the Marconi–EMI mobile camera control unit*

The outside broadcasts of the Coronation procession represented an outstanding technical achievement by the Marconi–EMI Television Company for it involved the transmission of wideband signals by both radio and cable links. Great credit was due to Shoenberg and his research team for making the broadcast possible, especially as they had been heavily engaged for the whole of 1936 on the installation and testing of their equipment at Alexandra Palace.

The BBC was, of course, delighted by the success of its outside broadcast and was keen, as was the trade, for television broadcasting to prosper, but until it achieved a modicum of popularity as evidenced by increasing sales of sets, the Corporation was unlikely to gain much by arguing a case for a separate television licence to cover expenditure on the broadcasts. There were two conflicting aspects to the development of the service: first, from the BBC's point of view, there was not sufficient material available for three hourly periods a day, and hence this limited their broadcasts of television to periods from 3.00 p.m. to 4.00 p.m. and from 9.00 p.m. to 10.00 p.m. on weekdays; and secondly, from the buying public's viewpoint, the price of a good set with a large screen was as much as a small car. At a time when a typical weekly wage was of the order of ten to twelve pounds a prospective purchaser had to consider whether television compared favourably, as a form of enjoyment, with motoring, for example.

Nevertheless, a solution had to be found and it was obvious that the impetus for breaking the deadlock had to be initiated by the Corporation with advice from the TAC. As early as February 1937 – only four months after the start of the service – Carpendale had told the TAC that earnest consideration was being given to the question of providing an additional hour's transmission in the afternoon:[9] however, there was the issue of production cost to be reckoned with and the present cost was already causing the Corporation grave concern. Lord Selsdon suggested that the extended programme might consist largely of film to keep costs as low as practical, but Carpendale was somewhat apprehensive about the effect on the reputation of the service if this idea were implemented. The practicability of including the transmission of records of several of the more successful television studio items was also advanced.

At the next meeting of the TAC the BBC's controller outlined the plans of the BBC for the additional broadcasts.[13] These were to be for trade demonstration purposes only and would consist of films, which to a considerable extent would be repeated daily, comprising a specially made film dealing with the television service, instructional films, and newsreels.

The revised programme timetable was as follows:[16]

1 23rd and 24th August
3 hours transmission of film with sound announcements from 11.30 a.m. to 12.30 p.m.; 3.00 p.m. to 4.00 p.m.; and from 9.00 p.m. to 10.00 p.m.

2 26th August to 4th September
3 hours transmission from 11.30 a.m. to 12.30 p.m.; 4.00 p.m. to 5.00 p.m.; and from 9.00 p.m. to 10.00 p.m. (these were to be actuality transmissions)

3 6th September *et seq.*

10.30 a.m. to 11.00 a.m.	cruciform pattern accompanied by tone
11.00 a.m. to 12.00 noon	television demonstration film and film magazine with appropriate sound – for trade purposes only
3.00 p.m. to 4.00 p.m.	normal transmission
9.00 p.m. to 10.00 p.m.	normal transmission

Unfortunately, notwithstanding the extra hour of television for retailers of sets, the RMA's auditors noted that as few as 183 television receivers had been sold for the quarter ending 30th September 1937 – about half the number for the previous quarter. Clearly the position was serious; unless some action could be taken to stimulate sales the service would either die out or it would have to be handed over to private enterprise and be supported by advertisement revenue. There was thus a necessity for more service hours, particularly in the evenings, and a requirement for a set costing approximately £30.

The BBC in October 1937[28] planned to present two separate hours of television on Sundays in 1938, and to improve the programmes and their layout by utilising the theatre at Alexandra Palace. It was also intended to develop further the outside broadcasts by the acquisition of an additional mobile unit at a cost of £40 000. The total extra cost of these proposals was estimated to be £100 000 for capital expenditure and £70 000 for running costs. All of this was a move in the right direction of course, but Lord Selsdon felt that further efforts should be made to extend the evening programmes to a total of $2\frac{1}{2}$ hours of viewing – 'say, $1\frac{1}{2}$ hours of studio and outside broadcasts and 1 hour of good films'. He understood that old films could be made suitable for television reproduction at a cost of £25 to £30 for 60 000 ft.

The use of old film had attracted the attention of L. H. Bedford, of A. C. Cossor Ltd., and after the Radiolympia Exhibition of 1936 he recorded some most helpful comments on the types of films which were satisfactory for television programmes.[29] He found that a great deal of judgement had to be exercised in the selection of film material. One film which was shown at the exhibition was 'It's love again' but Bedford considered it a complete failure simply because the print was far too dense. Consequently insufficient light passed through the film and the noise/signal ratio was very high. A less dense print was required and it may have been this feature Selsdon had in mind when he thought old films could be made suitable for television.

Bedford's perceptive observations are quite important for they indicate some of the problems which faced producers of television programmes in the early days of the service. Some of his remarks were:

1 'Cover to cover' – Although this film contained some excellent shots which came over extremely well, it was on the whole quite unsuitable, because some portions of the film were quite unintelligible even to those who had seen it some twenty times. Films containing rapidly changing scenes . . . are unsuitable because the detail obtainable with television at the present date is hardly sufficiently good to enable one to appreciate what is happening in the short time available.

2 'Show boat' – This film was ideal because there was plenty of movement and good music.

3 'Rembrandt' – This film was most unsuitable because it does not take advantage of movement and it must be remembered that movement adds considerably to the value of a television image . . . The film itself was too dense to come over well and gave in general poor results.

4 'When knights were bold' – This film also was too dense for successful results to be obtained and, during a good portion of the run, Jack Buchanan was too small to be seen.
5 'Poste haste' – This film suffered greatly from lack of movement, and its lack of value for television purpose was apparently realised, as the film was removed from the programme.
6 'First a doll' – Excellent in every way. Plenty of movement.

Here, then, in Bedford's report was an early indication that film studio techniques could not be taken over *in toto* into the world of television production. Still, some films were appropriate for the new medium and Selsdon's suggestion seemed to provide a reasonably cheap and satisfactory solution for an extension of the television service.

Certainly cost was of central importance in this matter for the expenditure on television for 1937 was estimated at £65 000 capital and £330 000 running expenses, of which £50 000 was found by the Corporation out of 1936 income. Any proposals, therefore, which increased the number of television broadcast hours without incurring a substantial extra cost were of significance to the BBC.[28]

Selsdon's idea, however, was rather negated by the difficulty of obtaining old films at any price. It was thought that about 30 could be procured from Gaumont-British, but this number was much too small to sustain an additional regular period of television. In any case later enquiries by the BBC to the cinema interests on this matter showed that the film companies were displaying the greatest unwillingness to co-operate, and by January 1938, Graves, the BBC's Deputy Director General, reported:[30] 'It has proved impossible to obtain English and American feature films. Recently, however, it has seemed likely that some good continental films (with English sub-titles) may be secured; these can be used as an occasional experiment and afford temporary relief and change from studio programmes.' Again it was felt that there were not enough of such films to make a special film hour broadcast outside the normal times.

There was another possibility: if the film interests were unwilling to loan or hire full length feature films to the BBC would they consider allowing trailers to be shown on television?[31] Rosen, of the Radio Manufacturers' Association, put this idea forward as he was under the impression that the television industry in America intended using trailers as a part of their programmes when they started a service. It was felt such an arrangement would be attractive to the cinema companies, and, if so, it might pave the way towards establishing friendlier relations between the BBC and the film industry. However, there were difficulties:

1 the speed of film was too great for successful television transmission and some of the trailers were short and poor in quality;
2 the present number of viewers was too small to make the proposition attractive to film proprietors – if the potential audience had been substantially

greater they might have been induced to provide trailers more suitable for television;

3 the showing of trailers might be thought to infringe the government's ban on sponsored programmes, although Phillips thought this arrangement was defensible.

From the film interests point of view, the BBC, with its new service, was a potential serious rival with whom it was felt they should not co-operate, at least until their position *vis-à-vis* the Corporation became clearer regarding any deleterious effect on the film industry. Still, the BBC had one likely rejoinder to the cinema companies stance: if it had the power it could ban the cinema reproduction of their programmes unless the industry could be persuaded to make films available for television.

The use of films would have alleviated the provision of extra television broadcast time in the short term, but in the medium term the BBC had to provide more studio programmes and outside broadcasts. By the end of 1937, approximately one year after the start of the service, the shortage of studio accommodation at Alexandra Palace made it impossible for the evening programmes to be extended to two hours.[32] Fortunately there were two ways which would allow extra studios to be established: first, with the removal of the Baird facilities and equipment from the London Station the BBC could proceed to re-equip the Baird studio; and secondly, there was at Alexandra Palace a theatre which could be converted into a studio. The first of these proposals was likely to take four or five months to complete, while the second was thought to require a year's work.

In addition to the time element inherent in these suggestions, there was also the question of cost. The modification of the theatre into a studio with a series of sets and a central control tower was estimated to cost some £90 000, and, moreover was unlikely to provide a long term answer to the Corporation's problems.[33] The following issues had to be considered:

1 the conversion of the theatre for studio use did not afford full value for the extra cost;
2 the existing layout of the station was inconvenient;
3 with the above improvements the Alexandra Palace station might meet the requirements of the television service adequately for about five years;
4 if the BBC stayed at Alexandra Palace, the Corporation might find themselves without sufficient accommodation or equipment to meet the demands of the service as it grew.

Contemplation of all these points suggested to some members of the BBC that the theatre should not be transformed, but instead an immediate search should be made for another site on which an entirely new television station, properly designed, and with provision for future additions, could be constructed. Such a new station would not of course be available for the transmission of television broadcasts for two or three years. Because of this fact and the need for an interim solution to the problem of studio accommodation, the Television Advisory Committee was inclined to the view that any scheme which involved the complete abandonment of the Alexandra Palace station, apart from the transmitting plant, in about three years time could not be recommended and that consequently the theatre should be restructured.

Concurrent with the above issue there was also the immediate problem of furthering the sales of receivers. With the non-availability of British and American films for an extra programme, other ideas had to be discussed which would make television more popular to the lay public. Several points were advanced which were felt to be helpful;

1 an improvement of the production quality of the existing programmes;
2 the amalgamation of some sound and television broadcasts;
3 the expansion of outside broadcasts;
4 the need to give an assurance about the stability of the 405-line standard;
5 the necessity of pursuing technical developments so as to utilise the full capacity of the high definition specification of the London Station;
6 the reduction of electrical interference.

On the quality of television programmes, the *Observer's* television correspondent said[34] in July 1937:

> During the first month or two . . . the programmes were not too good, in fact they were, with a few noteworthy exceptions, of very low grade. The performers were not worthy of a very great scientific achievement. Also the engineers were often too ambitious. Everything was in an experimental stage and faults must be forgiven and forgotten; but it is a fact that the poorness of the early pictures did a great deal of harm to the young art.

This was not an isolated opinion, for in December 1937, M. M. Macqueen, the chairman of a deputation from the Radio Manufacturers' Association to the TAC, observed[31] that 'apart from occasional "high spots", much of the programme material was "poor stuff".' He was not sure whether this was attributable to lack of imagination on the part of the BBC or to lack of finance, but he suspected the latter. Outside broadcasts, short plays, and 'Picture Page' were regarded as being among the best items. Second rate cabaret items were severely criticised and films generally were regarded as being very poor. He acknowledged detailed improvements of the BBC television service during the latter part of 1937 and certain outstanding items such as the Cenotaph ceremony and 'Journey's end', but maintained that programmes were lacking in real entertainment value. Macqueen emphasised the need for a definite assurance that some improvement would be effected if members of his association were to endeavour to reduce prices.

Actually the BBC was well aware of this type of criticism, for at the 45th meeting of the TAC in November 1937 Carpendale informed[35] the members of the Corporation's plans for a better service. Of the estimated £69 000 increase in expenditure on programmes for 1938 about £38 000 would be employed to increase programme hours and the remainder would be used for improvements in quality. Selsdon asked if this was a sufficient provision as he was of the opinion that in order to make television a national success and a financially self-supporting institution it was necessary to provide periodic outstanding programme items, featuring stars, which could be widely publicised.

Macqueen's reference to the Cenotaph ceremony highlighted the area where television programmes could be extended without the need for extra studio space. Outside broadcasts, using the versatile and reliable emitron cameras, had been a feature of the Corporation's services since its inception, and some of these had been quite excellent. The Coronation and Cenotaph ceremonies, for example, showed the immediate advantage of television compared to the film cinema – the instantaneous transmission of images of actual events. Additionally, the Cenotaph service, perhaps unwittingly, pointed the way ahead for better pictures.

Jonah Barrington, of the *Daily Express*, on 12th November 1937 (the day after Armistice Day), titled his news report:[36] 'Television is perfect for two minutes'. He continued:

> As cars, buses, lorries outside switched off their engines and came to rest, so did the crackling fade from the sound reception and the spots from the viewing screen, rather as if some unseen smudge had been wiped off a palette. For those two minutes the picture came to us clean, clear and steady – like a photograph. For the first time we were seeing what long-distance television would be like if engineers could solve the interference problem.

On the broadcast itself Barrington had this to say: 'The transmission touched great heights. Close-ups of the King . . . and of the Prime Minister came through splendidly. We should have liked more.'

Here then was an indication that the BBC could send out worthy programmes, but the immediate problem was to reduce the time intervals when such superlatives, as Barrington had used, could be employed.

At a later date the RMA's Television Development Sub-committee indicated how television programmes could be improved and communicated their views to the TAC. Many of their points seem obvious when set against present day television, but at the time they were made they were clearly necessary. Some of them were:[37]

1 The duplication of programmes in the afternoon and evening of the same day is to be deprecated.
2 The programme transmitted on a Saturday evening should, in general, be of the variety type and consist of a number of short items.
3 In the normal variety programme, the items included are often very

similar in character and insufficient variation is introduced, both as regards the artists employed and the type of material used.

4 Transmissions of drama have usually been of too ambitious a character, the type of play generally selected has been too morbid and it is felt that short, one-act plays of a more cheerful nature would be preferable. In this connection, it is felt that so large a use of 'horrific' subjects – whether it is in plays or films – is to be deprecated.

5 It is apparent from the transmissions from the studio that in some cases adequate rehearsal has not been possible and the difference between the slickness and perfection of performance which is achieved when sufficient rehearsal has been provided is most marked.

6 It is desired to emphasise the importance of the part played by the announcer. It is felt that the brightness of the television programme is, to a large extent, dependent upon him.

Fig. 19.5 *FA cup final. Television outside broadcast from Wembley Stadium, 30th April 1938*

7 The method of displaying television programmes in the *Radio Times* is not such as to give proper prominence to star items. It is suggested that a better use of headlines should be made and, further that more space should be devoted in the *Radio Times* as a whole to television news.

8 The method of putting out television news is not calculated to secure the widest possible notice in the press. For instance, the announcement that the Cup Final (Fig. 19.5) would be televised was first made on the television screen during a television programme, with the result that its news value, as an item of interest to the press, was destroyed at the outset.

9 Any tendency to give information concerning forthcoming television events as a 'scoop' to individual papers is to be deprecated. A positive suggestion is made for a weekly press visit to Alexandra Palace, preferably on Thursday, in order that publicity in the Sunday papers might be secured.

10 As far as possible, in television programmes, all announcements should be accompanied by some picture, either of the announcer or illustrating the announcement being made.

The RMA's perceptive and sensible comments showed that much could be done to improve programmes, notwithstanding the limited studio space available. In the absence of immediate extra studio space, the BBC could expand its service by making use of more outside broadcast transmissions and here the Corporation had already taken steps to implement this. They had ordered an additional mobile unit and planned to transmit a number of attractive outside broadcasts from Twickenham, Epsom and several other places. By 25th January 1938 Graves[30] was able to give some details of the BBC's plans in this field. Unfortunately the extra mobile unit was not expected until the summer, and as the existing mobile unit required appreciable time for setting up, testing and subsequent removal to other sites, outside broadcasts could not be given in all afternoon programmes – even assuming good lighting conditions. When the new unit arrived the BBC planned to provide outside broadcasts on Sundays as well as weekdays and to increase their number. The Corporation recognised that whatever developments might occur in studio programmes the future of television would depend very largely on the televising of outside activities. All of this was not to be achieved without some difficulty for 'considerable opposition had already been experienced and more was anticipated in connection with the televising of important sporting events under commercial auspices.

On the issue of amalgamating some of the sound and vision programmes, which had been put forward by Brown at the 45th TAC meeting[35] and which had also been advanced by the RMA, the association thought that the television service might be advertised by the televising of artists appearing in the ordinary sound programmes – the listeners being informed that the act was being televised. Carpendale referred to some rather serious difficulties regarding the implementation of this idea; possibly they related to the artists contracts.

The RMA was particularly keen to the marrying of sound and vision programmes and raised the matter on a number of occasions. Sir Frank Smith and Colonel Angwin also supported such a move and urged the BBC to experiment with the television transmission of sound broadcasts as a means of solving the accommodation problem.

Towards the end of March 1938 the Corporation broadcast its first programme on sound and television wavelengths. The RMA's view was vindicated for the programme was 'very good' and they recommended a further extension of this procedure. They stressed that items from the normal sound broadcasting programmes should be televised to as large an extent as possible, the general policy being that the two services should be merged and not kept distinct. As an example of what they had in mind, the RMA suggested that when 'Music Hall' was broadcast from St George's Hall, in the sound service, the programme could also be televised. The difficulty was that theatre managers did not like the idea and the BBC found that their hostility made it impossible to obtain for television the stars who appeared in sound programmes.

On the technical side of television broadcasting two issues had to be faced by the BBC, the TAC and the GPO. First, the introduction of legislation to decrease the level of electrical interference from internal combustion engines, medical equipment and other sources, and secondly, the need to give the general public an assurance regarding the stability, for several years, of the London Station's waveform specification.

Both the Marconi–EMI Television Company Ltd. and Baird Television Ltd. had carried out extended tests on the transmission characteristics of horizontally polarised and vertically polarised electromagnetic radiations having frequencies in the 6 m to 8 m waveband, and both companies had concluded that the decreased interference obtained with horizontally polarised waves was not sufficient to justify the 'inconvenience of having a horizontal aerial' at the receiver.

The BBC was particularly concerned about this matter as their OB mobile units had to operate frequently in electrically noisy environments, unlike the cameras at Alexandra Palace. The worst disturbances were caused by electrotherapy apparatus in hospitals, but although the Post Office was willing to give advice on remedial measures in a number of cases, the users of such apparatus were unwilling or unable to meet the cost.[33] At the beginning of 1938 the Post Office was powerless to enforce the suppression of interference as it had no legal powers, but in the new Wireless Telegraphy Bill, which it was drafting at that time, provisions empowering the Department in concert with the Electricity Commissioners to compel owners of interfering apparatus to adopt preventative measures were included.

The second point mentioned, touching on standards, was more amenable to an immediate solution. Following the adoption of the 405-line standard in January 1937, the TAC had issued a press announcement confirming that the standard would not be substantially altered before the end of 1938. Consequently, a prospective purchaser of an expensive set at the end of 1937 could not be given an assurance from his dealer that the receiver he intended buying would be of use after about twelve months. The *Wireless and Electrical Trader*, in an editorial[38] published in October 1937, urged the TAC to look into the matter afresh and let both the industry and the public know where they would be standing in a year's time. For the editor the obvious practical course was to leave the Alexandra Palace transmissions unchanged and to install improved high definition equipment at the provincial stations when these became available. 'The disadvantages of having the London transmission somewhat out of date for a year or two would be as nothing compared with the need for elaborate adjustment at the public's expense of present day instruments,' the editorial stated.

This statement was discussed at the TAC's meeting[28] in October 1937, but not withstanding their discussion on agenda items dealing with sales of sets and the development of the television service, the general feeling of the committee was that no action on the lines indicated in the editorial was required. Their advice seems rather strange when set against the poor sales record and the need for a guarantee to be given to buyers. The committee, which was normally forward looking and perceptive on television matters, was singularly unappreciative of the point raised by the trade journal. However, the RMA understood the position well, and, at their meeting with the TAC in December, Macqueen reaffirmed the view that an official pledge to the public on the issue would assist sales. With this corroboration, the committee agreed to the publication of a press notice[39] on the subject. Initially they were of the view that the standard should remain unchanged for a period of two years only from 1st January 1938, as 'it would be undesirable to specify a longer period in view of the possibility of technical developments', but at the next meeting Smith suggested that the time span should be three years and this was concurred.

Confirmation was given in the reply by the Postmaster General to a question, in the House of Commons, raised by a Captain Evans, on 1st February 1938: '. . . the public may therefore purchase television sets without any fear that they will become obsolete or require substantial alteration for a very considerable time to come'.[40]

Whatever the individual contributions of each of the proposals, outlined above, towards the sale of receivers, there was certainly no doubt about their effect in 1938. The RMA's figures showed that the numbers of sets sold, in the periods indicated, were:[41]

Quarter ended	
31st March 1938	582
30th June 1938	446
30th September 1938	954
Month ended	
31st October 1938	1896
30th November 1938	1571
31st December 1938	1745

Whereas 2121 television sets had been sold by the 31st December 1937, a year later the total number of receivers sold was 9315, and by August 1939 the figure was 18 999: television sales were advancing rapidly, only to be stopped by the outbreak of war.

Actually there was no need to change the transmission standard, for the 405-line system was capable of being developed to give much better pictures than those shown in 1937. The RMA referred to this in December 1937 but the chief engineer of the BBC did not think anything dramatic could be expected in the way of image betterment. Conceivably he may have been reluctant to spend money on this aspect of the service when so much more could be done with the limited funds at hand.

The RMA's enquiry seems to have been quite valid, for the American Telegraph and Telephone Company had recently given a laboratory demonstration of television transmission along cable on a standard of 240-lines and Angwin had been informed that the definition realised had been comparable with the results reached over the air by the 405-line transmission. A similar observation had been made by Post Office engineers in 1935 when films had been televised by the Marconi–EMI and by the Baird Television systems working on 405-lines and 240-lines, respectively. These facts did not show the 405-line system in a poor light, but rather highlighted the improvements which could take place based on the existing waveform specification, as technical progress was advanced in electronics. Shoenberg and Blumlein had stated, to the Technical Subcommittee of the TAC, their belief that the 405-line television demonstrations which they gave in 1936 did not represent the ultimate performance of the chosen standard: as Shoenberg informed the sub-committee, one reason for the choice of 405-lines in 1935 was that it enabled future picture quality to be raised without the necessity for an alteration of system parameters. Blumlein repeated this view in 1938.

A surprising feature of this subject of line standards concerned the number of lines per picture. When the 405-line system was announced in 1935 there had been a considerable outcry, against the use of such a high definition, by certain sections of the radio and television industry, but now in April 1938, C. O. Stanley, a member of the RMA,[42] felt able to tell the chairman of the TAC that opinion in some technical quarters was that 'the present definition was not sufficiently high, and that the question of a higher standard should be considered now, when the number of sets which would need to be adapted was small'. This was rather a *volte face* for a section of the industry and vindicated Shoenberg's 1935 judgement.

Moreover, the resolution of the received image was dependent on the receiving equipment: if this were poor then clearly good pictures could not be reproduced. If, for example, the bandwidth of the various circuits in a receiver were 1 MHz instead of 2·5–3·0 MHz, degradation would result. Both Ashbridge and Angwin[42] thought that 1937 receivers did not take full advantage of the definition offered by the 405-line standard, and the BBC's chief engineer noted that if any improvement was to be made the number of lines per picture would have to be at least doubled. Nevertheless, the point brought up by Stanley was important, for the TAC had to reflect on the standard which should be imposed at the end of the guaranteed period, and associated with it there was the effect of a greater band of frequencies on both free space and cable transmission to be examined. In retrospect Ashbridge's view on future line standards – about 811 lines per picture – seems to have been rather pessimistic bearing in mind current British practice.

The poor sales of receivers in 1937 undoubtedly had one beneficial effect on the television service; it contributed significantly to the improvement of television productions in 1938. Both the BBC together with the TAC had had to examine critically the quality of the Alexandra Palace programmes to determine whether there were aspects of production technique which ought to be improved. The RMA, in particular, played a most useful role when it adopted its watchful stance, and was thereby able to proffer advice on many aspects of television broadcasting.

Some of the comments made by the Television Development Sub-committee of the RMA on television programmes for the fortnight ending on the 5th April 1938 give some measure of the progress which the BBC was making in the new medium:[43]

1 The programmes transmitted showed, in general, a marked improvement both as regards entertainment value in the items transmitted and the technique of the transmission itself.
2 The crazy programme 'Nice work' and 'If you can get it' were specially well done and very attractive in character.
3 The transmission of the boat race was another good feature. The idea of putting over a film of the race in a later programme on the same day was particularly happy and indicates, clearly, that there is always a possibility in such topical events as this that something may be obtained which can be repeated by the use of film later on. As a first attempt in this style of transmission the item was a very good one.
4 Transmission of 'Henry IV' was another welcome item and Ernest Milton in the title role was particularly effective. The success of this indicates the wisdom of the policy of cultivating personality in television transmissions.
5 The cabaret transmissions are appreciated, but it is suggested that their attractiveness might be increased from time to time if it were possible to arrange for a 'high spot' say once a month without detracting from the normal high level.
6 'Picture Page' is considered to be a very popular programme and should be continued with an attempt to improve, as far as possible, upon the material employed in it. It would appear that there is some paucity of good material which it is hoped will be found possible to overcome.
7 It is recommended that during the summer months, once a week for, say, a period of six weeks, outside broadcasts from the zoo would be welcome.
8 Transmissions of comedy films such as Mickey Mouse should be used as much as possible in the afternoon shows, as they are of particular appeal to children.
9 It is again suggested that consideration should be given to the possibility of transmitting in the television programmes items from the 'In town tonight' broadcast series.

10 Two outside broadcasts of particular interest and merit were the McAvoy fight on the 7th April and the soccer international on the 9th April. In both cases the arrangements made were altogether adequate and the transmissions resulting were of a very high quality.

11 'Cabaret cruise no. 4' on 11th April was also a marked success and its presentation must have done a great deal of good, especially where demonstrations were taking place to prospective customers.

12 The experiment of showing a semi-educational film was interesting and it is considered good provided it is put over in such a manner that it stresses the interest rather than the educational aspect.

13 It is felt that the Pepler Masque is a type of entertainment which had such a small appeal that it should not be transmitted very frequently.

Fig. 19.6 *Oxford and Cambridge athletic sports at the White City Stadium, 1938. A super emitron camera is being used*

The programmes offered to British viewers, in a weekly schedule of approximately 18 hours, could be classified generally as outside broadcasts, studio produced programmes and televised films. Of these the most newsworthy, and the most controversial in regard to their relationships with other interests, were the outside broadcasts. From the inception of outside broadcasting on Coronation Day, mobile television units enabled a section of the general public to see a wide variety of events in public, social and sporting life – the Wimbledon tennis tournaments, the Lord Mayor's Show, the Cenotaph ceremony on Armistice Day, the Prime Minister alighting from his plane at Heston after his visits to Berchtesgaden and Munich, the Derby, the University Boat Race, the Cup Final, Trooping the Colour, the Test Matches from Lord's and the Oval, boxing matches such as those between McAvoy and Harvey, and Boon and Danahar and so on (Figs. 19.6 and 19.7).

Until the end of March 1939 the London Television Service was the only high definition service anywhere which provided regular programmes for viewing in the home. In the United States a public television service was inaugurated on the

Fig. 19.7 *Super emitron camera*

occasion of the opening, at the end of April 1939, of the World's Fair in New York. The programmes were radiated from the top of the Empire State Building by the National Broadcasting Company and from the Chrysler building by the Columbia Broadcasting System,[44] and were limited to about two hours per week. Germany started a regular service, using the Berlin transmitter, in March 1939, and in Paris the first public televised programme to be seen was shown at the Theatre Marigny on 2nd April 1939.[45]

Some idea of the cost of providing a television service may be gained from the fact that the capital expenditure on the London Station up to 30th September 1938, less depreciation written off, was nearly £126 000, and the revenue expenditure up to that date, including depreciation and programme, engineering and staff costs, was about £660 000. To meet the rapidly growing costs the Civil Estimates for the last financial year ending 31st March 1939 provided an extra grant of 15% of the net listeners' licence receipts, additional to the normal 75% handed over by the Treasury, to be earmarked specifically for television and foreign language broadcasts. This was estimated to produce £610 000, of which approximately two-thirds, or £406 000, would be required for television.[46]

The growth in expenditure was matched by an increasing public interest in television and in March 1939 *The Times* noted: '. . . the very people who set out to popularise television are astonished by the response of the public to their campaign'.

Fig. 19.8 *HMV model 901 pre-war television receiver – price 60 guineas*

Several facts contributed to this success. The public found that sets could be bought at reasonable prices and the service had improved to such an extent as to inspire confidence in its future well-being. By the summer of 1939 between 50 and 60 different models, produced by some 20 manufacturers, were available to the public at prices ranging upwards from £22 for an add-on vision unit to a mains-operated sound receiver. Prices for one-piece sets giving a picture of $7\frac{1}{2} \times 6$ inch were about £30 and for sets offering 10 inch × 8 inch pictures the cost was almost £40.[46] This was a much improved retail situation compared to that which had prevailed two years previously (Fig. 19.8).

However, in addition to all this it was the advertising campaigns, with local exhibitions in and around London, which had focused attention on television. During the 1939 television exhibition held at Selfridges no fewer than 10 000 people had attended each day of the first week in March, and the sale of sets averaged £500 per day throughout the first fortnight of the exhibition.[44]

Symptomatic of this increasing awareness was the encouraging response to the television questionnaire organised by the BBC. More than 4000 viewers sent in completed forms although the questionnaire was a somewhat 'formidable document' occupying three typewritten pages. The 13 questions were so subdivided that a viewer could record as many as 40 expressions of opinion on the radiated programmes. An analysis[47] showed that the preferences of the television audience for various items in the programmes were:

plays or variety relayed from theatres	93%
newsreels	93%
'Picture Page', the topical magazine	92%
cabaret	91%
outside broadcasts	89%
sporting events	88%
cartoon films	82%
news maps	74%
short films	56%
musical features	12%

The only ominous note to this general enthusiasm was the complaint that the BBC was employing too many foreign artists, but as Cock, the BBC's Director of Television commented: 'We are glad to have them because certain managers will not allow their British artists to be televised,' (Figs. 19.9 and 19.10).

In paragraph 48 of their report of January 1935, the Television Committee envisaged 'the ultimate establishment of a general television service in this country' and the relaying of television broadcasts by land-line or by wireless from one or more main transmitting stations to sub-stations in different parts of the country. It was clear that the creation of a network of stations to serve all but the most sparsely populated areas would be a gradual process which would occupy a good many years and require the building of at least 12 to 15 stations.

The Television Advisory Committee gave much thought to this matter and in the summer of 1939 drafted a report, for the attention of the Postmaster General, on the expansion of the service. Based on the technical experience which had been gained from more than two years working of the London Station and the stage of acceptance of television by the general public, it was appropriate for a plan of action to be formulated which would lead after six or seven years to the build up of a semi-national service; one which would bring television within reach of most of the densely populated areas of Great Britain.

With this objective the committee foresaw the foundation of four regional stations, each having about four times the power of the London television station, which would probably be located at or near Birmingham, Moorside Edge near Huddersfield, Westerglen near Falkirk and Clevedon near Bristol. It was contemplated, not unduly optimistically, to anticipate that approximately 25 million people would be within the effective range of one or other of these stations and that the first of them, at Birmingham, would be ready after the end of 1941. Two other stations might be constructed simultaneously and become operational in 1943, while the fourth regional station might follow in late 1944 or early 1945.

Fig. 19.9 *Picture Page no. 2. Entire personnel concerned in the first television magazine programme, including artists, interviewers, producer, stage manager, engineering and camera crew. (October 1936)*

Much attention was paid to the financing of the London and proposed regional stations, and, in the first instance, it was considered necessary to restrict the regional service to the relay of programmes initiated in London and thereby avoid the substantial costs of separate regional programmes. Assuming $17\frac{1}{2}$ hours a week of television – the 1939 allocation – the committee estimated an annual expenditure of £900 000 to £1 000 000, see Appendix 3.

As an essential preliminary to the extension of television to the provinces the TAC felt that certain developments associated with the London Station should be undertaken 'immediately' in order to make the station an adequate centre for the future national service. They anticipated that the expenditure for the station for 1939 would be approximately £576 000 (see Appendix 4), an increase of £154 000 or 36% over the estimate for 1938. The latter was £158 000 over the actual expenditure for 1937.

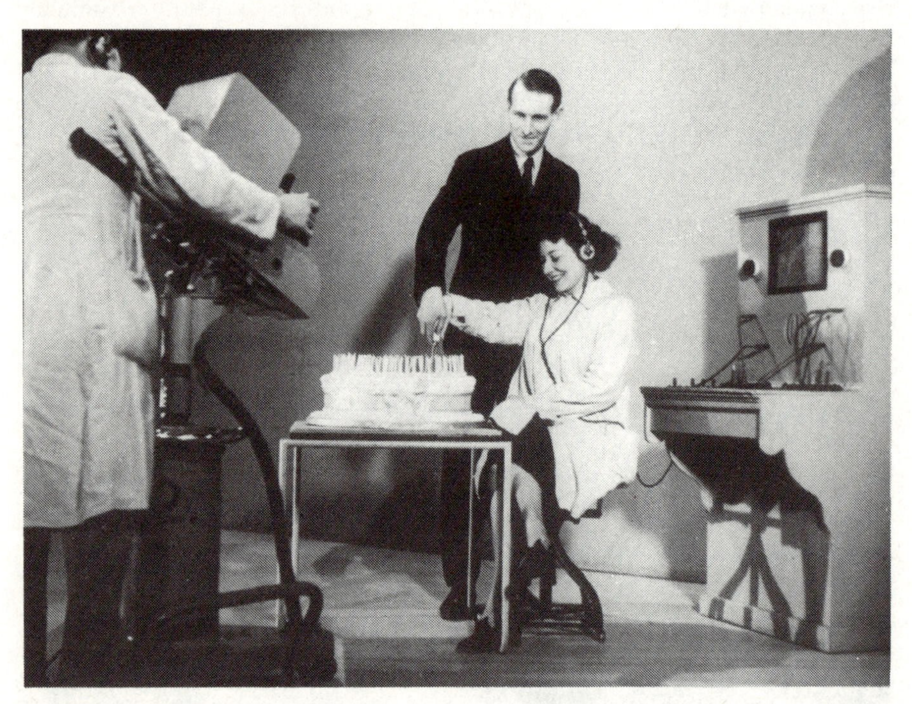

Fig. 10 *Picture Page – 100th edition. Cecil Madden and Joan Miller cutting the cake. (December 1937)*

However, all of this overlooked the possible effects of the ominous European political situation. When war was declared on 3rd September 1939 the Alexandra Palace broadcasts had already terminated. The closure caused some dismay among the general public and the BBC received many letters asking if it would be possible for the television transmissions to be restored. The Corporation's response to these requests was that the television programmes had been withdrawn for reasons affecting national security and the question of resumption could, in wartime conditions, be taken only by the government.

In Germany television broadcasts continued for several months after the commencement of hostilities and this fact led to questions being raised in the House of Commons in February and May 1940 about the prospect of the television transmissions being resumed. Essentially the arguments against a resumption were:

1 the frequency bands which had been allocated for television were required for defence purposes;
2 the technicians who were capable of providing a service were needed for war work;
3 the government had to conserve money and would not be justified in expending resources for a limited audience;
4 manufacturers of television sets were heavily engaged in supplying communications and radar equipment for the war effort.

These cogent points led to one conclusion: the London Station had to remain closed for the duration of the war.

References

1 Anon.: 'BBC Annual 1937' (BBC publications)
2 MARSLAND GANDER, L.: 'Professional touch needed', *Daily Telegraph*, 24th November 1936
3 Anon.: 'More syncopated dance music', *Evening News*, 10th December 1936
4 Anon.: 'Television to be brighter', *Daily Telegraph*, 21st December 1936
5 Anon.: 'Television receivers – reduction in prices', *The Times*, 9th February 1937
6 ASHBRIDGE, N.: letter to Lord Selsdon, 9th February 1937, Post Office bundle 5536
7 SELSDON, Lord.: letter to N. Ashbridge, 10th February 1937, Post Office bundle 5536
8 OSBORNE, G. M.: letter to J. H. Brebner, Press Controller, GPO, 19th February 1937, Post Office bundle 5536
9 Minutes of the 40th meeting of the Television Advisory Committee, 26th February 1937, Post Office bundle 5536
10 VARLEY ROBERTS, J.: memorandum to F. W. Phillips, 5th March 1937, Post Office bundle 5536
11 Anon.: 'Sales of television sets', a note, Post Office bundle 5536
12 Anon.: 'Radiolympia – television sets', a note, September 1937, Post Office bundle 5536
13 Minutes of the 41st meeting of the Television Advisory Committee, 22nd April 1937
14 VARLEY ROBERTS, J.: letter to Sir Stephen Tallents, BBC, 30th April 1937, Post Office bundle 5536
15 Anon.: 'Television to shut down for holidays', *Daily Herald*, 14th May 1937
16 Paper no. 48, Television Advisory Committee, Post Office bundle 5536
17 Minutes of Sir Noel Ashbridge, remarks made at a Television Advisory Committee, 19th December 1935, Post Office bundle 5536
18 Minutes of the 28th meeting of the Television Advisory Committee, 24th June 1936, Post Office bundle 5536
19 Minutes of the 24th meeting of the Television Advisory Committee, 19th December 1935, Post Office bundle 5536
20 Minutes of the 25th meeting of the Television Advisory Committee, 20th January 1937, Post Office bundle 5536
21 See British patents 452,713 (Dec. 1934), 452,772 (Feb. 1935)
22 Paper no. 35, Television Advisory Committee, Post Office bundle 5536
23 MILES, Lt. Col. W. G. H.: letter to F. W. Phillips, 23rd January 1936, Post Office bundle Post 33/5143, file 7
24 ASHBRIDGE, Sir Noel.: letter to Col. A. S. Angwin, 16th October 1936, Post Office bundle Post 33/5143
25 MILES, Lt. Col. W. G. H.: letter to Col. A. S. Angwin, 10th October 1936, Post Office bundle, Post 33/5143

26 Minutes of the 26th meeting of the Television Advisory Committee, 20th February 1936, Post Office bundle 5536
27 BBC announcement: 'Television in Coronation week', 22nd April 1937
28 Minutes of the 44th meeting of the Television Advisory Committee, 5th October 1937, Post Office bundle 5536
29 Paper no. 31, Television Advisory Committee, Post Office bundle 5536
30 Minutes of the 49th meeting of the Television Advisory Committee, 25th January 1938, Post Office bundle 5536
31 Minutes of the 46th meeting of the Television Advisory Committee, 7th December 1937, Post Office bundle 5536
32 Minutes of the 47th meeting of the Television Advisory Committee, 15th December 1937, Post Office bundle 5536
33 Minutes of the 48th meeting of the Television Advisory Committee, 14th January 1938, Post Office bundle 5536
34 Anon.: 'Television – a year's progress', *Observer*, 11th July 1937
35 Minutes of the 45th meeting of the Television Advisory Committee, 11th November 1937, Post Office bundle 5536
36 BARRINGTON, Jonah.: 'Television is perfect for two minutes', *Daily Express*, 12th November 1937
37 Paper no. 54, Television Advisory Committee, Post Office bundle 5536
38 Editorial: 'Television anniversary', *Wireless and Electrical Trader*, 2nd October 1937
39 RMA: proposed statement to the press, Post Office bundle 5536
40 Paper no. 55, Television Advisory Committee, Post Office bundle 5536
41 Minutes of the 57th meeting of the Television Advisory Committee, 18th January 1938, Post Office bundle 5536
42 Minutes of the 50th meeting of the Television Advisory Committee, 29th April, Post Office bundle 5536
43 Papers nos. 54A and 54B, Television Advisory Committee, Post Office bundle 5536
44 Anon.: 'The popularity of television', *The Times*, 7th March 1939
45 Anon.: 'Television in France', *The Times*, 3rd April 1939
46 LIVEING, E.: 'The progress of television', *Fortnightly Review*, June 1939, p. 654
47 Anon.: 'Televiewers like variety', *Daily Telegraph*, 28th June 1939

Appendixes

Appendix 1: List of witnesses and organisations represented (Television Committee)

Organisation	Witnesses
Baird Television Ltd.	Major A. G. Church A. G. D. West
A. C. Cossor Ltd.	W. R. Bullimore J. H. Thomas L. H. Bedford
Electric and Musical Industries Ltd.	A. Clark I. Shoenberg C. S. Agate A. D. Blumlein C. O. Browne G. E. Condliffe N. E. Davis S. J. Preston
Ferranti Ltd.	V. Z. de Ferranti A. Hall
General Electric Co. Ltd.	C. C. Patterson T. W. Heather
Plew Television Ltd.	Dr C. G. Lemon
Scophony Ltd.	S. Sagall G. W. Walton G. Wikkenhauser
British Broadcasting Corporation	Sir J. C. W. Reith
Newspaper Proprietors' Association	Col. the Hon. F. E. Lawson Sir Thomas McAra A. J. Polley F. W. Jarvis E. J. Robertson
Radio Manufacturers' Association	W. W. Burnham R. Milward Ellis

Popular Wireless and the *Wireless Constructor*
The Television Society

and
Sir William Jarratt
W. Barrie Abbott
J. Guibiansky
A. B. Storrar
R. W. Hughes
Dr J. H. T. Roberts
Dr C. Tierney
R. R. Poole
W. G. W. Mitchell

Appendix 2: Estimates of cost of providing and working the London Station (Television Committee Report)

1 Capital cost:	£
Television transmitter (10 kW)	
Sound transmitter (2 kW)	
complete with televising and sound recording	
equipment and power supplies	25 000
Allow for duplication of television equipment	
for use of two systems but with common transmitter	9500
Mast and aerial systems	3000
Site and building for transmitter	7000
	45 000
Allow for higher costs of first sets,	
development charges and contingencies, say 10%	4500
	49 000
say, £50 000	

2 Running costs per annum:	
a Engineering costs	£
Staff, operating and maintenance	2200
Power	1300
Valves, stores etc.	1600
Studio links, land-lines etc.	2400
	7500
Allow contingencies	500
	8000
b Allowance for amortisation	
Total capital expenditure	£50 000
Write off in four equal payments, say,	
£12 000 per annum	
c Programmes	
Estimate based on information supplied	
by the BBC but limited to programme as	
indicated in paragraphs 60 and 61 of report	£100 000 per annum

Total running costs per annum	£
a Engineering costs	8000
b Amortisation	12 000
c Programmes	100 000
Total	£120 000

For the period up to December 31st 1936	
approximately 18 months, estimated cost	£180 000

Appendix 3: Estimated cost of a television service from London and four regional stations

Revenue expenditure	London Station (three stations, including present theatre) £	Four regional stations	Total £
Programmes – artists, orchestra, etc.	151 000		151 000
Cable link	6 000	104 000	110 000
Engineering – power and maintenance	58 000	40 000	98 000
Premises – maintenance and alterations	40 000	4 000	44 000
Salaries and pension scheme	146 000	32 000	178 000
Depreciation	92 000	85 000	177 000
Overheads	50 000	27 000	77 000
Income tax	17 000	25 000	42 000
	560 000	317 000	877 000
		say	900 000

Capital expenditure			
Improvements to plant	85 000		85 000
Theatre equipment	90 000		90 000
Land, premises and plant		512 000	512 000
	175 000	512 000	687 000

Appendix 4: Estimate of the cost of the television service for 1939

Estimate 1938 £	Revenue expenditure	Service on existing 2 studios	Cost 1939 including conversion	Annual cost thereafter without increase in programme hours
142 000	Programmes, artists, etc.	157 000		
47 000	Engineering, power and maintenance	54 000		4 000
15 000	Premises – maintenance, minor alterations	19 000	4 000	14 000
25 000	Major alterations	7 000	50 000	
100 000	Salaries and pension scheme	136 000	2 000	10 000
55 000	Depreciation at 25% per annum	69 000	12 000	23 000
38 000	Overheads	45 000	7 000	5 000
	Income tax	13 000		4 000
422 000		500 000	75 000	60 000*
		575 000		

	Capital expenditure	
	Sundry improvements to plant	85 000
	Theatre equipment	90 000
		175 000

* Excluding income tax liability of £14 000 in 1940 in respect of 1939 alterations.

Index